Mucosal Immune Defense: Immunoglobulin A

Mucosal Immune Defense: Immunoglobulin A

Edited by

Charlotte Slayton Kaetzel

Professor of Microbiology, Immunology and Molecular Genetics
University of Kentucky, Lexington, Kentucky

Charlotte S. Kaetzel
Department of Microbiology
Immunology and Molecular Genetics
University of Kentucky
Lexington, KY 40536
USA
cskaet@uky.edu

ISBN-13: 978-0-387-72231-3 e-ISBN-13: 978-0-387-72232-0

Library of Congress Control Number: 2007934987

Printed on acid-free paper.

9 8 7 6 5 4 3 2 1

springer.com

Preface

Although the existence of a humoral "immune" system has been appreciated for millennia, it was not until 1890 that "antibodies" were identified as serum proteins capable of recognizing and neutralizing antigens with a high degree of specificity (von Behring and Kitasato, 1890). Nearly 50 years later, the advent of physicochemical techniques for analyzing the size and charge of serum proteins led to the proposal that antibodies comprised multiple isotypes (Tiselius and Kabat, 1939). The pioneering work of Heremans and colleagues (Carbonara and Heremans, 1963; Heremans, 1959; Heremans et al., 1959, 1963) demonstrated that a carbohydrate-rich antibody species found in the β-globulin fraction of human serum was distinct from the previously identified IgG and IgM isotypes; this new form of antibody was subsequently designated "IgA". Shortly thereafter, Tomasi and Zigelbaum (1963) demonstrated that IgA in external secretions, unlike serum IgA, consisted mainly of dimers of the basic immunoglobulin subunit. Further structural studies revealed that the IgA dimers in SIgA were linked to an additional glycoprotein of about 80 kDa, which was originally designated the "secretory piece" and is now called the secretory component (SC) (Tomasi et al., 1965). Because most of the serum IgA is monomeric, the question arose whether IgA dimers in SIgA were assembled from serum-derived monomeric IgA or were derived from locally synthesized dimeric IgA. Two landmark experiments demonstrated that the IgA in colostrum is synthesized by local plasma cells as an 11S dimer. Lawton and Mage (1969) examined the distribution of *b* locus light-chain allotypic markers in colostral IgA from heterozygous rabbits. If the SIgA were assembled from serum-derived monomeric IgA, one would expect to find a random assortment of the light-chain markers. However, immunoprecipitation with antiallotypic antibodies revealed that individual SIgA molecules contained either the *b4* or *b5* marker, but not both, suggesting that the IgA dimers were assembled within local plasma cells. Similar results were obtained by Bienenstock and Strauss (1970), who demonstrated that individual SIgA molecules from human colostrum contained either κ or λ light chains, but not both. The concept of local origin of SIgA was upheld by later studies in which transport of locally synthesized IgA into jejunal secretions (Jonard et al., 1984) and saliva (Kubagawa et al., 1987) was found to be significantly greater than transport of serum-derived IgA.

Further support for the model of local synthesis of polymeric IgA came with the discovery of the "joining" (J) chain, a peptide of about 15 kDa that was found to be a subunit of dimeric IgA and pentameric IgM isolated from colostrum (Halpern and Koshland, 1970; Mestecky et al., 1971). Subsequent studies demonstrated that the J-chain was expressed by a high percentage of IgA- and IgM-secreting plasma cells in mucosal tissues and exocrine glands (Brandtzaeg, 1974, 1983; Brandtzaeg and Korsrud, 1984; Crago et al., 1984; Korsrud and Brandtzaeg, 1980; Kutteh et al., 1982; Nagura et al., 1979) and that expression of the J-chain was correlated with *in vitro* binding of SC to immunocytes in tissue sections (Brandtzaeg, 1974, 1983; Brandtzaeg and Korsrud, 1984). Current evidence suggests that the J-chain is not obligatory for polymerization of IgA and IgM, rather that the presence of the J-chain is required for binding of polymeric IgA to SC.

It is now appreciated that IgA is the most abundant immunoglobulin isotype; its total daily synthesis exceeds that of all other isotypes combined. The predominance of IgA derives from the continuous production and transport of SIgA across the vast surfaces of mucosal epithelia, some 300–400 m^2 in adult humans. In fact, it has been estimated that 3 g of SIgA are transported daily into the intestines of the average adult (Conley and Delacroix, 1987; Mestecky et al., 1986). SIgA antibodies are the diplomats of the immune system, with the mission of maintaining homeostasis at mucosal surfaces. We receive the first members of this diplomatic corps from our mothers, in the form of SIgA antibodies in breast milk, which serve until our developing immune system produces its own envoys. SIgA antibodies can be found at every mucosal surface, where they enlist the aid of epithelial cells and a host of innate immune factors to negotiate with microbes, food antigens, and environmental substances. Microbes that agree not to breach the mucosal barrier and to work for the benefit of the host are welcomed as members of the commensal microbiota. Pathogens and noxious substances that threaten to invade the body proper are neutralized and deported through the closest body orifice. Only when diplomacy fails, because of overwhelming numbers of enemy combatants, microbial weapons of mass destruction, or defects in the IgA system, are the big guns of the adaptive immune response (such as IgG antibodies and effector T-cells) recruited to protect the host. The cost of the military option is collateral damage to the mucosal surface, with the risk of serious injury and even death. A healthy IgA system allows us to thrive in a world full of potential pathogens and to co-exist peacefully with hundreds of billions of commensal microorganisms.

Recent advances in human genomics, gene regulation, structural biology, cell signaling, and immunobiology have greatly enhanced our understanding of this important class of antibody. This volume is designed to serve as a reference for current knowledge of the biology of IgA and its role in mucosal immune defense and homeostasis. Topics include the structure of

IgA (Chapter 1), the development of IgA plasma cells (Chapter 2), epithelial transport of IgA and interaction with Fc receptors (Chapters 3 and 4), regulation of the IgA system (Chapter 5), biological roles of IgA, including newly discovered functions (Chapters 6–9), regional functions of IgA (Chapters 10–12), IgA-associated diseases (Chapter 13), and potential therapeutic applications for IgA (Chapters 14 and 15).

References

Bienenstock, J., and Strauss, H. (1970). Evidence for synthesis of human colostral γA as an 11S dimer. *J. Immunol.* 105:274–277.

Brandtzaeg, P. (1974). Presence of J chain in human immunocytes containing various immunoglobulin classes. *Nature* 252:418–420.

Brandtzaeg, P. (1983). Immunohistochemical characterization of intracellular J-chain and binding site for secretory component (SC) in human immunoglobulin (Ig)-producing cells. *Mol. Immunol.* 20:941–966.

Brandtzaeg, P., and Korsrud, F. R. (1984). Significance of different J chain profiles in human tissues: Generation of IgA and IgM with binding site for secretory component is related to the J chain expressing capacity of the total local immunocyte population, including IgG and IgD producing cells, and depends on the clinical state of the tissue. *Clin. Exp. Immunol.* 58:709–718.

Carbonara, A. O., and Heremans, J. F. (1963). Subunits of normal and pathological γ-1A-globulins. (β-2A-globulins). *Arch. Biochem. Biophys.* 102:137–143.

Conley, M. E., and Delacroix, D. L. (1987). Intravascular and mucosal immunoglobulin A: two separate but related systems of immune defense? *Ann. Intern. Med.* 106:892–899.

Crago, S. S., Kutteh, W. H., Moro, I., Allansmith, M. R., Radl, J., Haaijman, J. J., and Mestecky, J. (1984). Distribution of IgA1-, IgA2-, and J chain-containing cells in human tissues. *J. Immunol.* 132:16–18.

Halpern, M. S., and Koshland, M. E. (1970). Novel subunit in secretory IgA. *Nature* 228:1276–1278.

Heremans, J. F. (1959). Immunochemical studies on protein pathology. The immunoglobulin concept. *Clin. Chim. Acta* 4:639–646.

Heremans, J. F., and Schultz, H. E. (1959). Isolation and description of a few properties of the β2A-globulin of human serum. *Clin. Chim. Acta.* 4:96–102.

Heremans, J. F., Vaerman, J. P., and Vaerman, C. (1963). Studies on the immune globulins of human serum. II. A study of the distribution of anti-*Brucella* and anti-diphtheria antibody activities among γ-ss, γ-im and γ-1a-globulin fractions. *J Immunol.* 91:11–17.

Jonard, P. P., Rambaud, J. C., Vaerman, J. P., Galian, A., and Delacroix, D. L. (1984). Secretion of immunoglobulins and plasma proteins from the jejunal mucosa. Transport rate and origin of polymeric immunoglobulin A. *J. Clin. Invest.* 74:525–535.

Korsrud, F. R., and Brandtzaeg, P. (1980). Quantitative immunohistochemistry of immunoglobulin- and J-chain-producing cells in human parotid and submandibular salivary glands. *Immunol.* 39:129–140.

Kubagawa, H., Bertoli, L. F., Barton, J. C., Koopman, W. J., Mestecky, J., and Cooper, M. D. (1987). Analysis of paraprotein transport into the saliva by using anti-idiotype antibodies. *J. Immunol.* 138:435–439.

Kutteh, W. H., Prince, S. J., and Mestecky, J. (1982). Tissue origins of human polymeric and monomeric IgA. *J. Immunol.* 128:990–995.

Lawton, A. R., III, and Mage, R. G. (1969). The synthesis of secretory IgA in the rabbit. I. Evidence for synthesis as an 11S dimer. *J. Immunol.* 102:693–697.

Mestecky, J., Russell, M. W., Jackson, S., and Brown, T. A. (1986). The human IgA system: A reassessment. *Clin. Immunol. Immunopath.* 40:105–114.

Mestecky, J., Zikan, J., and Butler, W. T. (1971). Immunoglobulin M and secretory immunoglobulin A: Presence of a common polypeptide chain different from light chains. *Science* 171:1163–1165.

Nagura, H., Brandtzaeg, P., Nakane, P. K., and Brown, W. R. (1979). Ultrastructural localization of J chain in human intestinal mucosa. *J. Immunol.* 123:1044–1050.

Tiselius, A., and Kabat, E. A. (1939). An electrophoretic study of immune sera and purified antibody preparations. *J. Exp. Med.* 69:119–131.

Tomasi, T. B., Jr., Tan, E. M., Solomon, A., and Prendergast, R. A. (1965). Characteristics of an immune system common to certain external secretions. *J. Exp. Med.* 121:101–124.

Tomasi, T. B., Jr., and Zigelbaum, S. (1963). The selective occurence of γ-1A globulins in certain body fluids. *J Clin. Invest.* 42:1552–1560.

von Behring, E., and Kitasato, S. (1890). On the acquisition of immunity against diphtheria and tetanus in animals. *Deutsch. Med. Wochenschr.* 16:1145–1148.

Acknowledgments

The Editor gratefully acknowledges the hard work of all of the contributors to this volume, each of whom is a leader in the field of IgA immunology. It has been a pleasure to work with such an outstanding group of scientists. I also appreciate the invaluable editorial assistance of my colleague Dr. Maria Bruno. Finally, I wish to express my sincere thanks to all of the helpful people at Springer Publishing, especially Andrea Macaluso, who conceived the concept for this book, Lisa Tenaglia and Suji Prakash.

Charlotte Slayton Kaetzel
Lexington, Kentucky

Contents

Contributors

Laurent Boursier
Peter A. Gorer Department of Immunobiology, Kings College London School of Medicine at Guy's King's College and St. Thomas' Hospitals, Guy's Hospital, London, SE1 9RT, England, United Kingdom

Ranveig Braathen
Laboratory for Immunohistochemistry and Immunopathology (LIIPAT), Department and Institute of Pathology, Rikshospitalet University Hospital, N-0027 Oslo, Norway

Per Brandtzaeg
Laboratory for Immunohistochemistry and Immunopathology (LIIPAT), Department and Institute of Pathology, Rikshospitalet University Hospital, N-0027 Oslo, Norway

Maria E. C. Bruno
Department of Microbiology, Immunology and Molecular Genetics, University of Kentucky College of Medicine, Lexington, KY 40536, United States

Koteswara R. Chintalacharuvu
Department of Microbiology, Immunology and Molecular Genetics and the Molecular Biology Institute, University of California Los Angeles, Los Angeles, CA 90095, United States

Blaise Corthésy
Laboratoire de Recherche et Développement, Service d'Immunologie et d'Allergie, Centre Hospitalier Universitaire Vaudois, Lausanne, Switzerland

Jonathan D. Edgeworth
Department of Nephrology & Transplantation, Kings College London School of Medicine at Guy's King's College and St. Thomas' Hospitals, Guy's Hospital, London, SE1 9RT; and Department of Infection, Guy's & St. Thomas' NHS Foundation Trust, St. Thomas' Hospital, Lambeth Palace Road, London, SE1 7EH, England, United Kingdom

Lennart Hammarström
Department of Clinical Immunology, Huddinge University Hospital, Karolinska Institute, 86 Huddinge, Sweden

Finn-Eirik Johansen
Laboratory for Immunohistochemistry and Immunopathology (LIIPAT), Department and Institute of Pathology, Rikshospitalet University Hospital, N-0027 Oslo, Norway

Charlotte S. Kaetzel
Department of Microbiology, Immunology and Molecular Genetics, University of Kentucky College of Medicine, Lexington, KY 40536, United States

Charu Kaushic
Department of Pathology and Molecular Medicine, McMaster University, Hamilton, Ontario, Canada

Hiroshi Kiyono
Division of Mucosal Immunology, Department of Microbiology and Immunology, The Institute of Medical Science, The University of Tokyo and Core Research for Evolutional Science and Technology (CREST), Japan Science and Technology Corporation (JST), Tokyo, Japan

Jun Kunisawa
Division of Mucosal Immunology, Department of Microbiology and Immunology, The Institute of Medical Science, The University of Tokyo and Core Research for Evolutional Science and Technology (CREST), Japan Science and Technology Corporation (JST), Tokyo, Japan

Michael E. Lamm
Department of Pathology, Case Western Reserve University, Cleveland, OH 44106, United States

Nicholas J. Mantis
Division of Infectious Disease, Wadsworth Center, New York State Department of Health, Albany, NY 12208, United States

Jerry R. McGhee
Departments of Oral Biology and Microbiology, The Immunobiology Vaccine Center, University of Alabama at Birmingham, Birmingham AL 35294, United States

Jiri Mestecky
Departments of Microbiology and Medicine, University of Alabama at Birmingham, Birmingham, AL 35294, United States

Dennis W. Metzger
Center for Immunology and Microbial Disease, Albany Medical College, Albany, NY 12208, United States

Sherie L. Morrison
Department of Microbiology, Immunology and Molecular Genetics and the Molecular Biology Institute, University of California Los Angeles, Los Angeles, CA 90095, United States

H. Craig Morton
Laboratory for Immunohistochemistry and Immunopathology (LIIPAT), Department and Institute of Pathology, Rikshospitalet University Hospital, N-0027 Oslo, Norway; Present address: Institute of Marine Research, 5817 Bergen, Norway

Else Munthe
Laboratory for Immunohistochemistry and Immunopathology (LIIPAT), Department and Institute of Pathology, Rikshospitalet University Hospital, N-0027 Oslo, Norway

Armelle Phalipon
Laboratoire de Pathogénie Microbienne Moléculaire, Institut Pasteur, INSERM U389, Paris Cedex 15, France

Michael W. Russell
Departments of Microbiology and Immunology, and Oral Biology, Witebsky Center for Microbial Pathogenesis and Immunology, University at Buffalo, Buffalo, NY 14214, United States

Hilde Schjerven
Laboratory for Immunohistochemistry and Immunopathology (LIIPAT), Department and Institute of Pathology, Rikshospitalet University Hospital, N-0027 Oslo, Norway

Jo Spencer
Peter A. Gorer Department of Immunobiology, Kings College London School of Medicine at Guy's King's College and St. Thomas' Hospitals, Guy's Hospital, London, SE1 9RT, England, United Kingdom

Charles R. Wira
Department of Physiology, Dartmouth Medical School, Lebanon, NH 03756, United States

Jenny M. Woof
Division of Pathology and Neuroscience, University of Dundee Medical School, Ninewells Hospital, Dundee DD1 9SY, United Kingdom

Esther M. Yoo
Department of Microbiology, Immunology and Molecular Genetics and the Molecular Biology Institute, University of California Los Angeles, Los Angeles, CA 90095, United States

1
The Structure of IgA

Jenny M. Woof[1]

[1] Division of Pathology and Neuroscience, University of Dundee Medical School, Ninewells Hospital, Dundee DD1 9SY, United Kingdom

1.1. Introduction

1.1.1. Immunoglobulin A is the Most Abundant Ig in the Body

It is a surprising and often overlooked fact that the majority of the body's immunoglobulin (Ig) production is geared toward the IgA class. Indeed, the daily synthesis of IgA far outstretches the combined production of all the other Ig classes. Most IgA is produced in mucosa-associated tissue by large numbers of plasma cells in the mucosal subepithelium (Conley and Delacroix, 1987; Mestecky et al., 1991). The necessity for such intensive IgA production at the mucosa presumably reflects a critical requirement, at least in evolutionary terms, for immune protection of mucosal sites. The mucosal surfaces collectively have a huge surface area (~400 m^2 in the human adult) (Childers et al., 1989). They represent, by far, the largest area of contact between the immune system and the environment and can be considered an important point of exposure to inhaled and ingested pathogens.

1.1.2. Distribution and Molecular Forms of IgA

Immunoglobin A is the predominant antibody in the secretions that bathe mucosal surfaces such as the gastrointestinal, respiratory, and genitourinary tracts and in external secretions such as colostrum, milk, tears, and saliva. In addition, IgA is present in serum at concentrations of 2–3 mg/mL, making it the second most prevalent serum Ig after IgG. Although IgG is present at around five times greater concentration than IgA, it is metabolized about five times more slowly, suggesting that the production rates of serum IgA and IgG are similar.

Serum IgA derives from the bone marrow and is principally monomeric in form (Fig. 1.1). However, the IgA in secretions, termed secretory IgA (SIgA), is chiefly polymeric. These locally produced polymers are principally dimers (Fig. 1.1), with a small proportion of higher polymers such as trimers and tetramers.

1.2. The Structure of Monomeric IgA

1.2.1. Component Polypeptides

Like all antibodies, the basic monomer unit of IgA comprises two identical heavy chains and two identical light chains, folded up into globular domains and linked by disulfide bridges. Whereas the light chains are common to all Ig

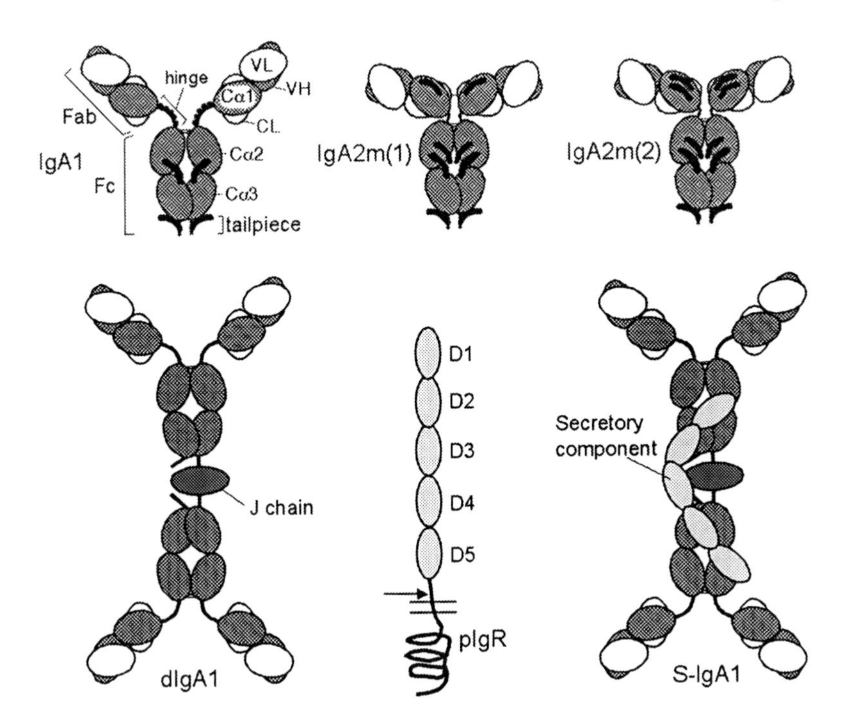

FIG. 1.1. Schematic representation of the monomeric forms of human IgA1, IgA2m(1) and IgA2m(2), the dimeric (dIgA1) and secretory (SIgA1) forms of IgA1, and pIgR. IgA heavy-chain domains are shown hatched and light-chain domains are shown in white. On the monomeric forms of IgA, O-linked sugars (on the IgA1 hinge) are shown as black circles, whereas N-linked oligosaccharides are shown as black disks. For clarity, the oligosaccharide moieties of pIgR, dIgA, and SIgA1 are not included. The J-chain is shown in dark gray. The extracellular Ig-like domains (D1 to D5) of pIgR are shown in pale gray. The approximate site at which pIgR is cleaved to yield secretory component is indicated by an arrow.

classes, the heavy chains (or α-chains) are unique to IgA. In the α-heavy-chain, the amino-terminal variable domain (VH) is followed by three constant domains, termed Cα1, Cα2, and Cα3, with a hinge region separating the Cα1 and Cα2 domains (Fig. 1.2). As with other Ig heavy-chain constant region genes, each domain is encoded by a separate exon (Fig. 1.2). Unlike some Igs, the IgA hinge region is not encoded by a separate exon or exons but is encoded at the 5' end of the Cα2 exon.

The α heavy chain possesses an 18-amino-acid extension at the carboxy-terminus termed the tailpiece, which is lacking in antibodies such as IgG and IgE, which exist solely as monomers. This structural feature is critical for polymer formation (see Sect 1.3) and is shared with IgM, the only other Ig class capable of polymerization.

The domains of the light and heavy chains each adopt a characteristic secondary structure called the immunoglobulin fold (Ig-fold) (Amzel and Poljak,

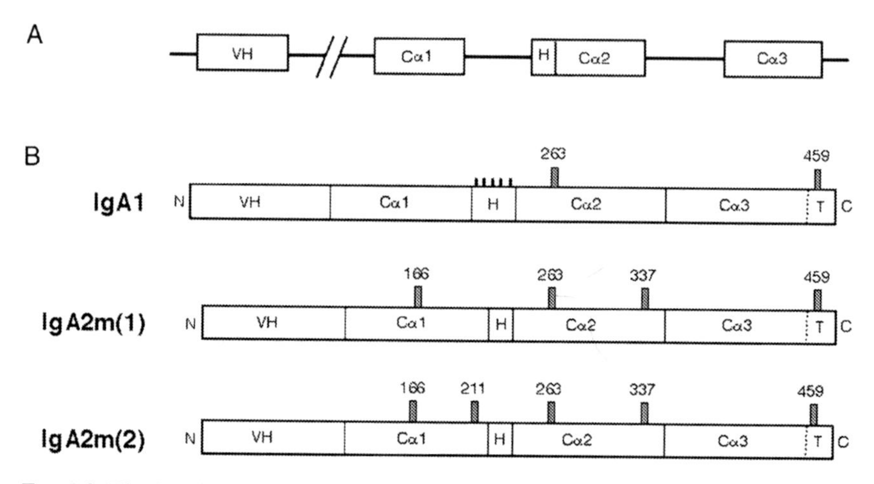

FIG. 1.2. The IgA heavy chain. **(A)** Gene structure showing exons (white boxes). The region encoding the variable domain becomes juxtaposed following splicing out of the intervening DNA during class switching. The hinge (H) is encoded within the same exon as the Cα2 domain. **(B)** Polypeptide structure of the heavy chains of IgA1, IgA2m(1) and IgA2m(2). The attachment sites of N-linked glycan structures (gray rectangles) are numbered using Bur IgA1 numbering. O-Linked glycans in the IgA1 hinge are shown as black rectangles. T = tailpiece.

1979). Although originally described in immunoglobulins, it has long been recognized that this same basic structural building block is present in large numbers of other proteins, making up the extensive Ig gene superfamily (Williams and Barclay, 1988). Comprising approximately 110 amino acids, the Ig-fold is essentially a sandwich of two β-sheets, each consisting of antiparallel β-strands, stabilized by a conserved disulfide bond.

1.2.2. Fab and Fc Regions

The heavy and light chains are arranged to form two antigen-binding (Fab) regions linked to the Fc region via the hinge region (Fig. 1.1). With the exception of the Cα2 domains, the domains are closely paired (VH with VL, Cα1 with CL, and Cα3 with Cα3), with extensive lateral van der Waals contacts and hydrogen bonds between the domains. These close interactions result in the removal of significant surface areas of the domains from contact with solvent (see Sect. 1.2.5.3). Some contribution to structural stability might also be afforded by interaction between neighboring domains on the same polypeptide.

The amino-terminal variable domains of the two Fab regions are responsible for recognition of antigen. The exquisite specificity of antigen binding is determined by the three-dimensional conformation of the antigen-binding site, peculiar to each antibody, formed from the association of three hypervariable loops or complementarity determining regions (CDRs) from the VH domain and three from the VL domain. The unique feature of the IgA Fab region is the Cα1 domain, because the VH, VL, and CL domains are present in all antibodies, regardless of class. Structural information on the Cα1 domain derives principally from the Fab structures of two mouse IgA myeloma proteins determined by X-ray crystallography (Satow et al., 1986; Suh et al., 1986). The elbow-bend angles between the variable and constant domains of the Fab regions were different in the two structures [133° in one (Satow et al., 1986) and 145° in the other (Suh et al., 1986)], in keeping with a degree of flexibility at this point.

The Fc region is critical for the effector function of IgA, mediated via interaction with various cell surface receptors. Binding of IgA immune complexes to such receptors essentially serves to link recognition of pathogens via the Fab regions of IgA with mechanisms capable of bringing about elimination of these foreign invaders. Receptors that bind to the Fc region of IgA include FcαRI present on phagocytes such as neutrophils, macrophages, and eosinophils, Fcα/μR on B-cells, and the polymeric immunoglobulin receptor (pIgR) on epithelial cells. The molecular details of receptor–IgA interactions along with examples of their perturbation by certain pathogens will be discussed in a later section (Sect. 1.5).

1.2.3. Human IgA Subclasses

Two subclasses of IgA, named IgA1 and IgA2, exist in humans. Products of separate genes, they exhibit several sequence differences along the length of their heavy-chain constant regions (Fig. 1.3). Two IgA subclasses are also present in most anthropoid apes (chimpanzee, gorilla, and gibbon) (Kawamura et al., 1992). Other mammals have only one IgA isotype, with the exception of the rabbit, which has genes for 13 IgA subclasses, of which 10 or 11 appear to be expressed (Burnett et al., 1989; Spieker-Polet et al., 1993).

The most marked difference between the human IgA subclasses is seen in the hinge region, where an insertion in IgA1 has produced a much more extended hinge than in IgA2. The IgA1 hinge, rich in Ser, Thr, and Pro residues, shares similarities with regions of mucin molecules and generally carries three to five O-linked glycan moieties (Mattu et al., 1998; Royle et al., 2003). A recent study suggests that 5–10% of serum IgA1 molecules carry six O-linked sugars in the hinge (Tarelli et al., 2004).

There are two well-characterized allotypic variants of IgA2, termed IgA2m(1) and IgA2m(2) (Figs. 1.1 and 1.3). A third IgA2 variant, IgA2n, has also been reported (Chintalacharuvu et al., 1994). Although the heavy-chain constant region sequences of the allotypes differ at a number of points along their length, it is the arrangement of disulfide bonds between polypeptides that represents the most obvious difference between them. In IgA2m(2), typical interchain disulfide bridges link the light and heavy chains, but these are

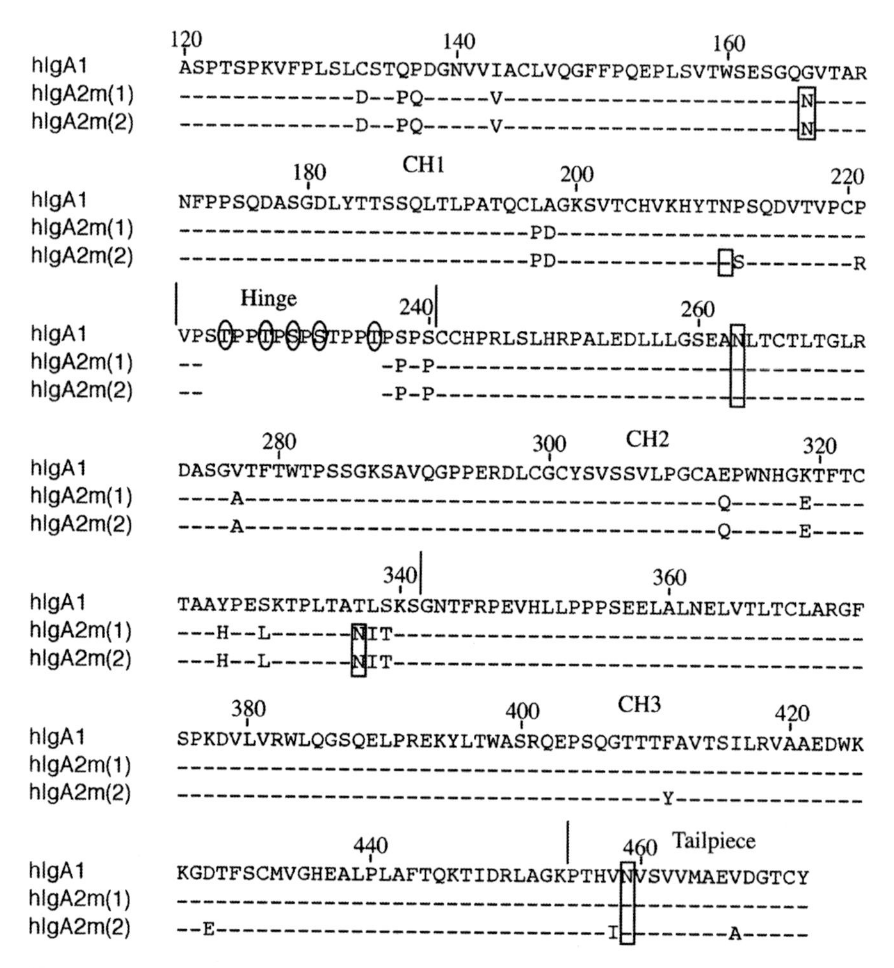

FIG. 1.3. Primary sequences of the constant regions of human IgA subclasses. Residues are numbered according to the Bur IgA1 numbering scheme. Dashes in the IgA2m(1) and IgA2m(2) sequences indicate identity to the IgA1 sequence at that residue. Spaces in the IgA2 sequences in the hinge region indicate an insertion in the IgA1 sequence at that point. Domain and other boundaries are indicated by vertical lines above the sequences. Attachment sites for N-linked sugars are boxed and those for O-linked sugars are circled.

generally missing in IgA2m(1). Instead, the light chains form disulfide bridges to each other and association with the heavy chains is stabilized through noncovalent interactions.

The relative contribution of each subclass to the IgA pool varies between different body compartments. In serum, the IgA1 subclass predominates (about 90% IgA1, 10% IgA2). In secretions, the ratio of the subclasses varies depending on the site, but it is generally more evenly balanced than in serum. Because the production is local, these proportions reflect the distribution of IgA1-secreting and IgA2-secreting cells in various tissues (Crago et al., 1984; Kett et al., 1986).

1.2.4. IgA Glycosylation

N-Linked carbohydrates are found attached to both IgA subclasses and make up 6–7% of total mass in IgA1 and 8–10% in IgA2, as assessed in myeloma proteins (Tomana et al., 1976). IgA1 has two N-linked glycans per heavy-chain constant region, attached to Asn263 in the Cα2 domain and Asn459 on the tailpiece (Figs. 1.1 and 1.2B). As well as these, the IgA2 subclass has further N-linked sugar moieties, the number depending on allotype. IgA2m(1) has two additional sugars, attached at Asn166 in the Cα1 domain and at Asn337 in the Cα2 domain. In addition to these, the IgA2m(2) allotype has a fifth N-linked sugar attached to Asn211 in the Cα1 domain (Fig. 1.3).

The composition of the IgA heavy-chain N-linked sugars has been analyzed in both serum IgA1 and SIgA (Field et al., 1994; Mattu et al., 1998; Royle et al., 2003). The carbohydrate chains are not a single entity but consist of a family of structures based on a mannosyl chitobiose core, represented in Figure 1.4. There is considerable heterogeneity in the number and type of terminal sugars (galactose and sialic acid) and in the level of fucosylation. Around 80% of the N-linked glycans of IgA1 are digalactosylated biantennary complex-type oligosaccharides. Less than 10% of the sugars are triantennary and there are few, if any, tetraantennary or extended structures.

As mentioned earlier, the hinge region of IgA1, but not that of IgA2, carries between three and six O-linked sugars attached to Thr and Ser residues (Field et al., 1994; Mattu et al., 1998; Tarelli et al., 2004) (Figs. 1.1, 1.2, and 1.3). These sugars are much smaller than the N-linked ones and are composed principally of *N*-acetyl galactosamine, galactose, and sialic acid. Again, there is significant heterogeneity, resulting in a family of structures that vary with respect to the presence or absence of galactose and sialic acid.

The roles of the oligosaccharides in various aspects of IgA function have been investigated. Mutation of one or both N-linked glycan attachment sites

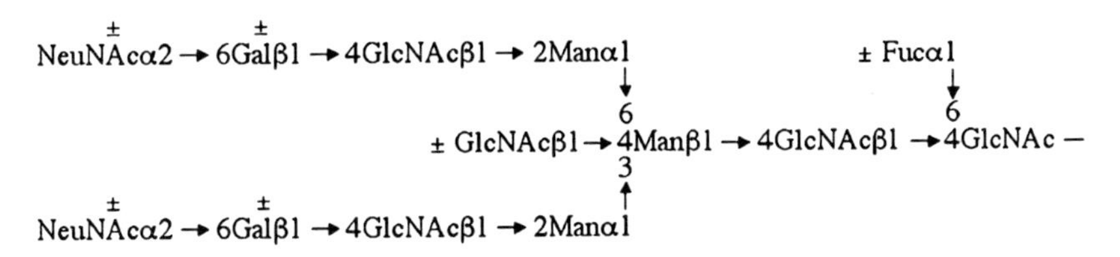

Fig. 1.4. Typical N-linked glycans of IgA heavy chains. Chain variants include the presence or absence (±) of bisecting *N*-acetyl glucosamine, fucose, galactose, and sialic acid. NeuNAc = *N*-acetyl neuraminic (sialic) acid; Gal = galactose; GlcNAc = *N*-acetyl glucosamine; Man = mannose; Fuc = fucose.

of human IgA1 had little effect on the assembly and secretion of monomeric IgA1 (Chuang and Morrison, 1997), whereas loss of the Cα2 domain N-linked glycan attached to Asn263 did not produce any detectable effect on binding to FcαRI (Mattu et al., 1998). Treatment of IgA1-producing cells with the inhibitor benzyl 2-acetamido-2-deoxy-α-D-galactopyranoside, which prevents elaboration of the O-linked sugars in the hinge beyond the first *N*-acetyl galactosamine residue, had no appreciable effect on IgA1 assembly and secretion (Gala and Morrison, 2002), indicating that the O-linked sugars play little role in IgA assembly.

1.2.5. Detailed Structural Information

1.2.5.1. Models Based on Neutron and X-ray Scattering

Techniques such as X-ray and neutron scattering have been used to gain insights into the structure of monomeric forms of IgA in solution by providing the basis for molecular models (Boehm et al., 1999; Furtado et al., 2004) (Table 1.1). A large array of models based on the known crystal structures of IgG Fab and Fc fragments were generated, each with a different hinge structure, and through a molecular dynamics curve fit search method, those structures most suited to fit the experimental scattering characteristics of the IgA antibody were selected. Models for human IgA1 and IgA2m(1) suggest that both subclasses adopt average T-shaped structures (Fig. 1.5). The T-shapes do not necessarily represent fixed, rigid structures but reflect averages of the different conformations presumed to be available to these molecules as a result of flexibility in their hinge regions.

TABLE 1.1. Details of sequence and structure database information for IgA and related molecules in humans.

Molecule	Nucleotide accession code[a]	Structure (PDB) accession code	Structure type	Structure details
IgA1	Cα region = J00220	1IGA	Molecular model	Intact antibody. Based on X-ray and neutron scattering
IgA1 Fc		1OW0	X-ray crystal	Fcα in complex with FcαRI
IgA2m(1)	Cα region = J00221	1R70	Molecular model	Intact antibody. Based on X-ray and neutron scattering
pIgR	NM_002644	1XED	X-ray crystal	Domain 1 only
FcαRI	X54150	1OVZ and 1UCT	X-ray crystal	Extracellular portion of receptor only
		1OW0	X-ray crystal	In complex with IgA1 Fc

[a] Some molecules have alternative accession codes. Sequence information is available from the Entrez Nucleotide database at http://www.ncbi.nlm.nih.gov/entrez/query.fcgi?db=Nucleotide, and structural information is available from the RCSB Protein Data Bank at http://www.rcsb.org/pdb/index.html.

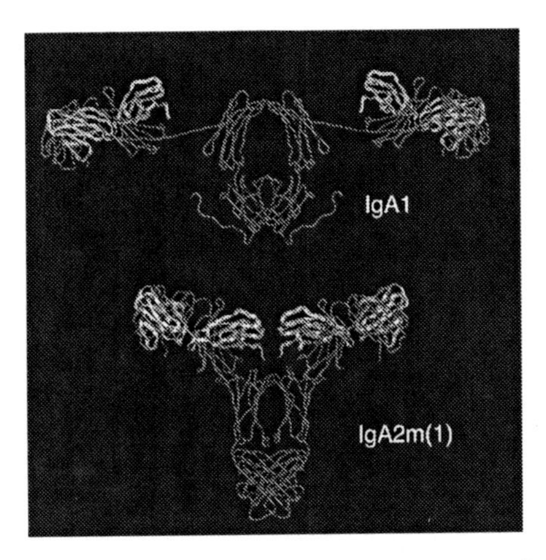

Fig. 1.5. Molecular models of human IgA1 and IgA2m(1) (using coordinates from PBD accession codes 1IGA and 1R70, respectively). Light chains are shown as white ribbons and heavy chains are shown as gray tubes.

1.2.5.2. Significance of Structural Differences Between the Human IgA Subclasses

The IgA1 subclass can be expected to have a significantly greater degree of flexibility than IgA2, due to the short hinge of the latter. The IgA2m(1) allotype is particularly constrained, due to the additional stricture of an inter-light-chain disulfide bridge at the hinge. Electron micrographs of human IgA2m(1) in complex with an anti-idiotype antibody indicate that this allotype, as a result of limited flexibility, is unable to form ring dimeric immune complexes (Roux et al., 1998). IgA1 was not tested for complex formation in this study, but earlier fluorescence polarization and spin-label experiments indicated that segmental flexibility in IgA1 is comparable to that seen in IgG (Dudich et al., 1980; Liu et al., 1981; Zagyansky and Gavrilova, 1974).

The molecular models derived from scattering analysis revealed that IgA1 has a much more extended reach between Fab regions than either IgG1 or IgA2. Thus, the Fab-to-Fab center-to-center distance is 8.2 nm for IgA2m(1) and 16.9 nm for IgA1 (Furtado et al., 2004). Hence, IgA1, to a much greater extent than IgG1 or IgA2, appears to be capable of attaining bivalent interaction with widely spaced antigen molecules. This property might lead to higher-avidity interactions with certain antigens and might be advantageous in the recognition of some pathogens. Overall, IgA1 would appear to increase the repertoire of possibilities within the humoral immune system for high-avidity antigen binding. This capacity might account for the persistence of this subclass after its relatively recent emergence in evolutionary time.

The extended hinge region in IgA1, as well as lending advantageous antigen-binding capabilities to the antibody, also confers increased susceptibility to proteolytic attack. Presumably, the hinge is more exposed because of the increased separation of the Fab arms from the Fc that its length enforces. The O-linked sugars attached to particular Ser and Thr residues along its length are thought to offer some protection, but, inevitably, the IgA1 hinge's enhanced accessibility results in increased possibilities for proteolysis. Indeed, as will be discussed later (Sect. 1.5.3), a number of important bacterial pathogens have evolved specific proteases, termed IgA1 proteases, which cleave in the hinge region of IgA1 (Plaut, 1983). The IgA2 subclass, with the much shorter hinge, is resistant to cleavage by these proteases.

1.2.5.3. X-ray Crystal Structure of IgA1 Fc

Our understanding of the structure of IgA took a significant step forward with the determination of the X-ray crystal structure of human IgA1 Fc in complex with the extracellular domains of human FcαRI (Herr et al., 2003) (Table 1.1). As yet, no crystal structure of uncomplexed IgA Fc is available, but the structure in the receptor complex provides many useful insights.

The IgA1 Fc structure is highly reminiscent of those of IgG Fc (comprising paired Cγ2 and Cγ3 domains) and IgE Fc (Cε3 and Cε4 domains), with a pseudotwofold symmetry and extensive noncovalent interactions between the C-terminal domain pair (Woof and Burton, 2004) (Fig. 1.6). This Cα3–Cα3 pairing makes a significant contribution to the stability of the Fc, burying 2061 Å^2 of surface area from solvent contact in IgA Fc. Interactions between the Cα2 and Cα3 domains also contribute to the overall stability by burying a further 1662 Å^2 of surface area per Fc. The upper portion of the Fc is further stabilized by disulfide bridges between the Cα2 domains. This inter-heavy-chain disulfide bridge arrangement, different from those in IgE and

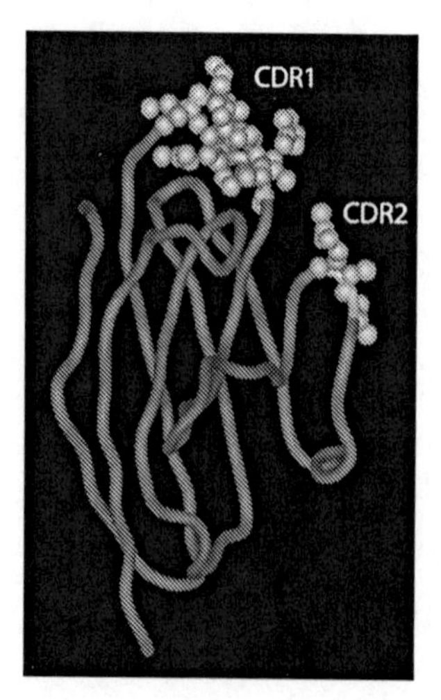

Fig. 1.6. X-ray crystal structure of domain 1 of human pIgR (using coordinates from PDB accession code 1XED). The atoms of residues implicated in binding to polymeric IgA (Thr27–Thr33 of CDR1 and Glu53–Gly54 of CDR2) are shown as white spheres.

IgG, ties the upper reaches of the domains together. Two disulfide bridges link Cys242 in one chain with Cys299 in the other. Additional bonds, either linking the two Cys241 residues or possibly between Cys241 and Cys301 on opposing chains, are presumed to be present. The truncated version of Fc in the crystal does not possess Cys241, so it is not possible as yet to irrefutably designate these final interchain disulfide bridges. Unlike IgG, there are no disulfide bridges in the hinge region, so in IgA, the two heavy chains must be free to flex independently on emerging from the Cα2 domains.

There are also differences between the Fc regions of IgA and those of IgG and IgE in terms of the positions occupied by N-linked sugar moieties. The glycans attached to Asn263 in the Cα2 domains of IgA are arranged over the outer surfaces of these domains, burying 930 $Å^2$ of domain surface area per Fc from solvent contact in the process. The sugar moieties make additional contacts with the Cα3 domains, burying a further 914 $Å^2$ of surface area per Fc. In contrast, the equivalent glycans in IgG and IgE are located between the domains corresponding to Cα2 (Cγ2 and Cε3 respectively), occupying the interior of the Fc. Interaction with these domains buries 1044 $Å^2$ (IgG) and 900 $Å^2$ (IgE) of surface area per Fc, but the glycans make no contact with the Cα3 domain equivalents (Cγ3 and Cε4).

Structural information on the tailpiece of IgA is still lacking because the IgA1 Fc fragment in the solved structure terminated at Lys454, immediately prior to the start of the tailpiece. Therefore, questions remain regarding potential linkages involving Cys471, the penultimate residue of the tailpiece (Fallgren-Gebauer et al., 1995; Prahl et al., 1971). For example, it has been postulated that disulfide linkages to another Cys in the heavy chain, such as Cys311 on the exterior surface of the Cα2 domain, might occur in monomeric IgA (Prahl et al., 1971).

Successful crystallization of three intact full-length IgG molecules, including human IgG1, has shown that, under appropriate conditions, crystal structures of these highly flexible molecules can be obtained (Saphire et al., 2002). However, as yet, there is no crystal structure for intact IgA, and our current understanding of the relative positioning of Fab and Fc regions relies on the low-resolution solution structural analysis described earlier (Table 1.1).

1.3. Dimeric IgA

1.3.1. J-Chain

Of all of the Ig classes, only IgA and IgM share the ability to polymerize through the linkage of multiple monomer units. IgA predominantly polymerizes into dimers, which are stabilized through covalent interaction with the joining (J) chain, a 15–16-kDa polypeptide also present in pentameric IgM (Koshland, 1985). Small proportions of larger IgA polymers, particularly tetramers, also form. Irrespective of the polymer size, only one molecule of the J-chain is incorporated (Halpern and Koshland, 1973; Zikan et al., 1986).

The J-chain, rich in acidic amino acids, does not resemble any other known protein. It comprises a single polypeptide containing eight Cys residues, six of which form conserved intrachain disulfide bridges (Cys12–Cys100, Cys17–Cys91, and Cys108–Cys133) (Bastian et al., 1992; Frutiger et al., 1992), and it is highly conserved (~70%) between species. Its presence has been demonstrated in a wide range of vertebrate species ranging from mammals, through birds and reptiles, to fish and amphibians (Kobayashi et al., 1973). Phylogenetic evidence indicates that the reported J-chain expression in earthworms is probably erroneous and might be accounted for by contamination with mammalian material (Hohman et al., 2003).

The J-chain carries an N-linked carbohydrate attached to Asn48 that accounts for about 8% of its molecular mass (Baenziger, 1979; Niedermeier et al., 1972). Recent analysis of the glycan composition of the J-chain released from SIgA revealed five major forms, chiefly sialylated biantennary complex structures, present in similar proportions (Royle et al., 2003).

The J-chain is expressed by antibody-producing cells and is incorporated into polymeric IgA or IgM shortly before or at the time of secretion (Moldoveanu et al., 1984; Parkhouse,1971). In the absence of IgA or IgM cosecretion, the J-chain is retained in the cell (Kutteh et al., 1983; Mosmann et al., 1978).

1.3.2. Structural Models of the J-Chain

The three-dimensional structure of the J-chain remains unresolved. Lack of significant sequence homology with other proteins has thwarted attempts to place the J-chain in a family of structurally related molecules. Nevertheless, models for the folding of the J-chain have been proposed on the basis of a number of predictive methods. A two-domain structure, with N-terminal β-sheets and C-terminal α-helical segments, has been proposed (Cann et al., 1982). Others have predicted a single domain β-barrel structure, akin to an Ig VL domain (Zikan et al., 1985). Once the precise disulfide bond-pairing arrangement was determined, an alternative two-domain model was postulated with an N-terminal β-barrel domain and a C-terminal domain comprising a mixture of α-helices and β-strands (Frutiger et al., 1992).

1.3.3. Structure of Dimeric IgA

1.3.3.1. Electron Microscopy

The earliest structural information on IgA was gleaned from electron microscopy of various dimeric IgA samples. Dimeric IgA derived from human myeloma patients was shown to have a double-Y shape, with some degree of flexibility (Bloth and Svehag, 1971; Dourmashkin et al., 1971; Munn et al., 1971). Earlier diagrams have tended to depict these double-Y shapes with overlapping Fc regions. However, analysis of the dimensions of the joined Fc regions (Fc–Fc) visible in these images indicates that overlap of the Fc regions is probably an inaccurate representation of the arrangement within the dimer. From the images, the Fab length was estimated to be about 70 Å, whereas the length of the joined Fc regions (Fc–Fc) lay in the range 125–155 Å. Because each Fc region is expected to be around 65–70 Å long, in keeping with estimates of the length of the Fc region of IgG at around 65 Å (Guddat et al., 1993), the observed Fc–Fc length is best explained by an end-to-end arrangement of the two Fc regions (Fig. 1.1). Overlap of the Fc regions might be expected to result in a much shorter Fc–Fc length than that observed. An extended arrangement was also predicted from hydrodynamic data from sedimentation and viscosity experiments (Björk and Lindh, 1974). Moreover, results from more recent mutagenesis experiments are consistent with the organization of the two Fc regions end-to-end (Krugmann et al., 1997).

1.3.3.2. Linkages Between IgA and the J-Chain

The two monomers and the J-chain of dimeric IgA are linked by disulfide bridges. Studies on myeloma proteins indicated that the penultimate residue of the tailpiece of the α-heavy-chain, Cys471, formed a disulfide bridge to the J-chain (Chapuis and Koshland, 1975; Mendez et al., 1973; Mestecky et al., 1974). Mutagenesis of the IgA tailpiece confirmed that Cys471 played a critical role in binding the J-chain in IgA dimerization (Atkin et al., 1996). On the J-chain, the second and third Cys residues, Cys14 and Cys68, were implicated in the interaction (Bastian et al., 1992; Mendez et al., 1973; Mole et al., 1976). A mutagenesis study confirmed that these residues were essential for disulfide bridge formation with dimeric IgA (Krugmann et al., 1997). The presence of N-linked sugars on both the tailpiece and the J-chain also appears to be important for correct dimer formation (Atkin et al., 1996; Krugmann et al., 1997). The domains of IgA Fc also contribute to dimer formation. The presence of the Cα2 domain drives more efficient dimer formation, whereas the presence of the Cα3 domain appears to contribute to the normal restriction of polymerization to dimer formation (Yoo et al., 1999). In fact, the J-chain might be partially obscured by parts of the Fc, because some J-chain epitopes are not accessible in polymeric IgA. However, the finding that cleavage of interchain disulfide bonds can be enough to release the J-chain from polymeric IgA suggests that there are only weak noncovalent interactions between the J-chain and the Fc region (Mestecky et al., 1972).

1.4. Secretory IgA

1.4.1. Polymeric Immunoglobulin Receptor/SC and Epithelial Transcytosis

Immunoglobin A that is destined for the mucosal secretions is produced locally by organized mucosal-associated lymphoid tissues. It is transported across the epithelium into the mucosal lumen by virtue of its interaction with the polymeric immunoglobulin receptor (pIgR) (Mostov et al., 1980). This receptor, which will be discussed in detail in Chapter 3, is expressed basolaterally on epithelial cells and specifically transports polymeric immunoglobulins [i.e., polymers of IgA (predominantly dimers) and, to a lesser extent, pentameric IgM]. Although pIgR is capable of transporting polymeric IgA and IgM at similar rates, the larger size of IgM restricts its diffusion to the receptor through the extracellular matrix and basement membrane so the smaller polymeric IgA molecule is transferred more efficiently (Natvig et al., 1997).

On binding, the pIgR–dimeric IgA complex is internalized and transcytosed through a series of vesicular compartments to the apical plasma membrane. At this point, the extracellular portion of the pIgR, comprising five Ig-like domains, is proteolytically cleaved to form the secretory component (SC). The formation of a disulfide bridge links SC covalently to dimeric IgA, and it is the complex of dimeric IgA and SC, termed SIgA, that is released into the secretions. SC might afford the antibody some protection against proteolytic degradation (Almogren et al., 2003; Crottet and Corthésy, 1998), and the carbohydrate residues on SC help to anchor SIgA to the mucus lining of the epithelium, thereby ensuring effective immune protection (Phalipon et al., 2002).

1.4.2. Available Structural Information on Secretory IgA

1.4.2.1. SC Domains and Their Implicated Roles

The pIgR polypeptide comprises an extracellular region of ~620 amino acids, a transmembrane region of 23 amino acids, and an intracellular tail of ~103 amino acids (Kaetzel and Mostov, 2005). The extracellular portion is responsible for ligand binding, and folds into five Ig-like domains with homology to Ig variable domains (termed D1 to D5 from the N-terminus), followed by a sixth non-Ig-like region that contains the site of cleavage to yield SC (Fig. 1.1). Each of the Ig-like domains is stabilized by one or more internal disulfide bridges (one each in D2, D3, and D4, two in D1, and three in D5).

The first three domains, D1, D2, and D3, are critical for the interaction of pIgR with polymeric human IgA, whereas D4 and D5 appear only to contribute to the affinity of the interaction (Norderhaug et al., 1999). In particular, D1 plays a key role in binding, with loops analogous to the CDRs of variable domains being implicated in the interaction (Bakos et al., 1993; Coyne et al., 1994). Thus, the binding of polymeric IgA to pIgR is initiated by noncovalent interactions between D1 of SC and the Fc region of IgA, followed by the formation of a disulfide bond between Cys467 in D5 of SC and Cys311 in the IgA Cα2 domain (Fallgren-Gebauer et al., 1995; Underdown et al., 1977). Direct interaction between the J-chain and SC is also required (Johansen et al., 2001).

A series of mutagenesis studies have thrown light on the region on human IgA Fc responsible for the noncovalent interaction with human pIgR. The Cα3 domain appears to play the major role (Braathen et al., 2002; Hexham et al., 1999), and residues 402–410 comprising an exposed loop within this domain have been particularly implicated (Hexham et al., 1999; White and Capra, 2002). Recent work suggests that residues Phe411, Val413, and Thr414 in this motif play key roles, with additional contributions from close-lying residue Lys377 and the interface loop Pro440–Phe443 (Lewis et al., 2005). Thus, motifs lying principally across the Cα2-proximal surface of the Cα3 domain have been identified as critical for pIgR binding.

1.4.2.2. EM Studies of Dimeric/Secretory IgA

Early electron microscopy (EM) studies revealed SIgA from colostrum to have a double-Y shape with similar dimensions to dimeric IgA (Bloth and Svehag, 1971), consistent with an end-to-end arrangement of IgA monomers within SIgA. The angle between the Fab arms was seen to vary widely, consistent with significant flexibility in the molecule. The addition of SC did not appear to have significant effects on the overall length of the joined Fc regions, perhaps suggesting that the domains of SC interact predominantly with the joined Fc regions but do not extend beyond their length (Fig. 1.1).

1.4.2.3. X-ray Crystal Structure of Domain 1 of SC

Two IgA monomers, the J-chain, and SC interact with each other through a multitude of noncovalent interactions and disulfide bridges to form the SIgA complex. Although the precise details of their arrangement must await further structural studies, the crystal structure of the first domain of SC has recently been solved (Hamburger et al., 2004) (Table 1.1). This domain, which has a critical role in mediating primary noncovalent interactions with IgA Fc, has a structure similar to that of Ig variable domains, but with differences in the loops that are analogous to the CDRs (Fig. 1.6). The CDR3 loop points away from the other CDR equivalents. Residues in CDR1 were implicated in IgA binding by earlier mutagenesis and peptide mapping studies. The crystal structure suggests that the key residues (Thr27–Thr33) lie in a conserved helical turn in the CDR1 loop (Fig. 1.6). The CDR2 loop (Glu53–Gly54) lies close in three-dimensional space and could contribute to the binding event. However, the positioning of CDR3 suggests that it would be difficult for a ligand contacting CDRs 1 and 2 to also contact CDR3 (Hamburger et al., 2004).

1.5. Relevance of the Structure of IgA to its Function

1.5.1. Molecular Basis of the IgA–FcαRI Interaction

Human FcαRI (CD89) is an important mediator of IgA effector function (see Chapter 4). Interaction of the receptor with IgA clustered on a pathogen surface triggers potent elimination processes such as phagocytosis, antibody-dependent cell-mediated cytotoxicity (ADCC), and release of activated oxygen species.

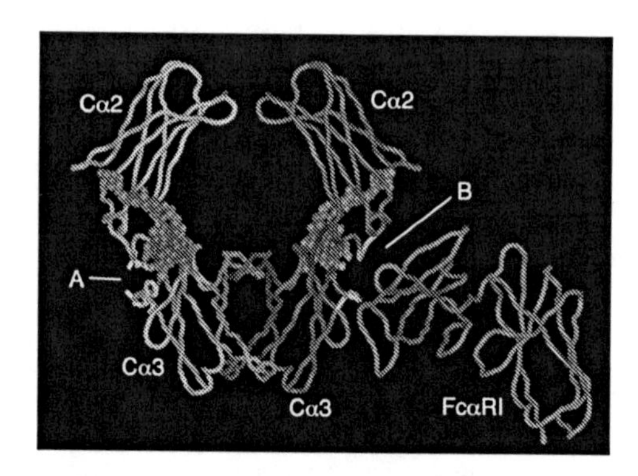

FIG. 1.7. X-ray crystal structure of the complex of IgA1 Fc, on the left, with the extracellular domains of FcαRI, on the right (PDB accession 1OW0). The atoms of the N-linked glycans attached to Asn263 of the Cα2 domains of IgA are shown as spheres. The receptor interacts with a region centered on two Fc interdomain loops, shown in white and indicated by B on the right heavy chain. The residues critical for interaction with streptococcal IgA-BPs are indicated by A and shown in white on the left heavy chain. The interaction site for the bacterial IgA-BPs overlaps with that for FcαRI.

The ligand-binding (α) chain of FcαRI, with an extracellular region comprising two Ig-like extracellular domains, binds both IgA1 and IgA2 via their Fc regions. Based on mutagenesis studies, an interaction site lying at the interface of the Cα2 and Cα3 domains of IgA was proposed (Carayannopoulos et al., 1996; Pleass et al., 1999). This site localization was confirmed by the resolution of the X-ray crystal structure of the complex of IgA1 Fc and the extracellular domains of FcαRI (Herr et al., 2003) (Fig. 1.7). Curiously, this interaction site differs from that on IgG and IgE for their specific Fc receptors, despite appreciable homology between both ligands and receptors (Woof and Burton, 2004). IgG Fc receptors (FcγR) all bind at a hinge-proximal site at the "top" of IgG Fc, and the high-affinity IgE receptor, FcεRI, binds at an equivalent site on IgE. Possibly the interdomain site on IgA Fc was favored by FcαRI as a result of steric restriction imposed by the particular arrangement of the hinge region and inter-heavy-chain disulfide bridges at the top of the Fc in IgA.

1.5.2. IgA Interaction with Bacterial IgA-Binding Proteins

Certain streptococci express surface proteins capable of binding specifically to human IgA. These so-called IgA-binding proteins (IgA-BPs) help the bacteria to subvert the protective IgA immune response. Examples of these proteins are Arp4 and Sir22 on group A *Streptococcus* (Frithz et al., 1989; Stenberg et al., 1994) and the unrelated β protein on group B *Streptococcus* (Héden et al., 1991). All three IgA-BPs interact with the interface between the Cα2 and Cα3 domains of IgA Fc (Pleass et al., 2001) (Fig. 1.7). This interaction site is essentially the same as that bound by FcαRI. Indeed, the IgA-BPs inhibit the binding of IgA to FcαRI and the triggering of a FcαRI-mediated respiratory burst (Pleass et al., 2001). Such a blockade presumably allows the bacteria to evade elimination mechanisms normally elicited by IgA through interaction with FcαRI. Recently, the SSL7 toxin from *Staphylococcus aureus* has been shown similarly to bind to the Cα2/Cα3 interface and competitively inhibit FcαRI binding (Wines et al., 2006), suggesting that this evasion strategy is used by a number of different bacteria.

1.5.3. Cleavage of the IgA1 Hinge by IgA1 Proteases

Differences within the hinge regions of IgA1 and IgA2 account for the differential susceptibility of the subclasses to cleavage by a group of highly specific proteolytic enzymes secreted by certain pathogenic bacteria. These enzymes, termed IgA1 proteases, each cleave at a specific Pro—Ser or Pro—Thr bond within the extended hinge region of IgA1 (Fig. 1.8). In contrast, IgA2 has

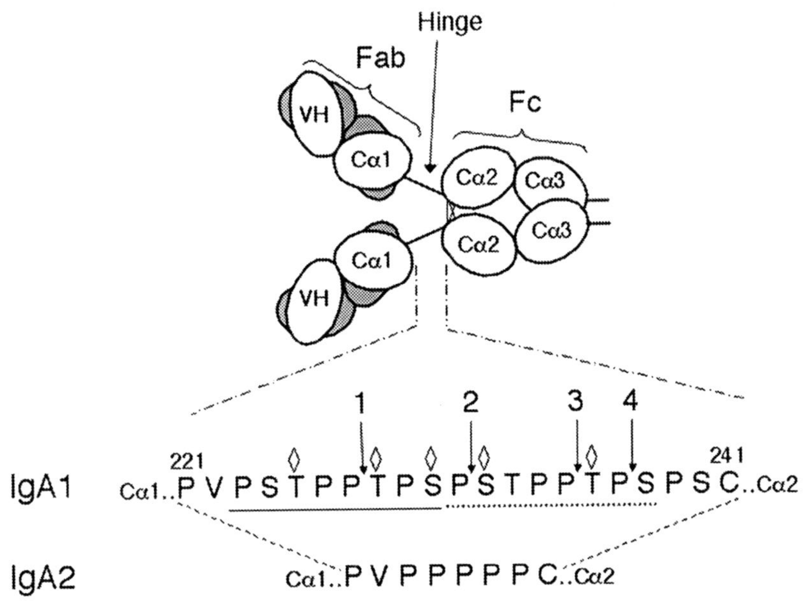

Fig. 1.8. Comparison of the hinge regions of human IgA1 and IgA2. The amino acid sequences of the hinge regions in the two subclasses are shown (Bur IgA1 numbering). The IgA1 hinge contains a duplicated octapeptide sequence (one repeat is underlined by a solid line, the other by a dotted line) that is missing from IgA2. Cleavage sites of bacterial IgA1 proteases within the IgA1 hinge region shown above the IgA1 sequence are numbered as follows: 1, *S. pneumoniae*, *S. sanguis*, *S. mitis*, and *S. oralis*; 2, *H. influenzae* type 1; 3, *N. meningitidis* type 2, *H. influenzae* type 2, *N. gonorrhoeae* type 2; 4, *N. meningitidis* type 1, and *N. gonorrhoeae* type 1. Attachment sites for O-linked glycans on IgA1 are indicated by diamonds above the sequence.

a much shorter hinge and remains resistant to cleavage. The bacteria that produce IgA1 proteases (e.g., *Neisseria meningiditis*, *N. gonorrhoeae*, *Haemophilus influenzae*, *Streptococcus pneumoniae*, and *S. sanguis* are responsible for a number of clinically important, and sometimes life-threatening, diseases (Plaut, 1983). Cleavage of IgA1 within the hinge region releases the Fc portion of the antibody; so although the Fab portions might still bind specific antigens on the bacterial surface, no Fc-mediated elimination mechanisms can be triggered. Access of intact Igs is also prevented. In this way, the bacterium evades the protection that would normally be provided by mucosal Igs.

The sequence requirements governing substrate recognition and cleavage by the various IgA1 proteases are now being teased out (Batten et al., 2003; Chintalacharuvu et al., 2003; Senior et al., 2000; Senior and Woof, 2005a, 2005b). The amino acid sequence of the susceptible region and its context within the protein as a whole are both important. It is clear that for optimal access and cleavage, the proteases require the susceptible bond to be suitably positioned relative to the Fc region (Senior and Woof, 2005b) and that, for at least some of the enzymes, elements within the Fc region also contribute to substrate recognition (Chintalacharuvu et al., 2003). Indeed, a recent study using site-directed mutagenesis to probe the role of the Fc found that IgA1 hinge cleavage can be influenced by particular regions in the CH3 domain (Senior and Woof, 2006). Thus, hinge cleavage by the *N. meningitidis* type 2 enzyme appears to require residues Pro440–Phe443, a motif previously shown to be critical for FcαRI interaction. In contrast, cleavage by the *H. influenzae* type 2 enzyme appears to depend on residues that also contribute to the binding site for pIgR (Senior and Woof, 2006). One might speculate that in order to stabilize interaction with IgA1 and achieve appropriate orientation for efficient cleavage of the hinge, some bacterial proteases interact with regions on IgA1 Fc that are particularly suited to protein–protein interactions and have been conserved because they form key interaction surfaces for host receptors.

1.6. Concluding Remarks

Our understanding of IgA structure continues to advance. However, many outstanding questions remain, particularly relating to the precise structural arrangement of the multiple polypeptides that make up dimeric IgA and SIgA. Further X-ray crystal information mighty ultimately unravel this complex puzzle. Given the growing interest in mucosal immunity and in the potential to use IgA therapeutically, structural information will continue to provide invaluable insights into the function of this important antibody class.

References

Almogren, A., Senior, B. W., Loomes L. M., and Kerr, M. A. (2003). Structural and functional consequences of cleavage of human secretory and human serum immunoglobulin

A1 by proteinases from *Proteus mirabilis* and *Neisseria meningitidis*. *Infect. Immun.* 71:3349–3356.
Amzel, L. M., and Poljak, R. J. (1979). Three-dimensional structure of immunoglobulins. *Annu. Rev. Biochem.* 48:961–997.
Atkin, J. D., Pleass, R. J., Owens, R. J., and Woof, J. M. (1996). Mutagenesis of the human IgA1 heavy chain tailpiece that prevents dimer assembly. *J. Immunol.* 157:156–159.
Baenziger, J. U. (1979). Structure of the oligosaccharide of human J chain. *J. Biol. Chem.* 254:4063–4071.
Bakos, M. A., Kurosky, A., Cwerwinski, E. W., and Goldblum, R. M. (1993). A conserved binding site on the receptor for polymeric Ig is homologous to CDR1 of Ig V_κ domains. *J. Immunol.* 151:1346–1352.
Bastian, A., Kratzin, H., Eckart, K., and Hilschmann, N. (1992). Intra- and interchain disulphide bridges of the human J chain in secretory immunoglobulin A. *Biol. Chem. Hoppe Seyler* 373:1255–1263.
Batten, M. R., Senior, B. W., Kilian, M., and Woof, J. M. (2003). Amino acid sequence requirements in the hinge of human immunoglobulin A1 (IgA1) for cleavage by streptococcal IgA1 proteases. *Infect. Immun.* 71:1462–1469.
Björk, I., and Lindh, E. (1974). Gross conformation of human secretory immunoglobulin A and its component parts. *Eur. J. Biochem.* 45:135–145.
Bloth, B., and Svehag, S. E. (1971). Further studies on the ultrastructure of dimeric IgA of human origin. *J. Exp. Med.* 133:1035–1042.
Boehm, M. K., Woof, J. M., Kerr, M. A., and Perkins S. J. (1999). The Fab and Fc fragments of IgA1 exhibit a different arrangement from that in IgG: a study by X-ray and neutron solution scattering and homology modelling. *J. Mol. Biol.* 286:1421–1447.
Braathen, R., Sorensen, V., Brandtzaeg, P., Sandlie, I., and Johansen, F. E. (2002). The carboxyl-terminal domains of IgA and IgM direct isotype-specific polymerization and interaction with the polymeric immunoglobulin receptor. *J. Biol. Chem.* 277:42755–42762.
Burnett, R. C., Hanly, W. C., Zhai, S. K., and Knight K. L. (1989). The IgA heavy chain gene family in rabbit: cloning and sequence analysis of 13 Cα genes. *EMBO J.* 8:4041–4047.
Cann, G. M., Zaritsky, A., and Koshland, M. E. (1982). Primary structure of the immunoglobulin J chain from the mouse. *Proc. Natl. Acad. Sci. USA* 79:6656–6660.
Carayannopoulos, L., Hexham, J. M., and Capra, J. D. (1996). Localization of the binding site for the monocyte immunoglobulin (Ig) A-Fc receptor (CD89) to the domain boundary between Cα2 and Cα3 in human IgA1. *J. Exp. Med.* 183:1579–1586.
Chapuis, R. M., and Koshland, M. E. (1975). Linkage and assembly of polymeric IgA immunoglobulins. *Biochemistry* 14:1320–1326.
Childers, N. K., Bruce, M. G., and McGhee, J. R. (1989). Molecular mechanisms of immunoglobulin A defense. *Annu. Rev. Immunol.* 43:503–536.
Chintalacharuvu, K. R., Chuang, P. D., Dragoman, A., Fernandez, C. Z., Qiu, J., Plaut, A. G., Trinh, K. R., Gala, F. A., and Morrison, S. L. (2003). Cleavage of the human immunoglobulin A1 (IgA1) hinge region by IgA1 proteases requires structures in the Fc region of IgA. *Infect. Immun.* 71:2563–2570.
Chintalacharuvu, K. R., Raines, M., and Morrison, S. L. (1994). Divergence of human α-chain constant region gene sequences: A novel recombinant α2 gene. *J. Immunol.* 152:5299–5304.

Chuang, P. D., and Morrison, S. L. (1997). Elimination of *N*-linked glycosylation sites from the human IgA1 constant region: Effects on structure and function. *J. Immunol.* 158:724–732.

Conley, M. E., and Delacroix, D. L. (1987). Intravascular and mucosal immunoglobulin A: Two separate but related systems of immune defense? *Ann. Intern. Med.* 106:892–899.

Coyne, R. S., Siebrecht, M., Peitsch, M. C., and Casanova, J. E. (1994). Mutational analysis of polymeric immunoglobulin receptor/ligand interactions. Evidence for the involvement of multiple complementarity determining region (CDR)-like loops in receptor domain I. *J. Biol. Chem.* 269:31620–31625.

Crago, S. S., Kutteh, W. H., Moro, I., Allansmith, M. R., Radl, J., Haaijman, J. J., and Mestecky, J. (1984). Distribution of IgA1-, IgA2-, and J chain-containing cells in human tissues. *J. Immunol.* 132:16–18.

Crottet, P., and Corthésy, B. (1998). Secretory component delays the conversion of secretory IgA into antigen-binding $F(ab')_2$: A possible implication for mucosal defence. *J. Immunol.* 161:5445–5453.

Dourmashkin, R. R., Virella, G., and Parkhouse, R. M. (1971). Electron microscopy of human and mouse myeloma serum IgA. *J. Mol. Biol.* 56:207–208.

Dudich, E. I., Dudich, I. V., and Timofeev, V. P. (1980). Fluorescence polarization and spin-label study of human myeloma immunoglobulins A and M. Presence of segmental flexibility. *Mol. Immunol.* 17:1335–1339.

Fallgren-Gebauer, E., Gebauer, W., Bastian, A., Kratzin, H., Eiffert, H., Zimmerman, B., Karas, M., and Hilschmann, N. (1995). The covalent linkage of the secretory component to IgA. *Adv. Exp. Med. Biol.* 371A:625–628.

Field, M. C., Amatayakul-Chantler, S., Rademacher, T. W., Rudd, P. M., and Dwek, R. A. (1994). Structural analysis of the N-glycans from human immunoglobulin A1: comparison of normal human serum immunoglobulin A1 with that isolated from patients with rheumatoid arthritis. *Biochem. J.* 299:261–275.

Frithz, E., Héden, L. O., and Lindahl, G. (1989). Extensive sequence homology between IgA receptor and M proteins in *Streptococcus pyogenes. Mol. Microbiol.* 3:1111–1119.

Frutiger, S., Hughes, G. J., Paquet, N., Luthy, R., and Jaton, J. C. (1992). Disulfide bond assignment in human J chain and its covalent pairing with immunoglobulin M. *Biochemistry* 31:12643–12647.

Furtado, P. B., Whitty, P. W., Robertson, A., Eaton, J. T., Almogren, A., Kerr, M. A., Woof, J. M., and Perkins S. J. (2004). Solution structure determination of monomeric human IgA2 by X-ray and neutron scattering, analytical ultracentrifugation and constrained modelling: A comparison with monomeric human IgA1. *J. Mol. Biol.* 338:921–941.

Gala, F. A., and Morrison, S. L. (2002). The role of constant region carbohydrate in the assembly and secretion of human IgD and IgA1. *J. Biol. Chem.* 277:29005-29011.

Guddat, L. W., Herron, J. N., Edmundson, A. B. (1993). Three-dimensional structure of a human immunoglobulin with a hinge deletion. *Proc. Natl. Acad. Sci. USA* 90:4271–4275.

Halpern, M. S., and Koshland, M. E. (1973). The stoichiometry of J chain in human secretory IgA. *J. Immunol.* 111:1653–1660.

Hamburger, A. E., West, A. P., Jr., and Bjorkman, P. J. (2004). Crystal structure of a polymeric immunoglobulin binding fragment of the human polymeric immunoglobulin receptor. *Structure (Camb.)* 12:1925–1935.

Héden, L. O., Frithz, E., and Lindahl, G. (1991). Molecular characterization of an IgA receptor from group B streptococci: Sequence of the gene, identification of a proline-rich region with unique structure and isolation of N-terminal fragments with IgA-binding capacity. *Eur. J. Immunol.* 21:1481–1490.

Herr, A. B., Ballister, E. R., and Bjorkman, P. J. (2003). Insights into IgA-mediated immune responses from the crystal structures of human FcαRI and its complex with IgA1-Fc. *Nature* 423:614–620.

Hexham, J. M., White, K. D., Carayannopoulos, L. N., Mandecki, W., Brisette, R., Yang, Y. S., and Capra, J. D. (1999). A human immunoglobulin (Ig)A Cα3 domain motif directs polymeric Ig receptor-mediated secretion. *J. Exp. Med.* 189: 747–752.

Hohman, V. S., Stewart, S. E., Rumfelt, L. L., Greenberg, A. S., Avila, D. W., Flajnik, M. F., and Steiner, L. A. (2003). J chain in the nurse shark: Implications for function in a lower vertebrate. *J. Immunol.* 170:6016–6023.

Johansen, F. E., Braathen, R., and Brandtzaeg, P. (2001). The J chain is essential for polymeric Ig receptor-mediated epithelial transport of IgA. *J. Immunol.* 167:5185–5192.

Kaetzel, C. S., and Mostov, K. E. (2005). Immunoglobulin transport and the polymeric immunoglobulin receptor. In: Mestecky, J., Bienenstock, J., Lamm, M. E., Mayer, L., McGhee, J. R., and Strober W. (eds.), *Mucosal Immunology*, 3rd ed. Elsevier/Academic Press, Amsterdam, pp. 211–250.

Kawamura, S., Saitou, N., and Ueda, S. (1992). Concerted evolution of the primate immunoglobulin α-gene through gene conversion. *J. Biol. Chem.* 267:7359–7367.

Kett, K., Brandtzaeg, P., Radl, J., and Haaijman, J. J. (1986). Different subclass distribution of IgA-producing cells in human lymphoid organs and various secretory tissues. *J. Immunol.* 136:3631–3635.

Kobayashi, K., Vaerman, J.-P., Bazin, H., Lebacq-Verheyden, A.-M., and Heremans, J. F. (1973). Identification of J-chain in polymeric immunoglobulins from a variety of species by cross-reaction with rabbit antisera to human J-chain. *J. Immunol.* 111:1590–1594.

Koshland, M. E. (1985). The coming of age of the immunoglobulin J chain. *Annu. Rev. Immunol.* 3:425–453.

Krugmann, S., Pleass, R. J., Atkin, J. D., and Woof, J. M. (1997). Structural requirements for assembly of dimeric IgA probed by site-directed mutagenesis of J chain and a cysteine residue of the α chain CH2 domain. *J. Immunol.* 159:244–249.

Kutteh, W. H., Moldoveanu, Z., Prince, S. J., Kulhavy, R., Alonso, F., and Mestecky, J. (1983). Biosynthesis of J chain in human lymphoid cells producing immunoglobulins of various isotypes. *Mol. Immunol.* 20:967–976.

Lewis, M. J., Pleass, R. J., Atkin, J. D., Batten, M. R., and Woof, J. M. (2005). Structural requirements for the interaction of human immunoglobulin A with the human polymeric immunoglobulin receptor. *J. Immunol.* 175:6694–6701.

Liu, B. M., Cheung, H. C., and Mestecky, J. (1981). Nanosecond fluorescence spectroscopy of human immunoglobulin A. *Biochemistry* 20:1997–2003.

Mattu, T. S., Pleass, R. J., Willis, A. C., Kilian, M., Wormald, M. R., Lellouch, A. C., Rudd, P. M., Woof, J. M., and Dwek, R. A. (1998). The glycosylation and structure of human serum IgA1, Fab and Fc regions and the role of N-glycosylation on Fcα receptor interactions. *J. Biol. Chem.* 273:2260–2272.

Mendez, E., Prelli, F., Frangione, B., and Franklin, E. C. (1973). Characterization of a disulphide bridge linking the J chain to the α chain of polymeric immunoglobulin A. *Biochem. Biophys. Res. Commun.* 55:1291–1297.

Mestecky, J., Kulhavy R., and Kraus, F. W. (1972). Studies on human secretory immunoglobulin A. II. Subunit structure. *J. Immunol.* 108:738–747.

Mestecky, J., Lue, C., and Russell, M. W. (1991). Selective transport of IgA: Cellular and molecular aspects. *Gastroenterol. Clin. North Am.* 20:441–471.

Mestecky, J., Schrohenloher, R. E., Kulhavy, R., Wright, G. P., and Tomana, M. (1974). Site of J chain attachment to human polymeric IgA. *Proc. Natl. Acad. Sci. USA* 71:544–548.

Mikoryak, C. A., Margolies, M. N., and Steiner, L. A. (1988). J chain in *Rana catesbiana* high molecular weight Ig. *J. Immunol.* 140:4279–4285.

Moldoveanu, Z., Egan, M. L., and Mestecky, J. (1984). Cellular origins of human polymeric and monomeric IgA: Intracellular and secreted forms of IgA. *J. Immunol.* 133:3156–3162.

Mole, J. E., Bhown, A. S., and Bennett, J. C. (1976). Sequence analysis of human J chain. Amino terminal location of a disulphide bond linking the immunoglobulin heavy chain. *Biochem. Biophys. Res. Commun.* 73:92–97.

Mosmann, T. R., Gravel, Y., Williamson, A. R., and Baumal, R. (1978). Modification and fate of J chain in myeloma cells in the presence and absence of polymeric immunoglobulin secretion. *Eur. J. Immunol.* 8:94–101.

Mostov, K. E., Kraehenbuhl, J. P., and Blobel, G. (1980). Receptor-mediated transcellular transport of immunoglobulin: synthesis of secretory component as multiple and larger transmembrane forms. *Proc. Natl. Acad. Sci. USA* 77:7257–7261.

Munn, E. A., Feinstein, A., and Munro, A. J. (1971). Electron microscope examination of free IgA molecules and of their complexes with antigen. *Nature* 231:527–529.

Natvig, I. B., Johansen, F. E., Nordeng, T. W., Haraldsen, G., and Brandtzaeg, P. (1997). Mechanism for enhanced external transfer of dimeric IgA over pentameric IgM: studies of diffusion, binding to the human polymeric Ig receptor, and epithelial transcytosis. *J. Immunol.* 159:4330–4340.

Niedermeier, W., Tomana, M., and Mestecky, J. (1972). The carbohydrate composition of J chain from human serum and secretory IgA. *Biochim. Biophys. Acta* 257:527–530.

Norderhaug, I. N., Johansen, F. E., Krajci, P., and Brandtzaeg, P. (1999). Domain deletions in the human polymeric Ig receptor disclose differences between its dimeric and pentameric IgM interaction. *Eur. J. Immunol.* 29:3401–3409.

Parkhouse, R. M. E. (1971). Immunoglobulin A biosynthesis. Intracellular accumulation of 7S subunits. *FEBS Lett.* 16:71–73.

Phalipon, A., Cardona, A., Kraehenbuhl, J. P., Edelman, L., Sansonetti, P. J., and Corthésy, B. (2002). Secretory component: a new role in secretory IgA-mediated immune exclusion *in vivo*. *Immunity* 17:107–115.

Plaut, A. G. (1983). The IgA1 proteases of pathogenic bacteria. *Annu. Rev. Microbiol.* 37:603–622.

Pleass, R. J., Areschoug, T., Lindahl, G., and Woof, J. M. (2001). Streptococcal IgA-binding proteins bind in the Cα2–Cα3 interdomain region and inhibit binding of IgA to human CD89. *J. Biol. Chem.* 276:8197–8204.

Pleass, R. J., Dunlop, J. I., Anderson, C. M., and Woof, J. M. (1999). Identification of residues in the CH2/CH3 domain interface of IgA essential for interaction with the human Fcα receptor (FcαR) CD89. *J. Biol. Chem.* 274:23508–23514.

Prahl, J. W., Abel, C. A., and Grey, H. M. (1971). Carboxy-terminal structure of the α chain of human IgA myeloma proteins. *Biochemistry* 10:1808–1812.

Roux, K. H., Strelets, L., Brekke, O. H., Sandlie, I., and Michaelsen, T. E. (1998). Comparisons of the ability of human IgG3 hinge mutants, IgM, IgE, and IgA2, to form small immune complexes: A role for flexibility and geometry. *J. Immunol.* 161:4083–4090.

Royle, L., Roos, A., Harvey, D. J., Wormald, M. R., van Gijlswijk-Janssen, D., Redwan, El-R. M., Wilson, I. A., Daha, M. R., Dwek, R. A., and Rudd, P. M. (2003). Secretory IgA N- and O-glycans provide a link between the innate and adaptive immune systems. *J. Biol. Chem.* 278:20140–20153.

Saphire, E. O., Stanfield, R. L., Crispin, M. D., Parren, P. W., Rudd, P. M., Dwek, R. A., Burton, D. R., and Wilson, I. A. (2002). Contrasting IgG structures reveal extreme asymmetry and flexibility. *J. Mol. Biol.* 319:9–18.

Satow, Y., Cohen, G. H., Padlan, E. A., and Davies, D. R. (1986). Phosphocholine binding immunoglobulin Fab McPC603. An X-ray diffraction study at 2.7 Å. *J. Mol. Biol.* 190:593–604.

Senior, B. W., Dunlop, J. I., Batten, M. R., Kilian, M., and Woof, J. M. (2000). Cleavage of a recombinant human immunoglobulin A2 (IgA2)–IgA1 hybrid antibody by certain bacterial IgA1 proteases. *Infect. Immun.* 68:463–469.

Senior, B. W., and Woof, J. M. (2005a). Effect of mutations in the human immunoglobulin A1 (IgA1) hinge on its susceptibility to cleavage by diverse bacterial IgA1 proteases. *Infect. Immun.* 73:1515–1522.

Senior, B. W., and Woof, J. M. (2005b). The influences of hinge length and composition on the susceptibility of human IgA to cleavage by diverse bacterial IgA1 proteases. *J. Immunol.* 174:7792–7799.

Senior, B. W., and Woof, J. M. (2006). Sites in the CH3 domain of human IgA1 that influence sensitivity to bacterial IgA proteases. *J. Immunol.* 177:3913–3919.

Spieker-Polet, H., Yam, P. C., and Knight, K. L. (1993). Differential expression of 13 IgA-heavy chain genes in rabbit lymphoid tissues. *J. Immunol.* 150:5457–5465.

Stenberg, L., O'Toole, P. W., Mestecky, J., and Lindahl, G. (1994). Molecular characterization of protein Sir, a streptococcal cell surface protein that binds both immunoglobulin A and immunoglobulin G. *J. Biol. Chem.* 269:13458–13464.

Suh, S. W., Bhat, T. N., Navia, M. A., Cohen, G. H., Rao, D. N., Rudikoff, S., and Davies, D. R. (1986). The galactan-binding immunoglobulin Fab J539: An X-ray diffraction study at 2.6-Å resolution. *Proteins* 1:74–80.

Tarelli, E., Smith, A. C., Hendry, B. M., Challacombe, S. J., and Pouria, S. (2004). Human serum IgA1 is substituted with up to six *O*-glycans as shown by matrix assisted laser desorption ionisation time-of-flight mass spectrometry. *Carbohydrate Res.* 339:2329–2335.

Tomana, M., Niedermeier, W., Mestecky, J., and Skvaril, F. (1976). The differences in carbohydrate composition between the subclasses of IgA immunoglobulins. *Immunochemistry* 13:325–328.

Underdown, B. J., DeRose, J., and Plaut, A. (1977). Disulfide bonding of secretory component to a single monomer subunit in human secretory IgA. *J. Immunol.* 118:1816–1821.

White, K. D., and Capra, J. D. (2002). Targeting mucosal sites by polymeric immunoglobulin receptor-directed peptides. *J. Exp. Med.* 196:551–555.

Williams, A. F., and Barclay, A. N. (1988). The immunoglobulin superfamily: Domains for cell surface recognition. *Annu. Rev. Immunol.* 6:381–405.

Wines, B. D., Willoughby, N., Fraser, J. D., and Hogarth, P. M. (2006). A competitive mechanism for staphylococcal toxin SSL7 inhibiting the leukocyte IgA receptor,

FcαRI, is revealed by SSL7 binding at the Cα2/Cα3 interface of IgA. *J. Biol. Chem.* 281:1389–1393.

Woof, J. M., and Burton, D. R. (2004). Human antibody–Fc receptor interactions illuminated by crystal structures. *Nat. Rev. Immunol.* 4:89–99.

Yoo, E. M., Coloma, M. J., Trinh, K. R., Nguyen, T. Q., Vuong, L. U., Morrison, S. L., and Chintalacharuvu K. R. (1999). Structural requirements for polymeric immunoglobulin assembly and association with J chain. *J. Biol. Chem.* 274:33771–33777.

Zagyansky, Y. A., and Gavrilova, E. M. (1974). Segmental flexibility of human myeloma immunoglobulins A. *Immunochemistry* 11:681–682.

Zikan, J., Mestecky, J., Kulhavy, R., and Bennett, J. C. (1986). The stoichiometry of J chain in human secretory dimeric IgA. *Mol. Immunol.* 23:541–544.

Zikan, J., Novotny, J., Trapane, T. L., Koshland, M. E., Urry, D. W., Bennett, J. C., and Mestecky, J. (1985). Secondary structure of the immunoglobulin J chain. *Proc. Natl. Acad. Sci. USA* 82:5905–5909.

2
IgA Plasma Cell Development

Jo Spencer[1], Laurent Boursier[1], and Jonathan D. Edgeworth[2]

2.1. Introduction

Evolution in biological systems is rarely wasteful; it involves both adaptation and conservation of resources. In this context especially, the quantity of IgA secreted onto mucosal surfaces and the cellular processes that generate it are all the more remarkable. Approximately 10^{10} plasma cells per meter of gut are situated in the diffuse connective tissue stroma between the epithelium and the muscularis mucosa referred to as the lamina propria (Fig. 2.1) (Brandtzaeg et al., 1999; Brandtzaeg and Pabst, 2004). These produce antibody, most of which is immumoglobin A (IgA), so that ~3–5g of IgA is actively transported each day into the lumen of the human gut (Conley and Delacroix, 1987). This secreted

[1] Peter A. Gorer Department of Immunobiology, Kings College London School of Medicine at Guy's King's College and St. Thomas' Hospitals, Guy's Hospital, London, SE1 9RT, United Kigdom
[2] Department of Nephrology & Transplantation, Kings College London School of Medicine at Guy's King's College and St. Thomas' Hospitals, Guy's Hospital, London, SE1 9RT, United Kingdom

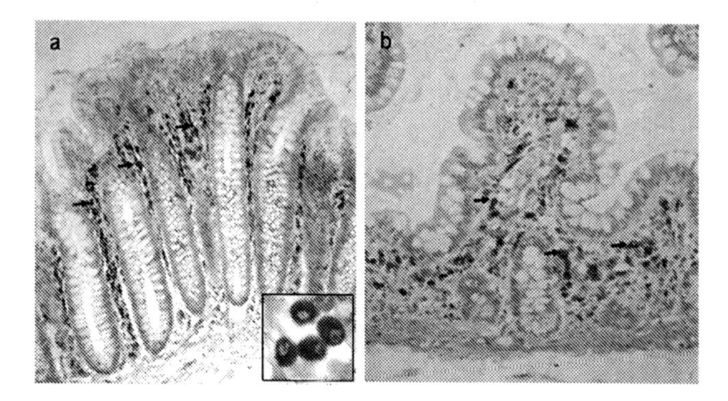

FIG. 2.1. Paraffin sections of **(a)** colon and **(b)** ileum stained with antibody to IgA. IgA plasma cells are visible in the lamina propria. Under higher magnification (inset), single IgA plasma cells can be identified by abundant oval cytoplasm, eccentric round nucleus with characteristic condensation of chromatin that resembles a clock face.

antibody has a critical role in maintaining homeostasis in an environment where the immune system and potentially proinflammatory bacterial stimuli are closely juxtaposed and separated by a single epithelial layer (Fagarasan et al., 2002). The aim of this chapter is to discuss the mechanisms that generate, diversify, and disseminate this extensive IgA-producing plasma cell population. There are considerable interspecies differences in mucosal lymphoid tissue that will be identified where relevant, but the final outcome in all species is the same: the production of the largest population of plasma cells in the body.

An early indication of local production of mucosal immunoglobulins was provided by Ogra and Karzon (1969), who noted specific antibody in secretions but not serum in response to human mucosal immunization with polio vaccine. The cellular basis for this partitioning of mucosal and systemic responses was identified by Gowans and Knight (1964), who described the "homing" of adoptively transferred and labeled immunoblasts from thoracic duct lymph of rats (which contains the lymphatic drainage from the gut), back to the gut. Craig and Cebra (1971) identified that commitment to IgA production was associated with the Peyer's patch (PP) B-cells. Although the dynamics of the responses were debated to some degree, it was generally agreed from studies of animal models at the time that PPs are a source of precursors of IgA plasma cells that enter the blood via the lymphatics and subsequently home back to the gut and that might proliferate locally in response to local antigenic challenge before terminal differentiation (Husband, 1982; Husband and Gowans, 1978).

Analysis of human plasma cells and their precursors has been advanced by studies of the non-germline-encoded junctional regions of the immunoglobulin variable region (IgV) genes, which are unique for each cell (or clone of cells) of B lineage. Microdissection of cells from different zones of tissue sections combined with IgV gene analysis identified clonally related cells in the germinal centers of PPs and the adjacent lamina propria, indicating the origin of plasma cells from PPs (Dunn-Walters et al., 1997). IgV gene analysis of single human intestinal IgA- and IgM-secreting plasma cells (Fischer and Kuppers, 1998) demonstrated a diverse population. However, evidence of clonally related cells within small areas of lamina propria and widespread plasma cell clones could be seen when cells using the smaller families of IgV segments were analyzed independently (Boursier et al., 1999, 2005; Holtmeier et al., 2000), consistent with wide, blood-borne dissemination of plasma cell precursors.

Human intestinal plasma cells share the same profile of markers of differentiation, longevity, and function as bone marrow plasma cells, suggesting that like bone marrow plasma cells, the majority are also long-lived terminally differentiated effector cells (Medina et al., 2002, 2003). Although the focus of this chapter is the production of IgA plasma cells, it should be noted that both IgM and IgA are mucosal immunoglobulins. Both can be polymeric through expression of the J-chain, which allows binding to the poly-Ig receptor and transport onto mucosal surfaces (Brandtzaeg et al., 1999). In humans, clonally related plasma cell variants using the μ– and α-heavy-chain can be detected in lamina propria, implying that IgM and IgA with the same specificity have the same origin and operate side by side (Boursier et al., 1999, 2005).

2.2. Gut-Associated Lymphoid Tissue

The term *gut-associated lymphoid tissue* (GALT) refers to all of the organized follicular lymphoid tissue beneath the epithelial surface of the intestinal tract and includes PP and isolated lymphoid follicles (ILFs) (Brandtzaeg and Pabst 2004). Although not anatomically part of the GALT, there is evidence that mesenteric lymph nodes are functionally linked to the GALT and, therefore, can be included in an assessment of mucosal plasma cell development. The broader term *mucosa-associated lymphoid tissue* (MALT) includes all organized lymphoid tissue associated with epithelia at all mucosal surfaces (Fig. 2.2). GALT is, by

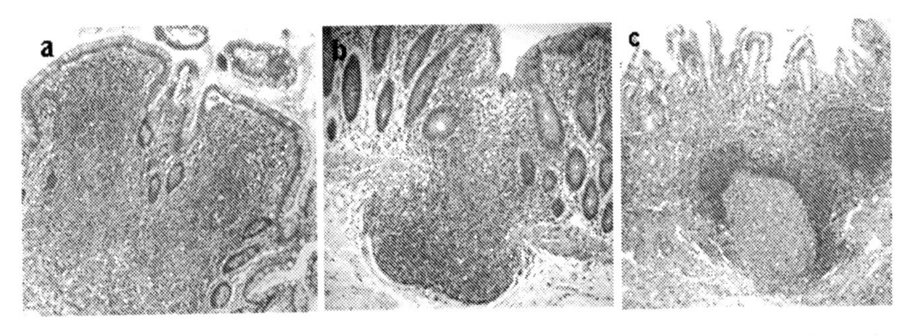

FIG. 2.2. Paraffin sections stained with hematoxylin and eosin of **(a)** a Peyer's patch, **(b)** an isolated lymphoid follicle from colon, and **(c)** an acquired lymphoid follicle in a stomach infected with *H. pylori,* all stained with Hematoxylin and Eosin. In each case, the follicular structure is unencapsulated and intimately associated with the adjacent follicle-associated epithelium.

far, the largest compartment of MALT and is responsible for generating far more IgA plasma cells than other components of MALT (Brandtzaeg et al., 1986). GALT is also the most thoroughly characterized and we will therefore focus on the properties of GALT specifically rather than MALT generally.

Gut-associated lymphoid tissue differs from organized lymphoid tissue in lymph nodes in that the tissue is not encapsulated; the lymphoid tissue at the borders of the follicle infiltrates diffusely between the crypts. There are no afferent lymphatics; the antigen enters the lymphoid tissue directly from the specialized follicle-associated epithelium (FAE), which differentiates from the adjacent crypt epithelial stem cells (Bhalla et al., 1982). This very close micro anatomic and functional association between the lymphoid tissue and the FAE is a defining characteristic of MALT (Anderle et al., 2005). The FAE generally lacks secretory cells (such as goblet cells), digestive enzymes, and the polymeric Ig receptor (Pappo and Owen, 1988; Sierro et al., 2000; Smith, 1985), but it expresses chemokines such as CCL9 and CCL20 involved in the recruitment of dendritic cells to the FAE microenvironment (Tanaka et al., 1999; Zhao et al., 2003). Probably the most notable cellular feature of the FAE are the "microfold" or "membranous" M-cells, which differentiate from epithelial stem cells. They comprise a variable proportion of the FAE in different species and can be identified by scanning electron microscopy by a lack of surface villi (Owen and Jones, 1974). M-Cells are highly active in transcytosis, taking up soluble and particulate antigens from the intestinal lumen, importantly including microbial antigens, which are passed into the subepithelial space. They are intimately associated with lymphocytes and dendritic cells at their basolateral surface, forming a unique lymphoepithelial microenvironment (Farstad et al., 1994). Unlike the lymphocytes infiltrating the villus epithelium, which are almost exclusively $CD8^+$ cells, B- and T-lymphocytes, including $CD4^+$ cells that express CD40L involved in cognate interactions, are present in the FAE. These lymphoepithelial structures, which develop in association with microbial colonization, are involved in the initial inductive events required for plasma cell generation and represent a front line of defense against intestinal pathogens. It has been suggested that the intimacy between the lymphoid tissue and the epithelium includes invaginations of the follicular germinal centers into the FAE (Yamanaka et al., 2001, 2003).

2.2.1. Peyer's Patches

Clusters of lymphoid follicles (5–200 in man) (Cornes, 1965) that are predominantly located in the terminal ileum are termed Peyer's patches (PPs) after Johann Conrad Peyer (1653–1712), who probably observed follicular aggregates macroscopically (Haubrich, 2005). In some species, including rats and mice, the PPs are clearly visible as bulbous protrusions from the serosal surface of the bowel. In other species, including adult humans, they are barely visible even from the mucosal surface with the naked eye, and original macroscopic quantification required fixation and staining of the mucosa. They might be visible in children when they are more numerous (Cornes, 1965; MacDonald et al., 1987).

Peyer's patches are constitutive and develop during fetal life, being identifiable with discrete B- and T-cell zones from ~20 weeks of gestation in man (Spencer et al., 1986b). From murine models, it is clear that organogenesis of PPs is dependent on the development of anlagen-containing $CD45^+$, $CD4^+$, and $CD3^-$ inducer cells. The functional expression of interleukin (IL) 7Rα, lymphotoxin (LT)αb, and CXCR5 by inducer cells is essential for the initiation of the developmental process (Finke et al., 2002; Nishikawa et al., 2003). Subepithelial clusters of cells with the phenotype of inducer cells are present in human fetal intestine from 11 weeks of gestation, which are probably equivalent to inducer cells identified in mice (Spencer et al., 1987). The development of PPs per se is therefore antigen independent, although they increase in number after birth, reaching maximum at 12 years and gradually declining in number through adult life (Cornes, 1965).

2.2.2. *Isolated Lymphoid Follicles*

In addition to clustered follicles of the PP, ILFs are distributed throughout the intestine. In the human small bowel, single follicles that closely resemble a single Peyer's patch follicle are present with a frequency of ~1 per 269 villi in the jejunum and 1 per 28 villi in the ileum along the mesenteric and antimesenteric regions of the circumference (Moghaddami et al., 1998). In mice, ILFs are identifiable as ~100–200 lymphoid clusters along the antimesenteric border of the small intestine (Hamada et al., 2002). Dukes and Bussey (1926) observed an average of three follicles per square centimeter of colon in man. In the colon, ILFs often have a flask-shaped appearance as they penetrate the muscularis mucosae and form a relatively narrow follicular structure as they approach the epithelium (Fig. 2.2).

Isolated lymphoid follicles are not visible to the naked eye and it is possible to underestimate the potential contribution of B-cell responses in the ILF to the lamina propria plasma cell population. Although PPs have traditionally been considered the main site for induction of IgA responses in the gut, recent evidence suggests that ILFs are also likely to contribute significantly and there are likely to be innate mechanisms driving their maturation (Kweon et al., 2005; Lorenz and Newberry, 2004; McDonald et al., 2005; Shikina et al., 2004). The fuller appreciation of the contribution of ILFs to IgA-mediated immunity is relatively recent and of particular importance because they are on the front line of the immune responses against pathogenic bacteria. *Salmonella typhimurium*, *Yersinia enterocolitica*, and *Shigella flexneri* can all invade via ILFs.

2.2.3. *Acquired MALT*

There are circumstances in which acquisition of MALT is clearly antigen dependent, such as that acquired in the stomach in response to *Helicobacter pylori* infection (Genta et al., 1993). This is particularly noticeable since the stomach does not normally contain lymphoid tissue (Fig. 2.2). Ironically, IgA produced in response to *H. pylori* infection appears not to be protective but enhances the infective process (Akhiani *et al.*, 2004, 2005).

2.3. B-Cell Compartments of GALT; Potential Plasma Cell Precursors

There are three major B-cell compartments in GALT, all of which are potential precursors of IgA plasma cells at different stages of development and activation: the germinal center, the mantle, and the marginal zone (Spencer et al., 1986a). The germinal center, which is situated centrally as in peripheral secondary lymphoid tissues, is the major site of proliferation of B-cells activated in the PPs. It has been suggested that the GC of PPs differ from those in peripheral lymphoid tissues in their lower expression of CD38, implying that the pathways to plasma cell development might differ in the gut and the periphery (Guilliano et al., 2001), As in peripheral lymphoid tissues, activated B-cells proliferate as centroblasts in the dark zones of germinal centers. During this phase of rapid proliferation, two characteristic features of the immunoglobulin secreted by plasma cells are thought to be acquired. First, a very high load of somatic mutations is introduced into the variable region genes and, second, class switch recombination to IgA takes place (Butcher et al., 1982; Liu et al., 1996). Both processes are dependent on expression of the activation-induced cytidine deaminase (AID) (Muramatsu et al., 2000). Further rounds of mutation might occur after the class-switch event. This chronology of events has been documented in germinal centers outside GALT (Liu et al., 1996), but the facts that, in man, variable region genes of IgM- and IgA-secreting plasma cells are both highly mutated and also that class-switch variants share a proportion of V region mutations support this view (Boursier et al., 2005). Two routes to IgA switch recombination of human B-cells have been described. The first is dependent on the switch factor transforming growth factor (TGF)-β. IL-10, IL-2, and IL-4 have also been implicated together with proliferative signals, including those provided by cross-linking of CD40 (Briere et al., 1995; Zan et al., 1998; Cazac and Roes, 2000; Defrance et al., 1992; Tangye et al., 2002; Zan et al., 1998). The second is dendritic cell associated through the surface receptors BAFF (Blys) and the proliferation inducing ligand APRIL and might be CD40 and therefore T-cell independent (Fayette et al., 1997; Litinskiy et al., 2002). Although dendritic cell-associated switch to IgA has been described *in vitro*, there has been a tendency to associate T-cell-dependent class switching with GALT. However, dendritic cells are abundant in GALT and it is possible that adaptive and innate routes to IgA class switch in plasma cell precursors might occur in GALT.

In conventional models of germinal center function, centroblasts mature to smaller, nonproliferating germinal center centrocytes that reexpress Ig (Pascual et al., 1994). At this stage, when IgV has been diversified by hypermutation, centrocytes undergo competitive selection for specificity for antigen that is retained in complexes on the follicular dendritic cells (Hardie et al., 1993). Germinal center formation in the peripheral lymphoid tissue is associated with B-cell responses to T-dependent antigens. T-dependent antigens are protein antigens or antigens with protein carriers that facilitate the cognate T/B interaction when B-cells solicit the help of specific T-cells through presentation of peptide. In an analysis of receptor signaling on germinal center formation, Casola et al. (2004) observed that germinal centers in GALT might be initiated by innate rather than adaptive processes. They observed that, as predicted, when B-cells did not express specific Ig, there was no cognate B/T interaction and germinal centers did not form in the peripheral lymphoid tissue. However, lack of B-cell specificity did not prevent germinal center formation in the PPs, although germinal center formation remained T-cell dependent and was also dependent on the bacterial flora, implying that B-cells in germinal centers of GALT can be driven by stimuli associated with innate rather than adaptive responses. The lack of dependence on B-cell specificity for germinal center formation in GALT might also be reflected in the distribution of mutations in IgV gene sequences and also the tolerance of such high loads of mutations. In an analysis of the distribution of mutations carried by intestinal B-cells, Dunn-Walters and Spencer (1998) compared the mutations in IgV genes that encode the expressed Ig with the distribution in alleles that were rearranged but not used. These unused rearrangements, which are carried by 40% of mature B lineage cells, acquire mutations due to the AID enzyme, but these are not selected by antigen. Strong biases in the distribution of mutations were observed, but these biases were largely the same in used and unused rearrangements and were considered to reflect the mechanism that introduced the mutations. There was a suggestion of selection for functional viability in the genes encoding the used Ig but not convincing selection for antigen specificity. In contrast, the imprint of antigen selection is more apparent in the Ig genes of peripheral B-cells with a known high affinity for antigen. It is possible that the lack of selection for B-cell specificity mighty remove the ceiling on the level of mutations the IgV gene can tolerate, generating a highly diversified repertoire and resulting in the characteristic high level of mutations characteristic of postfollicular intestinal B-cells and plasma cells. Although not exactly parallel, in that it is an antigen-independent process, somatic hypermutation of IgV genes in the PPs of sheep generates a diverse primary repertoire. The highly diversified B-cell receptor repertoire of potential plasma cell precursors generated in the GALT of other species, including rabbits, involves somatic hypermutation driven by the flora. This apparent absence of antigen selection raises questions as to the nature of the initiating commensal bacterial stimulus and the extent to which GALT generates specific IgA responses to intestinal pathogens (Lanning *et al.*, 2002, 2005; Reynaud *et al.*, 1995).

The germinal center is surrounded by a mantle of small, round, quiescent naïve B-cells, which, as migrant cells through GALT, have the potential to encounter stimulus and progress along the activation and differentiation pathways to plasma cell development. The mantle zone in PPs tends to be narrower in the context of the follicle as a whole compared to secondary follicles in the systemic immune system (Spencer et al., 1986a).

The mantle zone is surrounded by the marginal zone. The marginal zone extends up to and into the FAE. It might intersperse between the crypts at the edge of the follicle and sometimes infiltrates between the epithelial cells of the crypts (Spencer et al., 1986a). A subset of this population, predominantly the intraepithelial and subepithelial components, express the immunoglobulin superfamily receptor translocation-associated 1 (IRTA1) receptor, although the functional significance of this is not yet known (Falini et al., 2003). Analysis of Ig gene rearrangements isolated from marginal zone B-cells of PPs reveals the presence of cells with highly mutated Ig genes, indicative of GALT germinal center memory cells, which is consistent with studies using animal models (Dunn-Walters et al., 1996; Liu et al., 1988). Marginal zone B-cells of GALT are thought to be the benign equivalent of lymphomas of mucosa-associated lymphoid tissue (MALT lymphomas), which arise most commonly in the stomach as an aberrant component of the response to infection with *H. pylori*. MALT lymphomas have mutated IgV genes and continue to acquire mutations as the tumor progresses (Du et al., 1996). MALT lymphoma cells are often interspersed between the dendrites of follicular dendritic cells, suggesting an association between the tumor and the germinal center, and it is possible that they continue to acquire mutations following malignant transformation in the germinal center microenvironment. It is possible that this germinal-center-independent proliferation, but germinal center interaction reflects the properties of normal human marginal zone B-cells of GALT. We have observed proliferation of B-cells as isolated cells in the marginal zone of human GALT, although it is not clear whether this event is dependent on germinal centers at any stage (Boursier et al., 2005).

T-cell populations in ILFs and PPs resemble those in peripheral lymphoid tissue. They are predominantly $CD4^+$ (Spencer et al., 1986a). The majority are located in the T-cell zone on the serosal aspect of the follicles. In colonic ILFs, it is not unusual to see the T-cell zone entirely on the serosal side of the muscularis, although in serial sections, once the B-cell zone is sectioned, the lymphoid tissue punctures the muscularis extending toward the mucosa (O'Leary and Sweeney, 1986).

In addition to memory, the splenic marginal zone is known to be a site where T-independent responses are initiated. This was originally demonstrated in animal models, although splenectomized humans also have impaired responses to T-independent antigens (Amlot et al., 1985; Amlot and Hayes, 1985; Claassen et al., 1986). Consistent with this, marginal zone B-cells in mice express Toll-like receptor 4, which mediates innate responsiveness to lipopolysaccharide (LPS) (Viau and Zouali, 2005). Analysis of IgV genes from the microdissected GALT marginal zone revealed some unmutated sequences, implying that the IgD-negative cells from the marginal zone of human organized mucosal lymphoid tissue might not be exclusively germinal center derived (Dunn-Walters et al., 1996), although this observation requires substantiation.

2.4. Generation of IgA Plasma Cells Outside GALT

Gut-associated lymphoid tissue was considered to be the sole source of intestinal IgA plasma cells until Kroese et al. (1989) observed that the self-replenishing B1 B-cell population in the peritoneal cavity of mice contributed significantly to the IgA response—original estimates being around 50% contribution to the total population from the peritoneum. Murakami et al. (1994) identified a functional relationship between the peritoneum and the intestinal microenvironments, by demonstrating that orally administered LPS could activate B1 B-cells in both the lamina propria and the peritoneal cavity. Although current literature suggests that the contribution of the peritoneum to the lamina propria plasma cell population in mice might be substantially less than 50% peritoneal origin (Thurnheer et al., 2003), it is clear that in mice, peritoneal B1 B-cells contribute to the total IgA produced. It is now known that the IgA response in mice can be independent of costimulatory signals that mediate interaction between T-cells and antigen-presenting cells and that it has a different cytokine requirement to antibody production in the periphery (Gardby et al., 2003). There is evidence that the Ig derived from peritoneal precursors is encoded by somatically mutated IgV genes, despite the apparent lack of association with GALT (Bos et al., 1996). All of the above features of the peritoneal contribution to the IgA response relate to mice. In humans there is no known contribution to the intestinal response from peritoneal lymphocytes in terms of either response to intraperitoneal immunization (Lue et al., 1994), phenotypic analysis, or analysis of configuration of the Ig genes (Boursier et al., 2002).

Possibly related to the contribution of B1 B-cells to the plasma cell population is the recently described series of observations that suggest that IgA responses can be generated independently of GALT, in the lamina propria. Dendritic cells have been shown to extend processes through the epithelial layer to capture lumenal bacteria and might then go on to present derived antigens to naïve B-cells in the lamina propria (Macpherson and Uhr, 2004; Rescigno et al., 2001). Dendritic cells can directly activate B-cells and preferentially induce switching to IgA (Fayette et al., 1997; Litinskiy et al., 2002). Several routes to T-independent innate B-cell responses have been described. Macpherson et al. (2000) identified T-cell independent IgA responses to a variety of antigenic structures, including protein and LPS that was independent of GALT. IgA plasma cell development that is independent of organized lymphoid tissue (Gardby et al., 2003) and even independent of peripheral naive B-cells (Macpherson et al., 2001) have been described.

Having initially observed IgM$^+$ cells in lamina propria cells from mice lacking AID, Fagarasan et al. (2001) tested the hypothesis that these represented precursors of lamina propria IgA plasma cells detectable in wild-type animals by looking for indicators of local class switching. They observed circle-switch transcripts indicting recent class-switch events and AID expression. They also observed that intestinal stromal cells can support B-cell class-switch recombination. Although the detection of these factors clearly indicated local class-switch events, it

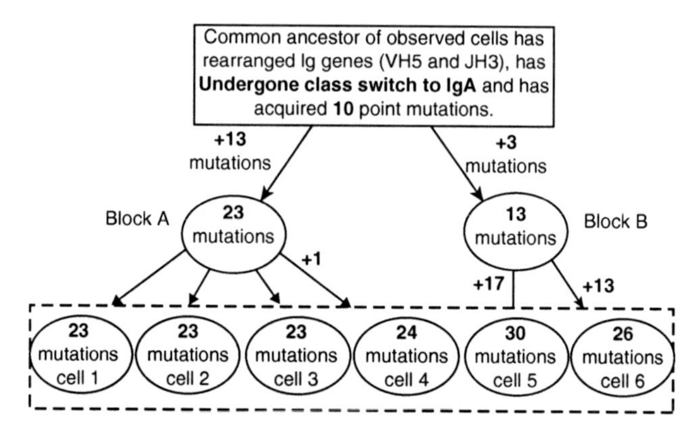

FIG. 2.3. A lineage tree analysis of the relationship between 6 IgA+ plasma cells (labeled cells 1–6, in the dotted box) isolated from blocks (A and B) of human colonic mucosa that were 5 cm apart in the origin sample of colon. The sequences of the heavy-chain genes of all 6 cells had several features in common; they had all switched to IgA and they had 10 shared somatic mutation events. They had then diversified by somatic hypermutation, as illustrated. By analysis of shared mutations, it is possible to deduce that cell 6 from block B was more closely related to cells 2–4 from block A than to cell 5 from block B. Likewise, cell 1 from block A was more closely related to cell 5 from block B than cells 2–4 from the same microenvironment in block A. This supports the concept of the wide dissemination of cells from distant common precursors.

was subsequently suggested that although PP cells were excluded from this study, ILFs might have contaminated the preparation and provide the alternative GALT-dependent source for IgA (Shikina et al., 2004). In studies deliberately avoiding ILF structures, no evidence of extrafollicular diversification has been observed in mice (Shikina et al., 2004) or human tissues (Boursier et al., 2005). Analysis of IgV gene sequences using lineage trees has demonstrated that plasma cells can be more closely linked to distant than local IgA or IgM plasma cells (Fig. 2.3). AID expression and cell division, which are required for somatic hypermutation and class-switch recombination, are not seen in human intestinal lamina propria. In addition, the phenotype of lamina propria B lineage cells is overwhelmingly that of plasma cells and their immediate precursors; rare memory cells only have been seen in one study (Boursier et al., 2005; Farstad et al., 2000).

2.5. Antigen-Specific IgA Responses to Mucosal Pathogens

The most recent literature on IgA plasma cell biology emphasizes the role of innate immunity in driving the production of IgA plasma cells and the protective properties of natural poly-specific IgA (Wijburg et al., 2006).

The specificity of IgA in general might therefore be of little importance to its biological function. However, it has been clear since the early days of mucosal immunity as a discipline that specific IgA responses can be generated in the gut (e.g., Ogra and Karzon, 1969). Specific responses to mucosally delivered protein antigens and killed whole bacteria usually given with the mucosal adjuvant cholera toxin have been studied for many years (Holmgren et al., 1993). Cholera toxin, the B-subunit (CTB) of which is a potent adjuvant presumably due to its binding of GM1 ganglioside on cell surfaces, has a complex effect on the mucosal immune system leading to upregulation of costimulatory molecules on T-cells (Gardby et al., 2003), preferential class switching to IgA (Xu-Amano et al., 1993), and the ability to stimulate long-lived (greater than 3 years) protective immunity (Clemens et al., 1986).

However, this compartmentalized view of nonspecific responses against commensal bacteria and specific responses against protein antigens with adjuvant must accommodate the fact that the dominant and specific protective immune response against many bacterial intestinal pathogens is directed against LPS. For example, studies of natural *Shigella flexneri* infection and the response to experimental vaccination in humans and animal models have together demonstrated a key role for intestinal IgA directed against LPS serotype-specific epitopes (DuPont et al., 1972; Formal et al., 1966; Mel et al., 1965; Phalipon et al., 1995). *S. flexneri* LPS serotypes commonly differ only in the positioning of one or more glucose molecules on a common backbone of repeating oligosaccharides, which indicates a high degree of antigenic specificity (Simmons and Romanowska, 1987). It is not known whether this IgA anti-LPS response is completely T-cell independent or whether it involves noncognate T-cell help as suggested in the experiments of Casola et al. (2004) or even involves cognate T-cell recognition of protein covalently attached to the LPS. The duration of protection against reinfection with the same *S. flexneri* serotype remains unclear but is probably less than a year, implying poor memory cell generation. There is also evidence that specific IgA responses are important in the protective response in nontyphoidal Salmonella infection (Michetti et al., 1992). The existence of over 1500 Salmonella LPS serotypes strongly implies selective pressure exerted by intestinal IgA.

2.6. Homing to the Lamina Propria

Precursors of antibody-producing cells migrate from GALT and mesenteric lymph node (MLN) via the lymphatics to the blood and subsequently home back to the gut. The return of cells to the gut is regulated by a combination of lymphocyte homing receptors and their endothelial ligands and mucosally derived chemokines and their receptors (Kunkel and Butcher, 2003; Salmi and Jalkanen, 2005). The integrin $\alpha 4\beta 7$ expressed by plasma cell precursors facilitates their homing to the mucosa, by binding to its endothelial ligand, the

Ig superfamily member MAdCAM-1 (Farstad et al., 1995). MAdCAM-1 is expressed on both the high endothelial venules of GALT and the flat endothelium of vessels in the lamina propria (Briskin et al., 1993). The chemokines CCL25 and CCL28 are both involved in attracting cells expressing their receptors CCR9 and CCR10, respectively, into the intestinal lamina propria. CCL28 is expressed by epithelium in large and small intestines, and cells expressing CCR10 are recruited to both sites (Kunkel et al., 2003; Pan et al., 2000). In contrast, CCL25 is produced predominantly by the small bowel endothelium from where it recruits IgA plasma cell precursors expressing CCR9 (Bowman et al., 2002).

The final stage in plasma cell development is terminal differentiation in the lamina propria, which involves, for example, loss of CD19 and gain of plasma cell-associated markers such as CD138, probably in response to locally secreted cytokines such as IL-6 (Goodrich and McGhee, 1998).

2.7. Concluding Remarks

Although there is currently debate over the obligatory involvement of GALT at some stage in the generation of intestinal IgA plasma cells, there is little doubt that innate stimuli are significant driving forces in their massive propagation. It is likely that the size and diversity of the IgA response is intimately associated with the "danger signals" from the intestinal flora. It is in this reactive background that specific IgA responses might be generated. Imposing specificity on a system geared to broad-spectrum polyreactivity is a major challenge for strategies to generate effective mucosal immunization.

References

Akhiani, A. A., Schon, K., Franzen, L. E., Pappo, J., and Lycke, N. (2004). *Helicobacter pylori*-specific antibodies impair the development of gastritis, facilitate bacterial colonization, and counteract resistance against infection. *J. Immunol.* 172:5024–5033.

Akhiani, A. A., Stensson, A., Schon, K., and Lycke, N. Y. (2005). IgA antibodies impair resistance against *Helicobacter pylori* infection: Studies on immune evasion in IL-10-deficient mice. IgA antibodies impair resistance against *Helicobacter pylori* infection: studies on immune evasion in IL-10-deficient mice. *J. Immunol.* 174:8144–8153.

Amlot, P. L., Grennan, D., and Humphrey, J. H. (1985). Splenic dependence of the antibody response to thymus-independent (TI-2) antigens. *Eur. J. Immunol.* 15:508–512.

Amlot, P. L., and Hayes, A. E. (1985). Impaired human antibody response to the thymus-independent antigen, DNP-Ficoll, after splenectomy. Implications for post-splenectomy infections. *Lancet* 1:1008–1011.

Anderle, P., Rumbo, M., Sierro, F., Mansourian, R., Michetti, P., Roberts, M. A., and Kraehenbuhl, J.-P. (2005). Novel markers of the human follicle-associated epithelium identified by genomic profiling and microdissection. *Gastroenterology* 129:321–327.

Bhalla, D. K., Owen, R. L., Bhalla, D. K., and Owen, R. L. (1982). Cell renewal and migration in lymphoid follicles of Peyer's patches and cecum:An autoradiographic study in mice. *Gastroenterology* 82:232–242.

Bos, N. A., Bun, J. C., Popma, S. H., Cebra, E. R., Deenen, G. J., van der Cammen, M. J., Kroese, F. G., and Cebra, J. J. (1996). Monoclonal immunoglobulin A derived from peritoneal B cells is encoded by both germ line and somatically mutated VH genes and is reactive with commensal bacteria. *Infect. Immun.* 64:616–623.

Boursier, L., Dunn-Walters, D. K., and Spencer, J. (1999). Characteristics of IgVH genes used by human intestinal plasma cells from childhood. *Immunology* 97:558–564.

Boursier, L., Farstad, I. N., Mellembakken, J. R., Brandtzaeg, P., and Spencer, J. (2002). IgVH gene analysis suggests that peritoneal B cells do not contribute to the gut immune system in man. *Eur. J. Immunol.* 32:2427–2436.

Boursier, L., Gordon, J. N., Thiagamoorthy, S., Edgeworth, J. D., and Spencer J. (2005). Human intestinal IgA response is generated in the organized gut-associated lymphoid tissue but not in the lamina propria. Human intestinal IgA response is generated in the organized gut-associated lymphoid tissue but not in the lamina propria. *Gastroenterology* 128:1879–1889.

Bowman, E. P., Kuklin, N. A., Youngman, K. R., Lazarus, N. H., Kunkel, E. J., Pan, J., Greenberg, H. B., and Butcher, E. C. (2002). The intestinal chemokine thymus-expressed chemokine (CCL25) attracts IgA antibody-secreting cells. *J. Exp. Med.* 195:269–275.

Brandtzaeg, P., Farstad, I. N., Johansen, F. E., Morton, H. C., Norderhaug, I. N., and Yamanaka, T. (1999). The B-cell system of human mucosae and exocrine glands. *Immunol. Rev.* 171:45–87.

Brandtzaeg, P., Kett, K., Rognum, T. O., Soderstrom, R., Bjorkander, J., Soderstrom, T., Petrusson, B., and Hanson, L. A. (1986). Distribution of mucosal IgA and IgG subclass-producing immunocytes and alterations in various disorders. *Monogr. Allergy* 20:179–194.

Brandtzaeg, P., and Pabst, R. (2004). Let's go mucosal: Communication on slippery ground. *Trends Immunol.* 25:570–577.

Briere, F., Defrance, T., Vanbervliet, B., Bridon, J.-M., Durand, I., Rousset, F., and Banchereau, J. (1995). Transforming growth factor β (TGFβ) directs IgA1 and IgA2 switching in human naive B cells. *Adv. Exp. Med. Biol.* 371A:21–26.

Briskin, M. J., McEvoy, L. M., and Butcher, E. C. (1993). MAdCAM-1 has homology to immunoglobulin and mucin-like adhesion receptors and to IgA1. *Nature* 363:461–464.

Butcher, E. C., Rouse, R. V., Coffman, R. L., Nottenburg, C. N., Hardy, R. R., and Weissman, I. L. (1982). Surface phenotype of Peyer's patch germinal center cells: Implications for the role of germinal centers in B cell differentiation. *J. Immunol.* 129:2698–2707.

Casola, S., Otipoby, K. L., Alimzhanov, M., Humme, S., Uyttersprot, N., Kutok, J. L., Carroll, M. C., and Rajewsky, K. (2004). B cell receptor signal strength determines B cell fate. *Nat. Immunol.* 5:317–327.

Cazac, B. B., and Roes, J. (2000). TGF-b receptor controls B cell responsiveness and induction of IgA *in vivo*. *Immunity* 13:443–451.

Claassen, E., Kors, N., Dijkstra, C.D., and van Rooijen, N. (1986). Marginal zone of the spleen and the development and localization of specific antibody-forming cells against thymus-dependent and thymus-independent type-2 antigens. *Immunology* 57:399–403.

Clemens, J. D., Sack, D. A., Harris, J. R., Chakraborty, J., Khan, M. R., Stanton, B. F., Kay, B. A., Khan, M. U., Yunus, M., Atkinson, W., and Holmgren, J. (1986). Field trial of oral cholera vaccines in Bangladesh. *Lancet* 328:124–127.

Conley, M. E., and Delacroix, D. L. (1987). Intravascular and mucosal immunoglobulin A: Two separate but related systems of immune defense? *Ann. Intern. Med.* 106:892–899.

Cornes, J. S. (1965) Number, size and distribution of Peyer's patches in the human small intestine. *Gut* 6:225–233.

Craig, S. W., and Cebra, J. J. (1971) Peyer's patches: An enriched source of precursors for IgA-producing immunocytes in the rabbit. *J. Exp. Med.* 134:188–200.

Defrance, T., Vanbervliet, B., Briere, F., Durand, I., Rousset, F., and Banchereau, J. (1992). Interleukin 10 and transforming growth factor b cooperate to induce anti-CD40-activated naive human B cells to secrete immunoglobulin A. *J. Exp. Med.* 175:671–682.

Du, M., Diss, T. C., Xu, C., Peng, H., Isaacson, P.G., and Pan, L. (1996). Ongoing mutation in MALT lymphoma immunoglobulin gene suggests that antigen stimulation plays a role in the clonal expansion. *Leukemia* 10:1190–1197.

Dukes, C., and Bussey, H. J. R. (1926). The number of lymphoid follicles of the human large intestine. *J. Pathol. Bacteria* 29:111–116.

Dunn-Walters, D. K., Isaacson, P. G., and Spencer, J. (1996). Sequence analysis of rearranged IgVH genes from microdissected human Peyer's patch marginal zone B cells. *Immunology* 88:618–624.

Dunn-Walters, D. K., Isaacson, P. G., and Spencer, J. (1997). Sequence analysis of human IgVH genes indicates that ileal lamina propria plasma cells are derived from Peyer's patches. *Eur. J. Immunol.* 27:463–467.

Dunn-Walters, D. K. and Spencer J. (1998). Strong intrinsic biases towards mutation and conservation of bases in human IgVH genes during somatic hypermutation prevent statistical analysis of antigen selection. *Immunology* 95:339–345.

DuPont, H. L., Hornick, R. B., Snyder, M. J., Libonati, J. P., Formal, S. B., and Gangarosa, E. J. (1972). Immunity in shigellosis. II. Protection induced by oral live vaccine or primary infection. *J. Infect. Dis.* 125:12–16.

Fagarasan, S., Kinoshita, K., Muramatsu, M., Ikuta, K., and Honjo, T. (2001). *In situ* class switching and differentiation to IgA-producing cells in the gut lamina propria. *Nature* 413:639–643.

Fagarasan, S., Muramatsu, M., Suzuki, K., Nagaoka, H., Hiai, H., and Honjo, T. (2002). Critical roles of activation-induced cytidine deaminase in the homeostasis of gut flora. *Science* 298:1424–1427.

Falini, B., Tiacci, E., Pucciarini, A., Bigerna, B., Kurth, J., Hatzivassiliou, G., Droetto, S., Galletti, B. V., Gambacorta, M., Orazi, A., Pasqualucci, L., Miller, I., Kuppers, R., Dalla-Favera, R., and Cattoretti, G. (2003). Expression of the IRTA1 receptor identifies intraepithelial and subepithelial marginal zone B cells of the mucosa-associated lymphoid tissue (MALT). *Blood* 102:3684–3692.

Farstad, I. N., Carlsen, H., Morton, H. C., and Brandtzaeg, P. (2000). Immunoglobulin A cell distribution in the human small intestine: Phenotypic and functional characteristics. *Immunology* 101:354–363.

Farstad, I. N., Halstensen, T. S., Fausa, O., and Brandtzaeg, P. (1994). Heterogeneity of M-cell-associated B and T cells in human Peyer's patches. *Immunology* 83:457–464.

Farstad, I. N., Halstensen, T. S., Lazarovits, A. I., Norstein, J., Fausa, O., and Brandtzaeg, P. (1995). Human intestinal B-cell blasts and plasma cells express the mucosal homing receptor integrin a4b7. *Scand. J. Immunol.* 42:662–672.

Fayette, J., Dubois, B., Vandenabeele, S., Bridon, J.-M., Vanbervliet, B., Durand, I., Banchereau, J., Caux, C., and Brière, F. (1997). Human dendritic cells skew isotype switching of CD40-activated naive B cells towards IgA1 and IgA2. *J. Exp. Med.* 185:1909–1918.

Finke, D., Acha-Orbea, H., Mattis, A., Lipp, M., and Kraehenbuhl, J. (2002). $CD4^{+}CD3^{-}$ cells induce Peyer's patch development: role of $\alpha 4\beta 1$ integrin activation by CXCR5. *Immunity* 17:363–373.

Fischer, M., and Kuppers, R. (1998). Human IgA- and IgM-secreting intestinal plasma cells carry heavily mutated VH region genes. *Eur. J. Immunol.* 28:2971–2977.

Formal, S. B., Kent, T. H., May, H. C., Palmer, A., Falkow, S., and LaBrec, E. H. (1966). Protection of monkeys against experimental shigellosis with a living attenuated oral polyvalent dysentery vaccine. *J. Bacteriol.* 92:17–22.

Gardby, E., Wrammert, J., Schon, K., Ekman, L., Leanderson, T., and Lycke, N. (2003). Strong differential regulation of serum and mucosal IgA responses as revealed in CD28-deficient mice using cholera toxin adjuvant. *J. Immunol.* 170:55–63.

Genta, R. M., Hamner, H. W., and Graham, D. Y. (1993). Gastric lymphoid follicles in *Helicobacter pylori* infection: Frequency, distribution, and response to triple therapy. *Hum. Pathol.* 24:577–583.

Goodrich, M. E., and McGee, D. W. (1998). Regulation of mucosal B cell immunoglobulin secretion by intestinal epithelial cell-derived cytokines. *Cytokine* 10:948–955.

Gowans, J. L., and Knight, E. J. (1964). The route of recirculation of lymphocytes in the rat. *Proc. R. Soc. Lond. B: Biol. Sci.* 159:257–282.

Guilliano, M. J., Foxx-Orenstein, A. E., and Lebman, D. A. (2001). The microenvironment of human Peyer's patches inhibits the increase in CD38 expression associated with the germinal center reaction. *J. Immunol.* 166:2179–2185.

Hamada, H., Hiroi, T., Nishiyama, Y., Takahashi, H., Masunaga, Y., Hachimura, S., Kaminogawa, S., Takahashi-Iwanaga, H., Iwanaga, T., Kiyono, H., Yamamoto, H., and Ishikawa, H. (2002). Identification of multiple isolated lymphoid follicles on the antimesenteric wall of the mouse small intestine. *J. Immunol.* 168:57–64.

Hardie, D. L., Johnson, G. D., Khan, M., and MacLennan I. C. (1993). Quantitative analysis of molecules which distinguish functional compartments within germinal centers. *Eur J Immunol.* 23:997–1004.

Haubrich, W. S. (2005). Peyer of Peyer's patches. *Gastroenterology* 129:85.

Holmgren, J., Lycke, N., and Czerkinsky, C. (1993). Cholera toxin and cholera B subunit as oral-mucosal adjuvant and antigen vector systems. *Vaccine* 11:1179–1184.

Holtmeier, W., Hennemann, A., and Caspary, W. F. (2000). IgA and IgM V(H) repertoires in human colon: Evidence for clonally expanded B cells that are widely disseminated. *Gastroenterology* 119:1253–1266.

Husband, A. J., (1982). Kinetics of extravasation and redistribution of IgA-specific antibody-containing cells in the intestine. *J. Immunol.* 128:1355–1359.

Husband, A. J. and Gowans, J. L. (1978). The origin and antigen-dependent distribution of IgA-containing cells in the intestine. *J. Exp. Med.* 148:1146–1160.

Kroese, F. G., Butcher, E. C., Stall, A. M., Lalor, P. A., Adams, S., and Herzenberg, L. A. (1989). Many of the IgA producing plasma cells in murine gut are derived from self-replenishing precursors in the peritoneal cavity. *Int. Immunol.* 1:75–78.

Kunkel, E. J., and Butcher, E. C. (2003). Plasma-cell homing. *Nat. Rev. Immunol.* 3:822–829.

Kunkel, E. J., Kim, C. H., Lazarus, N. H., Vierra, M. A., Soler, D., Bowman, E. P., and Butcher, E. C. (2003). CCR10 expression is a common feature of circulating and mucosal epithelial tissue IgA Ab-secreting cells. *J. Clin. Invest.* 111:1001–1010.

Kweon, M. N., Yamamoto, M., Rennert, P. D., Park, E. J., Lee, A. Y., Chang, S. Y., Hiroi, T., Nanno, M., and Kiyono, H. (2005). Prenatal blockage of lymphotoxin β receptor and TNF receptor p55 signaling cascade resulted in the acceleration of tissue genesis for isolated lymphoid follicles in the large intestine. *J. Immunol.* 174:4365–4372.

Lanning, D. K., Rhee, K. J., and Knight, K. L. (2005). Intestinal bacteria and development of the B-lymphocyte repertoire. *Trends Immunol.* 26:419–425.

Lanning, D., Zhu, X., Zhai, S. K., and Knight, K. L. (2000). Development of the antibody repertoire in rabbit: gut-associated lymphoid tissue, microbes, and selection. *Immunol Rev*. 175:214–228.

Litinskiy, M. B., Nardelli, B., Hilbert, D. M., He, B., Schaffer, A., Casali, P., and Cerutti, A. (2002). DCs induce CD40-independent immunoglobulin class switching through BLyS and APRIL. *Nat. Immunol*. 3:822–829.

Liu, Y.-J., Malisan, F., de Bouteiller, O., Guret, C., Lebecque, S., Banchereau, J., Mills, F. C., Max, E. E., and Martinez-Valdez, H. (1996). Within germinal centers, isotype switching of immunoglobulin genes occurs after the onset of somatic mutation. *Immunity* 4:241–250.

Liu, Y.-J., Oldfield. S., and MacLennan, I. C. (1988). Memory B cells in T cell-dependent antibody responses colonize the splenic marginal zones. *Eur. J. Immunol*. 18:355–362.

Lorenz, R. G., and Newberry, R. D. (2004). Isolated lymphoid follicles can function as sites for induction of mucosal immune responses. *Ann. NY Acad. Sci*. 1029:44–57.

Lue, C., van den Wall Bake, A. W., Prince, S. J., Julian, B. A., Tseng, M. L., Radl, J., Elson, C. O., and Mestecky J. (1994). Intraperitoneal immunization of human subjects with tetanus toxoid induces specific antibody-secreting cells in the peritoneal cavity and in the circulation, but fails to elicit a secretory IgA response. *Clin. Exp. Immunol*. 96:356–363.

McDonald, K. G., McDonough, J. S., and Newberry, R. D. (2005). Adaptive immune responses are dispensable for isolated lymphoid follicle formation: Antigen-naive, lymphotoxin-sufficient B lymphocytes drive the formation of mature isolated lymphoid follicles. *J. Immunol*. 174:5720–5728.

MacDonald, T. T., Spencer, J., Viney, J. L., Williams, C. B., and Walker-Smith, J. A. (1987). Selective biopsy of human Peyer's patches during ileal endoscopy. *Gastroenterology* 93:1356–1362.

Macpherson, A. J., Gatto, D., Sainsbury, E., Harriman, G. R., Hengartner, H., and Zinkernagel, R. M. (2000). A primitive T cell-independent mechanism of intestinal mucosal IgA responses to commensal bacteria. *Science* 288:2222–2226.

Macpherson, A. J., Lamarre, A., McCoy, K., Harriman, G. R., Odermatt, B., Dougan, G., Hengartner, H., and Zinkernagel, R. M. (2001). IgA production without μ or δ chain expression in developing B cells. *Nat. Immunol*. 2:625–631.

Macpherson, A. J., and Uhr T. (2004). Induction of protective IgA by intestinal dendritic cells carrying commensal bacteria. *Science* 303:1662–1665.

Medina, F., Segundo, C., Campos-Caro, A., González-García, I., and Brieva, J. A. (2002). The heterogeneity shown by human plasma cells from tonsil, blood, and bone marrow reveals graded stages of increasing maturity, but local profiles of adhesion molecule expression. *Blood* 99:2154–2161.

Medina, F., Segundo, C., Campos-Caro, A., Salcedo, I., Garcia-Poley, A., and Brieva, J. A. (2003). Isolation, maturational level, and functional capacity of human colon lamina propria plasma cells. *Gut* 52:383–389.

Mel, D. M., Terzin, A. L., and Vuksic, L. (1965). Studies on vaccination against bacillary dysentery. 3. Effective oral immunization against *Shigella flexneri* 2a in a field trial. *Bull. World Health Org*. 32:647–655.

Michetti, P., Mahan, M. J., Slauch, J. M., Mekalanos, J. J., and Neutra, M. R. (1992). Monoclonal secretory immunoglobulin A protects mice against oral challenge with the invasive pathogen *Salmonella typhimurium*. *Infect. Immun*. 60:1786–1792.

Moghaddami, M., Cummins, A., and Mayrhofer, G. (1998). Lymphocyte-filled villi: Comparison with other lymphoid aggregations in the mucosa of the human small intestine. *Gastroenterology* 115:1414–1425.

Muramatsu, M., Kinoshita, K., Fagarasan, S., Yamada, S., Shinkai, Y., and Honjo, T. (2000). Class switch recombination and hypermutation require activation-induced cytidine deaminase (AID), a potential RNA editing enzyme. *Cell* 102:553–563.

Murakami, M., Tsubata, T., Shinkura, R., Nisitani, S., Okamoto, M., Yoshioka, H., Usui, T., Miyawaki, S., and Honjo T. (1994). Oral administration of lipopolysaccharides activates B-1 cells in the peritoneal cavity and lamina propria of the gut and induces autoimmune symptoms in an autoantibody transgenic mouse. *J. Exp. Med.* 180:111–121.

Nishikawa, S., Honda, K., Vieira, P., and Yoshida, H. (2003). Organogenesis of peripheral lymphoid organs. *Immunol. Rev.* 195:72–80.

Ogra, P. L., and Karzon, D. T. (1969). Distribution of poliovirus antibody in serum, nasopharynx and alimentary tract following segmental immunization of lower alimentary tract with poliovaccine. *J. Immunol.* 102:1423–1430.

O'Leary, A. D., and Sweeney, E. C. (1986). Lymphoglandular complexes of the colon: Structure and distribution. *Histopathology* 10:267–283.

Owen, R. L., and Jones, A. L. (1974). Epithelial cell specialization within human Peyer's patches: An ultrastructural study of intestinal lymphoid follicles. *Gastroenterology* 66:189–203.

Pan, J., Kunkel, E. J., Gosslar, U., Lazarus, N., Langdon, P., Broadwell, K., Vierra, M. A., Genovese, M. C., Butcher, E. C., and Soler, D. (2000). A novel chemokine ligand for CCR10 and CCR3 expressed by epithelial cells in mucosal tissues. *J. Immunol.* 165:2943–2949.

Pappo, J., and Owen, R. L. (1988). Absence of secretory component expression by epithelial cells overlying rabbit gut-associated lymphoid tissue. *Gastroenterology* 95:1173–1177.

Pascual, V., Liu, Y.-J., Magalski, A., de Bouteiller, O., Banchereau, J., and Capra, J. D. (1994). Analysis of somatic mutation in five B cell subsets of human tonsil. *J. Exp. Med.* 180:329–339.

Phalipon, A., Kaufmann, M., Michetti, P., Cavaillon, J.-M., Huerre, M., Sansonetti, P., and Kraehenbuhl, J.-P. (1995). Monoclonal immunoglobulin A antibody directed against serotype-specific epitope of *Shigella flexneri* lipopolysaccharide protects against murine experimental shigellosis. *J. Exp. Med.* 182:769–778.

Rescigno, M., Urbano. M., Valzasina, B., Francolini, M., Rotta, G., Bonasio, R., Granucci, F., Kraehenbuhl, J.-P., and Ricciardi-Castagnoli, P. (2001). Dendritic cells express tight junction proteins and penetrate gut epithelial monolayers to sample bacteria. *Nat. Immunol.* 2:361–367.

Reynaud, C.-A., Garcia, C., Hein, W. R., and Weill, J.-C. (1995). Hypermutation generating the sheep immunoglobulin repertoire is an antigen-independent process. *Cell* 80:115–125.

Salmi, M., and Jalkanen, S. (2005). Lymphocyte homing to the gut: Attraction, adhesion, and commitment. *Immunol. Rev.* 206:100–113.

Shikina, T., Hiroi, T., Iwatani, K., Jang, M. H., Fukuyama, S., Tamura, M., Kubo, T., Ishikawa, H., and Kiyono, H. (2004). IgA class switch occurs in the organized nasopharynx- and gut-associated lymphoid tissue, but not in the diffuse lamina propria of airways and gut. *J Immunol.* 172:6259–6264.

Sierro, F., Pringault, E., Assman, P. S., Kraehenbuhl, J.-P., and Debard, N. (2000). Transient expression of M-cell phenotype by enterocyte-like cells of the follicle-associated epithelium of mouse Peyer's patches. *Gastroenterology* 119:734–743.

Simmons, D. A., and Romanowska, E. (1987). Structure and biology of *Shigella flexneri* O antigens. *J. Med. Microbiol.* 23:289–302.

Smith, M. W. (1985). Selective expression of brush border hydrolases by mouse Peyer's patch and jejunal villus enterocytes. *J. Cell. Physiol.* 124:219–225.

Spencer, J., Finn, T., and Isaacson, P. G. (1986a) Human Peyer's patches: An immunohistochemical study. *Gut* 27:405–410.

Spencer, J., MacDonald, T. T., Finn, T., and Isaacson, P. G. (1986b). The development of gut associated lymphoid tissue in the terminal ileum of fetal human intestine. *Clin. Exp. Immunol.* 64:536–543.

Spencer, J., MacDonald. T. T., and Isaacson. P. G. (1987). Heterogeneity of non-lymphoid cells expressing HLA-D region antigens in human fetal gut. *Clin. Exp. Immunol.* 67:415–424.

Tanaka, Y., Imai, T., Baba, M., Ishikawa, I., Uehira, M., Nomiyama, H., and Yoshie, O. (1999). Selective expression of liver and activation-regulated chemokine (LARC) in intestinal epithelium in mice and humans. *Eur. J. Immunol.* 29:633–642.

Tangye, S. G., Ferguson, A., Avery, D. T., Ma, C. S., and Hodgkin P. D. (2002). Isotype switching by human B cells is division-associated and regulated by cytokines. Isotype switching by human B cells is division-associated and regulated by cytokines. *J. Immunol.* 169:4298–4306.

Thurnheer, M. C., Zuercher, A. W., Cebra, J. J., and Bos, N. A. (2003). B1 cells contribute to serum IgM, but not to intestinal IgA, production in gnotobiotic Ig allotype chimeric mice. *J. Immunol.* 170:4564–4571.

Viau, M., and Zouali, M. (2005). B-lymphocytes, innate immunity, and autoimmunity. *Clin. Immunol.* 114:17–26.

Wijburg, O. L. C., Uren, T. K., Simpfendorfer, K., Johansen, F.-E., Brandtzaeg, P., and Strugnell, R.A. (2006). Innate secretory antibodies protect against natural *Salmonella typhimurium* infection. *J. Exp. Med.* 203:21–26.

Xu-Amano, J., Kiyono, H., Jackson, R. J., Staats, H. F., Fujihashi, K., Burrows, P. D., Elson, C. O., Pillai, S., and McGhee, J. R. (1993). Helper T cell subsets for immunoglobulin A responses: Oral immunization with tetanus toxoid and cholera toxin as adjuvant selectively induces Th2 cells in mucosa associated tissues. *J. Exp. Med.* 178:1309–1320.

Yamanaka, T., Helgeland, L., Farstad, I. N., Fukushima, H., Midtvedt, T., and Brandtzaeg, P. (2003). Microbial colonization drives lymphocyte accumulation and differentiation in the follicle-associated epithelium of Peyer's patches. *J. Immunol.* 170:816–822.

Yamanaka, T., Straumfors, A., Morton, H., Fausa, O., Brandtzaeg, P., and Farstad, I. N. (2001). M cell pockets of human Peyer's patches are specialized extensions of germinal centers. *Eur. J. Immunol.* 31:107–117.

Zan, H., Cerutti, A., Dramitinos, P., Schaffer, A., and Casali, P. (1998). CD40 engagement triggers switching to IgA1 and IgA2 in human B cells through induction of endogenous TGF-β: Evidence for TGF-β but not IL-10-dependent direct Sμ→Sα and sequential Sμ→Sγ, Sγ→Sα DNA recombination. *J. Immunol.* 161:5217–5225.

Zhao, X., Sato, A., Dela Cruz, C. S., Linehan, M., Luegering, A., Kucharzik, T., Shirakawa A. K., Marquez, G., Farber, J. M., Williams, I., and Iwasaki, A. (2003). CCL9 is secreted by the follicle-associated epithelium and recruits dome region Peyer's patch $CD11b^+$ dendritic cells. *J. Immunol.* 171:2797–2803.

3 Epithelial Transport of IgA by the Polymeric Immunoglobulin Receptor

Charlotte S. Kaetzel[1] and Maria E. C. Bruno[1]

[1] Department of Microbiology, Immunology and Molecular Genetics, University of Kentucky College of Medicine, Lexington KY 40536, USA.

3.1. Introduction

The mucosal surfaces lining the gastrointestinal, respiratory and genitourinary tracts are continuously bombarded by potentially infectious agents such as bacteria, viruses, fungi, and parasites, in addition to soluble dietary and environmental substances. The first line of specific immunological defense against these environmental antigens is secretory IgA (SIgA) (Brandtzaeg et al., 1997; Lamm, 1997), which is produced by selective transport of polymeric IgA (pIgA) across epithelial cells lining mucosal surfaces (Kaetzel, 2005; Kaetzel and Mostov, 2005; Norderhaug et al., 1999). The magnitude of this transport process is impressive; it has been estimated that ~3 g of SIgA are transported daily into the intestines of the average adult (Conley and Delacroix, 1987; Mestecky et al., 1986). Transport of polymeric immunoglobulins (IgA and, to a lesser extent, IgM) across mucosal epithelial cells is mediated by a transmembrane glycoprotein called the polymeric immunoglobulin receptor (pIgR).

Forty years ago, Tomasi et al. (1965) isolated SIgA and demonstrated that it comprised a dimer of IgA subunits, joined by a small polypeptide called the "J-chain," and covalently bound to a glycoprotein of about 80 kDa [originally designated the "secretory piece" and now called the secretory component (SC)]. Immunohistochemical studies with specific antibodies to SC demonstrated that it is synthesized not by plasma cells but by epithelial cells lining mucous membranes and exocrine glands (reviewed in Kaetzel and Mostov, 2005). The paradox was that SC was a soluble secretory protein, whereas one would expect that the receptor for transcytosis of IgA would be an integral membrane protein. Resolution of this paradox came with the discovery that SC is a proteolytic fragment of an integral membrane protein, named the polymeric immunoglobulin receptor (pIgR) (Mostov et al., 1980). Brandtzaeg and Prydz (1984) provided direct evidence for an integrated function of the J-chain and pIgR in the epithelial transport of immunoglobulins, by demonstrating that only polymeric IgA and IgM containing the J-chain could bind to the surface of human intestinal epithelial cells expressing pIgR.

The pathway of pIgR-mediated transport of pIgA across polarized epithelial cells has now been characterized thoroughly (reviewed in Kaetzel and Mostov, 2005; Mostov et al., 2003) (Fig. 3.1). The pIgR is synthesized as an integral membrane protein in the rough endoplasmic reticulum and then travels to the Golgi apparatus. In the last station of the Golgi, known as the trans-Golgi network (TGN), pIgR is sorted into vesicles that deliver it to the basolateral surface of the epithelial cell. At that surface, pIgR can bind to pIgA that is produced by plasma cells, most commonly found in the lamina propria underlying the epithelium. With or without bound pIgA, pIgR is

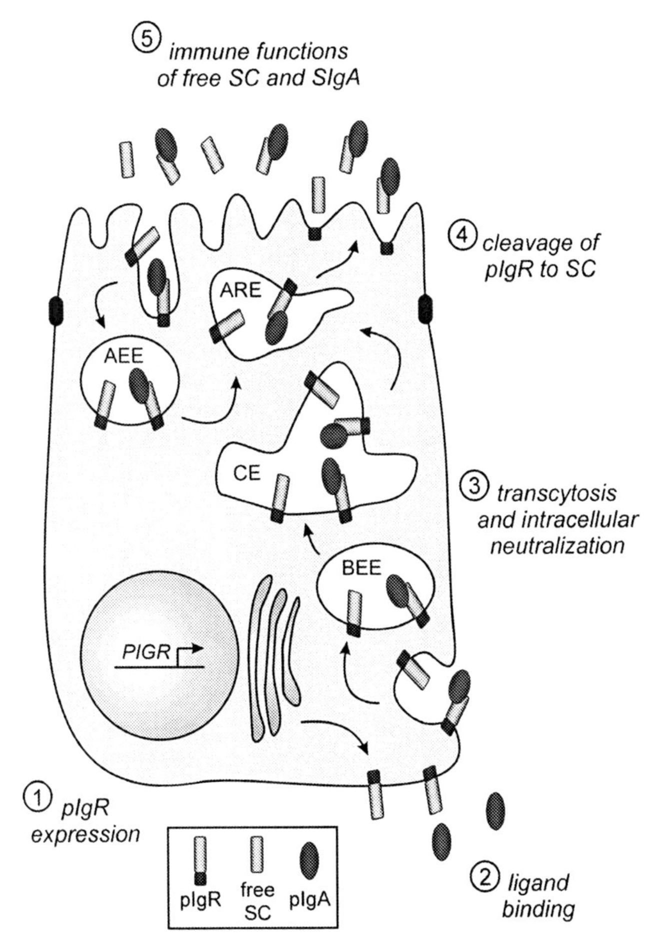

FIG. 3.1. Pathway of the pIgR through a polarized epithelial cell. A simplified epithelial cell is illustrated, with the apical surface at the top and the basolateral surface at the bottom. Newly synthesized pIgR is targeted to the basolateral surface, where ligand binding occurs. Following receptor-mediated endocytosis, ligand-bound or unoccupied pIgR is transported through a series of intracellular vesicles, in which neutralization of pathogens and antigens can take place. At the apical surface, pIgR is proteolytically cleaved to SC. At the mucosal surface and in external secretions, free SC and SIgA contribute to innate and adaptive immune defense. pIgA, polymeric IgA; BEE, basolateral early endosome; CE, common endosome; ARE, apical recycling endosome; AEE, apical early endosome.

endocytosed and delivered to basolateral early endosomes (BEEs). The receptor and ligand then move through a common endosomal compartment (CE) and are sorted into apical recycling endosomes (AREs) for delivery to the apical membrane. At this surface, cleavage of the extracellular ligand-binding portion of pIgR results in release of SC in free form or as part of the SIgA complex. A fraction of the pIgR at the apical surface might be reinternalized into apical early endosomes (AEEs) and then delivered back to the apical surface through the ARE. Mice with two disrupted alleles at the *Pigr* locus have markedly reduced IgA in external secretions, accompanied by elevated serum IgA (Johansen et al., 1999; Shimada et al., 1999), demonstrating that pIgR is required for transcytosis of pIgA across mucosal epithelia *in vivo*.

Current investigations into the immunobiology of the pIgR are focused in five major areas (highlighted in Fig. 3.1): regulation of pIgR expression in mucosal epithelial cells; binding of ligands to pIgR; transcytosis of pIgR and the role of pIgA in intracellular neutralization of pathogens and antigens; cleavage of pIgR to SC at the apical surface; and novel functions of free SC and SIgA at the epithelial surface and in external secretions. This chapter, along with other chapters in this volume, will highlight recent findings regarding the important role of pIgR in mucosal defense and homeostasis.

3.2. Regulation of pIgR Expression

The pIgR is unique among immunoglobulin Fc receptors in two ways: first, pIgR expression is restricted to mucosal and glandular epithelial cells and, second,, pIgR makes only one trip across these cells before being cleaved and released at the apical surface. Because of the 1:1 stoichiometry between SC and dimeric IgA in SIgA, one molecule of pIgR must be produced for every molecule of dimeric IgA that is transported across an epithelial cell. It follows that upregulation of pIgR expression would increase the capacity for epithelial transcytosis of IgA. Regulation of pIgR expression involves complex interactions among host-, microbial-, and environmental-derived factors, involving transcriptional and posttranscriptional mechanisms (reviewed in Johansen and Brandtzaeg, 2004; Kaetzel, 2005; Kaetzel and Mostov, 2005) (see also Chapter 5).

3.2.1. Structure of the Human PIGR Gene

3.2.1.1. Exon–Intron Structure

The National Center for Biotechnology Information has compiled data regarding the chromosomal location and exon–intron structure for *PIGR* genes from several mammalian and one avian species (www.ncbi.nlm.nih.gov/entrez/query.fcgi?db=gene). The human *PIGR* gene comprises 11 exons and 10 introns spanning 17,944 base pairs on chromosome 1q31-q41 (Gene ID: 5284) (Fig. 3.2A). *PIGR* genes of similar structure have been characterized on chromosome 1 of two primate species: *Pan troglodytes* (chimpanzee; Gene ID: 457685) and *Macaca mulatta* (Rhesus monkey; GeneID: 694638). Although significant divergence in sequence has occurred during evolution of the *PIGR* gene (see Sect. 3.3.1), the exon–intron structure appears to have been

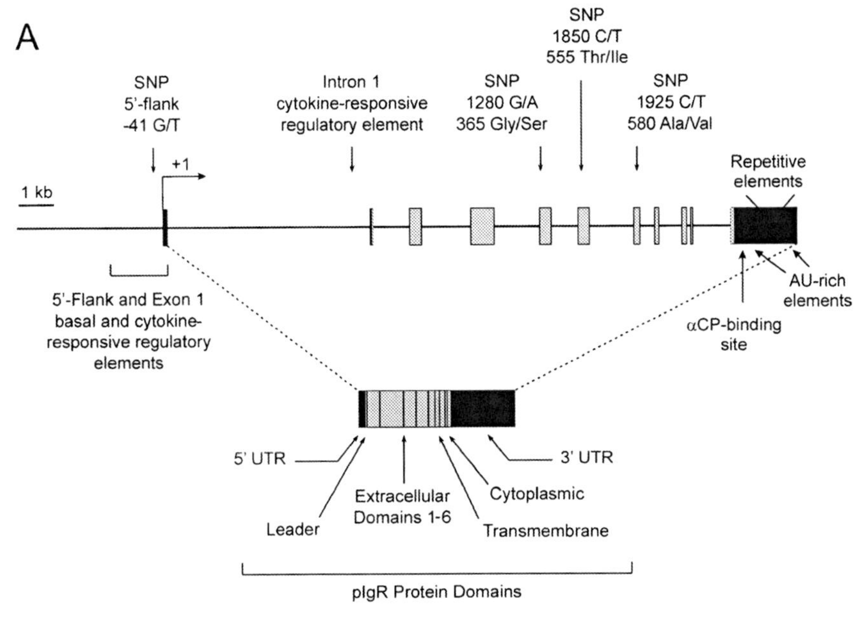

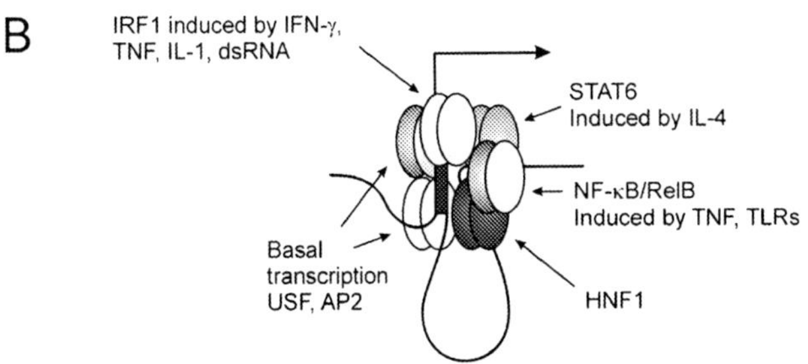

FIG. 3.2. Organization of the human polymeric immunoglobulin receptor (*PIGR*) gene locus and regulatory elements. **(A)** Eleven exons comprise the *PIGR* gene, encoding the 5′-UTR (black box), the coding region (gray boxes), and the 3′-UTR (solid box) of pIgR mRNA. Two regulatory regions containing binding sites for basal and cytokine-responsive transcription factors are designated. Also shown are the locations of four SNPs associated with this gene locus. A schematic of the human pIgR mRNA structure indicates the relative positions of exons encoding the 5′-UTR, protein domains, and 3′-UTR. **(B)** Working model for regulation of *PIGR* gene transcription by cytokines and microbial factors. See the text for a detailed discussion of potential interactions among transcription factors. AP, activator protein; HNF, hepatocyte nuclear factor; IRF, interferon regulatory factor; NF-κB, nuclear factor-kappaB; STAT, signal transducer and activator of transcription; USF, upstream stimulatory factor.

conserved across mammalian species, including *Bos taurus* (cow; chromosome 16q13, Gene ID: 281401), *Canis familiaris* (dog; chromosome 7, Gene ID: 474357), *Mus musculus* (mouse; chromosome 1, 68.2 cM, Gene ID: 18703), and *Rattus norvegicus* (rat; chromosome 13q13, Gene ID: 25046) and is also similar in an avian species, *Gallus gallus* (chicken; chromosome 26, Gene ID: 419848). The relationship of the exon–intron structure to untranslated and coding regions of pIgR mRNA is shown in Figure 3.2A.

3.2.1.2. Binding Sites for Transcription Factors

The proximal 5′-flanking region of the *PIGR* gene contains binding sites for transcription factors that regulate basal transcription of pIgR mRNA. Mutational analyses have demonstrated that an "E-Box" motif at position −71 of the human *PIGR* gene and −74 of the mouse *Pigr* gene is essential for basal promoter activity (Hempen et al., 2002; Johansen et al., 1998; Martín et al., 1998; Solorzano-Vargas et al., 2002). E-Box motifs with the sequence "CAC$^{G}/_{A}$TG" bind to transcription factors of the basic helix–loop–helix/leucine zipper family, including the Myc, upstream stimulatory factor (USF) and transcription factor E (TFE) subfamilies (Kiermaier et al., 1999; Lüscher and Larsson, 1999; Sirito et al., 1998). We have demonstrated that USF-1 and USF-2 but not c-Myc bind to the *PIGR* E-Box *in vitro* and *in vivo* and enhance promoter activity (Bruno et al., 2004). Activator protein-2 (AP2) binds to a site adjacent to the USF site and cooperatively enhances *PIGR* promoter activity (Hempen et al., 2002). Dysregulation of USF and AP2 expression in non-small-cell lung cancer has been shown to correlate with downregulation of pIgR expression relative to adjacent normal tissue (Khattar et al., 2005). Expression of pIgR is also decreased in adenomas and carcinomas of the colon (reviewed in Kaetzel and Mostov, 2005), through as-yet undiscovered mechanisms that might involve both transcriptional and posttranscriptional regulation (Traicoff et al., 2003).

Binding sites for transcription factors that are induced by cytokines, hormones, and microbial factors have been identified in the 5′-flanking, exon 1, and intron 1 of the *PIGR* gene. Putative regulatory elements in the 5′-flanking region include two potential sites for the cytokine-inducible transcription factor nuclear factor (NF)-κB (Takenouchi-Ohkubo et al., 2000) and sites for the androgen receptor (Claessens et al., 2001) in the human *PIGR* gene and a glucocorticoid-responsive element in the mouse *Pigr* gene (Li et al., 1999). An "interferon-sensitive response element" (ISRE) in exon 1, comprising the sequence "AGTTTCAGTTTT," is 100% conserved among species for which the complete nucleotide sequence of exon 1 is known, including human, chimpanzee, cow, dog, rat, and mouse. ISREs bind members of the interferon regulatory factor (IRF) family of transcription factors, which are induced by proinflammatory cytokines as well as microbial products (Honda and Taniguchi, 2006). A composite regulatory element, with binding sites for tissue-specific and cytokine-inducible transcription factors, has been identified in intron 1 of the human *PIGR* and mouse *Pigr* genes (Schjerven et al., 2000). The function of inducible regulatory elements in the control of pIgR mRNA transcription is discussed in Section 3.2.2 and in Chapter 5.

3.2.1.3. Elements in the 3′-Untranslated Region

Motifs that regulate the subcellular localization, stability, and translation of mRNA are often found in the untranslated regions (UTRs) of eukaryotic genes (Mignone et al., 2002). The complete sequences of the 5′-UTRs of human, rat, and mouse pIgR mRNA have been reported (Fodor et al., 1997; Martín et al., 1997; Piskurich et al., 1997), but their potential roles in the regulation of pIgR mRNA processing or translation have not been assessed. The complete 1793-nt sequence of the 3′-UTR of human pIgR mRNA has been determined, revealing the presence of two tandem *Alu* repeats, as well as elements that could affect mRNA processing and stability (Traicoff et al., 2003) (Fig. 3.2A). We have recently demonstrated that treatment of human intestinal epithelial cells with tumor necrosis factor (TNF) enhances the stability of pIgR mRNA (Bruno and Kaetzel, 2005). As expected, the 3′-UTR of chimpanzee pIgR mRNA is highly similar to that of human pIgR mRNA, including *Alu* repeats (GenBank accession number XM_514153). In contrast, homology in the 3′-UTR of pIgR mRNA among nonprimate mammals is quite low. The 1183-nt 3′-UTR of bovine pIgR mRNA (Kulseth et al., 1995; Verbeet et al., 1995) and the 1068-nt 3′-UTR of rabbit pIgR mRNA (Mostov et al., 1984) do not contain repetitive sequences or putative regulatory elements such as those found in primate pIgR mRNA. The 1416-nt 3′-UTR of rat pIgR mRNA is exceptional in that at least six unique alternatively processed forms have been identified (Koch et al., 1995). All of the pIgR mRNA transcripts contained unusual S1-nuclease-sensitive microsatellite elements consisting of multiple tandem d[GGA] and d[GAA] repeats, which are also found in mouse pIgR mRNA (Kushiro and Sato, 1997), but not in any other species yet studied. The microsatellite repeats in the 3′-UTRs of rodent pIgR mRNAs are located in approximately the same position as the *Alu* repeats in human pIgR mRNA, suggesting that this location might represent a genomic "hot spot" for insertion of repetitive elements. Attenuation of gene expression was observed when a fragment containing the microsatellite elements from rat pIgR mRNA was inserted into the 3′-UTR of a luciferase reporter gene, suggesting that it might act as a negative regulatory element (Aoki et al., 1997). The 3′-UTR is considerably shorter in pIgR mRNA from two marsupial species: 480-nt for the common brushtail possum (Adamski and Demmer, 1999) and 485-nt for the tammar wallaby (Taylor et al., 2002). The longer 3′-UTR of pIgR mRNA in eutherian mammals might represent an evolutionarily recent acquisition of novel mechanisms for regulating the transcription and/or stability of pIgR mRNA.

3.2.1.4. Polymorphisms in the Human *PIGR* Gene

As the Human Genome Project has expanded to include nucleotide sequences from diverse population groups, a number of single-nucleotide polymorphisms (SNPs) have been reported within the human *PIGR* gene locus

(National Center for Biotechnology Information, www.ncbi.nlm.nih.gov/projects/SNP/). Among SNPs with a frequency of heterozygosity of 0.1 or higher, four have been identified that might impact expression or function of pIgR (Fig. 3.2A). Allelic frequencies at these SNPs vary considerably among populations in different geographical locations, and two of these SNPs have been associated with human diseases (see below).

A SNP at position −41 (relative to the start site of transcription) falls within a previously identified binding site for the transcription factor AP2 (Hempen et al., 2002). The ancestral allele at this position (based on the sequence of the chimpanzee *PIGR* gene) is “G”; the impact of substitution of the minor “T” allele at that position on AP2 binding or *PIGR* transcription has not yet been determined. Based on 612 chromosomal sequences from 343 individuals of European, Asian, and African descent, heterozygosity at position −41 was estimated to be 0.408 ± 0.193 (http://www.ncbi.nlm.nih.gov/SNP/snp_ref.cgi?rs=2275529). Frequency of the ancestral “G” allele ranged from a high of about 0.9 for three different Asian populations to a low of 0.38 for a group from Sub-Saharan Africa and 0.39 for a group of African Americans. Significantly, the presence of the minor “T” allele was associated with increased risk for IgA nephropathy in a cohort of 389 Japanese patients and 465 controls (odds ratio = 1.60, $p = 0.00055$) (Obara et al., 2003). These investigators suggested that reduced expression of pIgR could impair epithelial transcytosis of IgA at mucosal surfaces, leading to increases in circulating IgA immune complexes that ultimately deposit in the renal glomeruli (see Chapter 13 for a discussion of the pathogenesis of IgA nephropathy). This interpretation was complicated by the fact that the −41 SNP was in linkage disequilibrium with another disease-associated SNP in the *PIGR* gene (see below). It is not known whether the relatively high frequency of the minor “T” allele predisposes individuals of African or African American ethnicity to increased risk for IgA nephropathy or whether additional risk factors in the Japanese population contributed to the disease association of the −41 *PIGR* SNP.

Another important SNP in the human *PIGR* gene is located within the coding region, at position 1925 of pIgR mRNA. Substitution of the ancestral “C” allele with the minor “T” allele causes a change in amino acid 580 from Ala to Val within domain 6, the linker peptide in which cleavage of pIgR to SC occurs (see Fig. 3.4). It is not known whether the Ala/Val substitution alters the structure of the linker peptide or alters the rate of cleavage of pIgR, which could affect the rate of epithelial transcytosis of IgA (see Sect. 3.6). Based on 554 chromosomal sequences from 277 individuals of European, Asian, and African descent, heterozygosity at position 1925 was estimated to be 0.536 ± 0.212 (http://www.ncbi.nlm.nih.gov/SNP/snp_ref.cgi?rs=291102). Frequency of the ancestral allele varied dramatically among ethnic groups, ranging from about 0.9 for Europeans and Asians to 0.15 for individuals from Sub-Saharan Africa (data on this SNP were not available for the cohort of African Americans). Two independent studies reported association of this SNP with diseases in which the role of pIgR in epithelial transcytosis of IgA might be relevant. Obara et al. (2003) reported that the presence of the minor "T" allele at position 1925 was associated with increased risk for IgA nephropathy in a Japanese population (odds ratio = 1.59, $p = 0.0003$). This SNP was in significant linkage disequilibrium (0.901) with the minor "T" allele at position −41 (see above), suggesting that one or the other (or both) of these SNPs contributed to increased risk of IgA nephropathy. Although no studies have been reported on the linkage of *PIGR* SNPs with IgA nephropathy in African populations, it is noteworthy that the prevalence of the minor alleles at both position −41 and 1925 is high in the African cohort. Another group reported a significant association of the *PIGR* 1925 SNP with Epstein–Barr virus (EBV)-associated nasopharyngeal cancer (Hirunsatit et al., 2003) (see Chapter 7 for a discussion of the role of pIgR and IgA in the pathogenesis of EBV infection). In contrast to its effect on susceptibility to IgA nephropathy, the presence of the minor "T" allele was associated with reduced risk of nasopharyngeal cancer in both Chinese (odds ratio = 2.35, $p < 0.05$) and Thai (odds ratio = 1.70, $p < 0.05$) populations. This study did not investigate a possible association between the −41 *PIGR* SNP and nasopharyngeal cancer. Although the association between *PIGR* SNPs and EBV-associated diseases has not been studied in African populations, it is noteworthy that the minor "T" allele dominated in the African cohort, suggesting selective pressure for mutation of the ancestral "C" allele. Clearly, it will be important to study the disease association of *PIGR* SNPs in individuals of varied ethnicity and to determine the effect of the Ala/Val substitution on the rate of pIgR cleavage in experimental systems.

Two additional SNPs have been reported that alter the amino acid sequence of pIgR, although no changes in pIgR function or disease associations have been linked to these SNPs. Substitution of the ancestral "C" allele at position 1280 of pIgR mRNA with the minor "T" allele results in a change from Gly to Ser at amino acid 365, in domain 4 of the extracellular region of pIgR protein. Heterozygosity at this position has been estimated at 0.489 ± 0.074 (http://www.ncbi.nlm.nih.gov/SNP/snp_ref.cgi?rs=2275531). In contrast to the SNPs at positions −41 and 1925, the ancestral allele predominated in African (frequency = 0.845) and African American populations (frequency = 0.770). Frequency of the ancestral allele was lower in individuals of European descent (0.575), perhaps suggesting selective pressure for mutation. Surprisingly, frequency of the minor "T" allele was 1.000 in Chinese and Japanese populations. It will be important to determine whether the substitution of Ser for Gly at this position alters function of pIgR and/or SC. Finally, a minor SNP with an estimated heterozygosity of 0.162 ± 0.234 has been reported at position 1850 of pIgR mRNA (http://www.ncbi.nlm.nih.gov/SNP/snp_ref.cgi?rs=7542760). Substitution of the ancestral "C" allele with the minor "T" allele results in a change from Thr to Ile at amino acid 555, in domain 5 of the extracellular region of pIgR protein. Frequency of the ancestral allele was 1.00 in European and Asian populations. Some evidence of mutation was found in African and African American populations, in which frequency of the ancestral allele was 0.720 and 0.717, respectively. It is not known whether this SNP is in linkage disequilibrium with the SNPs at positions −41 and 1925, in which frequency of the minor allele was also higher in African populations.

3.2.2. Factors That Regulate pIgR Expression

Because one molecule of pIgR must be synthesized for every molecule of IgA transported, the ability to control the cellular level of pIgR is crucial for IgA-mediated mucosal defense. A variety of host and microbial factors have been shown to regulate pIgR expression (Table 3.1) through complex and interconnected mechanisms.

TABLE 3.1. Regulators of pIgR expression.

Regulators	Nature of regulation
Cytokines	
IFN-γ	Upregulates pIgR in diverse mucosal epithelia; synergistic with TNF, IL-1, and IL-4
TNF	Upregulates pIgR in diverse mucosal epithelia; synergistic with IFN-γ, IL-1, and IL-4
IL-1	Upregulates pIgR in diverse mucosal epithelia; synergistic with IFN-γ and TNF
IL-4	Upregulates pIgR in diverse mucosal epithelia; synergistic with IFN-γ and TNF
Hormones	
Estradiol	Upregulates pIgR in human and rat endometrium Downregulates pIgR in rat mammary gland
Progesterone	Downregulates pIgR in human and rat endometrium and rabbit mammary gland
Androgens	Upregulate pIgR in male reproductive tract and lacrimal gland in rats, and in human breast and cervical cancer cell line
Glucocorticoids	Upregulate pIgR in human breast and colon cancer cell-lines, and rat liver and intestine Downregulate pIgR in rabbit mammary gland
Prolactin	Upgulates pIgR in rabbit and sheep mammary gland
Retinoic acid (vitamin A)	Enhances upregulation of pIgR by IL-4, IFN-γ, and TNF in a human intestinal epithelial cell line
Microbes/microbial factors	
Butyrate	Upregulates pIgR in a human intestinal epithelial cell line
Lipopolysaccharide	Upregulates pIgR in a human intestinal epithelial cell line
Escherichia coli	Upregulates pIgR in a human intestinal epithelial cell line
Bacterioides thetaiotaomicron	Colonization of germ-free mice with *B. thetaiotaomicron* upregulates intestinal expression of pIgR
Double-stranded RNA	Upregulates pIgR in a human intestinal epithelial cell line
Reovirus	Upregulates pIgR in a human intestinal epithelial cell line
Saccharomyces boulardii	Treatment of rats with *S. boulardii* upregulates intestinal expression of pIgR

3.2.2.1. Cytokines

The proinflammatory cytokines interferon (IFN)-γ, tumor necrosis factor (TNF), and interleukin (IL)-1, which are produced in response to a variety of bacterial and viral infections, play a key role in the upregulation of *PIGR* gene transcription. Treatment of the human intestinal epithelial cell line HT-29 with IFN-γ leads to phosphorylation and nuclear translocation of signal transducer and activator of transcription (STAT)-1, which induces transcription of the *IRF1* gene (Blanch et al., 1999; Kaetzel et al., 1997; Piskurich et al., 1997). IRF-1 enhances transcription of the *PIGR* gene by binding to a highly conserved ISRE in exon 1 (see Sect. 3.2.1.2). Inhibitors of tyrosine phosphorylation, which inhibit activation of STAT-1, potently inhibit the ability of IFN-γ to increase cellular pIgR levels in HT-29 cells (Denning, 1996). TNF and probably also IL-1 induce transcription of the *IRF1* gene by inducing nuclear translocation of NF-κB, which binds to an element adjacent to the STAT-1 site in the *IRF1* promoter (Pine, 1997). Surprisingly, inhibitors of NF-κB activation were found to inhibit IFN-γ-induced *PIGR* gene transcription partially in HT-29 cells (Ackermann and Denning, 2004). Although IFN-γ is only a weak activator of NF-κB, low levels of nuclear NF-κB could enhance *PIGR* gene transcription indirectly, by synergizing with STAT-1 to activate *IRF1* gene transcription, or directly, by binding to an NF-κB element in intron 1 of the *PIGR* gene (see below). Significantly, mice with two disrupted *Irf1* alleles were found to have a lower expression of pIgR mRNA in the liver and intestine than did wild-type mice (Blanch et al., 1999).

Binding of TNF and IL-1 to cell surface receptors initiates signal transduction pathways leading to the activation and nuclear translocation of the transcription factor NF-κB (Bonizzi and Karin, 2004; Dunne and O'Neill, 2003). Mutation of two NF-κB binding sites in the 5′-flanking region of the human *PIGR* gene caused a modest decrease in *PIGR* promoter activity in response to TNF (Takenouchi-Ohkubo et al., 2000). The NF-κB site within the intron 1 enhanceosome was found to cooperate with the IRF-1 site in exon 1 to mediate TNF-induced transcription of the human *PIGR* gene (Schjerven et al., 2001). Interestingly, the NF-κB site in intron 1 did not appear to cooperate with the NF-κB sites in the 5′-flanking region. These investigators subsequently reported that induction of *PIGR* transcription by TNF requires *de novo* synthesis of the RelB subunit of NF-κB (Schjerven et al., 2004), consistent with the relatively delayed kinetics of transcriptional activation. We have found that continuous stimulation of HT-29 cells with TNF for up to 3 weeks results in sustained upregulation of RelB and pIgR expression (Bruno and Kaetzel, 2005). TNF signaling is generally associated with the "classical" NF-κB pathway (i.e., rapid activation of RelA/p50 dimers and early transcription of genes involved in innate immune and inflammatory responses) (Bonizzi and Karin, 2004). The *RELB* gene is among the target genes of TNF-mediated RelA/p50 activation (Bren et al., 2001). Thus, activation of the classical NF-κB pathway by TNF in intestinal epithelial cells might indirectly induce

PIGR transcription by increasing steady-state levels of RelB protein. TNF might also induce synthesis of an as-yet unidentified cytokine or signaling molecule, which could then activate newly synthesized RelB through the "alternative" NF-κB pathway. Activation of the alternative NF-κB pathway by lymphotoxin and LIGHT signaling is crucial for the development of secondary lymphoid tissues, including Peyer's patches in the gut (reviewed in Schneider et al., 2004). Induction of *PIGR* transcription through the alternative NF-κB pathway, as a mechanism for enhanced transport of locally synthesized IgA, might represent an important bridge between innate and adaptive immune responses. In addition to its effects on the regulation of pIgR mRNA transcription, we have found that TNF enhances the stability of pIgR mRNA in a human intestinal epithelial cell line (Bruno and Kaetzel, 2005). It appears, therefore, that multiple mechanisms exist for maintaining high levels of pIgR expression during inflammatory responses.

A surprising aspect of the regulation of pIgR expression is cooperativity between the Th1-type cytokine IFN-γ and the Th2-type cytokine IL-4, the effects of which are usually antagonistic. IL-4 by itself, and in cooperation with IFN-γ, has been shown to upregulate pIgR mRNA and protein levels in human intestinal and respiratory epithelial cell lines (Ackermann et al., 1999; Denning, 1996; Loman et al., 1999; Phillips et al., 1990; Youngman et al., 1994). Insight into the mechanism of this coordinate regulation was provided by the discovery of a novel IL-4-inducible STAT-6 site in intron 1 of the human *PIGR* gene (Schjerven et al., 2000). As with the IRF-1 and NF-κB sites, the intron 1 STAT-6 site is conserved in the mouse *Pigr* gene. In HT-29 cells, the STAT-6 site was found to interact cooperatively with an adjacent site for the constitutively expressed hepatocyte nuclear factor (HNF)-1, providing a possible mechanism for tissue-specific cytokine inducibility of pIgR. Despite this progress, several mysteries remain regarding the mechanism of IL-4 regulation of pIgR expression. IL-4 was found to activate *PIGR* transcription with delayed kinetics, requiring *de novo* protein synthesis. However, nuclear translocation of activated STAT-6 occurred within minutes of IL-4 stimulation. These apparently contradictory observations led Schjerven et al. (2000) to postulate the existence of an unidentified IL-4-induced protein that cooperates with STAT-6 to form an IL-4 responsive enhancer element. An alternate explanation is suggested by the recent report that inhibitors of NF-κB activation significantly reduced IL-4-dependent upregulation of pIgR expression and *PIGR* transcriptional activity in HT-29 cells (Ackermann and Denning, 2004). Given the close proximity of the NF-κB/RelB element to the STAT-6 site in intron 1, it is possible that *de novo* synthesized RelB might synergize with STAT-6 to activate the IL-4- dependent enhancer element. To test this hypothesis, it will be important to determine whether IL-4 signaling induces RelB synthesis in intestinal epithelial cells. It will also be interesting to explore the possibility that related Th2 cytokines, such as IL-13, activate *PIGR* transcription.

A working model is proposed in Fig. 3.2B to unify the diverse data on transcriptional regulation of the human *PIGR* gene by cytokines. Signaling through multiple proinflammatory and Th2 cytokines, produced in mucosal tissues in response to infection or inflammation, leads to the synthesis and/or activation of IRF-1, NF-κB/RelB, and STAT-6. Binding elements for these transcription factors in exon 1 and intron 1 of the *PIGR* gene likely cooperate to form a cytokine-dependent "enhanceosome" (see Chapter 5). Recruitment of the enhanceosome to the *PIGR* promoter could enhance the activity of constitutive transcription factors such as USF and AP2 and could also facilitate interactions with the tissue-specific transcription factor HNF-1. Given the conservation of the IRF-1, NF-κB, and STAT-6 sites in the mouse *Pigr* gene, genetic approaches in mice could prove extremely useful to explore mechanisms of synergy among these transcription factors. Genetic approaches could also be used to determine the relative contributions of the classical and alternative NF-κB pathways in cytokine-induced *Pigr* transcription.

3.2.2.2. Hormones

Expression of pIgR is regulated by steroid and polypeptide hormones, in a cell-type-specific manner (Table 3.1). The effects of estrogen and progesterone are antagonistic in human and rat endometrium and, accordingly, the expression of pIgR varies during the estrous cycle (Bjercke and Brandtzaeg, 1993; Kaushic et al., 1995) (see Chapter 12). Binding sites for estrogen and progesterone receptors have not yet been identified in the *PIGR* gene, and the mechanisms by which they regulate pIgR expression are not known. Androgens have been shown to upregulate pIgR expression in male and female reproductive tissues, and two important binding sites for the androgen receptor (AR) have been characterized in the human *PIGR* gene (reviewed in Claessens et al., 2001). Although glucocorticoid receptors (GRs) and ARs often bind to common DNA elements, detailed studies demonstrated the androgen specificity of the AR elements in the human *PIGR* gene. A unique GR-responsive element (GRE) was identified in the 5′-flanking region of the mouse *Pigr* gene (Li et al., 1999). It was recently reported that dexamethasone treatment in a mouse model of asthma caused a significant decrease in pIgR protein levels in bronchoalveolar lavage fluid, suggesting negative regulation of pIgR expression by glucocorticoids (Zhao et al., 2006). The sequence of the mouse GRE shares significant homology with the orthologous region of the human *PIGR* gene, but its functional significance in glucocorticoid regulation of human pIgR expression has not yet been tested. The polypeptide hormone prolactin has been reported to upregulate pIgR expression in rabbit and sheep mammary gland, but its effects have not been evaluated in human or rodent tissues. Given the high concentration of free SC and SIgA in colostrum and milk (reviewed in Kaetzel and Mostov, 2005), it is likely that prolactin and other hormones regulate pIgR expression in mammary epithelial cells during pregnancy and lactation. Prolactin has been shown to stimulate

IRF-1 expression in several cell types through the Jak/Stat signaling pathway (Yu-Lee, 2001), and significant levels of IRF-1 protein have been detected in mouse mammary epithelial cells (Chapman et al., 2000). It is reasonable to speculate that IRF-1 plays a role in the maintenance of *PIGR* expression in the developing and lactating mammary gland, especially in the presence of other hormonal and immune stimuli.

3.2.2.3. Microbes

In addition to being regulated by host-derived factors, pIgR expression can be modulated directly by microbes and their metabolic products. Butyrate, a bacterial fermentation product and important energy source in the colon, has been reported to upregulate pIgR expression and to enhance the response to cytokines in HT-29 cells (Kvale and Brandtzaeg, 1995)(see Chapter 10). A role for commensal bacteria in pIgR regulation was suggested by the observation that pIgR expression was increased when germ-free mice were colonized with *Bacteroides thetaiotaomicron,* a prominent organism of the normal mouse and human intestinal microflora (Hooper et al., 2001). We found that pIgR mRNA levels were approximately sixfold higher in the colon than in the small intestine of C57BL/6 mice, consistent with a role for commensal bacteria in the maintenance of pIgR expression (Bruno and Kaetzel, unpublished observations). Furthermore, we found that treatment of mice with the antibiotic metronidazole, which targets anaerobic bacteria in the gut, resulted in a small but significant decrease in pIgR mRNA levels.

Changes in composition of the intestinal microflora might explain the recently reported roles for passive and adaptive immunity in the ontogeny of pIgR expression in mice (Jenkins et al., 2003). A direct role for intestinal bacteria in pIgR regulation suggests that the innate immune system might "prime" epithelial cells for transport of pIgA produced during the adaptive immune response. Host cells mediate innate immune responses to microbial components through Toll-like receptor (TLR) signaling (reviewed in Barton and Medzhitov, 2003; Takeda and Akira, 2003). Intestinal epithelial cells have been shown to express a wide variety of TLRs, the expression of which is upregulated during intestinal inflammation (Cario and Podolsky, 2000; Hausmann et al., 2002). We have recently reported that pIgR expression is upregulated in HT-29 cells by bacterial lipopolysaccharide (LPS), a ligand for TLR4, and double-stranded RNA, a ligand for TLR3 (Schneeman et al., 2005). Both of these TLR ligands were found to induce *de novo* synthesis of RelB, and activation of *PIGR* gene transcription was dependent on the NF-κB element in intron 1. Interestingly, double-stranded RNA, but not LPS, induced synthesis of IRF-1. Signaling through TLR3 and TLR4 has been found to upregulate synthesis of IRF-3, a related transcription factor that binds to the same ISRE motif as IRF-1 (Yamamoto et al., 2004). IRF-3 is known to cooperate with NF-κB to activate many target genes in the TLR pathway, and it is likely to interact with RelB in the context of the *PIGR* enhanceosome. It is noteworthy that

IL-1, a proinflammatory cytokine that upregulates pIgR expression, activates a signaling pathway that shares many elements of the TLR signaling pathways (Bowie and O'Neill, 2000). The observation that pathways for *PIGR* gene regulation by host and microbial factors appear to converge on the same enhanceosome provides evidence for the coordination between innate and adaptive immune responses in mucosal epithelia.

Upregulation of pIgR expression by LPS is significant in light of the finding that pIgR can participate in intracellular neutralization of LPS by epithelial cells (Fernandez et al., 2003) (see Chapters 7 and 8). Enhanced uptake and neutralization of LPS by IgA bound to pIgR might be an important mechanism by which intestinal inflammation is controlled in the presence of commensal bacteria. Similarly, upregulation of pIgR expression by double-stranded RNA might enhance antiviral immunity. It has recently been shown that reovirus, which has a double-stranded RNA genome, upregulates pIgR expression in HT-29 cells (Pal et al., 2005). Significantly, ultraviolet (UV)-inactivated reovirus induced a stronger increase in pIgR expression than did live virus, suggesting that viral components, but not viral replication, is required for induction of pIgR production. It is possible that viruses with single-stranded RNA or double-stranded DNA genomes might transiently produce double-stranded RNA species at sufficient levels to upregulate pIgR through the TLR3 pathway. Alternatively, these viruses might utilize other innate immune pathways to upregulate pIgR levels in epithelial cells. Signaling through various TLRs could also enhance pIgR expression by other classes of microorganisms, perhaps explaining the observation that pIgR expression was enhanced in the small intestine of rats treated with *Saccharomyces boulardii* (Buts et al., 1990). Increased expression of pIgR in response to a variety of probiotic organisms could contribute to increased secretion of anti-inflammatory SIgA in the intestine, thus providing broad-spectrum protection against pathogens and enteritis.

3.3. Structure and Function of pIgR

3.3.1. Functional Domains of pIgR

The pIgR protein consists of three major functional regions: an extracellular ligand-binding region, a short hydrophobic membrane-spanning domain, and a relatively long cytoplasmic tail (reviewed in Kaetzel and Mostov, 2005; Kaetzel, 2005) (Fig. 3.3). Five domains with homology to immunoglobulin variable domains comprise the ligand-binding portion of pIgR (see Sect. 3.4), which is connected by a loosely structured linker peptide (sometimes called domain 6) to the transmembrane domain. Cleavage of pIgR to form SIgA or free SC occurs within this linker peptide (see Sect. 3.6). The cytoplasmic tail of pIgR contains elements that interact with intracellular signaling proteins to regulate cellular trafficking (see Sect. 3.5).

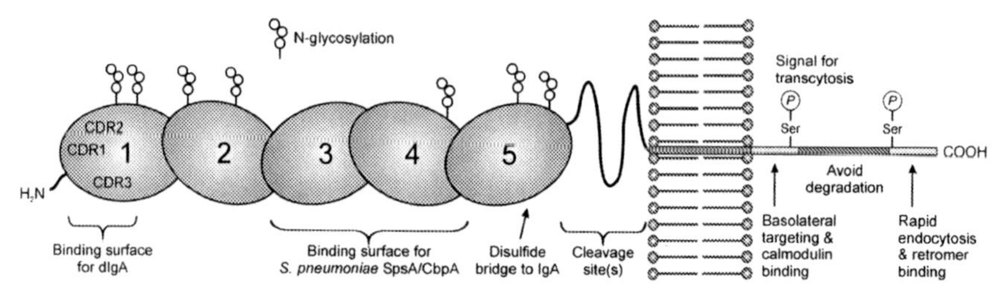

FIG. 3.3. Schematic structure of the human pIgR. The pIgR is a type I transmembrane protein, with an extracellular ligand-binding comprising five domains with homology to immunoglobulin variable regions. The three CDRs in domain 1 form a noncovalent binding surface for dimeric IgA (dIgA). During transcytosis, a disulfide bridge is formed between domain 5 of pIgR and the Fcα region of dimeric IgA. Peptide motifs in domains 3 and 4 cooperate to form a binding surface for the SpsA/CbpA protein of *S. pneumoniae*. A peptide of unknown structure links domain 5 to the membrane-spanning region and contains site(s) for proteolytic cleavage of pIgR to SC. Seven N-glycan residues on domains 1, 2, 4, and 5 contribute to innate immune functions of SC and might facilitate transcytosis of pIgR. The cytoplasmic domain of pIgR contains highly conserved signals for intracellular sorting, endocytosis and transcytosis.

3.3.2. Interspecies Conservation of pIgR Protein Structure and Function

The complete amino acid sequence of pIgR has been determined for 12 mammalian, 1 avian, and 1 amphibian species (See Fig. 3.4 for GenBank accession numbers and references therein). The recent reports of pIgR sequences from the chicken (*Gallus gallus*) (predicted from genomic sequence) and frog (*Xenopus laevis*) (Braathen et al., 2007) confirm earlier predictions that the ability to form complexes of polymeric antibodies with SC preceded mammalian evolution (reviewed in Peppard et al., 2005). Alignment of these sequences reveals a number of structural features that are highly conserved (Fig. 3.4). Within the five extracellular Ig-like domains, the Cys residues that form disulfide bonds to stabilize the characteristic "immunoglobulin fold" are invariant across species. An additional conserved Cys in domain 5 has been shown to form a disulfide bond with one of the α-heavy-chains of SIgA (Eiffert et al., 1984; Fallgreen-Gebauer et al., 1993). Interspecies homology is greatest in domain 1, especially within segments that have been shown to be critical for binding of IgA and IgM (see Sec. 3.4.1).

Domains 2 and 3 of pIgR are encoded by a single large exon in pIgR from mammalian species (see Fig. 3.2, exon 4), which is sometimes spliced out in rabbit pIgR mRNA (Deitcher and Mostov, 1986) but is invariantly included in pIgR transcripts from other mammalian species. Interestingly, there is no orthologue of domain 2 in chicken and frog pIgR, suggesting

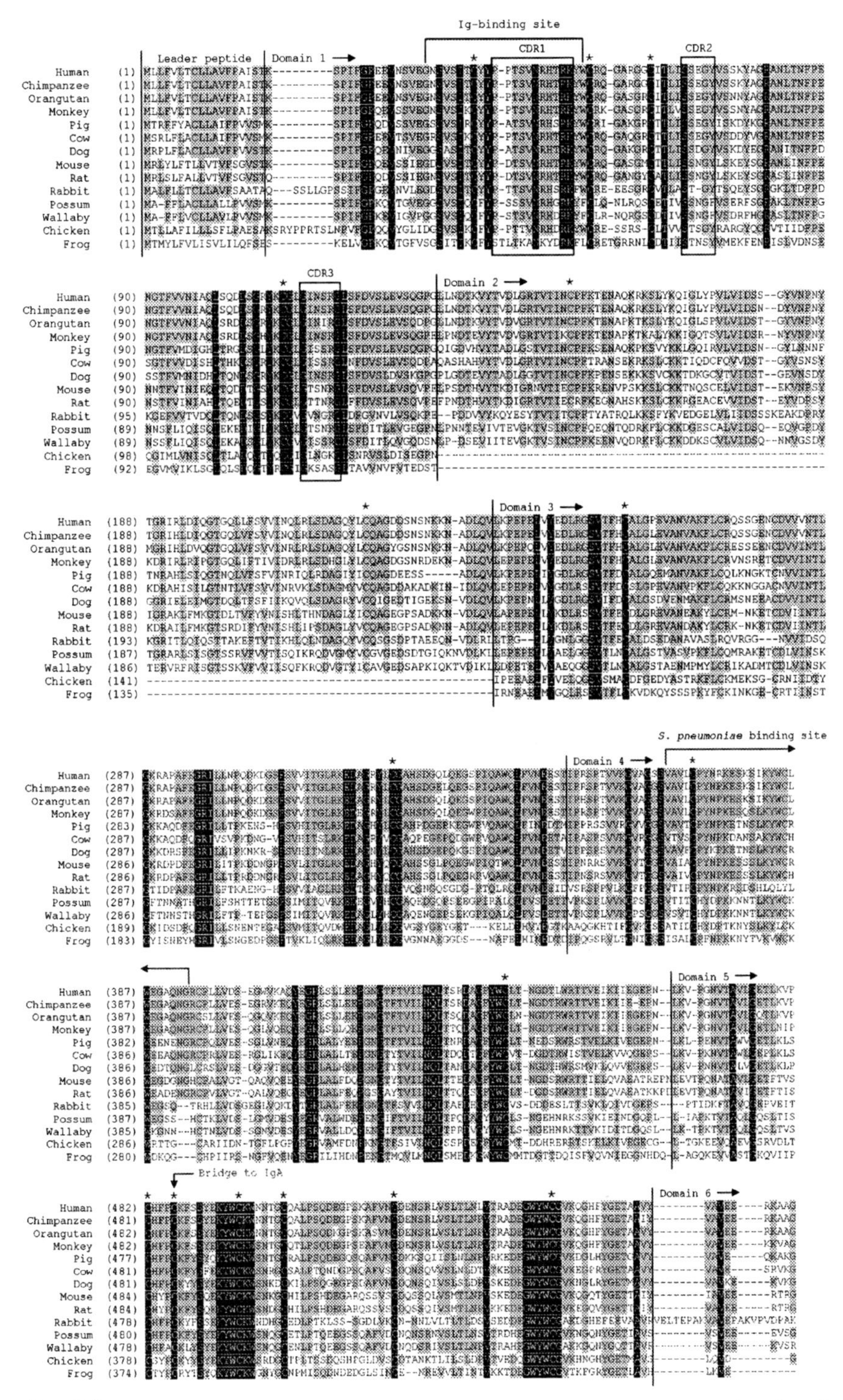

FIG. 3.4. (continued)

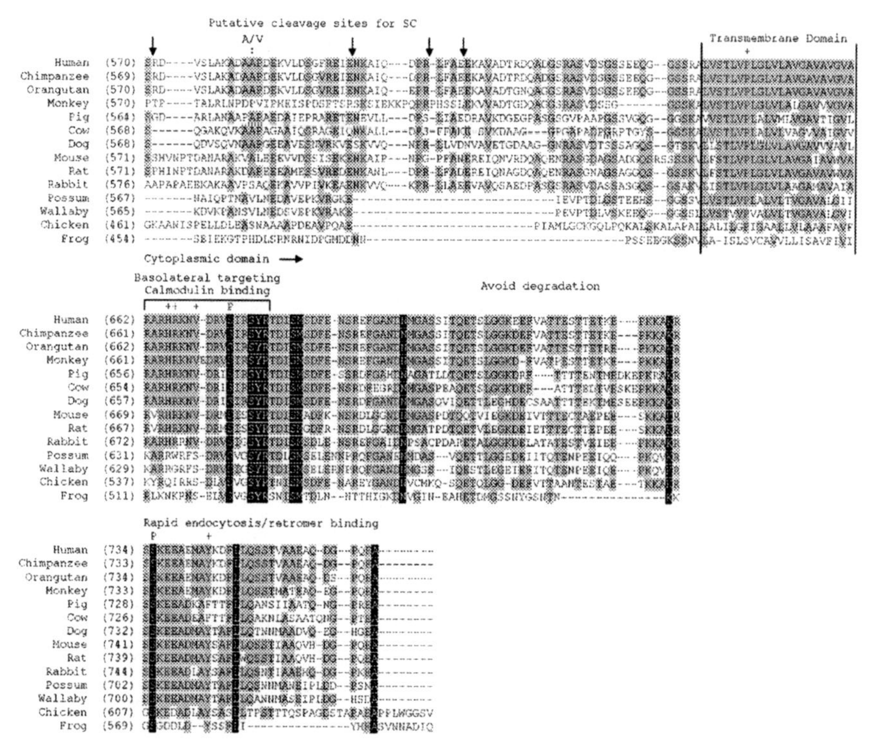

FIG. 3.4. Interspecies alignment of amino acid sequences of pIgR/SC. Black shading indicates sequence identity among all species, and gray shading indicates sequence identity among seven or more species. Asterisks denote conserved cysteine residues characteristic of immunoglobulin-like domains. Arrows in domain 6 indicate putative sites for cleavage of pIgR to SC. The Ala/Val polymorphism is noted at position 581 of human pIgR (amino acid 560 of the mature protein) (see Sect. 3.2.1). Conserved residues in the transmembrane and cytoplasmic domains associated with specific functions are marked with a "+" sign. The "P" symbol indicates conserved serine residues in the cytoplasmic domain that are phosphorylated during transcytosis. Accession codes for pIgR sequences: X73079, *Homo sapiens* (human); XM_514153, *Pan troglodytes* (chimpanzee); CR859163, *Pongo pygmaeus* (orangutan); XM_001083307, *Macaca mulatta* (rhesus monkey); AB032195, *Sus scrofa* (pig); X81371, *Bos taurus* (cow); XM_537133, *Canis familiaris* (dog); U06431, *Mus musculus* (mouse); X15741, *Rattus norvegicus* (rat), X00412, *Oryctolagus cuniculus* (rabbit); AF091137, *Trichosurus vulpecula* (silver-gray brushtail possum); AF317205, *Macropus eugenii* (tammar wallaby); XM_417977, *Gallus gallus* (chicken); EF079076, *Xenopus laevis* (frog).

that duplication and fusion of a smaller primordial exon 4 might have occurred during mammalian evolution. In support of this hypothesis, a cryptic splice site was found at the junction of domains 2 and 3 in exon 4 of the mouse *Pigr* gene (Bruce et al., 1999). Mutational analyses of exon 4 and surrounding intronic regions revealed the presence of multiple elements that promote inclusion of this unusually large exon (Bruce and Peterson, 2001). Motifs within domain 2 of pIgR have been shown to enhance binding of polymeric Igs (reviewed in Kaetzel and Mostov, 2005), providing a possible explanation for selective pressure for evolution of this "extra" extracellular domain in mammals (see Sect. 3.4.1).

The sixth extracellular domain of pIgR has a more random structure than the five immunoglobulin homology domains and is poorly conserved across species (Fig. 3.4). Proteolytic cleavage of pIgR within this domain leads to the release of SC from the apical surface of epithelial cells, either free or bound to SIgA or SIgM. Given the random structure and poor conservation of domain 6, it is likely that multiple proteases can cleave pIgR to SC (see Sect. 3.6).

The 23-amino-acid transmembrane region of pIgR is highly conserved among mammalian species, with differences involving only conservative substitutions of one hydrophobic amino acid for another (Fig. 3.4). The perfectly conserved Pro residue in the center of this core sequence might participate in signal transduction following ligand binding, by causing a bend in the α-helical structure of the membrane-spanning domain (Mostov et al., 1984). Although the sequence of the transmembrane domain of chicken pIgR is quite different from that of mammalian pIgRs, it includes the conserved Pro residue and varies at other residues by conservative substitutions. In contrast, the putative transmembrane region of frog pIgR is shorter, lacks the conserved Pro, and includes four polar amino acids (three Ser and one Cys). This striking difference in sequence suggests a possible divergence in pIgR function early in vertebrate evolution. The cytoplasmic domain of pIgR contains a number of highly conserved intracellular sorting signals that interact with cytoplasmic proteins to direct pIgR through the transcytotic pathway (see Sect. 3.5). Curiously, frog and chicken pIgR appear to have an eight-amino-acid extension at their C-termini relative to mammalian pIgRs.

3.4. Binding of Ligands to pIgR and SC

Membrane-bound pIgR and soluble SC have been shown to interact with a variety of physiological and pathophysiological ligands, through both peptide and carbohydrate-based motifs. Recent studies have considerably advanced our understanding of the structural elements that mediate ligand binding by pIgR and SC.

3.4.1. Polymeric IgA and IgM

The association of pIgR with polymeric IgA and IgM involves multiple structural elements that participate in both noncovalent and covalent bonds. Biochemical and mutagenesis studies have demonstrated that domain 1 of pIgR is both necessary and sufficient for binding of polymeric IgA and IgM, although in some species the other extracellular domains contribute to binding affinity (reviewed in Norderhaug et al., 1999; Kaetzel and Mostov, 2005). Given the sequence homology of pIgR domain 1 to immunoglobulin variable domains, it has been hypothesized that loop structures analogous to the complementarity-determining regions (CDRs) of antibodies might form the pIg-binding surface. The primary sequences of the CDRs in domain 1 are highly conserved among pIgR proteins, including the frog and chicken as well as mammalian species (Fig. 3.4). Most of our knowledge of the biochemical nature of pIgR–pIg interactions come from studies of human and rabbit pIgR. In the context of full-length rabbit pIgR, point mutations within CDR1, CDR2, or CDR3 were found to diminish pIgA binding (Coyne et al., 1994). "Swapping" of the putative CDR loops between human and rabbit pIgR demonstrated that these sequences were interchangeable with respect to binding of pIgA; however, CDR2 from human pIgR was required for binding of pIgM (Roe et al., 1999). This finding was consistent with decades-old observations that pIgR from some species (such as humans and cows) can bind and transport both pIgA and pIgM, whereas pIgR from other species (such as rodents and rabbits) transports pIgA but not pIgM (reviewed in Brandtzaeg and Johansen, 2001; Kaetzel and Mostov, 2005).

A major breakthrough in our understanding of the ligand-binding properties of pIgR came with the solving of the crystal structure of domain 1 from human pIgR (Hamburger et al., 2004) (see Chapter 1). As predicted from the sequence analysis, the overall structure of domain 1 was found to be very similar to that of immunoglobulin variable domains, including conservation of key residues that stabilize the β-pleated sheet structure of the "immunoglobulin fold." However, the structures of the CDR loops differ markedly from those found in antibody variable regions. The relatively long CDR1 loop in pIgR domain 1 contains a single helical turn (Ser28 through His32) comprising highly conserved amino acids. Previous studies had shown that mutations within this helix in CDR1 of rabbit pIgR abolished binding of pIgA (Coyne et al., 1994), emphasizing the contribution of this structure to the ligand-binding surface.

Surprisingly, the CDR2 loop in domain 1 was found to be only two amino acids in length, with a highly divergent amino acid at position 53 (Glu in human pIgR) followed by an invariant Gly at position 54. Given the key role of CDR2 in binding of human pIgR to IgM, it is significant that rabbit pIgR has a deletion at the position corresponding to the surface-exposed Glu53 residue. Notably, pIgR proteins that bind IgM (such as human and cow) have a charged residue at position 53, whereas those that bind pIgA exclusively have at that position a polar, uncharged residue (Asn in mouse and rat) or a deletion (rabbit) (Fig. 3.4). Experiments with chimeric "domain swap" immunoglobulins revealed that the C-terminal domains of human IgA (Cα3) and IgM (Cμ4) are required for J-chain binding, polymerization, and association with pIgR (Hexham et al., 1999). Chimeric IgM molecules in which the Cα3 domain of IgA was substituted for the Cμ4 domain of IgM were able to bind both rabbit and human pIgR, whereas native IgM bound only human pIgR (Braathen et al., 2002). Thus, the species specificity of pIgR binding appears to map to the C-terminal domain of IgM and the CDR2 loop of pIgR. It will be interesting to determine whether these patterns of discrimination of pIgR for pIgA versus pIgM hold for other species; pIgRs from primates, pigs, and dogs have negatively charged residues (Glu or Asp) at position 53, like the human and cow, whereas marsupial pIgRs have an Asn at this position, similar to mouse and rat pIgR. Chicken pIgR has a Ser residue at position 53 in CDR2, different from all known mammalian pIgR proteins (Wieland et al., 2004). This difference might be significant in that chicken IgA, like mammalian IgM, lacks a hinge region and has four C_H domains (reviewed in Peppard et al., 2005). Frogs, which produce IgM but not IgA, have the most divergent sequence within CDR2 of pIgR. Chicken and frog pIgRs are also unique in that they have a shorter extracellular region that lacks a domain orthologous to domain 2 of mammalian pIgRs (see Sect. 3.3.2). Modeling of the interactions between pIgR and pIgA or IgM from different species might shed light on the structural requirements for the receptor–ligand interface.

Unlike the CDR3 loop in immunoglobulin variable domains, the position of CDR3 in human pIgR was found to tilt away from CDR1 and CDR2 (Hamburger et al., 2004) (see Chapter 1). This unique conformation of CDR3 has two important consequences for the pIg-binding interface. First, the surface area comprising the three CDRs of pIgR is substantially larger than the antigen-binding surface of immunoglobulin variable domains. Second, the "tilted" orientation of CDR3 is consistent with the inability of pIgR domain 1 to form dimers, unlike the V_H-V_L dimers typical of antibody molecules.

Several other important findings regarding pIgR structure and function emerged from the studies of Hamburger et al. (2004). Biosensor binding experiments confirmed the specificity of isolated pIgR domain 1 for the J-chain containing pIgA. Domain 1 from pIgR bound native pIgA, but did not bind monomeric Fcα fragments, unlike another IgA receptor, FcαRI (CD89) (Herr et al., 2003) (see Chapter 4). Binding of native pIgR to pIgA or pentameric IgM requires the presence of a J-chain connecting the Ig subunits (see Chapter 5), and J-chain-deficient mice have diminished IgA levels in external secretions (Hendrickson et al., 1995., 1996; Lycke et al., 1999). The binding studies with isolated domain 1 suggest that structural determinants within this domain contribute to the specificity of pIgR for native pIgA, but the precise topology of the pIgR–pIgA interface and the role of the J-chain remain a mystery. The two N-glycan chains linked to domain 1 of pIgR (Fig. 3.2) do not appear to contribute to the interaction of pIgR with pIgA, because the affinities of binding of glycosylated versus nonglycosylated domain 1 to pIgA were found to be identical (Hamburger et al., 2004). Finally, it was noted that the sequences of the CDR loops in domain 1 are quite different from the corresponding structures in domains 2–5, consistent with the primary importance of domain 1 in noncovalent binding of pIgA.

3.4.2. *Streptococcus pneumoniae*

Streptococcus pneumoniae is a major human pathogen, causing diseases ranging from relatively mild otitis media to potentially fatal sepsis, pneumonia, and meningitis (Alonso de Velasco et al., 1995). Human pIgR and SC have been shown to bind specifically to a polymorphic surface protein from some strains of *S. pneumoniaie*, variously named SpsA, CbpA, or PspC (reviewed in Kaetzel, 2001; Kaetzel and Mostov, 2005). Mutational analysis defined a minimal SpsA/CbpA-binding fragment comprising ectodomains 3 and 4 of human pIgR (Elm et al., 2004; Lu et al., 2003). Further studies with "domain swap" chimeras of human and mouse pIgR (which does not bind *S. pneumoniae*) confirmed that the presence of both domains 3 and 4 from human pIgR were required for optimal *in vitro* binding to SpsA (Elm et al., 2004). A synthetic peptide corresponding to residues 349–375 from human pIgR domain 4 was found to inhibit adherence of pneumococci to human lung epithelial cells (Elm et al., 2004), suggesting that important contact residues reside within this site (noted as the "*S. pneumoniae* binding site" on Fig. 3.4). Interestingly, human pIgR contains two charged residues (Arg357 and Lys361) within the SpsA/CbpA-binding site where uncharged residues (Pro and Ser) are found at the corresponding positions of mouse pIgR. The Arg residue at position 357 of human pIgR appears to be a very recent evolutionary substitution, because all other known species (including nonhuman primates) have a Pro at this position. It will be interesting to determine whether these interspecies differences in pIgR sequence translate into different binding affinities for *S. pneumoniae,* a bacterium that has evolved within the human host.

3.4.3. *Substances That Bind the N-Glycans of SC*

Seven sites for N-glycosylation have been confirmed in the immunoglobulin-like ectodomains of human pIgR/SC (Eiffert et al., 1991; Hughes et al., 1999) (Fig. 3.3). The N-glycan chains have a unique structure that is heavily fucosylated and sialylated, similar to those of the antibacterial milk protein lactoferrin (Hughes et al., 1999; Royle et al., 2003). A variety of host-, pathogen-, and environment-derived substances with lectin-like activity have been shown to bind the N-glycans of SC, either in its free form or as part of SIgA (Table 3.2). It was recently reported that free SC from sputa of cystic fibrosis patients has an abnormal glycosylation pattern, relatively overfucosylated and undersialylated, and binds significantly less IL-8 than does free SC from normal sputa (Marshall et al., 2004). Thus, the specific composition of the

TABLE 3.2. Substances that bind pIgR and SC.

pIgR/SC Ligands	Biological function of pIgR/SC	Refs.
Substances that bind to peptide motifs of pIgR/SC		
Polymeric IgA and IgM	Epithelial transcytosis of mucosal antibodies Intracellular neutralization of pathogens and their by-products Protection of SIgA from proteolytic degradation	Reviewed in Norderhaug et al., 1999; Kaetzel and Mostov, 2005
IgA immune complexes	Excretion of antigens and pathogens across mucosal epithelia	Reviewed in Mazanec et al., 1993; Kaetzel and Mostov, 2005
S. pneumoniae SpsA/CbpA	Immune exclusion mediated by free SC Pneumococcal invasion mediated by cell surface pIgR	Kaetzel, 2001; Lu et al., 2003; Elm et al., 2004
Substances that bind to N-glycans of SC/SIgA		
Mucus	Immune exclusion of antigens and pathogens	Phalipon et al., 2002
IL-8	Inhibition of neutrophil chemotaxis	Marshall et al., 2001
Mac-1 (CD11b/CD18)	Increased affinity of FcαRI-SIgA	van Spriel et al., 2002
E. coli fimbrial lectins	Inhibition of adhesion to epithelial cells	Wold et al., 1990; Giugliano et al., 1995; Schroten et al., 1998; de Oliveira et al., 2001
H. pylori	Inhibition of adhesion to epithelial cells	Boren et al., 1993; Falk et al., 1993
Shigella spp.	Inhibition of adhesion to and invasion of epithelial cells	Willer et al., 2004
C. difficile toxin A	Inhibition of receptor binding	Dallas and Rolfe, 1998
Ricin	Reduction of toxicity at mucosal surfaces	Mantis et al., 2004

N-glycans likely influences the binding of SC and SIgA to lectins. Recent investigations of novel functions of SC and SIgA suggest that carbohydrate-mediated adherence to host and pathogenic factors plays an important role in innate immune defense (Phalipon and Corthesy, 2003) (see Chapter 8).

3.5. Transcytosis of IgA by pIgR

Polymeric IgR follows an unusually complex pathway through polarized epithelial cells, moving first to the basolateral surface and then to the apical surface (Fig. 3.1). Transcytosis of pIgR has proved to be an extremely useful model system to elucidate the pathways and mechanisms of polarized membrane traffic in epithelial cells (reviewed in Mostov et al., 2000, 2003; O'Brien et al., 2002). A circumferential tight junction divides the plasma membrane of each cell into two domains: the apical plasma membrane faces the lumen of the cavity, whereas the basolateral plasma membrane faces adjoining cells and the underlying basement membrane and connective tissue. These two domains of the plasma membrane differ in their protein and lipid compositions, reflecting the different functions served by the two surfaces in a polarized cell. Most of our understanding of pIgR trafficking has come from two experimental systems. The first is rat liver, where the expression of pIgR by hepatocytes allows transcytosis of pIgA from blood to bile (reviewed in Kaetzel and Mostov, 2005). In rat hepatocytes, pIgR is directed first to the sinusoidal surface, which is equivalent to the basolateral surface. pIgA in blood circulating through the hepatic sinusoids binds to pIgR and is transported to the bile canalicular surface, equivalent to the apical surface of polarized epithelial cells. At the canaliculus, pIgR is cleaved and free SC and SIgA are released into bile (reviewed in Brown and Kloppel, 1989). As a second experimental system, cloned cDNA encoding rabbit and human pIgR has been expressed in Madin–Darby canine kidney (MDCK) cells, which form a tight, well-polarized monolayer when cultured on permeable supports (Mostov and Deitcher, 1986; Tamer et al., 1995). pIgR expressed in MDCK cells follows the same transcytotic pathway as it does *in vivo*, and it serves as a useful model to study protein trafficking. An extensive series of deletions and point mutations have been made in the cytoplasmic domain of rabbit pIgR, which have uncovered several distinct sorting signals that control its intracellular trafficking (Figs. 3.3 and 3.4).

3.5.1. Basolateral Sorting Signal of pIgR

Integral proteins of the plasma membrane are synthesized in the rough endoplasmic reticulum and sent to the Golgi apparatus. As the proteins move through the last compartment of the Golgi, the TGN, they are sorted into vesicles that can deliver them to either the apical or basolateral surface of polarized epithelial cells (Mostov et al., 2000; Nelson and Yeaman, 2001). It was originally thought that sorting of transmembrane proteins to the basolateral surface was the default pathway, and did not require a specific sorting signal. However, studies with pIgR have shown that basolateral sorting requires a specific signal in the cytoplasmic domain (Casanova et al., 1991). The 17 amino acids that lie closest to the membrane comprise a signal that is necessary and sufficient for targeting of pIgR from the TGN to the basolateral surface (Fig. 3.4). When transplanted onto a heterologous reporter molecule, this peptide directs its delivery to the basolateral surface, indicating that the sorting signal can function autonomously (Casanova et al., 1991). Analysis of the basolateral sorting signal of rabbit pIgR by alanine scanning mutagenesis demonstrated that His675, Arg676, and Val679 were crucial for basolateral sorting (Aroeti et al., 1993; Reich et al., 1996) (noted with "+"

signs on Fig. 3.4). As would be expected from its important role in basolateral targeting, the 17-amino-acid sorting signal is highly conserved across species. The first eight amino acids proximal to the membrane, including the critical His, Arg, and Val residues, are nearly 100% conserved among placental mammals. The sequence of this octapeptide appears to have diverged at the eutherian–metatherian split in mammalian evolution, as it is quite different in pIgR from marsupial, avian, and amphibian species. It will be interesting to examine whether these orthologues of the eutherian octapeptide can also direct basolateral targeting. With the exception of an added Glu residue following the membrane-proximal octapeptide in monkey pIgR, the remaining nine amino acids of the basolateral sorting sequence are highly conserved across all species. Four of the nine residues are invariant, including a critical Ser that is phosphorylated upon ligand binding (see Sect. 3.4.3) and a Ser-Tyr-Arg tripeptide. A stretch of 20 amino acids beyond the "minimal" sorting signal is also highly conserved, suggesting an important function in pIgR trafficking.

3.5.2. *Regulation of pIgR Transcytosis*

Transcytosis of pIgR is regulated at multiple levels and provides an excellent model to study the regulation of membrane trafficking (reviewed in Mostov *et al.*, 2003; Rojas and Apodaca, 2002). Deletion of the C-terminal 30 amino acids of the cytoplasmic tail of rabbit pIgR dramatically reduced the rate of endocytosis from the basolateral surface of MDCK cells, suggesting that this region contains a signal for rapid endocytosis (Fig. 3.4). The tetrapeptide motif "YXXΦ," where Y = Tyr, X = any amino acid, and Φ = a bulky hydrophobic amino acid such as Phe, is known to facilitate clathrin-mediated endocytosis of transmembrane proteins by binding the adapter protein AP2 (Bonifacino and Traub, 2003). Significantly, the sequence "YXXΦ" is found near the C-terminus of pIgR in almost every species thus far sequenced (the Tyr in this motif is noted with a "+" in Fig. 3.4). The only exceptions are the pig and cow, in which the initial Tyr is substituted with Phe, and the chicken, in which Ser replaces the otherwise invariant Phe at the terminal "Φ." Mutation of Tyr753 in the "YXXΦ" motif of rabbit pIgR to Ser was found to reduce the rate of endocytosis from the basolateral surface of MDCK cells, confirming the importance of this endocytotic signal (Okamoto et al., 1992). Endocytosis of rabbit pIgR from the basolateral surface of MDCK cells was also found to be regulated by phosphorylation of Ser745 (Okamoto et al., 1994) (noted by "P" in Fig. 3.4). The importance of serine phosphorylation for endocytosis is supported by the fact that every known species of pIgR has a Ser residue at this position. It is possible that phosphorylation of Ser745 alters the folding of the cytoplasmic domain of pIgR, thereby exposing the downstream "YXXΦ" motif internalization signals to the endocytotic machinery. Ser745 is near the center of a highly conserved stretch of 12 amino acids (in every species except frog), suggesting that either the sequence or conformation of this region is important for phosphorylation and binding of this domain by cytoplasmic regulatory proteins.

Because sequences proximal to membrane serve to target pIgR to the basolateral surface, it follows that the basolateral sorting signal must somehow be "masked" in order for pIgR to be redirected to the apical surface. Mostov and colleagues have discovered two mechanisms that promote basolateral-to-apical transcytosis of pIgR. First, phosphorylation of Ser683, which is located in the 17-residue basolateral sorting signal (noted with "P" in Fig. 3.4), was found to enhance transcytosis of rabbit pIgR in MDCK cells (Casanova et al., 1990). Second, the calcium-regulated cytoplasmic protein calmodulin was found to bind to the basolateral targeting signal of rabbit pIgR expressed in MDCK cells and rat pIgR in liver endosomes (Chapin et al., 1996). These investigators proposed a model by which unphosphorylated pIgR is initially targeted to the basolateral surface via the basolateral sorting signal. After reaching that surface (or perhaps after endocytosis), pIgR is phosphorylated on Ser683 and/or binds calmodulin, thereby weakening the basolateral sorting signal and allowing pIgR to be transcytosed to the apical surface. Signaling pathways that promote phosphorylation of pIgR and activation of calmodulin would be expected to enhance the rate of transcytosis.

Although pIgR undergoes transcytosis in experimental epithelial cell systems in the absence of ligand, binding of pIgA has been shown to augment the rate of transcytosis of rabbit pIgR (Song et al., 1994). Interestingly, this stimulation did not depend on phosphorylation of Ser683. Binding of pIgA to pIgR apparently causes dimerization of the pIgR, although the evidence on this point is somewhat indirect (Singer and Mostov, 1998). Within 10 s of pIgA binding, several cytoplasmic proteins become tyrosine phosphorylated (Luton et al., 1998). The pIgR is not itself a tyrosine kinase nor is it phosphorylated on tyrosine, but somehow it acts to recruit p62*yes*, a nonreceptor tyrosine kinase of the Src family, to the plasma membrane (Luton et al., 1999). Mice with deletions in both alleles for p62*yes* exhibit a minor defect in basal transport of pIgA from blood to bile and exhibit a marked defect when challenged with a large bolus of intravenous pIgA (Luton et al., 1999). The direct substrates of p62*yes* are not known, but recruitment of p62*yes* to the plasma membrane was associated with downstream phosphorylation of phosphatidylinositol-specific phospholipase Cγ1 (PLCγ1). This enzyme causes hydrolysis of phosphatidylinositol-4,5-bis-phosphate (PIP2) to diacylglyceride (DAG) and inositol 1,4,5-trisphosphate (IP3). Release of DAG leads to activation of protein kinase C (PKC). Activation of PKC by phorbol ester, a DAG mimic, stimulates transcytosis of pIgR (Cardone et al., 1994), so it is likely that activation of PKC subsequent to pIgA binding also stimulates pIgR transcytosis (Cardone et al., 1996). Release of IP3 by hydrolysis of PIP2 causes the release of Ca^{2+} from intracellular stores and an increase in intracellular free Ca^{2+} ($[Ca^{2+}]_i$). Artificially increasing $[Ca^{2+}]_i$ rapidly stimulates transcytosis (Cardone et al., 1996; Luton and Mostov, 1999). Therefore, it is likely that the increase in $[Ca^{2+}]_i$ caused by pIgA binding to pIgR also stimulates transcytosis, perhaps through activation of calmodulin (Luton and Mostov, 1999).

Species differences have been noted in the stimulation of transcytosis by pIgA binding to pIgR. In addition to the above-described studies with rabbit pIgR exogenously expressed in MDCK cells, it has been demonstrated that pIgA stimulates blood-to-bile transcytosis of rat pIgR *in vivo* (Giffroy et al., 1998). In contrast, ligand-mediated stimulation of pIgR transcytosis was not observed in MDCK cells expressing human pIgR (Giffroy et al., 2001; Natvig et al., 1997). Binding of pIgA to the exogenously expressed human pIgR was found to trigger the IP3 signal; however, human pIgR appeared to be unable to respond to this signal (Giffroy et al., 2001). A teleological argument for differences in ligand-stimulated transcytosis can be made based on species differences in expression of pIgR by hepatocytes. A pIgR-mediated pathway for selective hepatobiliary transport of pIgA has been demonstrated in rats, mice, rabbits, chickens, and hamsters, but not in guinea pigs, sheep, dogs, or primates (reviewed in Delacroix et al., 1984; Kaetzel and Mostov, 2005; Moldoveanu et al., 1990). One can speculate that blood-to-bile transport of pIgA might require the ability to cope with large fluctuations in the amount of pIgA presented to the sinusoidal surface of the hepatocyte. Under these conditions, regulation of the rate of pIgR-mediated transcytosis by pIgA ligand would allow for rapid adjustment to the local ligand concentration. In contrast, pIgR-mediated transport of pIgA by nonhepatic epithelia might not require this type of regulation, perhaps because the level of pIgA presented to the basolateral surface does not vary as rapidly or dramatically. By this argument, species such as humans that do not transport pIgA across hepatocytes might have had no selective pressure to acquire or retain a mechanism for ligand-mediated control of the rate of pIgR transcytosis. It will be interesting to determine whether sequence differences in the cytoplasmic tail of pIgR confer differential sensitivity to ligand-mediated stimulation of transcytosis.

Recent research on pIgR transcytosis has focused on characterization of cytoplasmic proteins that regulate intracellular trafficking. Rab proteins are small GTPases that have been shown to regulate the formation of vesicles at the plasma membrane and the delivery of endocytosed proteins to multiple cellular locations (reviewed in Zerial and McBride, 2001). A novel function for a rab protein was revealed by the discovery that rab3b, in its GTP-bound state, interacts directly with the cytoplasmic domain of rabbit pIgR expressed in MDCK cells (Van Ijzendoorn et al., 2002). Bound rab3b appears to block transcytosis of pIgR, which then recycles to the basolateral surface. When pIgA binds to the pIgR, the rab3b hydrolyzes its GTP and dissociates from the pIgR, and the complex of pIgR and pIgA is transcytosed to the apical surface. In contrast to the inhibitory role of rab3b, the multimeric retromer complex has been shown to promote transcytosis of rabbit pIgR expressed in MDCK cells (Verges et al., 2004). Coimmunoprecipitation experiments with deletion mutants of pIgR revealed that the Vps35 component of retromer associates directly with the terminal 30-amino-acid segment of the pIgR cytoplasmic domain (Fig. 3.4). These investigators subsequently demonstrated that inhibition of phosphoinositide 3-kinase (PI3K) blocks the ability of retromer to facilitate pIgR transcytosis (Verges et al., 2006). Inhibition of PI3K reduced the membrane association of sorting nexins, components of the retromer complex that are involved in membrane deformation. Taken together, these results suggest that recruitment of the retromer complex to the cytoplasmic domain of pIgR initiates complex interactions with the intracellular sorting machinery.

Two classes of proteins have been shown to regulate transcytosis of pIgR in rat liver: p22, a novel Ca^{2+}-binding protein (Barroso et al., 1996); and several members of the rab family of small GTP-binding proteins (Jin et al., 1996). The rab3d protein was shown to colocalize with the pIgR in transcytotic vesicles, but not with the precursor form of pIgR associated with the secretory pathway and other Golgi markers (Larkin et al., 2000). Bile duct ligation caused rab3d to accumulate in the pericanalicular cytoplasm of hepatocytes, suggesting that it might play a role in the regulation of apically directed transcytosis. Other studies have implicated the cytoplasmic protein cellubrevin in regulating the apical transcytosis pathway in rat liver (Calvo et al., 2000). These investigators reported that pIgR coimmunoprecipitated with cellubrevin and that pIgA loading caused cellubrevin to redistribute into endosomal fractions enriched in transcytotic structures. Finally, it has been demonstrated that pIgR is a major calmodulin-binding protein in rat liver endosomes (Chapin et al., 1996; Enrich et al., 1996). Because calmodulin has been shown to bind to the cytoplasmic domain of rabbit pIgR and to regulate its transcytosis in MDCK cells (see above), it is likely to carry out a similar function in rat hepatocytes.

3.5.3. Transcytosis of IgA Immune Complexes

A novel function for pIgR was revealed by the discovery that this versatile receptor can transport complexes of pIgA and protein antigens from the lamina propria to the lumenal surface of mucosal epithelial cells (reviewed in Kaetzel and Mostov, 2005; Mazanec et al., 1993) (see Chapter 7). Thus, locally produced pIgA antibodies within the mucosa might serve to trap antigens derived from the environment, diet, and luminal microbiota or synthesized in the mucosal tissue during infections and to target these antigens for excretion. Antigens remain bound to the pIgA antibody throughout transcytosis, escaping lysosomal delivery and degradation, and are released along with SIgA at the apical surface. A mucosal IgA-mediated excretory immune system has been demonstrated *in vivo* using mice immunized mucosally with ovalbumin antigen (Robinson et al., 2001). pIgR has been shown to transport whole viruses and bacteria complexed to pIgA across epithelial layers (Gan et al., 1997; Yan et al., 2002), suggesting that mucosal IgA might "reroute" potential pathogens and prevent systemic infection. The ability of pIgR to bind and transport large pIgA-containing immune complexes (ICs) demonstrates that cross-linking of the Fab fragments with antigen does not appreciably alter binding of pIgR to the Fcα–J-chain segments of pIgA.

Hepatic clearance of IgA-containing ICs has been studied in humans and experimental animals in regard to systemic mechanisms for removal of circulating antigens. In humans (Rifai et al., 1989) and rodents (Brown et al., 1982, 1983; Harmatz et al., 1982; Phillips et al., 1988; Russell et al., 1981; Socken et al., 1981), intravenously administered IgA ICs or heat-aggregated IgA are cleared primarily by the liver, presumably via the same receptors that mediate uptake of IgA. In mice, pIgR-mediated hepatobiliary clearance of circulating IgA ICs has been shown to be involved in clearance of bacterial antigens absorbed through the intestine (Brown et al., 1984; Russell et al., 1983). Because human hepatocytes do not express significant levels of pIgR, it is likely that another receptor is involved in hepatic clearance of IgA IC—for example FcαRI (CD89) on Kupffer cells (Rifai and Mannik, 1984) (see Chapter 4). Because the systemic mechanisms for clearance of IgA IC are saturable (Rifai and Mannik, 1984; Russell et al., 1981), high concentrations of IgA IC in the circulation might lead to their deposition in extrahepatic tissues, as has been implicated in the pathogenesis of IgA nephropathy (see Chapter 13).

3.6. Cleavage of pIgR to SC

Despite considerable investigation and speculation during the past three decades, the mechanism by which pIgR is cleaved to SC remains a mystery. Cleavage has been demonstrated to occur within the unstructured peptide (sometimes called domain 6) that links the immunoglobulin-like domain 5 to the transmembrane region of pIgR (reviewed in Kaetzel and Mostov, 2005) (Figs. 3.3 and 3.4). Free SC purified from human colostrum (pooled from multiple women) was found to have a ragged C-terminus, varying from Ala550 to Lys559, with Ser552 as the dominant C-terminal residue (Eiffert et al., 1984). In a later study using colostrum from one woman, a single C-terminus of human SC at Arg585 was found (Hughes et al., 1997). Multiple C-termini in free SC in external secretions could result from trypsin-like cleavage of pIgR at Lys559 or Arg585 followed by additional exopeptidase cleavage. Prediction of a consensus cleavage site is hampered by the fact that the sequence and the length of the linker peptide is poorly conserved across species and that the C-terminal protein sequence is not available for free SC for species other than human. The identity of the protease that cleaves pIgR to SC is just as elusive. Cleavage of pIgR to SC in rat liver and MDCK cells is inhibited by leupeptin, suggesting a requirement for a cysteine protease (Breitfeld et al., 1989; Musil and Baenziger, 1987, 1988; Solari et al., 1989; Sztul et al., 1993). However, a candidate "SC protease" has never been identified. Furthermore, pIgR expressed ectopically in a variety of epithelial and nonepithelial cell types is efficiently cleaved to SC, suggesting that the protease (or proteases) are not cell-type-specific. Epithelial polarization appears to enhance cleavage

(Chintalacharuvu et al., 1991), but SC can be cleaved and released from non-polarized cells such as fibroblasts ectopically expressing pIgR (Asano et al., 2004; Deitcher et al., 1986).

The hypothesis most consistent with the data is that the relatively unstructured conformation, not the specific sequence, of the linker peptide exposes multiple potential cleavage sites to "nonspecific" proteases on the cell surface and in the extracellular milieu. Consistent with this hypothesis was the report that deletion of the cytoplasmic domain of human pIgR, ectopically expressed in baby hamster kidney fibroblasts, resulted in enhanced cleavage compared to full-length pIgR (Asano et al., 2004). These data suggest that association of the cytoplasmic domain of pIgR with intracellular proteins might constrain the conformation of the extracellular linker peptide, rendering it less susceptible to proteolytic cleavage. On the other hand, these investigators found that deletion of the linker peptide dramatically reduced cleavage of pIgR to SC. A small but detectable amount of cleavage was noted within extracellular domain 5 of this mutant pIgR, which due to its compact immunoglobulin-like structure would be expected to be less available to proteolytic cleavage than would the linker peptide. An important corollary to the "conformational" hypothesis is that cleavage of pIgR to SC is more efficient at the apical than the basolateral surface of polarized epithelial cells. The protein and lipid compositions of the apical and basolateral membranes differ substantially (Mostov et al., 2003), and it is reasonable to predict that pIgR might assume different conformations and/or encounter different proteases at the two surfaces. *In vivo*, the extracellular milieus at the apical and basolateral surfaces are quite different with respect to pH, mucous layer, and cellular infiltrate. At the apical surface, extracellular mucus-associated proteases as well as cell-associated proteases might participate in SC cleavage. Another mechanism for SC cleavage was suggested by the report that serine proteases released by activated neutrophils cleaved pIgR from the surface of human bronchial epithelial cells (Pilette et al., 2003). Thus, multiple proteases might participate in SC cleavage, particularly during inflammatory responses, consistent with the finding of multiple C-termini for free SC. Regardless of the source of the protease, it can be predicted that genetic polymorphisms that alter the sequence of the linker peptide might alter the susceptibility of pIgR to cleavage (see Sect. 3.2.1.4).

3.7. Immune Functions of Free SC and SIgA

3.7.1. Neutralization of Antigens and Pathogens by SIgA

Transcytosis of pIgA by pIgR promotes intracellular neutralization and transcellular excretion of antigens and pathogens and ensures continuous delivery of SIgA to the epithelial surface and external secretions. With each round of pIgA transport, epithelial cells "sacrifice" the extracellular domain of pIgR as cleaved SC, either free or complexed to pIgA (Fig. 3.1). It is reasonable to assume that the metabolic cost of synthesizing a new molecule of pIgR for each round of pIgA transport is compensated for by immune functions contributed by SC. Surface plasmon resonance-based binding experiments indicated that purified SIgA and pIgA antibodies had identical binding affinities for immobilized antigen (Lullau et al., 1996). Similarly, SIgA was found to be no more effective than pIgA in its ability to neutralize influenza virus *in vitro* (Renegar et al., 1998). However, the presence of SC has been shown to enhance both the stability and effector functions of pIgA. It has long been appreciated that bound SC protects the SIgA molecule from proteolytic degradation *in vitro* (Lindh, 1975; Mestecky et al., 1991; Renegar et al., 1998), and this protective mechanism has subsequently been demonstrated *in vivo* in the gastrointestinal tract (Chintalacharuvu and Morrison, 1997; Crottet and Corthesy, 1998) and the oral cavity (Ma et al., 1998). The presence of bound SC has also been shown to inhibit degradation of pIgA by neutrophil elastase, thus enhancing the effectiveness of humoral immunity in the respiratory tract (Pilette et al., 2003) (see Chapter 8).

3.7.2. Compromised Immune Functions in pIgR Knockout Mice

The generation of mice with two disrupted alleles in the *Pigr* gene (pIgR knockout mice) has provided an experimental model for determining the specific requirement for SIgA in mucosal homeostasis and protection against mucosal infections (Johansen et al., 1999; Shimada et al., 1999). pIgR knockout mice lack mucosal immunoglobulins and accumulate pIgA in the circulation at levels 100-fold higher than those in normal mice. Surprisingly, pIgR knockout mice were found to have about three times as many IgA-secreting plasma cells in the intestinal lamina propria as wild-type mice, suggesting that pIgR might contribute to mucosal B-cell homeostasis in addition to transporting mucosal pIgA (Uren et al., 2003). Therefore, both increased production of pIgA and lack of pIgR-mediated transcytosis appear to contribute to the elevated serum levels of pIgA in pIgR knockout mice. On the other hand, antigen-specific helper and cytotoxic T-cells responses appeared to be normal in pIgR knockout mice (Uren et al., 2003).

In comparison to wild-type mice, pIgR knockout mice have been shown to have reduced protection against infection with influenza A and B virus following intranasal immunization (Asahi et al., 2002; Asahi-Ozaki et al., 2004). Reduced levels of nasal IgA antibodies in pIgR knockout mice were also associated with inefficient cross-protection against related strains of influenza A or B. Defective protection against influenza infection in pIgR knockout mice could be attributed to a lack of intracellular neutralization as well as reduced immune exclusion by SIgA in nasal secretions. Other investigators observed no differences in induction of influenza virus-specific $CD8^+$

T-cells in pIgR knockout versus wild-type mice (Uren et al., 2003), confirming the crucial role for mucosal SIgA in protection against influenza infection.

Elucidation of the requirement for SIgA in protection against bacterial infections has proven to be more complex. Maaser et al. (2004) reported that naïve wild-type and pIgR knockout mice displayed identical kinetics of bacterial clearance following oral infection with *Citrobacter rodentium*, a murine pathogen that has been used as a model for enteropathogenic *Escherichia coli* in humans. Uren et al. (2005) found that previously vaccinated wild-type mice cleared *C. rodentium* more rapidly than did vaccinated pIgR knockout mice, but both mouse strains showed complete clearance by 9 days after oral challenge. These investigators also found that pIgR-mediated IgA transport was important for neutralization of a bacterial exotoxin. Vaccination with cholera toxin provided complete protection against subsequent challenge in wild-type mice, whereas vaccinated pIgR knockout mice were not protected and accumulated fluid in the small intestine in response to oral administration of cholera toxin. The importance of pIgR in protection against the intestinal pathogen *Salmonella typhimurium* appears to depend on the nature of infection and previous exposure. In one study, vaccinated wild-type and pIgR knockout mice were found to be equally resistant to challenge with *S. typhimurium*, despite dramatic differences in the titer of SIgA antibodies in intestinal secretions (Uren et al., 2005). However, another group reported that unimmunized pIgR knockout mice were more susceptible than wild-type mice to oral infection with low doses of *S. typhimurium* (Wijburg et al., 2006). These investigators further demonstrated that pIgR knockout mice were more susceptible to "natural" infection by the oral–fecal route with *S. typhimurium*, achieved by cohousing of infected and uninfected mice. The composition of the intestinal microbiota did not appear to differ substantially in wild-type and pIgR knockout mice (Sait et al., 2003). However, mice lacking pIgR expression were found to have increased levels of serum IgA specific for gut bacteria (Wijburg et al., 2006), suggesting that bacterial antigens had gained access to the systemic circulation. This finding is consistent with the role of pIgR in epithelial transport and excretion of pIgA complexed to bacteria (see Sect. 3.5.3). The importance of pIgR in the maintenance of intestinal homeostasis is highlighted by the recent report that pIgR knockout mice are more susceptible than wild-type mice to experimental colitis (Murthy et al., 2006). A detailed discussion of the role of IgA in intestinal homeostasis is presented in Chapter 10.

Studies with the respiratory pathogen *Streptococcus pneumoniae* revealed a very specific role for SIgA (Sun et al., 2004). These investigators found that vaccinated wild-type mice, but not pIgR knockout mice, were protected against infection with serotype 14 *S. pneumoniae*, which causes mucosal colonization but not systemic inflammation. In contrast, vaccinated wild-type and pIgR knockout mice were equally protected against lethal systemic infection with serotype 3 *S. pneumoniae*. These data suggest that the primary function of SIgA in protection against infection with *S. pneumoniae* is prevention of nasopharyngeal carriage, presumably by immune exclusion. In this context it is important to note that mouse pIgR, unlike human pIgR, does not bind to the SpsA/CbpA protein of *S. pneumoniae* (see above). Thus immune exclusion of this pathogen in mice is likely to be mediated by the antigen-specific pIgA moiety of SIgA. In humans, free SC and the SC moiety of SIgA might provide protection against nasopharyngeal colonization with *S. pneumoniae* by binding to SpsA/CbpA and blocking its association with pIgR on the surface of epithelial cells (reviewed in Kaetzel, 2001) (see Sect. 3.4.2).

The pIgR might have a role in clearance of intracellular as well as extracellular bacteria in the respiratory tract. Tjarnlund et al. (2006) found that immunized pIgR knockout mice were more susceptible than immunized wild-type mice to infection with *Mycobacterium bovis* bacillus Calmette-Guerin, based on higher bacterial loads and reduced production of both IFN-γ and TNF in the lungs. A detailed discussion of the role of pIgR and IgA in respiratory immunity can be found in Chapter 11).

A recent study suggests that pIgR and pIgA also coordinate immune defense against lumen-dwelling intestinal parasites (Davids et al., 2006). pIgR knockout mice were found to have higher parasite burdens and severer diarrhea than wild-type mice following experimental infection with *Giardia lamblia*.

3.7.3. Enhancement of Innate Immunity by SC

In addition to the functions of SIgA in antigen-specific immunity, free SC and SIgA have been shown to contribute in novel ways to the regulation of innate, "nonspecific" responses to pathogens (reviewed in Phalipon and Corthesy, 2003) (see Chapter 8). Many of these functions appear to result from binding of the unusual N-glycans of SC to bacterial and host factors (Table 3.2). Free SC has been shown to limit infection or reduce morbidity by binding to bacterial components such as *Clostridium difficile* toxin A (Dallas and Rolfe, 1998) and fimbriae of enterotoxigenic *E. coli* (de Oliveira et al., 2001). SIgA was shown to be more protective than pIgA in a mouse model of respiratory infection by *Shigella flexneri*, due to carbohydrate-dependent adherence of SIgA to the mucus lining of the epithelium of the upper airway (Phalipon et al., 2002). SC might also protect epithelial surfaces by reducing inflammation associated with host immune responses. Free SC has been shown to form a high-molecular-weight complex with IL-8 secreted by primary cultures of human bronchial epithelial cells, inhibiting its activity as a neutrophil chemoattractant (Marshall et al., 2001). Other investigators reported that monomeric and polymeric IgA triggered efficient phagocytosis of heat-killed *Neisseria meningitidis* by human neutrophils, but SIgA did not (Vidarsson et al., 2001). In contrast, SIgA was found to be more potent than serum IgA in stimulating degranulation and superoxide production by human eosinophils (Motegi and Kita, 1998; Motegi et al., 2000). Thus, the presence of SC in mucosal secretions, free or complexed to SIgA, might differentially modulate the host immune response to inflammatory stimuli.

3.8. Concluding Remarks

Emerging research on the complex biology of pIgR highlights the key role of this receptor in bridging innate and adaptive immune responses at mucosal surfaces (Fig. 3.5). Free SC, produced by transcytosis of pIgR in the absence of pIgA ligand, is an important component of innate antimicrobial defense. Commensal and pathogenic microorganisms can upregulate pIgR expression by signaling through Toll-like and probably other receptors, thus enhancing the innate immune response of SC and the capacity of epithelial cells to transport antigen-specific pIgA. Signaling pathways initiated by microorganisms also activate transcription of genes that encode chemokines, which recruit immune cells to the epithelium, and cytokines, which further amplify pIgR expression. In the adaptive phase of the immune response, additional

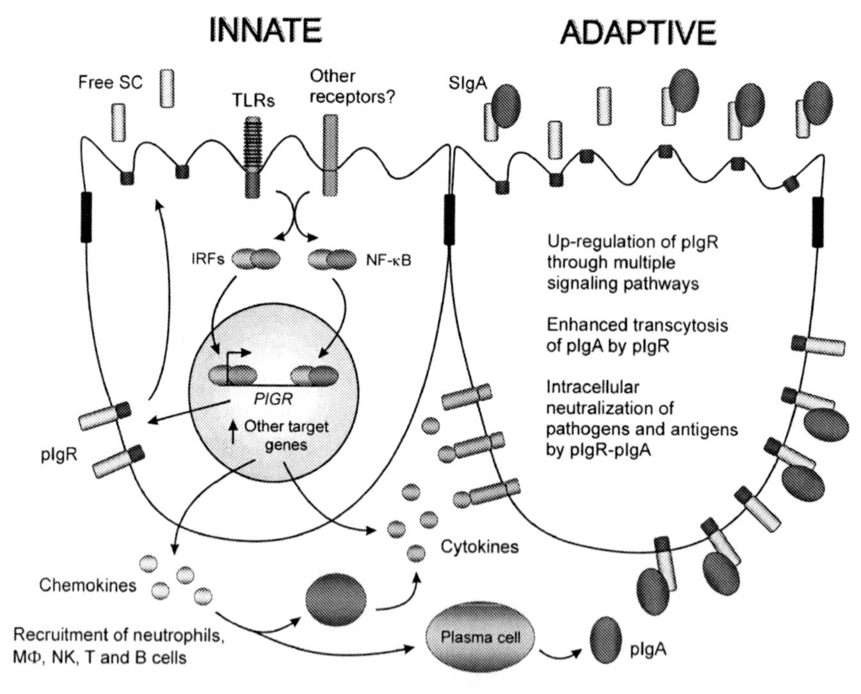

FIG. 3.5. The pIgR bridges innate and adaptive immunity. Signaling pathways initiated by pathogen-associated molecular patterns and host cytokines upregulate pIgR expression, leading to enhanced innate immune functions of SC and adaptive immune functions of SIgA. See the text for a detailed description of the signaling pathways and immune functions. TLRs, Toll-like receptors; IRFs, interferon regulatory factors; NF-κB, nuclear factor-kappaB; MΦ, macrophage; NK, natural killer cell; pIgA, polymeric IgA.

cytokines produced by immune cells contribute to the maintenance of high levels of pIgR expression, and local plasma cells secrete antigen-specific pIgA. Enhanced transcytosis of pIgA by pIgR facilitates intracellular neutralization of pathogens and antigens and the release of high levels of free SC and SIgA into external secretions. A key feature of this pathway is the presence of multiple checkpoints to minimize potentially damaging inflammatory responses. For example, signaling through TLR4 in intestinal epithelial cells can upregulate *PIGR* gene transcription while minimizing induction of proinflammatory genes. Synergistic effects of low levels of proinflammatory and Th2 cytokines can maximize pIgR expression while minimizing inflammatory damage to the epithelium. Intracellular neutralization of LPS by pIgR–pIgA can limit access of this inflammatory mediator to cells in the lamina propria. Thus, the polymeric immunoglobulin receptor plays a central role in innate and adaptive immune defense and the maintenance of homeostasis at mucosal surfaces that are constantly exposed to environmental pathogens, antigens, and commensal microorganisms.

Acknowledgments. Research in Dr. Kaetzel's laboratory is supported by the Crohn's & Colitis Foundation of America and the National Institutes of Health.

References

Ackermann, L. W., and Denning, G. M. (2004). Nuclear factor-κB contributes to interleukin-4- and interferon-dependent polymeric immunoglobulin receptor expression in human intestinal epithelial cells. *Immunology* 111:75–85.

Ackermann, L. W., Wollenweber, L. A., and Denning, G. M. (1999). IL-4 and IFN-γ increase steady state levels of polymeric Ig receptor mRNA in human airway and intestinal epithelial cells. *J. Immunol.* 162:5112–5118.

Adamski, F. M., and Demmer, J. (1999). Two stages of increased IgA transfer during lactation in the marsupial, *Trichosurus vulpecula* (brushtail possum). *J. Immunol.* 162:6009–6015.

Alonso de Velasco, E., Verheul, A. F., Verhoef, J., and Snippe, H. (1995). *Streptococcus pneumoniae*: Virulence factors, pathogenesis, and vaccines. *Microbiol. Rev.* 59:591–603.

Aoki, T., Koch, K. S., and Leffert, H. L. (1997). Attenuation of gene expression by a trinucleotide repeat-rich tract from the terminal exon of the rat hepatic polymeric immunoglobulin receptor gene. *J. Mol. Biol.* 267:229–236.

Aroeti, B., Kosen, P. A., Kuntz, I. D., Cohen, F. E., and Mostov, K. E. (1993). Mutational and secondary structural analysis of the basolateral sorting signal of the polymeric immunoglobulin receptor. *J. Cell Biol.* 123:1149–1160.

Asahi, Y., Yoshikawa, T., Watanabe, I., Iwasaki, T., Hasegawa, H., Sato, Y., Shimada, S., Nanno, M., Matsuoka, Y., Ohwaki, M., Iwakura, Y., Suzuki, Y., Aizawa, C., Sata, T., Kurata, T., and Tamura, S. (2002). Protection against influenza virus infection in polymeric Ig receptor knockout mice immunized intranasally with adjuvant-combined vaccines. *J. Immunol.* 168:2930–2938.

Asahi-Ozaki, Y., Yoshikawa, T., Iwakura, Y., Suzuki, Y., Tamura, S., Kurata, T., and Sata, T. (2004). Secretory IgA antibodies provide cross-protection against infection with different strains of influenza B virus. *J. Med. Virol.* 74:328–335.

Asano, M., Takenouchi-Ohkubo, N., Matsumoto, N., Ogura, Y., Nomura, H., Suguro, H., and Moro, I. (2004). Multiple cleavage sites for polymeric immunoglobulin receptor. *Immunology* 112:583–589.

Barroso, M., Bernd, K. K., DeWitt, N. D., Chang, A., Mills, K., and Sztul, E. S. (1996). A novel Ca^{2+}-binding protein, p22, is required for constitutive membrane traffic. *J. Biol. Chem.* 271:10,183–10,187.

Barton, G. M., and Medzhitov, R. (2003). Toll-like receptor signaling pathways. *Science* 300:1524–1525.

Bjercke, S., and Brandtzaeg, P. (1993). Glandular distribution of immunoglobulins, J chain, secretory component, and HLA-DR in the human endometrium throughout the menstrual cycle. *Hum. Reprod.* 8:1420–1425.

Blanch, V. J., Piskurich, J. F., and Kaetzel, C. S. (1999). Cutting edge: Coordinate regulation of IFN regulatory factor-1 and the polymeric Ig receptor by proinflammatory cytokines. *J. Immunol.* 162:1232–1235.

Bonifacino, J. S., and Traub, L. M. (2003). Signals for sorting of transmembrane proteins to endosomes and lysosomes. *Annu. Rev. Biochem.* 72:395–447. Bonizzi, G., and Karin, M. (2004). The two NF-kB activation pathways and their role in innate and adaptive immunity. *Trends Immunol.* 25:280–288.

Boren, T., Falk, P., Roth, K. A., Larson, G., and Normark, S. (1993). Attachment of *Helicobacter pylori* to human gastric epithelium mediated by blood group antigens. *Science* 262:1892–1895.

Bowie, A., and O'Neill, L. A. (2000). The interleukin-1 receptor/Toll-like receptor superfamily: Signal generators for pro-inflammatory interleukins and microbial products. *J. Leuk. Biol.* 67:508–514.

Braathen, R., Hohman, V. S., Brandtzaeg, P., and Johansen, F. E. (2007) Secretory antibody formation: Conserved binding interactions between J chain and polymeric Ig receptor from humans and amphibians. *J.Immunol.* 178:1589–1597.

Braathen, R., Sorensen, V., Brandtzaeg, P., Sandlie, I., and Johansen, F. E. (2002). The carboxy-terminal domains of IgA and IgM direct isotype-specific polymerization and interaction with the polymeric immunoglobulin receptor. *J. Biol. Chem.* 277:42,755–42,762.

Brandtzaeg, P., Berstad, A. E., Farstad, I. N., Haraldsen, G., Helgeland, L., Jahnsen, F. L., Johansen, F. E., Natvig, I. B., Nilsen, E. M., and Rugtveit, J. (1997). Mucosal immunity: A major adaptive defence mechanism. *Behring Inst. Mitt.* 98:1–23.

Brandtzaeg, P., and Johansen, F. E. (2001). Confusion about the polymeric Ig receptor. *Trends Immunol.* 22:545–546.

Brandtzaeg, P., and Prydz, H. (1984). Direct evidence for an integrated function of J chain and secretory component in epithelial transport of immunoglobulins. *Nature* 311:71–73.

Breitfeld, P. P., Harris, J. M., and Mostov, K. E. (1989). Postendocytotic sorting of the ligand for the polymeric immunoglobulin receptor in Madin-Darby canine kidney cells. *J. Cell Biol.* 109:475–486.

Bren, G. D., Solan, N. J., Miyoshi, H., Pennington, K. N., Pobst, L. J., and Paya, C. V. (2001). Transcription of the RelB gene is regulated by NF-κB. *Oncogene* 20:7722–7733.

Brown, T. A., Russell, M. W., Kulhavy, R., and Mestecky, J. (1983). IgA-mediated elimination of antigens by the hepatobiliary route. *Fed. Proc.* 42:3218–3221.

Brown, T. A., Russell, M. W., and Mestecky, J. (1984). Elimination of intestinally absorbed antigen into the bile by IgA. *J. Immunol.* 132:780–782.

Brown, T. A., Russell, M. W., and Mestecky, J. (1982). Hepatobiliary transport of IgA immune complexes: Molecular and cellular aspects. *J. Immunol.* 128:2183–2186.

Brown, W. R., and Kloppel, T. M. (1989). The liver and IgA: Immunological, cell biological and clinical implications. *Hepatology* 9:763–784.

Bruce, S. R., Kaetzel, C. S., and Peterson, M. L. (1999). Cryptic intron activation within the large exon of the mouse polymeric immunoglobulin receptor gene: Cryptic splice sites correspond to protein domain boundaries. *Nucleic Acids Res.* 27:3446–3454.

Bruce, S. R., and Peterson, M. L. (2001). Multiple features contribute to efficient constitutive splicing of an unusually large exon. *Nucleic Acids Res.* 29:2292–2302.

Bruno, M. E. C., and Kaetzel, C. S. (2005). Long-term exposure of the HT-29 human intestinal epithelial cell-line to TNF causes sustained up-regulation of the polymeric Ig receptor and pro-inflammatory genes through transcriptional and post-transcriptional mechanisms. *J. Immunol.* 174:7278–7284.

Bruno, M. E. C., West, R. B., Schneeman, T. A., Bresnick, E. H., and Kaetzel, C. S. (2004). Upstream stimulatory factor but not c-Myc enhances transcription of the human polymeric immunoglobulin receptor gene. *Mol. Immunol.* 40:695–708.

Buts, J.-P., Bernasconi, P., Vaerman, J.-P., and Dive, C. (1990). Stimulation of secretory IgA and secretory component of immunoglobulins in small intestine of rats treated with *Saccharomyces boulardii. Dig. Dis. Sci.* 35:251–256.

Calvo, M., Pol, A., Lu, A., Ortega, D., Pons, M., Blasi, J., and Enrich, C. (2000). Cellubrevin is present in the basolateral endocytic compartment of hepatocytes and follows the transcytotic pathway after IgA internalization. *J. Biol. Chem.* 275:7910–7917.

Cardone, M. H., Smith, B. L., Mennitt, P. A., Mochly-Rosen, D., Silver, R. B., and Mostov, K. E. (1996). Signal transduction by the polymeric immunoglobulin receptor suggests a role in regulation of receptor transcytosis. *J. Cell Biol.* 133:997–1005.

Cardone, M. H., Smith, B. L., Song, W., Mochly-Rosen, D., and Mostov, K. E. (1994). Phorbol myristate acetate-mediated stimulation of transcytosis and apical recycling in MDCK cells. *J. Cell Biol.* 124:717–727.

Cario, E., and Podolsky, D. K. (2000). Differential alteration in intestinal epithelial cell expression of Toll-like receptor 3 (TLR3) and TLR4 in inflammatory bowel disease. *Infect. Immun.* 68:7010–7017.

Casanova, J. E., Apodaca, G., and Mostov, K. E. (1991). An autonomous signal for basolateral sorting in the cytoplasmic domain of the polymeric immunoglobulin receptor. *Cell* 66:65–75.

Casanova, J. E., Breitfeld, P. P., Ross, S. A., and Mostov, K. E. (1990). Phosphorylation of the polymeric immunolglobulin receptor required for its efficient transcytosis. *Science* 248:742–745.

Chapin, S. J., Enrich, C., Aroeti, B., Havel, R. J., and Mostov, K. E. (1996). Calmodulin binds to the basolateral targeting signal of the polymeric immunoglobulin receptor. *J. Biol. Chem.* 271:1336–1342.

Chapman, R. S., Duff, E. K., Lourenco, P. C., Tonner, E., Flint, D. J., Clarke, A. R., and Watson, C. J. (2000). A novel role for IRF-1 as a suppressor of apoptosis. *Oncogene* 19:6386–6391.

Chintalacharuvu, K. R., and Morrison, S. L. (1997). Production of secretory immunoglobulin A by a single mammalian cell. *Proc. Natl. Acad. Sci. U S A* 94:6364–6368.

Chintalacharuvu, K. R., Piskurich, J. F., Lamm, M. E., and Kaetzel, C. S. (1991). Cell polarity regulates the release of secretory component, the epithelial receptor for polymeric immunoglobulins, from the surface of HT-29 colon carcinoma cells. *J. Cell. Physiol.* 148:35–47.

Claessens, F., Verrijdt, G., Schoenmakers, E., Haelens, A., Peeters, B., Verhoeven, G., and Rombauts, W. (2001). Selective DNA binding by the androgen receptor as a mechanism for hormone-specific gene regulation. *J. Steroid Biochem. Mol. Biol.* 76:23–30.

Conley, M. E., and Delacroix, D. L. (1987). Intravascular and mucosal immunoglobulin A: Two separate but related systems of immune defense? *Ann. Intern. Med.* 106:892–899.

Coyne, R. S., Siebrecht, M., Peitsch, M. C., and Casanova, J. E. (1994). Mutational analysis of polymeric immunoglobulin receptor/ligand interactions. Evidence for the involvement of multiple complementarity determining region (CDR)-like loops in receptor domain I. *J. Biol. Chem.* 269:31,620–31,625.

Crottet, P., and Corthesy, B. (1998). Secretory component delays the conversion of secretory IgA into antigen-binding competent F(ab')2: A possible implication for mucosal defense. *J. Immunol.* 161:5445–5453.

Dallas, S. D., and Rolfe, R. D. (1998). Binding of *Clostridium difficile* toxin A to human milk secretory component. *J. Med. Microbiol.* 47:879–888.

Davids, B. J., Palm, J. E. D., Housley, M. P., Smith, J. R., Andersen, Y. S., Martin, M. G., Hendrickson, B. A., Johansen, F. E., Svard, S. G., Gillin, F. D., and Eckmann, L. (2006). Polymeric immunoglobulin receptor in intestinal immune defense against the lumen-dwelling protozoan parasite *Giardia. Jr. Immunol.* 177:6281–6290.

Deitcher, D. L., and Mostov, K. E. (1986). Alternate splicing of rabbit polymeric immunoglobulin receptor. *Mol. Cell. Biol.* 6:2712–2715.

Deitcher, D. L., Neutra, M. R., and Mostov, K. E. (1986). Functional expression of the polymeric immunoglobulin receptor from cloned cDNA in fibroblasts. *J. Cell Biol.* 102:911–919.

Delacroix, D. L., Furtado-Barreira, G., Rahier, J., Dive, C., and Vaerman, J.-P. (1984). Immunohistochemical localization of secretory component in the liver of guinea pigs and dogs versus rats, rabbits, and mice. *Scand. J. Immunol.* 19:425–434.

Denning, G. M. (1996). IL-4 and IFN-γ synergistically increase total polymeric IgA receptor levels in human intestinal epithelial cells. Role of protein tyrosine kinases. *J. Immunol.* 156:4807–4814.

de Oliveira, I. R., de Araujo, A. N., Bao, S. N., and Giugliano, L. G. (2001). Binding of lactoferrin and free secretory component to enterotoxigenic *Escherichia coli. FEMS Microbiol. Lett.* 203:29–33.

Dunne, A. and O'Neill, L. A. (2003). The interleukin-1 receptor/Toll-like receptor superfamily: Signal transduction during inflammation and host defense. *Sci. STKE* 2000:re3.

Eiffert, H., Quentin, E., Decker, J., Hillemeir, S., Hufschmidt, M., Klingmuller, D., Weber, M. H., and Hilschmann, N. (1984). The primary structure of the human free secretory component and the arrangement of the disulfide bonds. *Hoppe-Seyler's Z. Physiol. Chem.* 365:1489–1495.

Eiffert, H., Quentin, E., Wiederhold, M., Hillemeir, S., Decker, J., Weber, M., and Hilschmann, N. (1991). Determination of the molecular structure of the human free secretory component. *Biol. Chem. Hoppe-Seyler* 372:119–128.

Elm, C., Braathen, R., Bergmann, S., Frank, R., Vaerman, J. P., Kaetzel, C. S., Chhatwal, G. S., Johansen, F. E., and Hammerschmidt, S. (2004). Ectodomains 3 and 4 of human polymeric immunoglobulin receptor (hpIgR) mediate invasion of *Streptococcus pneumoniae* into the epithelium. *J Biol. Chem.* 279:6296–6304.

Enrich, C., Jäckle, S., and Havel, R. J. (1996). The polymeric immunoglobulin receptor is the major calmodulin-binding protein in an endosome fraction from rat liver enriched in recycling receptors. *Hepatology* 24:226–232.

Falk, P., Roth, K. A., Boren, T., Westblom, T. U., Gordon, J. I., and Normark, S. (1993). An *in vitro* adherence assay reveals that Helicobacter pylori exhibits cell lineage-specific tropism in the human gastric epithelium. *Proc. Natl. Acad. Sci. USA* 90:2035–2039.

Fallgreen-Gebauer, E., Gebauer, W., Bastian, A., Kratzin, H. D., Eiffert, H., Zimmermann, B., Karas, M., and Hilschmann, N. (1993). The covalent linkage of secretory component to IgA. Structure of sIgA. *Biol. Chem. Hoppe-Seyler* 374:1023–1028.

Fernandez, M. I., Pedron, T., Tournebize, R., Olivo-Marin, J. C., Sansonetti, P. J., and Phalipon, A. (2003). Anti-inflammatory role for intracellular dimeric immunoglobulin A by neutralization of lipopolysaccharide in epithelial cells. *Immunity* 18:739–749.

Fodor, E., Feren, A., and Jones, A. (1997). Isolation and genomic analysis of the rat polymeric immunoglobulin receptor gene terminal domain and transcriptional control region. *DNA Cell Biol.* 16:215–225.

Gan, Y.-J., Chodosh, J., Morgan, A., and Sixbey, J. W. (1997). Epithelial cell polarization is a determinant in the infectious outcome of immunoglobulin A-mediated entry by Epstein-Barr virus. *J. Virol.* 71:519–526.

Giffroy, D., Courtoy, P. J., and Vaerman, J. P. (2001). Polymeric IgA binding to the human pIgR elicits intracellular signalling, but fails to stimulate pIgR-transcytosis. *Scand. J. Immunol.* 53:56–64.

Giffroy, D., Langendries, A., Maurice, M., Daniel, F., Lardeux, B., Courtoy, P. J., and Vaerman, J. P. (1998). *In vivo* stimulation of polymeric Ig receptor transcytosis by circulating polymeric IgA in rat liver. *Int. Immunol.* 10:347–354.

Giugliano, L. G., Ribeiro, S. T., Vainstein, M. H., and Ulhoa, C. J. (1995). Free secretory component and lactoferrin of human milk inhibit the adhesion of enterotoxigenic *Escherichia coli. J. Med. Microbiol.* 42:3–9.

Hamburger, A. E., West, A. P., Jr., and Bjorkman, P. J. (2004). Crystal structure of a polymeric immunoglobulin binding fragment of the human polymeric immunoglobulin receptor. *Structure* 12:1925–1935.

Harmatz, P. R., Kleinman, R. E., Bunnell, B. W., Bloch, K. J., and Walker, W. A. (1982). Hepatobiliary clearance of IgA immune complexes formed in the circulation. *Hepatology* 2:328–333.

Hausmann, M., Kiessling, S., Mestermann, S., Webb, G., Spottl, T., Andus, T., Scholmerich, J., Herfarth, H., Ray, K., Falk, W., and Rogler, G. (2002). Toll-like receptors 2 and 4 are up-regulated during intestinal inflammation. *Gastroenterology* 122:1987–2000.

Hempen, P. M., Phillips, K. M., Conway, P. S., Sandoval, K. H., Schneeman, T. A., Wu, H. J., and Kaetzel, C. S. (2002). Transcriptional regulation of the human polymeric Ig receptor gene: Analysis of basal promoter elements. *J. Immunol.* 169:1912–1921.

Hendrickson, B. A., Conner, D. A., Ladd, D. J., Kendall, D., Casanova, J. E., Corthesy, B., Max, E. E., Neutra, M. R., Seidman, C. E., and Seidman, J. G. (1995). Altered hepatic transport of immunoglobulin A in mice lacking the J chain. *J. Exp. Med.* 182:1905–1911.

Hendrickson, B. A., Rindisbacher, L., Corthesy, B., Kendall, D., Waltz, D. A., Neutra, M. R., and Seidman, J. G. (1996). Lack of association of secretory component with IgA in J chain-deficient mice. *J. Immunol.* 157:750–754.

Herr, A. B., Ballister, E. R., and Bjorkman, P. J. (2003). Insights into IgA-mediated immune responses from the crystal structures of human FcalphaRI and its complex with IgA1-Fc. *Nature* 423:614–620.

Hexham, J. M., White, K. D., Carayannopoulos, L. N., Mandecki, W., Brisette, R., Yang, Y. S., and Capra, J. D. (1999). A human immunoglobulin (Ig)A $C_{\alpha}3$ domain motif directs polymeric Ig receptor-mediated secretion. *J. Exp. Med.* 189:747–752.

Hirunsatit, R., Kongruttanachok, N., Shotelersuk, K., Supiyaphun, P., Voravud, N., Sakuntabhai, A., and Mutirangura, A. (2003). Polymeric immunoglobulin receptor polymorphisms and risk of nasopharyngeal cancer. *BMC Genet.* 4:3.

Honda, K., and Taniguchi, T. (2006). IRFs: master regulators of signalling by Toll-like receptors and cytosolic pattern-recognition receptors. *Nat. Rev. Immunol.* 6:644–658.

Hooper, L. V., Wong, M. H., Thelin, A., Hansson, L., Falk, P. G., and Gordon, J. I. (2001). Molecular analysis of commensal host-microbial relationships in the intestine. *Science* 291:881–884.

Hughes, G. J., Frutiger, S., Savoy, L. A., Reason, A. J., Morris, H. R., and Jaton, J. C. (1997). Human free secretory component is composed of the first 585 amino acid residues of the polymeric immunoglobulin receptor. *FEBS Lett.s* 410:443–446.

Hughes, G. J., Reason, A. J., Savoy, L., Jaton, J., and Frutiger-Hughes, S. (1999). Carbohydrate moieties in human secretory component. *Biochim. Biophys. Acta* 1434:86–93.

Jenkins, S. L., Wang, J., Vazir, M., Vela, J., Sahagun, O., Gabbay, P., Hoang, L., Diaz, R. L., Aranda, R., and Martín, M. G. (2003). Role of passive and adaptive immunity in influencing enterocyte-specific gene expression. *Am. J. Physiol.: Gastrointest. Liver Physiol.* 285:G714–G725.

Jin, M., Saucan, L., Farquhar, M. G., and Palade, G. E. (1996). Rab1a and multiple other Rab proteins are associated with the transcytotic pathway in rat liver. *J. Biol. Chem.* 271:30,105–30,113.

Johansen, F. E., Bosloven, B. A., Krajci, P., and Brandtzaeg, P. (1998). A composite DNA element in the promoter of the polymeric immunoglobulin receptor regulates its constitutive expression. *Eur. J. Immunol.* 28:1161–1171.

Johansen, F. E., and Brandtzaeg, P. (2004). Transcriptional regulation of the mucosal IgA system. *Trends Immunol.* 25:150–157.

Johansen, F. E., Pekna, M., Norderhaug, I. N., Haneberg, B., Hietala, M. A., Krajci, P., Betsholtz, C., and Brandtzaeg, P. (1999). Absence of epithelial immunoglobulin A transport, with increased mucosal leakiness, in polymeric immunoglobulin receptor/secretory component-deficient mice. *J. Exp. Med.* 190:915–922.

Kaetzel, C. S. (2001). Polymeric Ig receptor: defender of the fort or Trojan horse? *Curr. Biol.* 11:R35–R38.

Kaetzel, C. S. (2005). The polymeric immunoglobulin receptor: Bridging innate and adaptive immune responses at mucosal surfaces. *Immunol. Rev.* 206:83–99.

Kaetzel, C. S., Blanch, V. J., Hempen, P. M., Phillips, K. M., Piskurich, J. F., and Youngman, K. R. (1997). The polymeric immunoglobulin receptor: structure and synthesis. *Biochem. Soc. Trans.* 25:475–480.

Kaetzel, C. S., and Mostov, K. (2005). Immunoglobulin transport and the polymeric immunoglobulin receptor. In: Mestecky, J., Bienenstock, J., Lamm, M., Strober, W., McGhee, J. and Mayer, L. (eds.), *Mucosal Immunology*, 3rd ed., Academic Press, San Diego, pp. 211–250.

Kaushic, C., Richardson, J. M., and Wira, C. R. (1995). Regulation of polymeric immunoglobulin A receptor messenger ribonucleic acid expression in rodent uteri: Effect of sex hormones. *Endocrinology* 136:2836–2844.

Khattar, N. H., Lele, S. M., and Kaetzel, C. S. (2005). Down-regulation of the polymeric immunoglobulin receptor in non-small cell lung carcinoma: correlation with dysregulated expression of the transcription factors USF and AP2. *J. Biomed. Sci.* 12:65–77.

Kiermaier, A., Gawn, J. M., Desbarats, L., Saffrich, R., Ansorge, W., Farrell, P. J., Eilers, M., and Packham, G. (1999). DNA binding of USF is required for specific E-box dependent gene activation *in vivo. Oncogene* 18:7200–7211.

Koch, K. S., Gleiberman, A. S., Aoki, T., Leffert, H. L., Feren, A., Jones, A. L., and Fodor, E. J. (1995). Discordant expression and variable numbers of neighboring GGA- and GAA-rich triplet repeats in the 3′ untranslated regions of two groups of messenger RNAs encoded by the rat polymeric immunoglobulin receptor gene. *Nucleic Acids Res.* 23:1098–1112.

Kulseth, M. A., Krajci, P., Myklebost, O., and Rogne, S. (1995). Cloning and characterization of two forms of bovine polymeric immunoglobulin receptor. *DNA Cell Biol.* 14:251–256.

Kushiro, A., and Sato, T. (1997). Polymeric immunoglobulin receptor gene of mouse: sequence, structure and chromosomal location. *Gene* 204:277–282.

Kvale, D., and Brandtzaeg, P. (1995). Constitutive and cytokine induced expression of HLA molecules, secretory component, and intercellular adhesion molecule-1 is modulated by butyrate in the colonic epithelial cell line HT-29. *Gut* 36:737–742.

Lamm, M. E. (1997). Interaction of antigens and antibodies at mucosal surfaces. *Annu. Rev. Microbiol.* 51:311–340.

Larkin, J. M., Woo, B., Balan, V., Marks, D. L., Oswald, B. J., LaRusso, N. F., and McNiven, M. A. (2000). Rab3D, a small GTP-binding protein implicated in regulated secretion, is associated with the transcytotic pathway in rat hepatocytes. *Hepatology* 32:348–356.

Li, T. W., Wang, J., Lam, J. T., Gutierrez, E. M., Solorzano-Vargus, R. S., Tsai, H. V., and Martín, M. G. (1999). Transcriptional control of the murine polymeric IgA receptor promoter by glucocorticoids. *Am. J. Physiol* 276:G1425–G1434.

Lindh, E. (1975). Increased resistance of immunoglobulin A dimers to proteolytic degradation after binding of secretory component. *J. Immunol.* 1975:284–286.

Loman, S., Jansen, H. M., Out, T. A., and Lutter, R. (1999). Interleukin-4 and interferon-γ synergistically increase secretory component gene expression, but are additive in stimulating secretory immunoglobulin A release by Calu-3 airway epithelial cells. *Immunology* 96:537–543.

Lu, L., Lamm, M. E., Li, H., Corthesy, B., and Zhang, J. R. (2003). The human polymeric immunoglobulin receptor binds to *Streptococcus pneumoniae* via domains 3 and 4. *J. Biol. Chem.* 278:48,178–187.

Lullau, E., Heyse, S., Vogel, H., Marison, I., von Stockar, U., Kraehenbuhl, J. P., and Corthesy, B. (1996). Antigen binding properties of purified Immunoglobulin A and reconstituted secretory immunoglobulin A antibodies. *J. Biol. Chem.* 271:16,300–309.

Lüscher, B., and Larsson, L. G. (1999). The basic region/helix-loop-helix/leucine zipper domain of Myc proto-oncoproteins: Function and regulation. *Oncogene* 18:2955–2966.

Luton, F., Cardone, M. H., Zhang, M., and Mostov, K. E. (1998). Role of tyrosine phosphorylation in ligand-induced regulation of transcytosis of the polymeric Ig receptor. *Mol. Biol. Cell* 9:1787–1802.

Luton, F., and Mostov, K. E. (1999). Transduction of basolateral-to-apical signals across epithelial cells: Ligand-stimulated transcytosis of the polymeric immunoglobulin receptor requires two signals. *Mol. Biol. Cell* 10:1409–1427.

Luton, F., Verges, M., Vaerman, J. P., Sudol, M., and Mostov, K. E. (1999). The SRC family protein tyrosine kinase p62*yes* controls polymeric IgA transcytosis *in vivo. Mol. Cell* 4:627–632.

Lycke, N., Erlandsson, L., Ekman, L., Schon, K., and Leanderson, T. (1999). Lack of J chain inhibits the transport of gut IgA and abrogates the development of intestinal antitoxic protection. *J. Immunol.* 163:913–919.

Ma, J. K. C., Hikmat, B. Y., Wycoff, K., Vine, N. D., Chargelegue, D., Yu, L., Hein, M. B., and Lehner, T. (1998). Characterization of a recombinant plant monoclonal secretory antibody and preventive immunotherapy in humans. *Nature Med.* 4:601–606.

Maaser, C., Housley, M. P., Iimura, M., Smith, J. R., Vallance, B. A., Finlay, B. B., Schreiber, J. R., Varki, N. M., Kagnoff, M. F., and Eckmann, l. (2004). Clearance of *Citrobacter rodentium* requires B cells but not secretory immunoglobulin A (IgA) or IgM antibodies. *Infect. Immun.* 72:3315–3324.

Mantis, N. J., Farrant, S. A., and Mehta, S. (2004). Oligosaccharide side chains on human secretory IgA serve as receptors for ricin. *J. Immunol.* 172:6838–6845.

Marshall, L. J., Perks, B., Bodey, K., Suri, R., Bush, A., and Shute, J. K. (2004). Free secretory component from cystic fibrosis sputa displays the cystic fibrosis glycosylation phenotype. *Am. J. Respir. Crit Care Med.* 169:399–406.

Marshall, L. J., Perks, B., Ferkol, T., and Shute, J. K. (2001). IL-8 released constitutively by primary bronchial epithelial cells in culture forms an inactive complex with secretory component. *J. Immunol.* 167:2816–2823.

Martín, M. G., Gutierrez, E. M., Lam, J. T., Li, T. W., and Wang, J. (1997). Genomic cloning and structural analysis of the murine polymeric receptor (pIgR) gene and promoter region. *Gene* 201:189–197.

Martín, M. G., Wang, J., Li, T. W., Lam, J. T., Gutierrez, E. M., Solorzano-Vargas, R. S., and Tsai, A. H. (1998). Characterization of the 5′-flanking region of the murine polymeric IgA receptor gene. *Am. J. Physiol.* 275:G778–G788.

Mazanec, M. B., Nedrud, J. G., Kaetzel, C. S., and Lamm, M. E. (1993). A three-tiered view of the role of IgA in mucosal defense. *Immunol. Today* 14:430–435.

Mestecky, J., Lue, C., and Russell, M. W. (1991). Selective transport of IgA: Cellular and molecular aspects. *Gastroenterol. Clin. North Am.* 20:441–471.

Mestecky, J., Russell, M. W., Jackson, S., and Brown, T. A. (1986). The human IgA system: a reassessment. *Clin. Immunol. Immunopathol.* 40:105–114.

Mignone, F., Gissi, C., Liuni, S., and Pesole, G. (2002). Untranslated regions of mRNAs. *Genome Biol.* 3:0004.1–0004.10.

Moldoveanu, Z., Moro, I., Radl, J., Thorpe, S. R., Komiyama, K., and Mestecky, J. (1990). Site of catabolism of autologous and heterologous IgA in non-human primates. *Scand. J. Immunol.* 32:577–583.

Mostov, K., Su, T., and ter Beest, M. (2003). Polarized epithelial membrane traffic: conservation and plasticity. *Nat. Cell Biol.* 5:287–293.

Mostov, K. E., and Deitcher, D. L. (1986). Polymeric immunoglobulin receptor expressed in MDCK cells transcytoses IgA. *Cell* 46:613–621.

Mostov, K. E., Friedlander, M., and Blobel, G. (1984). The receptor for transepithelial transport of IgA and IgM contains multiple immunoglobulin-like domains. *Nature* 308:37–43.

Mostov, K. E., Kraehenbuhl, J.-P., and Blobel, G. (1980). Receptor-mediated transcellular transport of immunoglobulin: synthesis of secretory component as multiple and larger transmembrane forms. *Proc. Natl. Acad. Sci. USA* 77:7257–7261.

Mostov, K. E., Verges, M., and Altschuler, Y. (2000). Membrane traffic in polarized epithelial cells. *Curr. Opin. Cell Biol.* 12:483–490.

Motegi, Y., and Kita, H. (1998). Interaction with secretory component stimulates effector functions of human eosinophils but not of neutrophils. *J Immunol* 161:4340–4346.

Motegi, Y., Kita, H., Kato, M., and Morikawa, A. (2000). Role of secretory IgA, secretory component, and eosinophils in mucosal inflammation. *Int. Arch. Allergy Immunol.* 122(Suppl 1):25–27.

Murthy, A. K., Dubose, C. N., Banas, J. A., Coalson, J. J., and Arulanandam, B. P. (2006). Contribution of polymeric immunoglobulin receptor to regulation of intestinal inflammation in dextran sulfate sodium-induced colitis. *J Gastroenterol. Hepatol.* 21:1372–1380.

Musil, L. S., and Baenziger, J. U. (1987). Cleavage of membrane secretory component to soluble secretory component occurs on the cell surface of rat hepatocyte monolayers. *J. Cell Biol.* 104:1725–1733.

Musil, L. S., and Baenziger, J. U. (1988). Proteolytic processing of rat liver membrane secretory component. Cleavage activity is localized to bile canalicular membranes. *J. Biol. Chem.* 263:15,799–808.

Natvig, I. B., Johansen, F. E., Nordeng, T. W., Haraldsen, G., and Brandtzaeg, P. (1997). Mechanism for enhanced external transfer of dimeric IgA over pentameric IgM. Studies of diffusion, binding to the human polymeric Ig receptor, and epithelial transcytosis. *J. Immunol.* 159:4330–4340.

Nelson, W. J., and Yeaman, C. (2001). Protein trafficking in the exocytic pathway of polarized epithelial cells. *Trends Cell Biol.* 11:483–486.

Norderhaug, I. N., Johansen, F. E., Schjerven, H., and Brandtzaeg, P. (1999). Regulation of the formation and external transport of secretory immunoglobulins. *Crit Rev. Immunol.* 19:481–508.

Obara, W., Iida, A., Suzuki, Y., Tanaka, T., Akiyama, F., Maeda, S., Ohnishi, Y., Yamada, R., Tsunoda, T., Takei, T., Ito, K., Honda, K., Uchida, K., Tsuchiya, K., Yumura, W., Ujiie, T., Nagane, Y., Nitta, K., Miyano, S., Narita, I., Gejyo, F., Nihei, H., Fujioka, T., and Nakamura, Y. (2003). Association of single-nucleotide polymorphisms in the polymeric immunoglobulin receptor gene with immunoglobulin A nephropathy (IgAN) in Japanese patients. *J. Hum. Genet.* 48:293–299.

O'Brien, L. E., Zegers, M. M., and Mostov, K. E. (2002). Opinion: Building epithelial architecture: insights from three-dimensional culture models. *Nat. Rev. Mol. Cell Biol.* 3:531–537.

Okamoto, C. T., Shia, S.-P., Bird, C., Mostov, K. E., and Roth, M. G. (1992). The cytoplasmic domain of the polymeric immunoglobulin receptor contains two internalization signals that are distinct from its basolateral sorting signal. *J. Biol. Chem.* 267:9925–9932.

Okamoto, C. T., Song, W., Bomsel, M., and Mostov, K. E. (1994). Rapid internalization of the polymeric immunoglobulin receptor requires phosphorylated serine 726. *J. Biol. Chem.* 269:15,676–682.

Pal, K., Kaetzel, C. S., Brundage, K., Cunningham, C., and Cuff, C. F. (2005) Regulation of polymeric immunoglobulin receptor (pIgR) expression by reovirus. *J. Gen. Virol.* 86:2347–2357.

Peppard, J. V., Kaetzel, C. S., and Russell, M. W. (2005). Phylogeny and comparative physiology of IgA. In: Mestecky, J., Bienenstock, J., Lamm, M., Strober, W., McGhee, J. and Mayer, L. (eds.), *Mucosal Immunology*, 3rd ed., Academic Press, San Diego, pp. 195–210.

Phalipon, A., Cardona, A., Kraehenbuhl, J. P., Edelman, L., Sansonetti, P. J., and Corthesy, B. (2002). Secretory component: a new role in secretory IgA-mediated immune exclusion *in vivo. Immunity* 17:107–115.

Phalipon, A., and Corthesy, B. (2003). Novel functions of the polymeric Ig receptor: Well beyond transport of immunoglobulins. *Trends Immunol.* 24:55–58.

Phillips, J. O., Everson, M. P., Moldoveanu, Z., Lue, C., and Mestecky, J. (1990). Synergistic effect of IL-4 and IFN-γ on the expression of polymeric Ig receptor (secretory component) and IgA binding by human epithelial cells. *J. Immunol.* 145:1740–1744.

Phillips, J. O., Komiyama, K., Epps, J. M., Russell, M. W., and Mestecky, J. (1988). Role of hepatocytes in the uptake of IgA and IgA-containing immune complexes in mice. *Mol. Immunol.* 25:873–879.

Pilette, C., Ouadrhiri, Y., Dimanche, F., Vaerman, J. P., and Sibille, Y. (2003). Secretory component is cleaved by neutrophil serine proteinases but its epithelial production is increased by neutrophils through NF-κB- and p38 mitogen-activated protein kinase-dependent mechanisms. *Am. J. Respir. Cell Mol. Biol.* 28:485–498.

Pine, R. (1997). Convergence of TNFα and IFNγ signalling pathways through synergistic induction of IRF-1/ISGF-2 is mediated by a composite GAS/κB promoter element. *Nucleic Acids Res.* 25:4346–4354.

Piskurich, J. F., Youngman, K. R., Phillips, K. M., Hempen, P. M., Blanchard, M. H., France, J. A., and Kaetzel, C. S. (1997). Transcriptional regulation of the human polymeric immunoglobulin receptor gene by interferon-γ. *Mol. Immunol.* 34:75–91.

Reich, V., Mostov, K., and Aroeti, B. (1996). The basolateral sorting signal of the polymeric immunoglobulin receptor contains two functional domains. *J. Cell Sci.* 109:2133–2139.

Renegar, K. B., Jackson, G. D., and Mestecky, J. (1998). *In vitro* comparison of the biologic activities of monoclonal monomeric IgA, polymeric IgA, and secretory IgA. *J. Immunol.* 160:1219–1223.

Rifai, A., and Mannik, M. (1984). Clearance of circulating IgA immune complexes is mediated by a specific receptor on Kupffer cells in mice. *J. Exp. Med.* 160:125–137.

Rifai, A., Schena, F. P., Montinaro, V., Mele, M., D'Addabbo, A., Nitti, L., and Pezzullo, J. C. (1989). Clearance kinetics and fate of macromolecular IgA in patients with IgA nephropathy. *Lab. Invest.* 61:381–388.

Robinson, J. K., Blanchard, T. G., Levine, A. D., Emancipator, S. N., and Lamm, M. E. (2001). A mucosal IgA-mediated excretory immune system *in vivo. J. Immunol.* 166:3688–3692.

Roe, M., Norderhaug, I. N., Brandtzaeg, P., and Johansen, F. E. (1999). Fine specificity of ligand-binding domain 1 in the polymeric Ig receptor: importance of the CDR2-containing region for IgM interaction. *J. Immunol.* 162:6046–6052.

Rojas, R., and Apodaca, G. (2002). Immunoglobulin transport across polarized epithelial cells. *Nat. Rev. Mol. Cell Biol.* 3:944–955.

Royle, L., Roos, A., Harvey, D. J., Wormald, M. R., van Gijlswijk-Janssen, D., Redwan, E., Wilson, I. A., Daha, M. R., Dwek, R. A., and Rudd, P. M. (2003). Secretory IgA

N- and O-glycans provide a link between the innate and adaptive immune systems. *J. Biol. Chem.* 278:20,140–20,153.

Russell, M. W., Brown, T. A., Claflin, J. L., Schroer, K., and Mestecky, J. (1983). Immunoglobulin A-mediated hepatobiliary transport constitutes a natural pathway for disposing of bacterial antigens. *Infect. Immun.* 42:1041–1048.

Russell, M. W., Brown, T. A., and Mestecky, J. (1981). Role of serum IgA: hepatobiliary transport of circulating antigen. *J. Exp. Med.* 153:968–976.

Sait, L., Galic, M., Strugnell, R. A., and Janssen, P. H. (2003). Secretory antibodies do not affect the composition of the bacterial microbiota in the terminal ileum of 10-week-old mice. *Appl. Environ. Microbiol.* 69:2100–2109.

Schjerven, H., Brandtzaeg, P., and Johansen, F. E. (2000). Mechanism of IL-4-mediated up-regulation of the polymeric Ig receptor: role of STAT6 in cell type-specific delayed transcriptional response. *J. Immunol.* 165:3898–3906.

Schjerven, H., Brandtzaeg, P., and Johansen, F. E. (2001). A novel NF-κB/Rel site in intron 1 cooperates with proximal promoter elements to mediate TNF-α-induced transcription of the human polymeric Ig receptor. *J. Immunol.* 167:6412–6420.

Schjerven, H., Tran, T. N., Brandtzaeg, P., and Johansen, F. E. (2004). *De novo* synthesized RelB mediates TNF-induced up-regulation of the human polymeric Ig receptor. *J. Immunol.* 173:1849–1857.

Schneeman, T. A., Bruno, M. E. C., Schjerven, H., Johansen, F. E., Chady, L., and Kaetzel, C. S. (2005). Regulation of the polymeric Ig receptor by signaling through Toll-like receptors 3 and 4: linking innate and adaptive immune responses. *J. Immunol.* 175:376–384.

Schneider, K., Potter, K. G., and Ware, C. F. (2004). Lymphotoxin and LIGHT signaling pathways and target genes. *Immunol. Rev.* 202:49–66.

Schroten, H., Stapper, C., Plogmann, R., Kohler, H., Hacker, J., and Hanisch, F. G. (1998). Fab-independent antiadhesion effects of secretory immunoglobulin A on S-fimbriated *Escherichia coli* are mediated by sialyloligosaccharides. *Infect. Immun.* 66:3971–3973.

Shimada, S., Kawaguchi-Miyashita, M., Kushiro, A., Sato, T., Nanno, M., Sako, T., Matsuoka, Y., Sudo, K., Tagawa, Y., Iwakura, Y., and Ohwaki, M. (1999). Generation of polymeric immunoglobulin receptor-deficient mouse with marked reduction of secretory IgA. *J. Immunol.* 163:5367–5373.

Singer, K. L., and Mostov, K. E. (1998). Dimerization of the polymeric immunoglobulin receptor controls its transcytotic trafficking. *Mol. Biol. Cell* 9:901–915.

Sirito, M., Lin, Q., Deng, J. M., Behringer, R. R., and Sawadogo, M. (1998). Overlapping roles and asymmetrical cross-regulation of the USF proteins in mice. *Proc. Natl. Acad. Sci. USA* 95:3758–3763.

Socken, D. J., Simms, E. S., Nagy, B. R., Fisher, M. M., and Underdown, B. J. (1981). Secretory component-dependent hepatic transport of IgA antibody-antigen complexes. *J. Immunol.* 127:316–319.

Solari, R., Schaerer, E., Tallichet, C., Braiterman, L. T., Hubbard, A. L., and Kraehenbuhl, J.-P. (1989). Cellular location of the cleavage event of the polymeric immunoglobulin receptor and fate of its anchoring domain in the rat hepatocyte. *Biochem. J.* 257:759–768.

Solorzano-Vargas, R. S., Wang, J., Jiang, L., Tsai, H. V., Ontiveros, L. O., Vazir, M. A., Aguilera, R. J., and Martín, M. G. (2002). Multiple transcription factors in 5′-flanking region of human polymeric Ig receptor control its basal expression. *Am. J. Physiol. Gastrointest.:Liver Physiol.* 283:G415–G425.

Song, W., Bomsel, M., Casanova, J., Vaerman, J.-P., and Mostov, K. (1994). Stimulation of transcytosis of the polymeric immunoglobulin receptor by dimeric IgA. *Proc. Natl. Acad. Sci. USA* 91:163–166.

Sun, K., Johansen, F. E., Eckmann, l., and Metzger, D. W. (2004). An important role for polymeric Ig receptor-mediated transport of IgA in protection against Streptococcus pneumoniae nasopharyngeal carriage. *J. Immunol.* 173:4576–4581.

Sztul, E., Colombo, M., Stahl, P., and Samanta, R. (1993). Control of protein traffic between distinct plasma membrane domains. Requirement for a novel 108,000 protein in the fusion of transcytotic vesicles with the apical plasma membrane. *J. Biol. Chem.* 268:1876–1885.

Takeda, K., and Akira, S. (2003). Toll receptors and pathogen resistance. *Cell. Microbiol.* 5:143–153.

Takenouchi-Ohkubo, N., Takahashi, T., Tsuchiya, M., Mestecky, J., Moldoveanu, Z., and Moro, I. (2000). Role of nuclear factor-κB in the expression by tumor necrosis factor-α of the human polymeric immunoglobulin receptor (plgR) gene. *Immunogenetics* 51:289–295.

Tamer, C. M., Lamm, M. E., Robinson, J. K., Piskurich, J. F., and Kaetzel, C. S. (1995). Comparative studies of transcytosis and assembly of secretory IgA in Madin-Darby canine kidney cells expressing human polymeric Ig receptor. *J. Immunol.* 155:707–714.

Taylor, C. L., Harrison, G. A., Watson, C. M., and Deane, E. M. (2002). cDNA cloning of the polymeric immunoglobulin receptor of the marsupial Macropus eugenii (tammar wallaby). *Eur. J. Immunogenet.* 29:87–93.

Tjarnlund, A., Rodriguez, A., Cardona, P. J., Guirado, E., Ivanyi, J., Singh, M., Troye-Blomberg, M., and Fernandez, C. (2006). Polymeric IgR knockout mice are more susceptible to mycobacterial infections in the respiratory tract than wild-type mice. *Int. Immunol.* 18:807–816.

Tomasi, T. B., Jr., Tan, E. M., Solomon, A., and Prendergast, R. A. (1965). Characteristics of an immune system common to certain external secretions. *J. Exp. Med.* 121:101–124.

Traicoff, J. L., De Marchis, L., Ginsburg, B. L., Zamora, R. E., Khattar, N. H., Blanch, V. J., Plummer, S., Bargo, S. A., Templeton, D. J., Casey, G., and Kaetzel, C. S. (2003). Characterization of the human polymeric immunoglobulin receptor (*PIGR*) 3′UTR and differential expression of *PIGR* mRNA during colon tumorigenesis. *J. Biomed. Sci.* 10:792–804.

Uren, T. K., Johansen, F. E., Wijburg, O. L., Koentgen, F., Brandtzaeg, P., and Strugnell, R. A. (2003). Role of the polymeric Ig receptor in mucosal B cell homeostasis. *J. Immunol.* 170:2531–2539.

Uren, T. K., Wijburg, O. L., Simmons, C., Johansen, F. E., Brandtzaeg, P., and Strugnell, R. A. (2005). Vaccine-induced protection against gastrointestinal bacterial infections in the absence of secretory antibodies. *Eur. J. Immunol.* 35:180–188.

Van Ijzendoorn, S. C., Tuvim, M. J., Weimbs, T., Dickey, B. F., and Mostov, K. E. (2002). Direct interaction between Rab3b and the polymeric immunoglobulin receptor controls ligand-stimulated transcytosis in epithelial cells. *Dev. Cell* 2:219–228.

van Spriel, A. B., Leusen, J. H., Vile, H., and van De Winkel, J. G. (2002). Mac-1 (CD11b/CD18) as accessory molecule for Fc alpha R (CD89) binding of IgA. *J. Immunol.* 169:3831–3836.

Verbeet, M. P., Vermeer, H., Warmerdam, G. C. M., De Boer, H. A., and Lee, S. H. (1995). Cloning and characterization of the bovine polymeric immunoglobulin receptor-encoding cDNA. *Gene* 164:329–333.

Verges, M., Luton, F., Gruber, C., Tiemann, F., Reinders, L. G., Huang, L., Burlingame, A. L., Haft, C. R., and Mostov, K. E. (2004). The mammalian retromer regulates transcytosis of the polymeric immunoglobulin receptor. *Nat. Cell Biol.* 6:763–769.

Verges, M., Sebastian, I., and Mostov, K. E. (2007). Phosphoinositide 3-kinase regulates the role of retromer in transcytosis of the polymeric immunoglobulin receptor. *Exp. Cell Res.* 313:707–718.

Vidarsson, G., Der Pol, W. L., van Den Elsen, J. M., Vile, H., Jansen, M., Duijs, J., Morton, H. C., Boel, E., Daha, M. R., Corthesy, B., and van De Winkel, J. G. (2001). Activity of human IgG and IgA subclasses in immune defense against Neisseria meningitidis serogroup B. *J. Immunol.* 166:6250–6256.

Wieland, W. H., Orzaez, D., Lammers, A., Parmentier, H. K., Verstegen, M. W., and Schots, A. (2004). A functional polymeric immunoglobulin receptor in chicken (*Gallus gallus*) indicates ancient role of secretory IgA in mucosal immunity. *Biochem. J.* 380:669–676.

Wijburg, O. L., Uren, T. K., Simpfendorfer, K., Johansen, F. E., Brandtzaeg, P., and Strugnell, R. A. (2006). Innate secretory antibodies protect against natural Salmonella typhimurium infection. *J. Exp. Med.* 203:21–26.

Willer, E. M., Lima, R. L., and Giugliano, L. G. (2004). *In vitro* adhesion and invasion inhibition of *Shigella dysenteriae, Shigella flexneri* and *Shigella sonnei* clinical strains by human milk proteins. *BMC Microbiol.* 4:18.

Wold, A. E., Mestecky, J., Tomana, M., Kobata, A., Ohbayashi, H., Endo, T., and Eden, C. S. (1990). Secretory immunoglobulin A carries oligosaccharide receptors for Escherichia coli type 1 fimbrial lectin. *Infect. Immun.* 58:3073–3077.

Yamamoto, M., Takeda, K., and Akira, S. (2004). TIR domain-containing adaptors define the specificity of TLR signaling. *Mol. Immunol* 40:861–868.

Yan, H., Lamm, M. E., Bjorling, E., and Huang, Y. T. (2002). Multiple functions of immunoglobulin A in mucosal defense against viruses: An *in vitro* measles virus model. *J. Virol.* 76:10,972–10,979.

Youngman, K. R., Fiocchi, C., and Kaetzel, C. S. (1994). Inhibition of IFN-γ activity in supernatants from stimulated human intestinal mononuclear cells prevents up-regulation of the polymeric Ig receptor in an intestinal epithelial cell-line. *J. Immunol.* 153:675–681.

Yu-Lee, L. (2001). Stimulation of interferon regulatory factor-1 by prolactin. *Lupus* 10:691–699.

Zerial, M., and McBride, H. (2001). Rab proteins as membrane organizers. *Nat. Rev. Mol. Cell Biol.* 2:107–117.

Zhao, J., Yeong, L. H., and Wong, W. S. (2006). Dexamethasone alters bronchoalveolar lavage fluid proteome in a mouse asthma model. *Int. Arch. Allergy Immunol.* 142:219–229.

4
Fc Receptors for IgA

H. Craig Morton[1]

4.1. Introduction

Immunoglobulin A (IgA) is, by far, the most abundant immunoglobulin produced in humans and is also the most heterogeneous (Kerr, 1990; Woof and Mestecky, 2005) (see Chapter 1). Human serum IgA, produced by plasma cells in the bone marrow, lymph nodes, and spleen, is mainly monomeric (mIgA) and constitutes ~15–20% of the total serum Ig pool. However, IgA

[1] Laboratory of Immunohistochemistry and Immunopathology (LIIPAT), Institute of Pathology, Rikshospitalet University Hospital, 0027 Oslo, Norway. Present address: Institute of Marine Research, Nordnes 5817, Bergen, Norway.

predominates at the mucosa, as 80–90% of mucosal plasma cells produce this isotype (Brandtzaeg et al., 1999). In addition to IgA, mucosal plasma cells also express a small polypeptide called the joining chain (J-chain) that directs the assembly of dimers and larger polymers [collectively called polymeric IgA (pIgA)] (Johansen et al., 2000).

Following its secretion by plasma cells, and in order to fulfill its role as an important component of the human adaptive immune system, IgA interacts with a number of specific cellular receptors (Table 4.1). One of the most well characterized, and the primary focus of this chapter, is the IgA Fc receptor (FcR) expressed by cells of the myeloid lineage (called CD89 or FcαRI). Binding of IgA-coated targets to CD89 on myeloid cells can trigger a wide variety of cellular effector functions, including phagocytosis, antibody-dependent cell-mediated cytotoxicity (ADCC), and the synthesis and release of cytokines and other inflammatory mediators. Another IgA FcR, critical for mucosal defense, is the polymeric immunoglobulin receptor (pIgR) (Kaetzel, 2005). The pIgR is a sacrificial receptor responsible for binding pIgA and transporting it into external secretions. The pIgA that has been transcytosed via this pathway is called secretory IgA (SIgA), and a portion of the pIgR [called secretory component (SC)] remains associated with the SIgA molecule (see Sect. 4.3.1). SIgA is thus the primary mediator of humoral immunity at mucosal surfaces, where it neutralizes pathogens and foreign antigens in a process known as immune exclusion. The pIgR will only be mentioned briefly here, but its structure and function are discussed extensively in Chapter 3. Three other IgA receptors—the asiaolglycoprotein receptor (ASGP-R), the transferrin receptor (TfR, CD71), and Fcα/μR—have been identified in humans. The ASGP-R is expressed in the liver and is thought to be involved in IgA catabolism. TfR has only recently been recognized as an IgA receptor and may be involved in the pathogenesis of the immune complex (IC) disease IgA nephropathy (IgAN). Fcα/μR is related to the pIgR and likewise binds both IgA and IgM. Its expression in human mesangial cells may also suggest a role for Fcα/μR in the pathogenesis of IgAN.

This chapter will discuss current and emerging knowledge of the five IgA receptors listed above and in Table 4.1. Interested readers are also directed to several other excellent reviews of this subject (Morton and Brandtzaeg, 2001; Monteiro and van de Winkel, 2003; Otten and van Egmond, 2004; van Egmond et al., 2001; Woof et al., 2005).

4.2. The Human Myeloid IgA FcR (CD89/FcαRI)

4.2.1. Protein Structure

CD89 is a type I transmembrane glycoprotein with an extracellular region composed of two Ig-like domains and a short cytoplasmic tail devoid of recognized signaling motifs (Maliszewski et al., 1990). CD89 can associate with

TABLE 4.1. Structures and characteristics of IgA receptors.

Receptor	FcαR1	Polymeric Ig receptor (pIgR)	Fcα/μR	Asialoglycoprotein receptor (ASGP-R)	Transferrin receptor (TfR)
Receptor structure[a]	D1 D2 ITAM FcRγ	D2 D1 D3 D4 D5	D1	CLECT	HD AD PLD
Type	Type I	Type I	Type I	Type II	Type II
CD designation	CD89	—	—	—	CD71
HUGO designation	*FCAR*	*PIGR*	*FCAMR*	*ASGR1*	*TFRC*
Reference sequences					
mRNA	NM 02000	NM 002644	NM 032029	NM 001671	NM 003234
Protein	NP 01991	NP 002635	NP 114418	NP 001662	NP 003225
Structures	1OVZ (CD89) 1UCT (CD89) 1OW0 (CD89+Fcα)	1XED (D1 only)	N.D.	N.D.	1CX8 (TfR) 1SUV (TfR+Tf) 1DE4 (TfR+HFE)
Chromosomallocation	19q13.4	1q31-q41	1q32.1	17p13.2	3q29
Ligands	mIgA, pIgA	pIgA, IgM	IgA, IgM	Desialylated glycoproteins	Transferrin, HFE, pIgA1

[a]D, domain; CLECT, C-type lectin domain; HD, helical domain; AD, apical domain; PLD, protease-like domain, HFE, hemochromatosis protein; N.D., not done.

the FcR γ-chain (FcRγ_2), a specialized signaling dimer with two cytoplasmic immunoreceptor tyrosine-based activation motifs (ITAMs) in its cytoplasmic domains. Co-expression of FcRγ_2 is not required for surface expression of CD89 in transfectants. However, when CD89 transgenic (CD89Tg) mice were crossed with FcRγ_2 knockout mice, surface expression of CD89 was abolished (van Egmond et al., 1999). The ability of CD89 to trigger cell-mediated effector functions like phagocytosis, respiratory burst, and cytokine release is critically dependent on its association with FcRγ_2 (Morton et al., 1995; van Egmond et al., 1999).

CD89 associates with FcRγ_2 via a basic arginine residue in its transmembrane (TM) domain (Morton et al., 1995). A recent study demonstrated that lateral transfer of this positive charge in an Arg209Leu/Met210Arg mutant of human CD89 completely abrogated FcRγ_2-dependent signaling (Bakema et al., 2006). It is now understood that the possession of a basic TM residue, which facilitates association with signaling molecules like FcRγ_2, is a characteristic of activating receptors. This charged-based mechanism for the pairing of ligand-binding chains with specialized signaling molecules appears to be an evolutionary successful approach that is utilized by numerous activating receptors expressed on hematopoietic cells (Feng et al., 2005).

Recently, the crystal structure of the ectodomain of CD89 has been solved (Ding et al., 2003; Herr et al., 2003a). The two Ig-like domains of CD89 are orientated at ~90° to one another, and this orientation most closely resembles that seen in the leukocyte Ig-like receptor B1 (LILRB1, also called LIR-1/CD85j) and the killer cell Ig-like receptors (KIRs). However, the relative orientation of the extracellular Ig-like domains of CD89 is opposite to that previously observed for the other Ig-superfamily FcRs such as FcγRIIa, FcγRIIb, FcγRIII, and FcεRI (Herr et al., 2003a; Woof and Burton 2004).

4.2.2. Evolution

It is now apparent that CD89 has followed a slightly different evolutionary pathway to the other Ig-superfamily FcRs. CD89 is actually more closely related to receptors encoded by genes located on a region of chromosome 19, at position 19q13.4, known as the leukocyte receptor complex (LRC) (Martin et al., 2002). In contrast, the genes of the other Ig-superfamily FcRs are located on chromosome 1 (Hulett and Hogarth, 1994). Interestingly, the genes encoding the pIgR and Fcα/μR are also located on chromosome 1 (see Sects. 4.3.1 and 4.3.2).

In addition to the CD89 gene (*FCAR*), the LRC includes genes for the LILRs, KIRs, NKp46 (also called natural cytotoxicity receptor 1, NCR1), platelet glycoprotein VI (GPVI), and leukocyte-associated Ig-like receptor-1 and -2 (LAIR-1 and LAIR-2). Although they display a wide variety of functions, LRC-encoded receptors display a relatively high level of sequence and structural homology, strongly suggesting that they have evolved from a common ancestor (Nikolaidis et al., 2005; Volz et al., 2001). Although the closest relatives of the LRC-encoded proteins are the Ig-superfamily FcRs, genomic analysis suggests that the ancestors of these two gene families diverged before the separation of birds and mammals (Nikolaidis et al., 2005). To date, CD89 orthologues have been described in chimpanzees (Morton et al., 2005), macaques (Rogers et al., 2004), cattle (Morton et al., 2004), horses (Morton et al., 2005), and rats (Maruoka et al., 2004). Although IgA is also found in birds, an avian orthologue of CD89 has not been described.

In evolutionary terms, $FcR\gamma_2$ and other homologous signaling adapter molecules (i.e., CD3ζ, DAP10, and DAP12) are much older than the activating receptors with which they associate (Abi-Rached and Parham, 2005). Therefore, this suggests that when activating receptors possessing a polar TM residue arose during evolution, they were able to use pre-existing signaling pathways. In addition, these ancient signaling molecules might also have influenced the evolution of activating receptors by selecting for variants with which they were able to associate (Abi-Rached and Parham, 2005; Feng et al., 2005).

4.2.3. Genomics

FCAR consists of five exons spanning ~12-kb (de Wit et al., 1995) and lies at the telomeric end of the LRC close to *NCR1* and *GPVI*. On the centromeric side of *FCAR* lie two clusters of genes encoding the KIRs, LILRs, LAIR-1 and LAIR-2. Synteny mapping has revealed a region on mouse chromosome 7 that appears to represent the murine LRC. A murine homologue of *FCAR* is not found in this region (or anywhere else in the mouse genome), thus explaining why a CD89 homologue has not been identified in mice (Martin et al., 2002).

Characterization of the *FCAR* promoter has mapped elements responsible for the myeloid specific expression of CD89 to a 259-bp region directly prior to the translation initiation site (Shimokawa et al., 2000). The transcription factors CCAAT/enhancer-binding protein α (C/EBPα) and GA-binding protein (GABP) have been shown to bind to sites within this core promoter and to regulate CD89 transcription (Shimokawa and Ra, 2003).

Transcription of *FCAR* gives rise to a variety of differently spliced mRNA transcripts. A full-length cDNA clone encoding CD89 was first isolated by Maliszewski and co-workers in 1990 (Maliszewski et al., 1990), and numerous alternatively spliced transcripts have since been described (Morton and Brandtzaeg, 2001; Monteiro and van de Winkel, 2003). At least one of these splice variants (FcαRb) might encode a soluble form of CD89 (van Dijk et al., 1996). The relationship between FcαRb and soluble forms of CD89, which have been detected in serum covalently linked to IgA, is unknown (Launay et al., 2000; van der Boog et al., 2002).

Recent genetic analysis has identified several single-nucleotide polymorphisms (SNPs) within *FCAR*. The activity of *FCAR* promoter constructs carrying T alleles at two polymorphic sites (positions −311 and −142 relative to the ATG translation initiation codon) was reduced approximately twofold when compared to those carrying C alleles (Shimokawa et al., 2000). One Japanese study has since shown that the frequencies of the −311C/C and −142C/C genotypes is significantly increased in IgAN patients (Tsuge et al., 2001). These authors recently reported that these SNPs were also associated with increased risk for chronic hepatitis C (Watanabe et al., 2006). However, a second study of a different population of Japanese IgAN patients failed to find a significant association between these SNPs and IgAN (Narita et al., 2001). Four other SNPs (−340G/A, 363A/G, 376G/A, and 844A/G) have been described in *FCAR* (Jasek et al., 2004), and the 363A/A genotype has been linked to a susceptibility to aggressive periodontitis (Kaneko et al., 2004).

4.2.4. Distribution

CD89 is expressed at high levels by circulating neutrophils and monocytes (Hamre et al., 2003; Monteiro et al., 1992, 1993). Eosinophils express much lower levels of CD89 compared to neutrophils, but CD89 expression is reported to be increased on the eosinophils of patients suffering from allergic rhinitis and/or asthma (Monteiro et al., 1993).

In contrast to blood monocytes, intestinal macrophages resident in the lamina propria do not express CD89. These macrophages still retain effective phagocytic and bactericidal activity but do not secrete pro-inflammatory cytokines and lack receptors for IgG (FcγRs), lipopolysaccharide (LPS) (CD14), complement (CR3, CR4, C5aR), interleukin (IL)-2 (CD25), and IL-3 (CD123) (Smith et al., 2005). Thus, intestinal macrophages appear ideally adapted to promoting the anti-inflammatory environment of the gut (see below) while still retaining effective host defense functions. In contrast to gut macrophages, varying levels of CD89 expression have been detected on several other macrophage populations, including those from the peritoneal cavity, the lungs (alveolar macrophages), and the liver (Kupffer cells) (Hamre et al., 2003; Ouadrhiri et al., 2002; Patry et al., 1996; van Egmond et al., 2000).

Similar to the above-described situation, the expression of CD89 by different dendritic cell (DC) populations is still controversial. Monocyte-derived DCs (Mo-DCs) generated *in vitro* retain low levels of CD89 (Geissmann et al., 2000; Heystek et al., 2002), but the situation *in vivo* is less clear. Although Langerhans cells do not express CD89 *in vivo* (Geissmann et al., 2000; Hamre et al., 2003), expression was detected on both DR^+ and DR^- cells in the dermis (Geissmann et al., 2000). As these cells were also $CD68^+$, it is unclear whether they were “true” DCs or macrophages. However, functionally, this distinction might be superfluous, as skin DCs and macrophages can display similar phenotypic characteristics, especially during inflammation (Kiekens et al., 2001). However, even if DCs *in vivo* lack functional levels of CD89, they have been reported to be capable of binding and internalizing SIgA at least partly via the mannose scavenger receptor (Heystek et al., 2002).

4.2.5. IgA Binding

The interaction between IgA and CD89 is quite distinct from that between IgG and IgE and their respective FcRs. CD89 binds IgA via its N-terminal membrane-distal domain 1 (D1) (Morton et al., 1999; Wines et al., 1999, 2001), whereas FcγRIIIb (CD16) and FcεRI bind their ligands via residues located in the membrane proximal domain (D2) and near the D1–D2 interface (Woof and Burton 2004). The FcR-binding sites within the Ig molecules themselves are also quite distinct. CD89 binds IgA at the Cα2–Cα3 interface, whereas FcγRs bind to the lower hinge region at the amino terminal of the IgG Cγ2 domain, and FcεRI binds IgE at the analogous Cε2–Cε3 linker region (Woof and Burton, 2004).

Results from earlier mutagenesis experiments have been confirmed and extended by resolution of the crystal structure of CD89 in complex with the Fc region of IgA (Herr et al., 2003a; Wines et al., 1999, 2001). The IgA-binding site of CD89 involves residues in the B-C loop, the D strand, the D-E loop, and the F-G loop of D1. The CD89–IgA interface is composed of a central hydrophobic core flanked by several charged residues (Herr et al., 2003a; Woof et al., 2005). Previously, it was shown that mutation of CD89 residues Tyr-35 (in the B-C loop) and Arg-82 (in the F-G loop) to alanine abolished IgA binding (Wines et al., 2001). The molecular basis for this ablation is revealed by the crystal structure of the complex, which shows that Tyr-35 lies at the center of the hydrophobic patch on CD89 and forms a potential hydrogen bond to the Leu-257 of IgA. Similarly, Arg-82 forms a hydrogen bond with Leu-256.

A further unique characteristic of CD89-IgA binding is the stoichiometry of the interaction. A conformational change occurs in the Fc regions of IgG and IgE after interaction with one FcR, preventing the binding of a second FcR to the other heavy chain (Kato et al., 2000). However, recent structural and biochemical analysis have shown that the receptor-binding sites on both IgA heavy chains remain accessible and are each able to bind one CD89 molecule (Herr et al., 2003a, 2003b). This unique 2:1 stoichiometry raises some interesting questions regarding our understanding of FcR signaling, as it is usually assumed that FcR-mediated cellular effector functions can be triggered by as few as two antibody molecules bound to an antigen (Segal et al., 1977). Therefore, several potential explanations (which are not mutually exclusive) have been put forward to explain why soluble mIgA molecules in serum do not trigger CD89-mediated cellular activation (Herr et al., 2003a). First, the cytoplasmic regions of the CD89 molecules might be too far apart to trigger downstream signaling. Second, the relatively high concentration of IgA in serum might favor the formation of 1:1 complexes, which would presumably prevail until displaced by larger multivalent immune complexes. Finally, the lateral movement of CD89 within the cell membrane might be restricted by elements of the cytoskeleton that might need to undergo rearrangement before CD89 can aggregate and signal.

Interestingly, a combination of these last two theories might help to explain some previous observations regarding the ability of cytokines to modulate IgA binding by neutrophils and eosinophils. Nearly 20 years ago it was shown that granulocyte monocyte-colony stimulating factor (GM-CSF) and granulocyte-colony stimulating factor (G-CSF) could rapidly switch the binding of IgA to neutrophils from low to high affinity without increasing receptor expression and that this switch was associated with the enhancement of IgA-mediated phagocytosis (Weisbart et al., 1988). Similarly, IgA binding by eosinophils has been shown to be modulated by a cytokine-induced "inside-out" signaling pathway involving phosphatidylinositol 3 kinase (PI 3-kinase) (Bracke et al., 2000, 2001). Reduced IgA binding was also seen following disruption of the cytoskeleton (Bracke et al., 2001). Furthermore, whereas truncation of the cytoplasmic domain of CD89 or mutation of serine 263 (S263) to alanine resulted in constitutive high-affinity binding, mutation of S263 to aspartate (mimicking the phosphorylated state) reduced the affinity of CD89 for IgA (Bracke et al., 2001). Taking these results and their own observations on the stoichiometry of the binding interaction into account, Herr et al. (2003b) have proposed a model for the cytokine-induced increase in CD89–IgA affinity. They proposed that, in unstimulated cells, CD89 is phosphorylated on S263 and its lateral movement in the plasma membrane is restricted via cytoskeletal interactions. Two receptors are therefore prevented from binding to a single IgA molecule. However, following cytokine stimulation, a signal is transmitted via PI 3-kinase, which results in dephosphorylation of S263 causing either a redistribution of CD89 molecules or a change in their orientation, thus allowing two receptors to bind one IgA molecule. In this way, the bivalent binding of one IgA with two CD89 molecules would produce an increase in avidity and, therefore, a higher apparent affinity without increasing receptor expression.

More recently, Pasquier et al. (2005) presented evidence that the 1:1 complexes formed between CD89 and IgA might actually play a more active role than previously thought. Their experiments showed that binding of mIgA or anti-CD89 Fab fragments to CD89 could inhibit activating responses triggered by the cross-linking of other FcRs for IgG or IgE. Thus, these authors suggested that their observations might help to explain the inhibitory effect of IgA that has been seen in some experimental systems (Kerr 1990). Further studies into this interesting phenomenon are clearly needed.

A recent study utilized surface plasmon resonance to investigate the association of mIgA versus pIgA with CD89 (Oortwijn et al., 2007). These authors concluded that the initial association of mIgA and pIgA with CD89 is similar, whereas mIgA dissociates more rapidly than does pIgA. Given the large excess of circulating mIgA over pIgA in humans, mIgA should compete effectively with pIgA for binding to CD89, thus preventing receptor aggregation and consequent cellular activation.

4.2.6. Signaling

Aggregation of CD89/FcRγ_2 by IgA complexes or anti-CD89 monoclonal antibodies (mAbs) triggers their redistribution into detergent-insoluble lipid rafts (Lang et al., 1999, 2002). Such lipid rafts have been shown to provide protection to the Src family PTK Lyn from dephosphosphorylation and inactivation by a transmembrane tyrosine phosphatase. In the raft environment, Lyn is able to phosphorylate the FcRγ_2 ITAMs (Gulle et al., 1998). Syk is then recruited to the phosphorylated ITAMs and subsequently becomes activated by phosphorylation (probably also by Lyn). In some studies, phosphorylation of the tec family PTK Btk has also been observed (Lang et al., 1999; Launay et al., 1998). The signal transduction cascade continues via numerous adaptor proteins (Grb2, Shc, SHIP, CrkL, Cbl, SLP-76), resulting in the recruitment of the GTPase Sos to the complex. Sos converts GDP-RAS to GTP-RAS, which subsequently activates the Raf-1/MEK/MAP kinase and PI 3-kinase signaling pathways (Park et al., 1999).

However, not all CD89 molecules appear to associate with FcRγ_2 (Launay et al., 1999; Saito et al., 1995). Neutrophils, monocytes, and the monocytic cell line U937 apparently express two forms of the receptor: CD89 alone and CD89/FcRγ_2. Although both forms of CD89 bound IgA with similar affinity and IgA complexes were endocytosed with similar kinetics, the intracellular fate of the internalized complexes differed (Launay et al., 1999). Experimental data suggest that IgA complexes endocytosed via CD89 alone might be recycled, whereas those internalized via CD89/FcRγ_2 are degraded and sorted for antigen presentation (Launay et al., 1999). Therefore, signaling via FcRγ_2 is not essential for endocytosis and presumably as-yet uncharacterized motifs within the cytoplasmic tail of CD89 are required to trigger this process.

The role of FcRγ_2 signaling in CD89-mediated antigen presentation was investigated by transfecting the A20 B cell line with CD89 and either wild-type FcRγ_2 or FcRγ_2 in which the ITAM motif was mutated (Lang et al., 2001; Lang and Lang, 2006). These investigators found that cells expressing wild-type but not mutant FcRγ_2 presented CD89-targeted OVA antigen to OVA-specific T-cell hybridomas in the context of MHC class II. Cross-linking of CD89 with soluble IgA–OVA complexes resulted in translocation of phosphatidylinositol-dependent protein kinase 1 and protein kinase Bα to MHC II peptide-loading compartments, a mechanism that appeared to link FcRγ_2 signaling with efficient presentation of OVA peptides by MHC class II. A possible role for CD89/FcRγ_2-mediated presentation of IgA-linked antigen by DCs was suggested by the report that CD89 expression was upregulated during *in vitro* maturation of human myeloid DCs isolated from peripheral blood (Pasquier et al., 2004). These investigators further demonstrated that CD89 cross-linking by IgA–antigen complexes resulted in upregulation of MHC class II and costimulator expression. However, a recent study demonstrated that presentation of CD89/IgA-targeted antigen by DCs from CD89 transgenic mice was inefficient (Otten et al, 2006). In conclusion, various cell types appear to be capable of processing IgA-bound antigens internalized via CD89/FcRγ_2; however, the *in vivo* significance of this phenomenon remains to be demonstrated.

4.2.7. Effector Functions of CD89 and the Role of IgA

The IgA system in humans is relatively compartmentalized, with mIgA and SIgA being differentially distributed between the systemic and mucosal compartments, respectively. The high concentration of mIgA in serum is a feature unique to humans and other primates, as other mammals have only low levels of pIgA in serum. Thus, the function of human serum IgA and its relationship with SIgA has been the source of debate for many years (Conley and Delacroix 1987).

In the gut (and other mucosal areas), where SIgA predominates, the immune system is under tight control (Sansonetti, 2004). Here, under normal conditions, inflammation is actively avoided and, instead, tolerogenic signals predominate. Therefore, the primary function of SIgA is considered to be the protection of the mucosae by the immune exclusion of commensal and pathogenic bacteria. In this way, SIgA protects against the possibility of a damaging inflammatory reaction at these delicate areas. The observation that macrophages resident in the gut lamina propria are still phagocytically active but do not secrete inflammatory cytokines and have down-regulated many of their activating receptors (including CD89; see Sect. 4.2.4) supports the concept of a non-inflammatory gut environment (Smith et al., 2005). It has recently been shown that SIgA is apparently unable to trigger phagocytosis by Kupffer cells *in vivo* or neutrophils *in vitro* (van Egmond et al., 2000; Vidarsson et al., 2001). Although the molecular basis for these observations is still unclear, they appear to define a novel anti-inflammatory role for SIgA.

Serum IgA, however, is able to trigger phagocytosis by neutrophils and monocytes (Morton and Brandtzaeg, 2001). Serum IgA-coated bacteria were also phagocytosed by CD89$^+$ Kupffer cells in G-CSF-treated CD89Tg mice (van Egmond et al., 2000). Therefore, it has been proposed that bacteria that are able to invade the gut mucosae enter the portal circulation, where they become opsonized with serum IgA and are subsequently phagocytosed by activated Kupffer cells (Otten and van Egmond, 2004; van Egmond et al., 2000). This observation appears to support earlier speculation that the role of serum IgA is to reinforce the first line of defense provided by SIgA at the mucosae (Conley and Delacroix, 1987).

In contrast to phagocytosis, serum IgA and SIgA are both able to trigger respiratory burst activity by neutrophils and monocytes (Monteiro and van de Winkel, 2003; Morton and Brandtzaeg, 2001; Otten and van Egmond, 2004). The ability of SIgA to trigger neutrophil respiratory burst was shown to be dependent on the coexpression of CR3 (Mac-1, CD11b/CD18) (van Spriel et al., 2002). Experiments suggested a direct interaction between CR3 and SC, but no detectable association between CR3 and CD89 was observed.

Recently, the therapeutic potential of CD89 against tumor cells has been demonstrated both with the use of recombinant IgA and with bispecific antibodies (Dechant et al., 2002; Valerius et al., 1997). Moreover, recombinant reagents, targeted to CD89, have been shown to effectively trigger neutrophil-mediated effector functions against several pathogenic microorganisms, including *Candida albicans*, *Bordetella pertussis*, *Streptococcus pneumoniae*, and *Neisseria meningitidis* (Hellwig et al., 2001; Valerius et al., 1997; van der Pol W.L. et al., 2000; Vidarsson et al., 2001).

4.3. Alternative IgA Receptors

4.3.1. The Polymeric Immunoglobulin Receptor (pIgR)

The pIgR [also known as the membrane secretory component (SC)] is expressed at high levels on the basolateral membrane of secretory epithelial cells (see Chapter 3). Here, the pIgR binds J-chain-containing pIgA or IgM molecules, which are then endocytosed and transported through the cell to the apical membrane, where the pIgR is cleaved and the complex released into the external secretions. The fragment of pIgA remaining bound is called the SC and the released Igs are referred to as SIgA or SIgM. Interestingly, unoccupied receptors are also transcytosed, resulting in the release of free SC at the apical surface, suggesting that this protein might have a biological role distinct from its association with IgA (see Chapter 8).

The pIgR is a heavily glycosylated transmembrane protein, containing five extracellular Ig-like domains. The pIgR gene (*PIGR*) is located on chromosome 1, close to the Fcα/μR gene (see Sect. 4.3.2). The pIgR initially binds to pIgA or IgM via a noncovalent interaction mediated by a specific motif in D1 of the pIgR. For IgA, but not IgM, a covalent disulfide bond between D5 of the pIgR and the Cα2 domain of IgA is formed during transcytosis and provides extra stabilization for SIgA.

At mucosal surfaces, immune exclusion by SIgA antibodies provides a first line of defense against pathogens. IgA undergoing pIgR-mediated transcytosis has also been implicated in the intracellular neutralization of LPS and virus and also in the transport of Ags (possibly even whole bacteria) out of the lamina propria and into the lumen (Johansen and Brandtzaeg, 2004; Phalipon and Corthesy 2003) (see Chapter 7). However, it appears that some microorganisms might in fact be able to exploit their ability to bind to the pIgR to infect epithelial cells (Kaetzel, 2001), although whether this actually happens *in vivo* is still unclear (Phalipon and Corthesy, 2003).

4.3.2. Fcα/μR

Fcα/μR is a transmembrane glycoprotein with one extracellular Ig-like domain that is able to bind both IgA and IgM (Shibuya et al., 2000; McDonald et al., 2002). Its ligand-binding site consists of a motif within its single

Ig-like domain that is highly homologous to the ligand-binding motif of the pIgR (Shibuya and Honda, 2006). The Fcα/μR gene (*FCAMR*) is located close to the pIgR gene (*PIGR*) on human and mouse chromosome 1, suggesting common ancestry (Shimizu et al., 2001). In the mouse, flow cytometry showed that Fcα/μR was expressed by the majority of B-cells and macrophages, but not by T-cells, natural killer (NK) cells, or granulocytes (Shibuya et al., 2000). Fcα/μR mRNA has been detected in primary human mesangial cells, which has led to speculation that this receptor might be involved in the kidney deposition of IgA-containing immune complexes (IgA-ICs) in IgAN (see Sects. 4.3.3 and 4.4.2).

4.3.3. Transferrin Receptor (CD71)

Transferrin receptor (TfR) is a disulfide-linked homodimeric type II transmembrane receptor that binds two proteins critical for iron metabolism: transferrin (Tf) and the hereditary hemochromatosis protein (HFE). Recently, it has been shown that TfR is also able to bind IgA (Moura et al., 2001). TfR has been reported to bind only human IgA1 (not IgA2 or SIgA), and it binds pIgA1 better than mIgA1 (Moura et al., 2004). Current evidence suggests that the TfR interacts mainly with the O-linked hinge region carbohydrate moieties of pIgA1, but that the N-linked sugar chains might also be involved (Moura et al., 2004). Thus, it has been suggested that TfR might represent the receptor for IgA1 O-linked glycans previously identified on human T-cells (Monteiro and van de Winkel, 2003; Rudd et al., 1994; Swenson et al., 1998). Significantly, TfR has been found to be expressed on human mesangial cells (HMCs) and cell lines. More importantly, TfR expression has been shown to be increased on HMCs in renal biopsies from IgAN patients and to colocalize with IgA deposits (Haddad et al., 2003; Moura et al., 2001). TfR has also been shown to preferentially bind the aberrantly glycosylated IgA1, which is a characteristic of IgAN (Moura et al., 2004) (see Sect. 4.4.2). Furthermore, HMCs stimulated with pIgA1 upregulate TfR expression, begin to proliferate, and secrete IL-6 and transforming growth factor (TGF)-β (Moura et al., 2005). Together, these results suggest that TfR is an important IgA FcR on HMCs and, as such, might be involved in the initiation of renal damage thought to be triggered by IgA-IC deposition in the kidney.

4.3.4. Asialoglycoprotein Receptor

The liver has been identified as the major site of IgA catabolism in humans. Here, IgA binds, in a calcium-dependent manner, to the ASGP-R expressed on the surface of hepatocytes. The ASGP-R binds the exposed terminal Gal or GalNAc residues of desialylated IgA, resulting in its internalization and eventual delivery to lysosomes for degradation (Stockert, 1995). Thus, ASGP-R is important for the regulation of the serum levels of IgA.

4.3.5. Other IgA Receptors

Many studies have noted that human T- and B-cells express receptors for IgA (Kerr et al., 1995; Morton et al., 1996). However, because CD89 is not expressed on these cell types, other novel IgA receptors, such as the recently described Fcα/μR and TfR, might explain some of these earlier reports.

A receptor for IgA has been described on several intestinal epithelial cell lines (Kitamura et al., 2000). These cells were shown not to express CD89, and the receptor was shown to be distinct from pIgR, as it was able to bind monomeric IgA. The possibility that IgA was binding to the ASGP-R was also discounted.

M-cells are specialized epithelial cells able to transport mucosal Ags across the intestinal epithelium and deliver them to the underlying Ag-presenting cells. Both human and murine M-cells have been reported to specifically bind IgA. The murine M-cell IgA receptor was shown to bind IgA with or without SC and recognizes a site spanning domains Cα1 and Cα2 of IgA (Mantis et al., 2002) (see Chapter 9).

Natural killer cells were found to specifically bind human pIgA and SIgA, but not to express CD89 or mannose receptor (Mota et al., 2003). The binding was not inhibited by L-fucose, D-galactose, D-glucose, D-mannose, or *N*-acetyl-D-glucosamine, suggesting that NK cells bind IgA via its protein backbone, not via sugar moieties present on IgA or SC.

4.4. IgA and IgA Receptors in Disease

4.4.1. CD89 Dysfunction

As mentioned earlier, eosinophils from some allergic patients express higher levels of CD89 than those from normal individuals (Monteiro et al., 1993). As atopic asthmatics have been shown to have elevated levels of specific IgA in sputum against both allergens and bacterial antigens (Nahm et al., 1998), this might suggest a role for IgA and CD89 in the pathogenesis of atopic allergy and extrinsic asthma.

CD89 has also been proposed to be important for the removal of potentially harmful IgA-ICs from the circulation via endocytosis. In diseases such as IgAN, Sjogren's syndrome, alcoholic liver cirrhosis, and human immunodeficiency virus (HIV) infection, which are characterized by high serum concentrations of pIgA and increased levels of circulating IgA-ICs, decreased CD89 expression levels and/or endocytotic rates have been noted (Grossetete et al., 1995, 1998; Monteiro et al., 1995; Silvain et al., 1995). Failure to clear IgA-ICs is proposed to lead to their deposition in the kidneys (see Sect. 4.4.2), where they are associated with inflammation and chronic tissue damage.

4.4.2. *IgA Nephropathy (IgAN)*

IgAN is the most common form of primary glomerulonephritis worldwide. This disease is characterized by the deposition of granular IgA-ICs in the glomerular mesangium, and approximately half of IgAN patients have elevated levels of IgA1 and/or IgA1-ICs in serum. Current evidence suggests that the basic abnormality in IgAN lies not in the kidney but, rather, involves a defect within the IgA immune system itself (van der Boog et al., 2005). Several studies have now shown that there is an under-galactosylation of the O-linked carbohydrate moieties in the IgA1 hinge region and that this might contribute to the formation and reduced clearance of IgA1-ICs (van der Boog et al., 2005). Recently, it was proposed that a soluble form of CD89 could be released into the serum of IgAN patients following interaction with abnormal IgA and thus contribute to the formation of IgA-ICs (Launay et al., 2000). Evidence in support of this theory was obtained from a CD89 transgenic mouse model that spontaneously developed IgAN-like symptoms (Launay et al., 2000). However, because the presence of IgA-CD89-ICs do not appear to be specific for IgAN (van der Boog et al., 2003) and the ability of CD89 to bind murine IgA is disputed, the potential involvement of CD89 in the pathogenesis of IgAN is still controversial (van der Boog et al., 2004).

Elevated serum levels of IgA1 and/or IgA1-ICs are not sufficient to cause mesangial deposition, and it is likely that a specific receptor for IgA is involved (Gomez-Guerrero et al., 2002). The identity of the mesangial cell IgA receptor has been hotly disputed for many years, but studies from several groups have now excluded CD89, ASGP-R, and pIgR (van der Boog et al., 2005). Similarly, although Fcα/μR mRNA is expressed in mesangial cells, this receptor does not seem to be involved in IgA deposition (van der Boog et al., 2005). Recently, however, an upregulation of TfR on mesangial cells of IgAN patients has been described (Moura et al., 2001). In addition, TfR has been shown to bind abnormally glycosylated IgA1 and IgA1-ICs more efficiently than normal IgA (Monteiro 2005; van der Boog et al., 2005). Taken together, these data suggest that TfR might turn out to be the elusive mesangial cell IgA receptor and thus be involved in the pathogenesis of IgAN.

4.5. Concluding Remarks

This chapter summarizes current knowledge concerning the structure and function of the five recognized types of IgA receptors, with special emphasis on the activating receptor CD89. The interplay between IgA and its receptors plays a critical role in immune defense in both the systemic and mucosal compartments. However, dysfunction of the IgA system has also been implicated in the pathogenesis of a number of diseases, especially IgAN. Future studies should help to elucidate the biological role of the newly described IgA receptors Fcα/μR and TfR and to further increase our understanding of the function of CD89 *in vivo*.

References

Abi-Rached, L., and Parham, P. (2005). Natural selection drives recurrent formation of activating killer cell immunoglobulin-like receptor and Ly49 from inhibitory homologues. *J Exp. Med.* 201:1319–1332.

Bakema, J. E., de Haij, S., den Hartog-Jager, C. F., Bakker, J., Vidarsson, G., van Egmond, M., van de Winkel, J. G., and Leusen, J. H. (2006). Signaling through mutants of the IgA receptor CD89 and consequences for Fc receptor γ-chain interaction. *J. Immunol.* 176:3603–3610.

Bracke, M., Lammers, J. W., Coffer, P. J., and Koenderman, L. (2001). Cytokine-induced inside-out activation of FcαR (CD89) is mediated by a single serine residue (S263) in the intracellular domain of the receptor. *Blood* 97:3478–3483.

Bracke, M., Nijhuis, E., Lammers, J. W., Coffer, P. J., and Koenderman, L. (2000). A critical role for PI 3-kinase in cytokine-induced Fcα-receptor activation. *Blood* 95:2037–2043.

Brandtzaeg, P., Farstad, I. N., Johansen, F. E., Morton, H. C., Norderhaug, I. N., and Yamanaka, T. (1999). The B-cell system of human mucosae and exocrine glands. *Immunol. Rev.* 171:45–87.

Conley, M. E., and Delacroix, D. L. (1987). Intravascular and mucosal immunoglobulin A: two separate but related systems of immune defense? *Ann. Intern. Med.* 106:892–899.

Dechant, M., Vidarsson, G., Stockmeyer, B., Repp, R., Glennie, M. J., Gramatzki, M., van de Winkel, J. G., and Valerius, T. (2002). Chimeric IgA antibodies against HLA class II effectively trigger lymphoma cell killing. *Blood* 100:4574–4580.

de Wit, T. P., Morton, H. C., Capel, P. J., and van de Winkel, J. G. (1995). Structure of the gene for the human myeloid IgA Fc receptor (CD89). *J. Immunol.* 155:1203–1209. Ding, Y., Xu, G., Yang, M., Yao, M., Gao, G. F., Wang, L., Zhang, W., and Rao, Z. (2003). Crystal structure of the ectodomain of human FcαRI. *J. Biol. Chem.* 278:27,966–27,970.

Feng, J., Garrity, D., Call, M. E., Moffett, H., and Wucherpfennig, K. W. (2005). Convergence on a distinctive assembly mechanism by unrelated families of activating immune receptors. *Immunity* 22:427–438.

Geissmann, F., Launay, P., Pasquier, B., Lepelletier, Y., Leborgne, M., Lehuen, A., Brousse, N., and Monteiro, R. C. (2000). A subset of human dendritic cells expresses IgA Fc receptor (CD89), which mediates internalization and activation upon cross-linking by IgA complexes. *J. Immunol.* 166:346–352.

Gomez-Guerrero, C., Suzuki, Y., and Egido, J. (2002). The identification of IgA receptors in human mesangial cells: in the search for "Eldorado." *Kidney Int.* 62:715–717.

Grossetete, B., Launay, P., Lehuen, A., Jungers, P., Bach, J. F., and Monteiro, R. C. (1998). Down-regulation of Fcα receptors on blood cells of IgA nephropathy patients: Evidence for a negative regulatory role of serum IgA. *Kidney Int.* 53:1321–1335.

Grossetete, B., Viard, J. P., Lehuen, A., Bach, J. F., and Monteiro, R. C. (1995). Impaired Fcα receptor expression is linked to increased immunoglobulin A levels and disease progression in HIV-1-infected patients. *AIDS* 9:229–234.

Gulle, H., Samstag, A., Eibl, M. M., and Wolf, H. M. (1998). Physical and functional association of FcαR with protein tyrosine kinase Lyn. *Blood* 91:383–391.

Haddad, E., Moura, I. C., Arcos-Fajardo, M., Macher, M. A., Baudouin, V., Alberti, C., Loirat, C., Monteiro, R. C., and Peuchmaur, M. (2003). Enhanced expression of the CD71 mesangial IgA1 receptor in Berger disease and Henoch-Schonlein nephritis: association between CD71 expression and IgA deposits. *J Am. Soc. Nephrol.* 14:327–337.

Hamre, R., Farstad, I. N., Brandtzaeg, P., and Morton, H. C. (2003). Expression and modulation of the human immunoglobulin A Fc receptor (CD89) and the FcR γ chain on myeloid cells in blood and tissue. *Scand. J. Immunol.* 57:506–516.

Hellwig, S. M., van Spriel, A. B., Schellekens, J. F., Mooi, F. R., and van de Winkel, J. G. (2001). Immunoglobulin A-mediated protection against Bordetella pertussis infection. *Infect. Immun.* 69:4846–4850.

Herr, A. B., Ballister, E. R., and Bjorkman, P. J. (2003a). Insights into IgA-mediated immune responses from the crystal structures of human FcαRI and its complex with IgA1-Fc. *Nature* 423:614–620.

Herr, A. B., White, C. L., Milburn, C., Wu, C., and Bjorkman, P. J. (2003b). Bivalent binding of IgA1 to FcαRI suggests a mechanism for cytokine activation of IgA phagocytosis. *J. Mol. Biol.* 327:645–657.

Heystek, H. C., Moulon, C., Woltman, A. M., Garonne, P., and van Kooten, C. (2002). Human immature dendritic cells efficiently bind and take up secretory IgA without the induction of maturation. *J. Immunol.* 168:102–107.

Hulett, M. D., and Hogarth, P. M. (1994). Molecular basis of Fc receptor function. *Adv. Immunol.* 57:1–127.

Jasek, M., Obojski, A., Manczak, M., Wisniewski, A., Winiarska, B., Malolepszy, J., Jutel, M., Luszczek, W., and Kusnierczyk, P. (2004). Are single nucleotide polymorphisms of the immunoglobulin A Fc receptor gene associated with allergic asthma? *Int. Arch. Allergy Immunol.* 135:325–331.

Johansen, F. E., Braathen, R., and Brandtzaeg, P. (2000). Role of J chain in secretory immunoglobulin formation. *Scand. J. Immunol.* 52:240–248.

Johansen, F. E., and Brandtzaeg, P. (2004). Transcriptional regulation of the mucosal IgA system. *Trends Immunol.* 25:150–157.

Kaetzel, C. S. (2001). Polymeric Ig receptor: Defender of the fort or Trojan horse? *Curr. Biol.* 11:R35–R38.

Kaetzel, C. S. (2005). The polymeric immunoglobulin receptor: Bridging innate and adaptive immune responses at mucosal surfaces. *Immunol. Rev.* 206:83–99.

Kaneko, S., Kobayashi, T., Yamamoto, K., Jansen, M. D., van de Winkel, J. G., and Yoshie, H. (2004). A novel polymorphism of FcαRI (CD89) associated with aggressive periodontitis. *Tissue Antigens* 63:572–577.

Kato, K., Fridman, W. H., Arata, Y., and Sautes-Fridman, C. (2000). A conformational change in the Fc precludes the binding of two Fcγ receptor molecules to one IgG. *Immunol. Today* 21:310–312.

Kerr, M. A. (1990). The structure and function of human IgA. *Biochem. J.* 271:285–296.

Kerr, M. A., Stewart, W. W., Bonner, B. C., Greer, M. R., Mackenzie, S. J., and Steele, M. G. (1995). The diversity of leucocyte IgA receptors. *Contrib. Nephrol.* 111:60–65.

Kiekens, R. C., Thepen, T., Oosting, A. J., Bihari, I. C., van de Winkel, J. G., Bruijnzeel-Koomen, C. A., and Knol, E. F. (2001). Heterogeneity within tissue-specific macrophage and dendritic cell populations during cutaneous inflammation in atopic dermatitis. *Br. J Dermatol.* 145:957–965.

Kitamura, T., Garofalo, R. P., Kamijo, A., Hammond, D. K., Oka, J. A., Caflisch, C. R., Shenoy, M., Casola, A., Weigel, P. H., and Goldblum, R. M. (2000). Human intestinal epithelial cells express a novel receptor for IgA. *J. Immunol.* 164:5029–5034.

Lang, M. L., Chen, Y. W., Shen, L., Gao, H., Lang, G. A., Wade, T. K., and Wade, W. F. (2002). IgA Fc receptor (FcαR) cross-linking recruits tyrosine kinases, phosphoinositide kinases and serine/threonine kinases to glycolipid rafts. *Biochem. J.* 364:517–525.

Lang, G. A., and Lang, M. L. (2006). Protein kinase Bα is required for vesicle trafficking and class II presentation of IgA Fc receptor (CD89)-targeted antigen. *J. Immunol.* 176:3987–3994.

Lang, M. L., Shen, L., Gao, H., Cusack, W. F., Lang, G. A., and Wade, W. F. (2001). Fcα receptor cross-linking causes translocation of phosphatidlyinositol-dependent protein kinase 1 and protein kinase Bα to MHC class II peptide-loading-like compartments. *J. Immunol.* 166:5585–5593.

Lang, M. L., Shen, L., and Wade, W. F. (1999). γ-Chain dependent recruitment of tyrosine kinases to membrane rafts by the human IgA receptor FcαR. *J. Immunol.* 163:5391–5398.

Launay, P., Grossetete, B., Arcos-Fajardo, M., Gaudin, E., Torres, S. P., Beaudoin, L., Patey-Mariaud, D. S., Lehuen, A., and Monteiro, R. C. (2000). Fcα receptor (CD89) mediates the development of Immunoglobulin A (IgA) nephropathy (Berger's disease). Evidence for pathogenic soluble receptor-IgA complexes in patients and CD89 transgenic mice. *J. Exp. Med.* 191:1999–2010.

Launay, P., Lehuen, A., Kawakami, T., Blank, U., and Monteiro, R. C. (1998). IgA Fc receptor (CD89) activation enables coupling to syk and Btk tyrosine kinase pathways: Differential signaling after IFN- γ or phorbol ester stimulation. *J. Leuk. Biol.* 63:636–642.

Launay, P., Patry, C., Lehuen, A., Pasquier, B., Blank, U., and Monteiro, R. C. (1999). Alternative endocytic pathway for immunoglobulin A Fc receptors (CD89) depends on the lack of FcRγ association and protects against degradation of bound ligand. *J. Biol. Chem.* 274:7216–7225.

Maliszewski, C. R., March, C. J., Schoenborn, M. A., Gimpel, S., and Shen, L. (1990). Expression cloning of a human Fc receptor for IgA. *J. Exp. Med.* 172:1665–1672.

Mantis, N. J., Cheung, M. C., Chintalacharuvu, K. R., Rey, J., Corthesy, B., and Neutra, M. R. (2002). Selective adherence of IgA to murine Peyer's patch M cells: Evidence for a novel IgA receptor. *J. Immunol.* 169:1844–1851.

Martin, A. M., Kulski, J. K., Witt, C., Pontarotti, P., and Christiansen, F. T. (2002). Leukocyte Ig-like receptor complex (LRC) in mice and men. *Trends Immunol.* 23:81–88.

Maruoka, T., Nagata, T., and Kasahara, M. (2004). Identification of the rat IgA Fc receptor encoded in the leukocyte receptor complex. *Immunogenetics* 55:712–716.

McDonald, K. J., Cameron, A. J., Allen, J. M., and Jardine, A. G. (2002). Expression of Fc α/μ receptor by human mesangial cells: A candidate receptor for immune complex deposition in IgA nephropathy. *Biochem. Biophys. Res. Commun.* 290:438–442.

Monteiro, R. C. (2005). New insights in the pathogenesis of IgA nephropathy. *Nefrologia* 25(Suppl. 2):82–86.

Monteiro, R. C., Cooper, M. D., and Kubagawa, H. (1992). Molecular heterogeneity of Fcα receptors detected by receptor-specific monoclonal antibodies. *J. Immunol.* 148:1764–1770.

Monteiro, R. C., Grossetete, B., Nguyen, A. T., Jungers, P., and Lehuen, A. (1995). Dysfunctions of Fcα receptors by blood phagocytic cells in IgA nephropathy. *Contrib. Nephrol.* 111:116–122.

Monteiro, R. C., Hostoffer, R. W., Cooper, M. D., Bonner, J. R., Gartland, G. L., and Kubagawa, H. (1993). Definition of immunoglobulin A receptors on eosinophils and their enhanced expression in allergic individuals. *J. Clin. Invest.* 92:1681–1685.
Monteiro, R. C., and van de Winkel, J. G. (2003). IgA Fc Receptors. *Annu. Rev. Immunol.* 21:177–204.
Morton, H. C., and Brandtzaeg, P. (2001). CD89: the human myeloid IgA Fc receptor. *Arch. Immunol. Ther. Exp.* (*Warsz.*) 49:217–229.
Morton, H. C., Pleass, R. J., Storset, A. K., Brandtzaeg, P., and Woof, J. M. (2005). Cloning and characterization of equine CD89 and identification of the CD89 gene in chimpanzees and rhesus macaques. *Immunology* 115:74–78.
Morton, H. C., Pleass, R. J., Storset, A. K., Dissen, E., Williams, J. L., Brandtzaeg, P., and Woof, J. M. (2004). Cloning and characterization of an immunoglobulin A Fc receptor from cattle. *Immunology* 111:204–211.
Morton, H. C., van den Herik-Oudijk, I. E., Vossebeld, P., Snijders, A., Verhoeven, A. J., Capel, P. J., and van de Winkel, J. G. (1995). Functional association between the human myeloid immunoglobulin A Fc receptor (CD89) and FcR γ chain. Molecular basis for CD89/FcR γ chain association. *J. Biol. Chem.* 270:29,781–29,787.
Morton, H. C., van Egmond, M., and van de Winkel, J. G. (1996). Structure and function of human IgA Fc receptors (FcαR). *Crit. Rev. Immunol.* 16:423–440.
Morton, H. C., van Zandbergen, G., van Kooten, C., Howard, C. J., van de Winkel, J. G., and Brandtzaeg, P. (1999). Immunoglobulin-binding sites of human FcαRI (CD89) and bovine Fcγ2R are located in their membrane-distal extracellular domains. *J. Exp. Med.* 189:1715–1722.
Mota, G., Manciulea, M., Cosma, E., Popescu, I., Hirt, M., Jensen-Jarolim, E., Calugaru, A., Galatiuc, C., Regalia, T., Tamandl, D., Spittler, A., and Boltz-Nitulescu, G. (2003). Human NK cells express Fc receptors for IgA which mediate signal transduction and target cell killing. *Eur. J. Immunol.* 33:2197–2205.
Moura, I. C., Arcos-Fajardo, M., Gdoura, A., Leroy, V., Sadaka, C., Mahlaoui, N., Lepelletier, Y., Vrtovsnik, F., Haddad, E., Benhamou, M., and Monteiro, R. C. (2005). Engagement of transferrin receptor by polymeric IgA1: Evidence for a positive feedback loop involving increased receptor expression and mesangial cell proliferation in IgA Nephropathy. *J Am. Soc. Nephrol.* 16:2667–2676.
Moura, I. C., Arcos-Fajardo, M., Sadaka, C., Leroy, V., Benhamou, M., Novak, J., Vrtovsnik, F., Haddad, E., Chintalacharuvu, K. R., and Monteiro, R. C. (2004). Glycosylation and size of IgA1 are essential for interaction with mesangial transferrin receptor in IgA nephropathy. *J Am. Soc. Nephrol.* 15:622–634.
Moura, I. C., Centelles, M. N., Arcos-Fajardo, M., Malheiros, D. M., Collawn, J. F., Cooper, M. D., and Monteiro, R. C. (2001). Identification of the transferrin receptor as a novel immunoglobulin (Ig)A1 receptor and its enhanced expression on mesangial cells in IgA nephropathy. *J. Exp. Med.* 194:417–425.
Nahm, D. H., Kim, H. Y., and Park, H. S. (1998). Elevation of specific immunoglobulin A antibodies to both allergen and bacterial antigen in induced sputum from asthmatics. *Eur. Respir. J.* 12:540–545.
Narita, I., Goto, S., Saito, N., Sakatsume, M., Jin, S., Omori, K., and Gejyo, F. (2001). Genetic polymorphisms in the promoter and 5′ UTR region of the Fcα receptor (CD89) are not associated with a risk of IgA nephropathy. *J. Hum. Genet.* 46:694–698.

Nikolaidis, N., Klein, J., and Nei, M. (2005). Origin and evolution of the Ig-like domains present in mammalian leukocyte receptors: Insights from chicken, frog, and fish homologues. *Immunogenetics* 57:151–157.

Oortwijn, B. D., Roos, A., van der Boog, P. J., Klar-Mohamad, N., van Remoortere, A., Deelder, A. M., Daha, M. R., and van Kooten, C. (2007). Monomeric and polymeric IgA show a similar association with the myeloid FcαRI/CD89. *Mol. Immunol.* 44:996–973.

Otten, M.A., Groenveld, I., van de Winkel, J. G., and van Egmond, M. (2006). Inefficient antigen presentation via the IgA Fc receptor (FcαRI) on dendritic cells. *Immunobiology* 211:503–510.

Otten, M. A., and van Egmond, M. (2004). The Fc receptor for IgA (FcαRI, CD89). *Immunol. Lett.* 92:23–31.

Ouadrhiri, Y., Pilette, C., Monteiro, R. C., Vaerman, J. P., and Sibille, Y. (2002). Effect of IgA on respiratory burst and cytokine release by human alveolar macrophages: role of ERK1/2 mitogen-activated protein kinases and NF-κB. *Am. J Respir. Cell Mol. Biol.* 26:315–332.

Park, R. K., Izadi, K. D., Deo, Y. M., and Durden, D. L. (1999). Role of src in the modulation of multiple adaptor proteins in FcαRI oxidant signaling. *Blood* 94:2112–2120.

Pasquier, B., Launay, P., Kanamaru, Y., Moura, I. C., Pfirsch, S., Ruffie, C., Henin, D., Benhamou, M., Pretolani, M., Blank, U., and Monteiro, R. C. (2005). Identification of FcαRI as an inhibitory receptor that controls inflammation: Dual role of FcRγ ITAM. *Immunity* 22:31–42.

Pasquier, B., Lepelletier, Y., Baude, C. O., and Monteiro, R. C. (2004). Differential expression and function of IgA receptors (CD89 and CD71) during maturation of dendritic cells. *J. Leuk. Biol.* 76:1134–1141.

Patry, C., Sibille, Y., Lehuen, A., and Monteiro, R. C. (1996). Identification of Fcα receptor (CD89) isoforms generated by alternative splicing that are differentially expressed between blood monocytes and alveolar macrophages. *J. Immunol.* 156:4442–4448.

Phalipon, A., and Corthesy, B. (2003). Novel functions of the polymeric Ig receptor: Well beyond transport of immunoglobulins. *Trends Immunol.* 24:55–58.

Rogers, K. A., Scinicariello, F., and Attanasio, R. (2004). Identification and characterization of macaque CD89 (immunoglobulin A Fc receptor). *Immunology* 113:178–186.

Rudd, P. M., Fortune, F., Patel, T., Parekh, R. B., Dwek, R. A., and Lehner, T. (1994). A human T-cell receptor recognizes 'O'-linked sugars from the hinge region of human IgA1 and IgD. *Immunology* 83:99–106.

Saito, K., Suzuki, K., Matsuda, H., Okumura, K., and Ra, C. (1995). Physical association of Fc receptor γ chain homodimer with IgA receptor. *J. Allergy Clin. Immunol.* 96:1152–1160.

Sansonetti, P. J. (2004). War and peace at mucosal surfaces. *Nat. Rev. Immunol.* 4:953–964.

Segal, D. M., Taurog, J. D., and Metzger, H. (1977). Dimeric immunoglobulin E serves as a unit signal for mast cell degranulation. *Proc. Natl. Acad. Sci. USA* 74:2993–2997.

Shibuya, A., and Honda, S. (2006). Molecular and functional characteristics of the Fcα/μR, a novel Fc receptor for IgM and IgA. *Springer Semin. Immun.* 28: 377–382.

Shibuya, A., Sakamoto, N., Shimizu, Y., Shibuya, K., Osawa, M., Hiroyama, T., Eyre, H. J., Sutherland, G. R., Endo, Y., Fujita, T., Miyabayashi, T., Sakano, S., Tsuji, T., Nakayama, E., Phillips, J. H., Lanier, L. L., and Nakauchi, H. (2000). Fcα/μ receptor mediates endocytosis of IgM-coated microbes. *Nat. Immunol.* 1:441–446.

Shimizu, Y., Honda, S., Yotsumoto, K., Tahara-Hanaoka, S., Eyre, H. J., Sutherland, G. R., Endo, Y., Shibuya, K., Koyama, A., Nakauchi, H., and Shibuya, A. (2001). Fcα/μ receptor is a single gene-family member closely related to polymeric immunoglobulin receptor encoded on chromosome 1. *Immunogenetics* 53:709–711.

Shimokawa, T., and Ra, C. (2003). C/EBP α and Ets protein family members regulate the human myeloid IgA Fc receptor (FcαR, CD89) promoter. *J. Immunol.* 170:2564–2572.

Shimokawa, T., Tsuge, T., Okumura, K., and Ra, C. (2000). Identification and characterization of the promoter for the gene encoding the human myeloid IgA Fc receptor (FcαR, CD89). *Immunogenetics* 51:945–954.

Silvain, C., Patry, C., Launay, P., Lehuen, A., and Monteiro, R. C. (1995). Altered expression of monocyte IgA Fc receptors is associated with defective endocytosis in patients with alcoholic cirrhosis. Potential role for IFN-γ. *J. Immunol.* 155:1606–1618.

Smith, P. D., Ochsenbauer-Jambor, C., and Smythies, L. E. (2005). Intestinal macrophages: unique effector cells of the innate immune system. *Immunol. Rev.* 206:149–159.

Stockert, R. J. (1995). The asialoglycoprotein receptor: Relationships between structure, function, and expression. *Physiol. Rev.* 75:591–609.

Swenson, C. D., Patel, T., Parekh, R. B., Tamma, S. M., Coico, R. F., Thorbecke, G. J., and Amin, A. R. (1998). Human T cell IgD receptors react with O-glycans on both human IgD and IgA1. *Eur. J. Immunol.* 28:2366–2372.

Tsuge, T., Shimokawa, T., Horikoshi, S., Tomino, Y., and Ra, C. (2001). Polymorphism in promoter region of Fcα receptor gene in patients with IgA nephropathy. *Hum. Genet.* 108:128–133.

Valerius, T., Stockmeyer, B., van Spriel, A. B., Graziano, R. F., van den Herik-Oudijk, I. E., Repp, R., Deo, Y. M., Lund, J., Kalden, J. R., Gramatzki, M., and van de Winkel, J. G. (1997). FcαRI (CD89) as a novel trigger molecule for bispecific antibody therapy. *Blood* 90:4485–4492.

van der Boog, P. J., de Fijter, J. W., van Kooten, C., Van Der, H. R., van Seggelen, A., van Es, L. A., and Daha, M. R. (2003). Complexes of IgA with FcαRI/CD89 are not specific for primary IgA nephropathy. *Kidney Int.* 63:514–521.

van der Boog, P. J., van Kooten, C., de Fijter, J. W., and Daha, M. R. (2005). Role of macromolecular IgA in IgA nephropathy. *Kidney Int.* 67:813–821.

van der Boog, P. J., van Kooten, C., van Zandbergen, G., Klar-Mohamad, N., Oortwijn, B., Bos, N. A., van Remoortere, A., Hokke, C. H., de Fijter, J. W., and Daha, M. R. (2004). Injection of recombinant FcαRI/CD89 in mice does not induce mesangial IgA deposition. *Nephrol. Dial. Transplant.* 19:2729–2736.

van der Boog, P. J., van Zandbergen, G., de Fijter, J. W., Klar-Mohamad, N., van Seggelen, A., Brandtzaeg, P., Daha, M. R., and van Kooten, C. (2002). FcαRI/CD89 circulates in human serum covalently linked to IgA in a polymeric state. *J. Immunol.* 168:1252–1258.

van der Pol W. L., Vidarsson, G., Vile, H. A., van de Winkel, J. G., and Rodriguez, M. E. (2000). Pneumococcal capsular polysaccharide-specific IgA triggers efficient neutrophil effector functions via FcαRI (CD89). *J. Infect. Dis.* 182:1139–1145.

van Dijk, T. B., Bracke, M., Caldenhoven, E., Raaijmakers, J. A., Lammers, J. W., Koenderman, L., and de Groot, R. P. (1996). Cloning and characterization of FcαRb, a novel Fcα receptor (CD89) isoform expressed in eosinophils and neutrophils. *Blood* 88:4229–4238.

van Egmond, M., Damen, C. A., van Spriel, A. B., Vidarsson, G., van Garderen, E., and van de Winkel, J. G. (2001). IgA and the IgA Fc receptor. *Trends Immunol.* 22:205–211.

van Egmond, M., van Garderen, E., van Spriel, A. B., Damen, C. A., van Amersfoort, E. S., van Zandbergen, G., van Hattum, J., Kuiper, J., and van de Winkel, J. G. (2000). FcαRI-positive liver Kupffer cells: Reappraisal of the function of immunoglobulin A in immunity. *Nat. Med.* 6:680–685.

van Egmond, M., van Vuuren, A. J., Morton, H. C., van Spriel, A. B., Shen, L., Hofhuis, F. M., Saito, T., Mayadas, T. N., Verbeek, J. S., and van de Winkel, J. G. (1999). Human immunoglobulin A receptor (FcαRI, CD89) function in transgenic mice requires both FcR γ chain and CR3 (CD11b/CD18). *Blood* 93:4387–4394.

van Spriel, A. B., Leusen, J. H., Vile, H., and van de Winkel, J. G. (2002). Mac-1 (CD11b/CD18) as accessory molecule for FcαR (CD89) binding of IgA. *J. Immunol.* 169:3831–3836.

Vidarsson, G., Der Pol, W. L., van Den Elsen, J. M., Vile, H., Jansen, M., Duijs, J., Morton, H. C., Boel, E., Daha, M. R., Corthesy, B., and van de Winkel, J. G. (2001). Activity of human IgG and IgA subclasses in immune defense against *Neisseria meningitidis* serogroup B. *J. Immunol.* 166:6250–6256.

Volz, A., Wende, H., Laun, K., and Ziegler, A. (2001). Genesis of the ILT/LIR/MIR clusters within the human leukocyte receptor complex. *Immunol. Rev.* 181:39–51.

Watanabe, A., Shimokawa, T., Moriyama, M., Komine, F., Amaki, S., Arakawa, Y., and Ra, C. (2006). Genetic variants of the IgA Fc receptor (FcαR, CD89) promoter in chronic hepatitis C patients. *Immunogenetics* 58:937–946.

Weisbart, R. H., Kacena, A., Schuh, A., and Golde, D. W. (1988). GM-CSF induces human neutrophil IgA-mediated phagocytosis by an IgA Fc receptor activation mechanism. *Nature* 332:647–648.

Wines, B. D., Hulett, M. D., Jamieson, G. P., Trist, H. M., Spratt, J. M., and Hogarth, P. M. (1999). Identification of residues in the first domain of human Fcα receptor essential for interaction with IgA. *J. Immunol.* 162:2146–2153.

Wines, B. D., Sardjono, C. T., Trist, H. H., Lay, C. S., and Hogarth, P. M. (2001). The interaction of FcαRI with IgA and its implications for ligand binding by immunoreceptors of the leukocyte receptor cluster. *J. Immunol.* 166:1781–1789.

Woof, J. M., and Burton, D. R. (2004). Human antibody-Fc receptor interactions illuminated by crystal structures. *Nat. Rev. Immunol.* 4:89–99.

Woof, J. M., and Mestecky, J. (2005). Mucosal immunoglobulins. *Immunol. Rev.* 206:64–82.

Woof, J. M., van Egmond, M., and Kerr, M. A. (2005). Fc receptors. In: Mesteck, J., Lamm, M. E., Strober, W., Bienenstock, J., McGhee, J. R., and Mayer, L. (eds), *Mucosal Immunology*, 3rd ed. Elsevier/Academic Press, Amsterdam, pp 251–265.

5 Regulation of the Mucosal IgA System

Finn-Eirik Johansen[1], Ranveig Braathen[1], Else Munthe[1], Hilde Schjerven[1], and Per Brandtzaeg[1]

[1]Laboratory of Immunohistochemistry and Immunopathology (LIIPAT), Department and Institute of Pathology, Rikshospitalet University Hospital, 0027 Oslo, Norway.

5.1. Introduction

The vulnerable mucosae are persistently exposed to airborne, ingested, and sexually transmitted potentially harmful agents. First-line protection of the mucosal surface area, encompassing about 400 m^2 in an adult human, depends on active export of secretory IgA (SIgA) and, to some extent, secretory IgM (SIgM) antibodies (Brandtzaeg et al., 1999c) (Fig. 5.1). This secretory immune system is distinct from the systemic counterpart and is operational earlier in ontogeny. Production of SIgA relies on a predominant class switch to the IgA isotypes and a high expression level of the joining (J) chain in mucosal plasma cells (PCs) to ensure adequate production of polymeric IgA (pIgA; mostly dimers) (Brandtzaeg, 1974; Brandtzaeg et al., 1999a). These processes depend on transcriptional activation of the IgA germline ($C_H\alpha$) genes and the J-chain gene. Active epithelial export of pIgA requires abundant expression of the epithelial polymeric Ig receptor (pIgR), because this transport protein is sacrificed with the delivery of its cargo (see Chapter 3). The crucial importance of the J-chain and the pIgR for epithelial transport of pIgA is highlighted by the absence of SIgA in knockout mice deficient for either of these genes (Hendrickson et al., 1995; Johansen et al., 1999; Lycke et al., 1999; Shimada et al., 1999).

5.2. Discovery of the Mucosal IgA System

Immunoglobulin A was identified as the predominant Ig class in various secretions in the early 1960s, by Tomasi and colleagues (reviewed in Hanson and Brandtzaeg, 1993; Tomasi, 1992). The discovery of the striking abundance of IgA-producing PCs in human secretory tissues was followed by findings that exported IgA in secretions contained an additional polypeptide absent from serum IgA. This additional polypeptide was first called "secretory piece"—now named secretory component (SC) or, in its membrane form, pIgR (Hanson, 1961; Tomasi et al., 1965) (see Chapter 3).

The J-chain was independently discovered as an inherent part of pIgA and pentameric IgM by Halpern and Koshland (1970) and by Mestecky et al. (1971), respectively, and was the first sign of shared functional features between these two Ig classes. A spurt of research on mucosal tissues and secretions followed. Decisive for the common transport model first proposed by Brandtzaeg in the early 1970s (Brandtzaeg, 1974, 1977, 1985; Brandtzaeg and Prydz, 1984) were three observations: (1) SC existed in a transmembrane form expressed by epithelial cells; (2) free SC could bind J-chain-containing pIgA and pentameric IgM specifically and with high affinity; and (3) the Golgi zone in epithelial cells contained unoccupied (newly synthesized) SC, whereas apical transport vesicles contained IgA-bound SC. Thus, membrane SC acts as a receptor for polymeric immunoglobulins (i.e., pIgR), translocating both pIgA and pentameric IgM to exocrine fluids by active and selective transepithelial transport. Although the generation of SIgA and SIgM is analogous, the secretory immune system

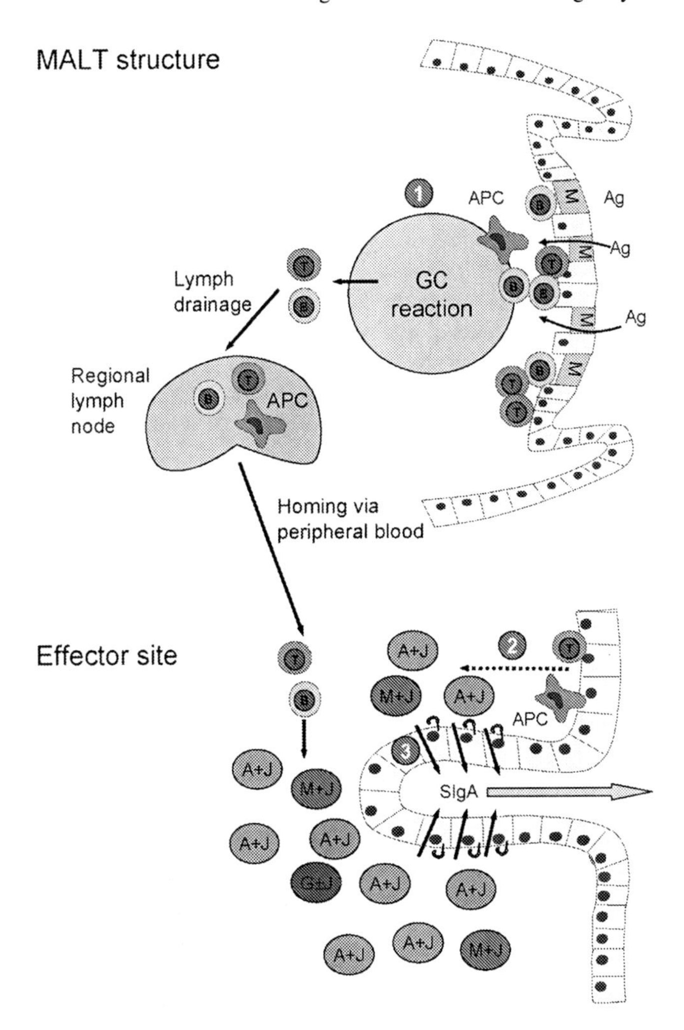

FIG. 5.1. Schematic steps in secretory IgA generation. (1) Naïve mucosal B-cells are induced in mucosa-associated lymphoid tissue (MALT) such as ileal Peyer's patches and isolated lymphoid follicles in the large bowel. Germinal center (GC) B-cells undergo somatic hypermutation and class-switch recombination before exit via lymph to regional lymph nodes and further homing via the thoracic duct and peripheral blood to mucosal effector sites such as the intestinal lamina propria. (2) Mucosal plasmablasts receive local signals (broken arrow) for terminal differentiation and express high levels of the J-chain; most of them are therefore committed to production of dimeric IgA (A + J), and some to production of pentameric IgM (M + J). Most of the rare mucosal IgG-producing cells also express the J-chain, although in a nonassociated form (IgG ± J). (3) Secretory epithelial cells express the polymeric Ig receptor (pIgR) basolaterally, thereby facilitating the export of dimeric IgA (and pentameric IgM) to provide SIgA (and SIgM) antibodies for defense functions. Ag = antigen, APC = antigen-presenting cell.

is often referred to as the mucosal IgA system because this isotype normally dominates at all mucosal effector sites.

5.3. Compartmentalization of the Mucosal IgA System

The mucosal immune system can be divided into two functionally distinct arms (Fig. 5.1) (see also Chapter 2). The inductive sites consist of organized mucosa-associated lymphoid tissue (MALT) and lymph nodes (LNs), where antigens sampled from mucosal surfaces stimulate naïve T- and B-lymphocytes. The effector sites consist of connective tissue stroma and secretory epithelium, where B-lymphocytes terminally differentiate into PCs that efficiently produce pIgA for export by pIgR-mediated transcytosis (see Chapter 3).

5.3.1. Mucosal Inductive Sites

The inductive sites for mucosal immunity are organized lymphoid structures with B-cell follicles containing germinal centers, intervening T-cell areas, and a variety of antigen-presenting cells (APCs) (Brandtzaeg and Johansen, 2005). Such sites of the gut include Peyer's patches (PPs), the appendix, and isolated lymphoid follicles (ILFs), whereas the unpaired nasopharyngeal tonsil (often called adenoids) and the paired palatine tonsils constitute inductive sites in the airways. All these lymphoid structures lack afferent lymphatics and sample exogenous antigens directly from mucosal surfaces through a characteristic follicle-associated epithelium (FAE), which contains "membrane" or "microfold" (M) cells (Kraehenbuhl and Neutra, 2000). These specialized thin cells have been shown to be effective in the uptake of especially microorganisms and other particulate antigens. A more detailed description of PPs and their function is found in Chapter 2.

Mucosal inductive sites can be subdivided after the anatomical location in the body. The gut is the largest antibody-producing organ in humans and mice, responding to the enormous stimulus of antigens from food products and the commensal microbiota. Inhaled antigens induce immune responses primarily in the nasopharynx-associated lymphoid tissue (NALT), whereas in the rat, rabbit, and guinea pig, the bronchus-associated lymphoid tissue (BALT) also plays a significant role (Brandtzaeg and Johansen, 2005). Mammary, salivary, and lacrimal glands, the urogenital tract, and the middle ear also participate in mucosal immunity. The role of IgA in mucosal immune responses in the respiratory and reproductive tracts is described in Chapters 11 and 12.

After primary stimulation of mucosal inductive sites, B-cells exit through draining lymphatics and local regional LNs into the blood circulation and "home" to mucosal effector sites for further maturation to PCs that produce pIgA. B-Cells activated in one inductive site might seed distant effector sites, and the term "common mucosal immune system" refers to the possibility of providing secretory antibodies at all mucosal surfaces of the body (Brandtzaeg and Johansen, 2005). Interestingly, however, the homing of B-cells activated in one mucosal microenvironment is not uniform to all mucosal surfaces, and lymphocytes preferentially migrate back to the region where they were originally stimulated (Brandtzaeg and Johansen, 2005; Johansen et al., 2005; Salmi and Jalkanen, 2005). This compartmentalization within the integrated mucosal immune system is supported by regionalized secretory immunity obtained after local immunization (reviewed in Holmgren and Czerkinsky, 2005). Thus, nasal antigen challenge will preferentially induce an immune response in the upper airways and saliva but, surprisingly, to some extent also in the female genital tract. Feeding of antigens (oral route) preferentially induces an immune response in the intestine but to some extent also in salivary and lactating mammary glands. Rectal antigen uptake preferentially induces an immune response in the large intestine and to some extent in the female genital tract.

5.3.2. Mucosal Effector Sites for Secretory Immunity

The effector sites, where B-cells terminally differentiate into mainly pIgA-producing PCs, consist of diffusely organized connective tissue, including the lamina propria (LP) of various mucosae, stroma of exocrine glands, and the secretory epithelia responsible for IgA export (Brandtzaeg and Johansen, 2005). The mucosa of the adult human gut has a surface area of about 300 m^2, whereas the airway epithelium covers ~100 m^2, compared with only 1.5 m^2 of skin. These mucosal surfaces are generally covered by a physically vulnerable monolayered epithelium, which is persistently exposed to airborne, ingested, and sexually transmitted agents, including numerous noxious substances and potential pathogenic microorganisms. In fact, most pathogens use the mucosae as their portal of entry. In addition to providing a physical barrier to the environment, the epithelium is equipped with several other innate defense mechanisms to prevent attachment, colonization, and possible damage by noxious agents. Many of these mechanisms cooperate with a first-line adaptive immune system provided by SIgA and SIgM (Brandtzaeg and Johansen, 2005). Secretory antibodies constitute the hallmark of mucosal immunity and are generated by the concerted action of two distinct cell types: pIgA-producing PCs and pIgR-expressing secretory epithelial cells.

The human gastrointestinal and naso-oropharyngeal tracts harbor the largest number and highest diversity of microorganisms in the body, with up to 10^{12}–10^{14} bacteria per gram of colonic tissue (Mackie et al., 1999). This enormous antigenic load warrants the need for an abundant export of SIgA, which in the human gut is calculated to be at least 40 mg/kg body weight or ~3 g per day (Conley and Delacroix, 1987). Production of this amount of SIgA requires a high number of pIgA-producing cells; in fact, more than 80% of all human PCs are located in the gastrointestinal tract and most of these are pIgA producers (Brandtzaeg et al., 1999a).

The microenvironments at different mucosal effector sites influence the relative level of B-cells expressing various Ig classes. In the gut lumen, the proximal small intestine contains mostly antigens derived from food, compared with the mainly commensal bacterial composition found in the distal ileum and large intestine. In humans, there are two isotypes of IgA, namely IgA1 and IgA2. PCs that produce IgA1 dominate in the systemic immune system, tonsils, and nasal mucosa (with ~ 90%), followed by the gastric mucosa, lacrimal glands, small intestine, and salivary and mammary glands, whereas IgA2-producing PCs dominate in the colon (65% IgA2) (Brandtzaeg and Johansen, 2005). Secretory antibodies to lipopolysaccharide (LPS) are generally of the IgA2 isotype, whereas protein antigens stimulate predominantly IgA1 production, although the molecular events underlying preferential IgA1 or IgA2 responses remain unclear (Mestecky and Russell, 1986; Tarkowski et al., 1990).

5.3.3. Homing of B-Cells to Mucosal Effector Sites

The homing of memory/effector lymphocytes depends on cell surface expression of chemokine receptors and several adhesion molecules that bind complementary chemokines and tissue-specific counterreceptors on vascular endothelial cells (Kunkel et al., 2003). Homing to the gut LP relies mainly on strong surface expression of $\alpha 4\beta 7$ and expression of CCR9 or CCR10 (see below), whereas a combined expression of $\alpha 4\beta 1$, L-selectin, and CCR7 appears to direct B-cells to the upper respiratory and uterine cervical mucosae (Johansen et al., 2005). Memory/effector B-lymphocytes involved in the systemic antibody response express mainly L-selectin and little $\alpha 4\beta 7$ (Quiding-Jarbrink et al., 1997) and are presumably guided to the bone marrow by expression of CXCR3 and CXCR4 (Odendahl et al., 2005). The counterreceptor for the $\alpha 4\beta 7$ integrin is the unmodified mucosal addressin cellular adhesion molecule-1 (MAdCAM-1) expressed apically on endothelial cells of the intestinal LP microvasculature (Brandtzaeg et al., 1999b; Butcher and Picker, 1996). By contrast, the $\alpha 4\beta 1$ integrin binds vascular cell adhesion molecule (VCAM)-1, expressed by the endothelium in human bronchial and nasal mucosa (Bentley et al., 1993; Jahnsen et al., 1995).

The microenvironments of the mucosal inductive sites and draining LNs regulate integrin expression pattern on lymphocytes, thus imprinting the homing capacity of the memory/effector cells (Campbell and Butcher, 2002). This imprinting is at least partially provided by regional dendritic cells (DCs). Such cells isolated from mesenteric lymph nodes (MLNs) and PPs are unique in the capacity to increase the expression of the mucosal homing receptor $\alpha 4\beta 7$ integrin on lymphocytes (Brandtzaeg and Johansen, 2005). Acquisition of such gut-homing properties apparently depends on retinoic acid derived by oxidative conversion from vitamin A, an enzymatic process preferentially mediated by the regional DCs (Iwata et al., 2003).

To attract the appropriately activated lymphocytes, expression of integrin counterreceptors on the microvascular endothelium is regulated by cell-intrinsic and microenvironmental factors at the mucosal effector sites. Additionally, local production of specific chemokines that activate integrins on the lymphocytes and promote chemotaxis directs selective extravasation. In the gut, expression of different chemokines in the small and large intestines explains the selective recruitment of lymphocytes activated by the oral and rectal route (Brandtzaeg and Johansen, 2005; Kiyono and Fukuyama, 2004). The chemokine CCL25 (TECK) is selectively produced by the crypt epithelium in the small intestine and attracts lymphocytes expressing CCR9 (Kunkel et al., 2000; Pabst et al., 2004). During an immune response to orally fed antigens, DCs in PPs and MLNs imprint lymphocytes with a high expression of CCR9 and $\alpha 4\beta 7$ integrin, combined with a downregulation of L-selectin (Johansson-Lindbom et al., 2003; Mora et al., 2003), thus directing these cells to the small intestine. In the large intestine, CCL28 (MEC) expression appears to be a decisive cue for attracting IgA^{+} plasmablasts that express high levels of CCR10 as well as $\alpha 4\beta 7$ (Hieshima et al., 2004; Kunkel et al., 2003).

5.3.4. Ig Class Distribution at Different Effector Sites

Antibodies constitute the effectors of humoral immune responses, and all five human Ig classes are involved in the protection of mucosal surfaces. Although IgG is the major systemic antibody class, IgA is nevertheless the predominant antibody class of the body on a biosynthetic basis—emphasizing the priority of the immune system to protect the mucosae. The estimated rate of biosynthesis of IgA is 66 mg/kg of body weight daily, compared to 34 mg of IgG and 7.9 mg of IgM (reviewed in Manz et al., 2005). In serum, the IgG subclasses predominate, making up 85% of total serum antibodies, whereas monomeric IgA accounts for 7–15% and ~5% is IgM, mainly pentameric. IgD and IgE contribute little to serum antibodies, with only 0.3% and 0.02%, respectively, due to low synthetic rates and short half-lives (Manz et al., 2005). In comparison, IgA-producing PCs dominate at all mucosal effector sites (Fig. 5.2A) (Brandtzaeg *et al.*, 1999a).

In the normal gut, IgM-producing PCs normally constitute a relative small and variable fraction in adults, whereas IgG-producing PCs are even less represented (Fig 5.2A). Nevertheless, SIgM-mediated mucosal protection is of significance, particularly in the neonatal period and in individuals with selective IgA deficiency—the most common immune disorder in humans. Only occasional IgD- and IgE-producing PCs are encountered in the gastrointestinal tract, whereas IgD-producing cells often constitute a significant fraction in the upper aerodigestive tract, particularly in IgA deficiency (Brandtzaeg et al., 1999c; Norderhaug et al., 1999b; Woof and Mestecky, 2005).

In accordance with the abundant local production of pIgA and its efficient pIgR-mediated export, SIgA is the major antibody class in most external secretions (Fig. 5.2B). However, in addition to local Ig production by PCs resident in the LP or stroma of exocrine glands, the Ig content in secretions might also be influenced by the relative serum concentrations and the transfer efficiency across endothelial and epithelial barriers. In this regard, it is of interest that the neonatal Fcγ receptor (FcRn), which is expressed both

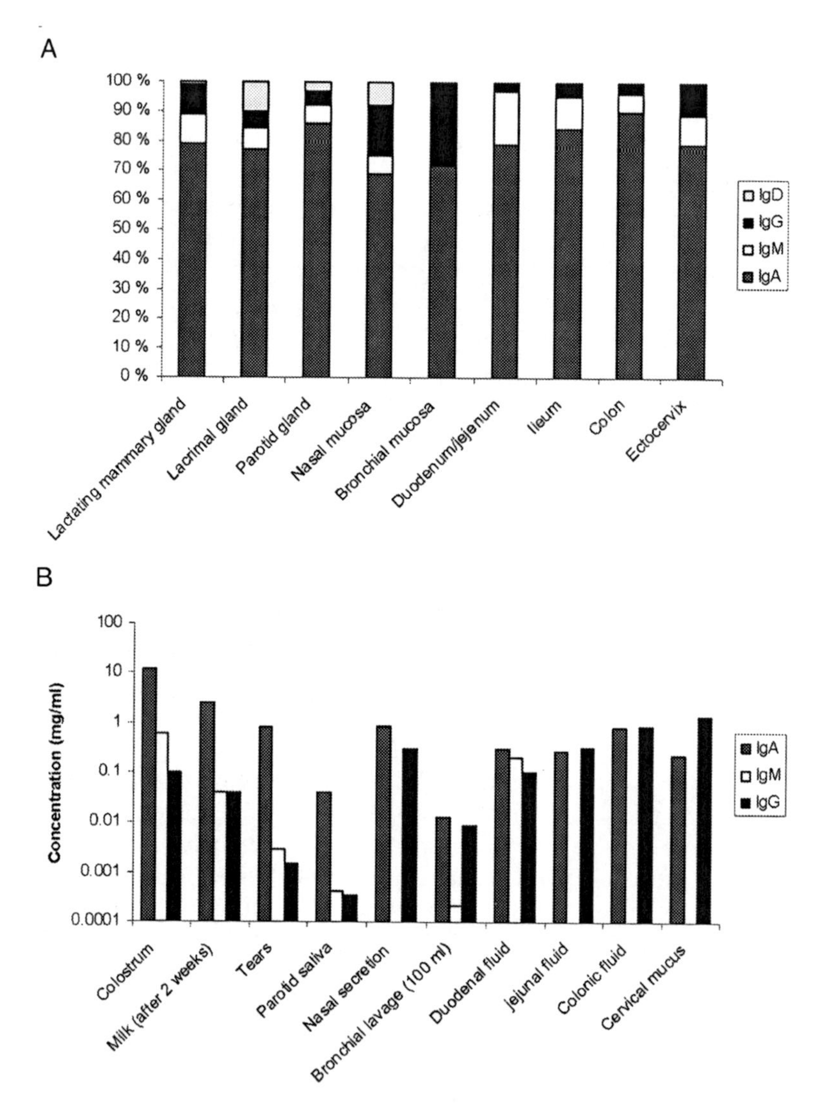

FIG. 5.2. Isotype distribution of mucosal plasma cells and immunoglobulins. **(A)** Percentage of Ig-producing plasma cells of various isotypes in the lamina propria at different mucosal effector sites in healthy human subjects. IgE-producing plasma cells make up less than 1% in the normal situation. **(B)** Concentrations of immunoglobulins sampled at different anatomical locations. Note logarithmic vertical axis. Data in (A) and (B) are based on papers reviewed in Norderhaug et al., 1999b.

on endothelial cells and gut epithelial cells, has recently been demonstrated to mediate transport of IgG across these barriers (Dickinson et al., 1999; Yoshida et al., 2004). In addition, this receptor functions to extend the half-life of IgG by directing endocytosed IgG away from lysosomal catabolism (Ghetie and Ward, 2000). Furthermore, paracellular bulk transfer of serum proteins also contributes to the Ig levels seen in secretions, particularly when the mucous membranes are irritated (Persson et al., 1998).

The relative levels of Ig classes in mucosal tissues are changed in several diseases. IgE is normally found at low concentrations in respiratory and gastrointestinal secretions. However, in subjects suffering from chronic allergic diseases such as asthma, allergic rhinitis, and food allergy, IgE concentrations in secretions might be significantly elevated. In inflammatory bowel disease (IBD), there is substantial increase in IgG, partly due to leakage from serum and partly because of significantly increased local production (Brandtzaeg et al., 2005).

5.4. Unique Features of Polymeric Igs

5.4.1. Antibody-Mediated Mucosal Defense

The vulnerable mucosae must carefully regulate the intensity of an immune response and associated effector mechanisms to combat efficiently the pathogen and at the same time avoid adverse effects that jeopardize the epithelial barrier function. SIgA is ideally suited for mucosal protection because it principally represents a noninflammatory combat system. Furthermore, active transcellular export of pIgA antibodies from the LP to the lumen allows such antibodies to operate in three distinct compartments (Mazanec et al., 1993) (see Chapter 7): (1) removal of antigens from LP (Kaetzel et al., 1991; Robinson *et al.,* 2001); (2) neutralization of virus or toxins within mucosal epithelial cells (Fujioka et al., 1998; Fernandez et al., 2003; Mazanec et al., 1992, 1995); and (3) immune exclusion at the luminal face to prevent epithelial colonization or penetration by microbes (Brandtzaeg and Johansen, 2005). In the secretions, the extracellular domain of the cleaved pIgR remains as bound SC incorporated in SIgA and SIgM after their release to the lumen, or it exists as free SC from cleavage of unoccupied receptor (Brandtzaeg, 1985). Free SC might exert certain innate immune functions in addition to stabilizing the noncovalently bound SC in SIgM (Brandtzaeg, 1975; Phalipon and Corthésy, 2003) (see Chapter 8). Covalently bound SC in SIgA confers extra stability to antibodies of this class and enhances their mucophilic properties (Phalipon and Corthésy, 2003). Serum-derived or locally produced IgG will also contribute to mucosal defense, but antibodies of this class might provide a proinflammatory reinforcement of the epithelial barrier function (Brandtzaeg and Tolo, 1977), although complement regulatory factors produced by mucosal epithelia probably counteracts such adverse effects (Berstad and Brandtzaeg, 1998).

5.4.2. Regulation of J-Chain Expression

Expression of the J-chain is a characteristic of human mucosal B-cells, but there is limited information regarding its local regulatory factors (Brandtzaeg and Johansen, 2005). In the mouse, the J-chain is a commonly used marker for activated B-cells, and its expression is induced by treatments that cross-link the antigen receptor, in addition to various cytokines such as interleukin (IL)-2, IL-5, and IL-6 (Randall et al., 1998; Tigges et al., 1989). Several promoter elements and transcription factors important for antigen-induced *J-chain* transcription in the mouse have been identified. In addition, IL-2-induced chromatin remodeling of the *J-chain* locus is necessary for its expression (Kang et al., 1998; Wallin et al., 1999). The best characterized promoter elements important for regulation of the murine *J-chain* gene have been designated JA, JB, JC, and JE (Lansford et al., 1992; Rinkenberger et al., 1996; Wallin et al., 1998) (Fig. 5.3A). The transcription factors B-cell myocyte enhancer factor 2-related nuclear factor (B-MEF2), PU.1, and upstream stimulatory factor (USF)-1 positively regulate J-chain transcription by binding the JA, JB, and JE elements, respectively (Rao et al., 1998; Shin and Koshland, 1993; Wallin et al., 1999).

B-Cell lineage-specific activator protein (BSAP; encoded by the *pax-5* gene) binds the JC element and thereby represses transcription from the J-chain promoter, apparently by preventing the binding of the activators USF-1 and B-MEF2 (Wallin et al., 1999). PU.1 is also involved in BSAP-mediated repression of J-chain transcription by recruiting the corepressor Groucho-4 (Linderson *et al.*, 2004) (Fig. 5.3A). Upon terminal maturation of murine B-cells, BSAP is subjected to repression by B-lymphocyte-induced maturation protein 1 (Blimp-1) that binds a target site in the *Pax-5* promoter (Lin et al., 2002), thereby relieving the inhibitory effects.

In mice, J-chain expression is only initiated during the last stages of antigen-driven B-cell differentiation, whereas in humans, J-chain transcription is active already during early stages of B-cell development (Bertrand et al., 1996). However, regulation of J-chain expression in the human mucosal immune system has apparently not been studied in any detail. Binding sites for BSAP, USF (E-Box), PU.1, and B-MEF-2 are conserved in the human promoter, but their functionality has not been reported.

5.4.3. IgA Class Switching

Production of antibodies with divergent effector functions is achieved through class-switch recombination (CSR), which occurs at the Ig heavy-chain constant region (C_H) gene locus after antigen activation in secondary lymphoid organs (Fig. 5.3B). This genetic event requires the enzymatic activity of several cell-type-specific proteins and general DNA repair enzymes, and it alters the isotype expression from IgM to downstream loci (reviewed in Manis et al., 2002). There are probably multiple IgA-inducing switch factors, but transforming growth factor (TGF)-β is the best recognized and most studied exogenous stimulus that promotes IgA expression when presented together with B-cell activation signals (Austin et al., 2003). Class switching requires that the DNA be made accessible to the enzymatic switch machinery, and this is achieved by activating transcription from the intronic germline (GL) α1 and GL α2 promoters (for IgA1 and IgA2, respectively).

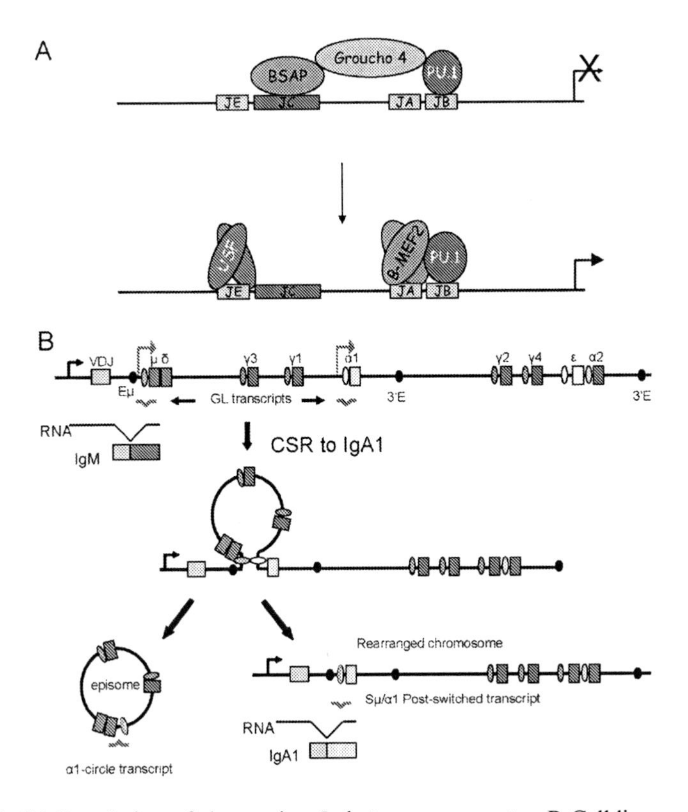

FIG. 5.3. **(A)** Regulation of the murine *J-chain* gene promoter. B-Cell lineage-specific activator protein (BSAP) acts as a repressor of J-chain transcription by recruitment of Groucho-4 and preventing DNA binding of upstream stimulatory factor (USF) and B-cell myocyte enhancer factor 2-related nuclear factor (B-MEF2). In terminally differentiated B-cells, BSAP expression is reduced and the *J-chain* gene is thus poised for activation. **(B)** Class-switch recombination (CSR) from IgM to IgA1 at the Ig heavy-chain locus. The variable region of the Ig heavy chain is assembled from the variable (V), diversity (D), and joining (J) gene segments by VDJ recombination during B-cell development. Transcription across the locus is driven by a promoter upstream of the rearranged VDJ segment (bent black arrow), which facilitates the synthesis of a μ-heavy-chain. The latter associates with a light chain, thereby forming a complete IgM molecule, which is displayed on the surface of the B-cell as part of the B-cell receptor. Secondary isotypes are produced by CSR and is initiated by germline (GL) transcription of the $C\mu$ and the $C\alpha 1$ genes (for switch to IgA1; bent gray arrows). Products of recombination at the C_H locus are the rearranged chromosome and an episomal circle, from which circle transcripts are derived. Cytokines stimulate transcription (bent gray arrows) through the C_H gene and determine the Ig isotype to which the B-cell will switch. The constant region of downstream isotypes are denoted by their corresponding Greek letters and the 3′ enhancers (3′E), which influence GL transcription and thereby the CSR, are indicated below the schematic chromosome.

Multiple transcription factor binding sites required for TGF-β activation of the mouse GL α promoter have been identified (Lin et al., 1992; Shi et al., 2001; Shi and Stavnezer, 1998; Zhang and Derynck, 2000). Transcription from the GL α promoter is also subjected to repression by BSAP, which is critical for B-cell lineage commitment and development but extinguished in terminally differentiated B-cells. Ectopic expression of BSAP repressed GL α transcription (and hence switching to IgA), whereas ectopic expression of this factor did not repress switching to other isotypes (Qiu and Stavnezer, 1998).

Interestingly, BSAP is also a major regulator of J-chain transcription (see above). Nuclear factor (NF)-κB activation is important for CSR, but an NF-κB site in the GL α promoter does not appear to be important for transcriptional activation of the gene (Shi et al., 2001). CD40, LPS, or other NF-κB-activating polyclonal stimuli therefore appear to have a role in CSR distinct from transcriptional activation. Recent data have suggested that soluble factors of the tumor necrosis factor (TNF) superfamily, such as BAFF or APRIL, that are produced by MALT DCs could play such a role (Castigli et al., 2005a, 2005b; Cerutti et al., 2005; Litinskiy et al., 2002).

5.4.4. Role of J-Chain and Heavy Chains in SIgA Formation

The heavy chains of IgA and IgM (the α- and μ-chains, respectively) contain a unique 18-amino-acid carboxyl terminal extension known as the secretory tailpiece (stp) required for appropriate polymerization and J-chain incorporation. A cysteine in the penultimate position in the stp forms a disulfide bridge between two heavy-chain monomeric units or between the heavy chain and the J-chain (Johansen et al., 2000). The interaction between pIgR and the polymeric Igs apparently utilizes motifs both in the heavy chains and in the J-chain (Johansen et al., 2000). Domain-swap chimeras between IgA and IgM have revealed that the carboxyl-terminal domain (Cα3 and Cμ4, respectively) dictates the size of the polymers formed (Braathen et al., 2002). This Ig domain also appears to be critical for pIgR binding because hybrid Ig molecules containing Cμ4 behaved like pentameric IgM and did not interact with rabbit pIgR, whereas hybrid Ig molecules containing Cα3 did (see below)—irrespective of the origin of the other Ig domains. Furthermore, a recombinant IgG with the Cα3 as an extra terminal domain could bind to SC (Chintalacharuvu et al., 2001).

Peptide display technology has been used to investigate the binding site for pIgR on IgA. Peptides with structural homology to two different loops in the Cα3 domain were able to bind to pIgR when present on phage particles (Hexham et al., 1999; White and Capra, 2002). One of these loops is three amino acids longer in IgA than in IgG and was critical for the binding to pIgR because dimeric IgA mutated in this loop was unable to bind to pIgR. Whether the corresponding loop in IgM, which is one amino acid shorter than in IgA (but two amino acids longer than in IgG), is involved in the interaction between pIgR and pentameric IgM remains to be determined.

The J-chain present in pIgA and pentameric IgM makes these two antibody isotypes unique. Production of this short (15 kDa) glycosylated polypeptide can be considered a hallmark of human mucosal PCs (Brandtzaeg et al., 1999c). The J-chain is required both for appropriate polymerization of IgA and IgM and for their affinity for pIgR (Johansen et al., 2000). The importance of this peptide for SIgA and SIgM formation appears to warrant its expression in all mucosal PCs although it is only incorporated into IgA and IgM polymers and is apparently degraded when coexpressed with other isotypes (Mosmann et al., 1978).

J-chain genes are found in lower vertebrates such as sharks (Hohman et al., 2003), which has IgM but no IgA. Curiously, it appears to be deleted in bony fish, although a functional pIgR gene was recently described in blowfish (Hamuro et al., 2007). Although the J-chain is required for appropriate polymerization of IgM and IgA present in serum (Erlandsson et al., 1998), the most distinguishable feature of J-chain$^{-/-}$ mice is their lack of secretory Igs (Hendrickson et al., 1995; Lycke et al., 1999).

The mature human J-chain peptide contains 137 amino acids, including 8 cysteines, which form 3 intrachain disulfide bridges and 2 bridges to the stp cysteines of the α-chain or the μ-chain (Johansen et al., 2000). The role of the J-chain in IgA polymerization and in the pIgA–pIgR interaction has been investigated by mutational analysis (Johansen et al., 2001; Krugmann et al., 1997). Two of the three intrachain disulfide bonds were found to be dispensable for IgA polymerization but essential for SIgA formation. Similarly, even the smallest carboxyl-terminal deletion abrogated SIgA formation although polymerization was intact—suggesting that the structural requirements of the J-chain for pIgR interaction are more stringent than the requirements for pIgA production (Johansen et al., 2001). Thus, although the J-chain might have an origin independent of the mucosal immune system, its structural properties are finely tuned to promote both polymerization of IgA and IgM and their export to the mucosal surface.

5.5. Structure and Ligand-Binding Properties of the pIgR

The pIgR is a type I transmembrane protein. The extracellular region contains five Ig-like domains (D1 to D5), a sixth flexible domain, and an endoproteolytic cleavage site that facilitates the apical epithelial release of SIgA and SIgM. After cleavage, pIgR ectodomain remains associated with SIgA and SIgM as the so-called bound SC.

A strong targeting signal in the cytoplasmic tail ensures basolateral delivery of nascent pIgR. Here, the receptor complexes with its ligands, and both ligand-bound and unoccupied receptors are constitutively endocytosed and transported to the apical domain by vesicular trafficking (Mostov and Kaetzel, 1999). The rate of this transport, known as transcytosis, can be stimulated by ligand binding in some species, but such stimulation apparently does not take place in humans (Brandtzaeg and Johansen, 2001; Giffroy et al., 2001; Natvig et al., 1997). See Chapter 3 for a detailed discussion of pIgR-mediated IgA transcytosis.

The mode of pIgR–pIg interaction and the isotype specificity (or preference) of the pIgR have diverged among mammalian species (Norderhaug et al., 1999a, 1999b; Roe et al., 1999). In humans, the pIgR binds pentameric IgM with a higher affinity than pIgA (Brandtzaeg, 1977; Roe et al., 1999). Furthermore, only D1 is required for pentameric IgM binding, whereas D2 and/or D3 are additionally required for a stable interaction between human pIgA and pIgR (Norderhaug et al., 1999a; Roe et al., 1999).

Although the initial interaction between pIgR and pIgA is somewhat weaker than that of pentameric IgM and pIgR, the former complex is stabilized by a disulfide bridge between D5 and Cα2, causing SC to be covalently bound to IgA in SIgA (Chintalacharuvu et al., 1994). This covalent attachment of SC is important for the stability and mucosal anchoring of SIgA and is, therefore, an important function of the pIgR beyond antibody export (Phalipon and Corthésy, 2003).

In rodents, the affinity of the pIgR for pentameric IgM is significantly reduced, and in rabbits, this interaction is negligible. Surprisingly, the pIgR–pIgA interaction in rabbits relies exclusively on D1, a binding mode reminiscent of that between human pIgR and pentameric IgM (Coyne et al., 1994). The crystal structure of D1 of the human pIgR was recently reported (Hamburger et al., 2004). As previously predicted (Bakos et al., 1993), it conformed to the general Ig fold, but the regions corresponding to complementarity-determining regions (CDRs) showed several unusual features. CDR1 contains a single α-helical turn rather than the extended structure with a hairpin loop seen in most variable domains (Chothia and Lesk, 1987). This region is highly conserved among different species and intimately involved in ligand binding both for pIgA and pentameric IgM (with the human receptor). A reinterpretation of mutational analysis of the binding site (Coyne et al., 1994) revealed that a hydrogen bond between Arg34, the first residue in the β-strand following CDR1, and Asn97 in CDR3 was essential for pIgA binding, probably because this bond is required to stabilize the pIg-binding site. Similarly, Lys35 stabilizes CDR1, whereas several of the neighboring amino acids have the potential to form direct noncovalent bonds with structures in pIgA or pentameric IgM.

A co-crystal of human pIgR D1 and J-chain-containing dimeric IgA (Fc region would probably suffice) will be required to ascertain the identity of interacting amino acids on both the ligand and receptor sides. Such knowledge would also contribute to our understanding of the differential interactions that various molecular forms of IgA might make with the IgA Fc receptor (CD89) (see Chapter 4).

5.6. Epithelial Expression of the pIgR

The pIgR is constitutively expressed by mucosal epithelial cells at quite high levels in the small and large bowels and is also expressed in lacrimal, salivary, and lactating mammary glands as well as in the lung, kidney, pancreas, and endometrium (Brandtzaeg, 1985; Norderhaug et al., 1999b). One notable difference between rodents and rabbits versus humans is that only the former species show significant pIgR expression in hepatocytes (Brandtzaeg, 1985). The consequence of this difference is that rabbits and rodents obtain most of their proximal intestinal SIgA content from hepatobiliary transfer of circulating pIgA, whereas in humans, nearly all of the SIgA (~98%) present in the gut lumen is derived from local production (Conley and Delacroix, 1987). These biological differences might lead to stronger compartmentalization of secretory antibody responses in humans than in rabbits and rodents, which would have to be considered when evaluating experimentally the effect of mucosal vaccines.

In addition to its constitutive expression, pIgR is transcriptionally upregulated by a number of mediators such as cytokines, hormones, and bacterial products that activate innate immunity (reviewed in Johansen and Brandtzaeg, 2004 ; Kaetzel, 2005; Norderhaug et al., 1999b). Several laboratories have studied its transcriptional regulation in human and murine cell lines. Consistent with the constitutive expression of the *PIGR* gene in the human intestinal adenocarcinoma cell line HT-29, histones bound to the promoter are acetylated in the basal state (Bruno et al., 2004 and our unpublished observations)). In humans, a distinct epithelial expression of pIgR is seen at 20 weeks of gestation (Brandtzaeg et al., 1991), but in rodents it is not seen until the time of weaning (Buts and Delacroix, 1985; Jenkins et al., 2003; Martin et al., 1998), which suggests dependency on exogenous stimuli. The mechanistic explanation of this species difference remains unknown.

5.6.1. Transcriptional Regulation of the pIgR

Transient transfection assays have identified several DNA elements that modulate the expression of the *PIGR* gene (see also Chapter 3). Binding of nuclear factors to these regulatory elements has been thoroughly investigated in pIgR-positive cell lines, such as the human colonic adenocarcinoma-derived cell line HT-29. Constitutive *PIGR* transcription depends primarily on an E-box located about 75 bp upstream of the transcription start site (Hempen et al., 2002; Johansen et al., 1998; Martin et al., 1998). The E-box is a target for the basic helix–loop–helix–leucine zipper family of transcription factors, and the members USF-1 and USF-2 have been shown to bind specifically to the *PIGR* E-box *in vitro* (Hempen et al., 2002; Martin et al., 1998). Furthermore, overexpression of USF-1 or USF-2 activated expression of chimeric *PIGR*-luciferase reporter genes (Bruno et al., 2004; Martin et al., 1998). Other DNA elements in the proximal promoter that might modulate constitutive transcription of the *PIGR* gene include an AP-2 element and an inverted repeat motif that binds a yet unidentified nuclear factor (Hempen et al., 2002; Johansen et al., 1998).

In intestinal and airway mucosae, pIgR expression and, hence, pIgA export has been suggested to depend not only on constitutive expression of the receptor but also on transcriptional enhancement by proinflammatory and immunoregulatory cytokines such as TNF, IL-1, interferon (IFN)-γ and IL-4 (reviewed in Kaetzel, 2005). In the female genital tract, lacrimal glands, and lactating mammary glands, however, regulated pIgR expression appears to depend on steroid hormones rather than cytokines (Norderhaug et al., 1999b) (see Chapter 12). In this regard, a glucocorticoid response element has been identified in the first exon of the human *PIGR* gene and an androgen response element in the upstream promoter of the murine *PIGR* gene (Haelens et al., 1999; Li et al., 1999).

The level of pIgR expression is modulated in mucosal disorders of the human intestine. In celiac disease, where the immunopathology is driven by gluten-reactive Th1 cells, epithelial expression of pIgR is upregulated mainly by IFN-γ (Brandtzaeg et al., 1992). In IBD, on the other hand, there is a more mixed pIgR expression despite increased mucosal levels of IFN-γ and TNF, both of which are positive regulators of pIgR expression (Macdonald and Monteleone, 2005) (see below). In regions of overt epithelial dysplasia, pIgR is downregulated—perhaps due to an altered cellular equilibrium between positive-acting and negative-acting transcription factors that bind to the E-box in the pIgR promoter. USF-1 is a ubiquitous transcription factor with high affinity to this element that exerts a positive effect on *PIGR* transcription (Bruno et al., 2004; Martin et al., 1998). However, in dysplastic or neoplastic epithelial cells, Myc-Max heterodimers might predominate. In accordance with this expression pattern of E-box-binding proteins, USF-1 and USF-2 were shown to transactivate a pIgR-derived reporter gene, whereas c-Myc had the opposite effect (Bruno et al., 2004). *In vivo,* downregulation of USF expression in human lung carcinomas was found to be associated with decreased levels of pIgR mRNA (Khattar et al., 2005). Therefore, it appears that *PIGR* belongs to a set of genes with noncanonical E-boxes that are downregulated in IBD and colorectal cancer.

5.6.2. Cytokine-Mediated Upregulation of pIgR Transcription

Transcription of the *PIGR* gene is upregulated slowly in response to IFN-γ, TNF, or IL-4—probably because this upregulation is largely dependent on *de novo* protein synthesis (Fig. 5.4). Interestingly, for both TNF and IL-4, nascent factors and latent, rapidly activated factors cooperate to mediate transcriptional induction of the *PIGR* gene (see below). Importantly, cytokines provide an immunoregulatory link between enhanced SIgA export and increased local pIgA production during low-grade inflammation or infection (Brandtzaeg et al., 1992).

An interferon-stimulated response element (ISRE) in the first exon and two in the proximal promoter region have been implicated in transcriptional activation of the *PIGR* gene in response to IFN-γ and TNF (Kaetzel et al., 1997; Piskurich et al., 1997; Blanch et al., 1999). The exon 1 ISRE has been shown to bind both interferon regulatory factor (IRF)-1 and IRF-2. The expression of these factors is induced by both IFN-γ and TNF, which supports the role of these two proinflammatory cytokines in upregulated pIgR expression

(Brandtzaeg et al., 1992). Nevertheless, the relatively low level of induction mediated by these DNA elements in reporter gene assays (compared with the level of induction of the endogenous gene) suggests that other DNA elements also contribute to regulation mediated by IFN-γ and TNF. However, no other DNA elements have been directly demonstrated to support IFN-γ induction of the *PIGR* gene, although this cytokine might be of primary importance for maintaining adequate levels of pIgR to meet the need for pIgA export.

5.6.3. A Cytokine-Responsive Enhancer in the First Intron of the PIGR Gene

We have identified an enhancer region in the large (5.7 kb) first intron of the human *PIGR* gene essential for induction by IL-4, which also contributes significantly to TNF responsiveness (Johansen and Brandtzaeg, 2004; Schjerven *et al.*, 2000, 2001, 2003, 2004). The minimal enhancer for IL-4-induced transcription is a 250-bp region located more than 4 kb downstream of the transcription start site and containing at least seven target elements for different DNA-binding factors (Johansen and Brandtzaeg, 2004). Central to this enhancer is a binding site for signal transducer and activator of transcription (STAT)-6.

Like the endogenous gene, IL-4 activation of a heterologous reporter gene containing the intronic IL-4-responsive enhancer depended on *de novo* protein synthesis (Schjerven et al., 2000). However, IL-4 stimulation of HT-29 cells caused rapid activation of STAT-6 without the need for new protein synthesis. Therefore, other events than STAT-6 activation and binding to the *PIGR* intronic enhancer must take place for IL-4-induced transcription of the gene to occur. Interestingly, ectopic activation of STAT-6 in HT-29 cells was sufficient to upregulate *PIGR* transcription (Schjerven et al., 2000), demonstrating that other signal transduction pathways from the IL-4 receptor were not required. STAT-6 activation could not activate the pIgR-derived reporter gene in fibroblasts, suggesting that induction of a cell-type-specific factor was required.

Functional analysis of the other six putative regulatory elements revealed minor roles in regulation of *PIGR* transcription in HT-29 cells (Fig. 5.5). Mutation of the binding site for hepatocyte nuclear factor (HNF)-1, a tissue-specific transcription factor (TF) whose expression pattern largely overlaps with that of pIgR, reduced the level of IL-4 activation of the reporter gene from about sixfold to about threefold (Schjerven et al., 2003). HNF-1α and HNF-1β present in nuclear extracts from HT-29 cells could bind the pIgR intronic HNF-1 site, which supported an involvement of this factor in pIgR regulation. The identity of the factors binding to the remaining elements required for optimal IL-4 activation remains unknown. Two of these elements, spaced 40 and 60 bp downstream of the consensus STAT-6 site, constitute STAT-6 half-sites and might be subjected to cooperative binding by multiple STAT-6 homodimers, a mechanism of cooperation demonstrated for other STATs (Xu et al., 1996).

Overlapping the intronic IL-4-responsive enhancer are DNA elements required for TNF-mediated pIgR upregulation. An NF-κB site was crucial

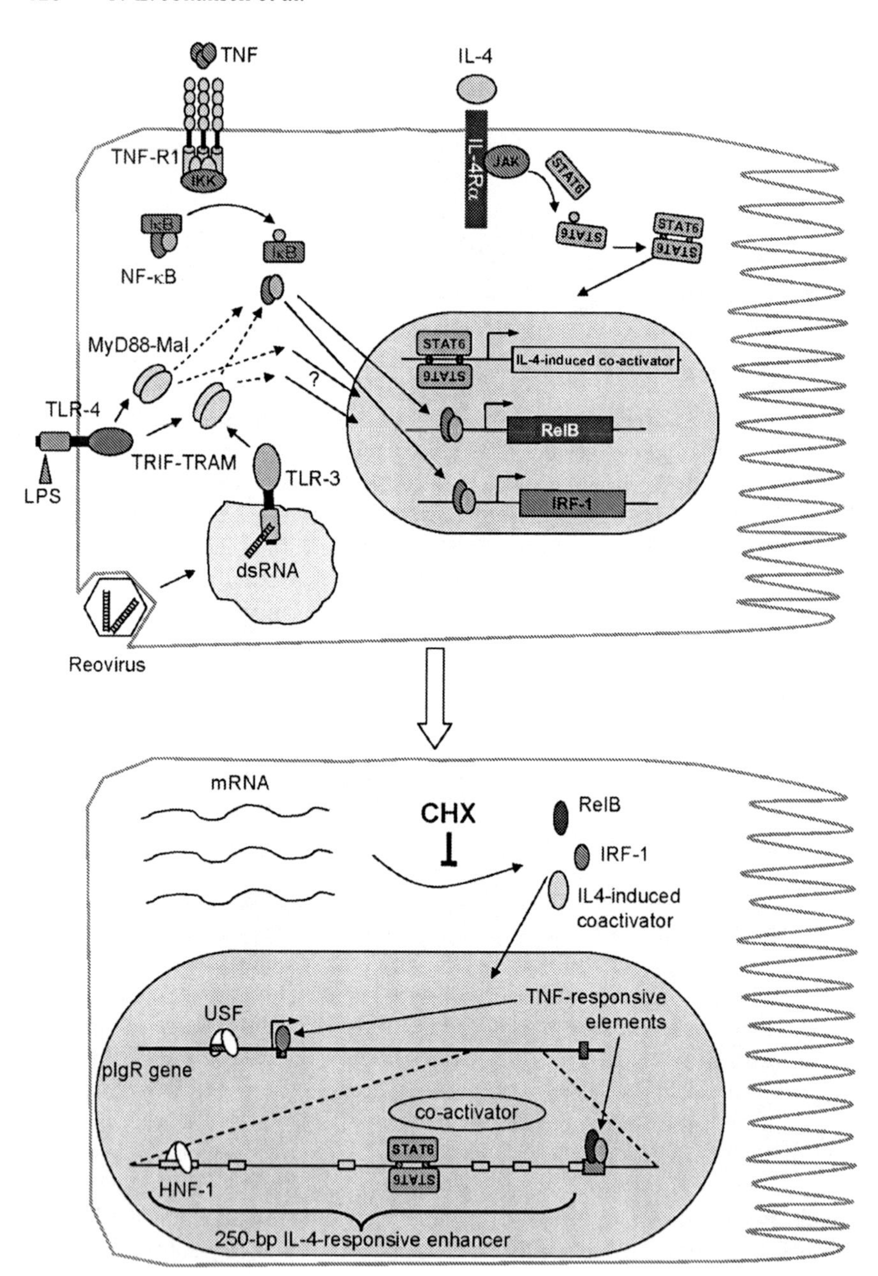
TNF
IL-4
TNF-R1
IKK
IL-4Rα
JAK
STAT6
IκB
NF-κB
MyD88-Mal
IL-4-induced co-activator
?
TLR-4
RelB
LPS
TRIF-TRAM
TLR-3
IRF-1
dsRNA
Reovirus
mRNA
CHX
RelB
IRF-1
IL4-induced
coactivator
TNF-responsive
elements
USF
pIgR gene
co-activator
HNF-1
250-bp IL-4-responsive enhancer

for efficient TNF-mediated *PIGR* gene induction, but adjacent sequences were also required. Nevertheless, induction by TNF was poor when these elements were transferred to a heterologous promoter, because cooperation with promoter-proximal elements was required for the full effect of this cytokine (Schjerven et al., 2001).

Interleukin-4 synergizes with IFN-γ and TNF to activate pIgR expression in HT-29 cells (Denning, 1996; Kvale and Brandtzaeg, 1995; Nilsen et al., 1999; Youngman et al., 1994). The short distance between the IL-4-responsive elements and the NF-κB site provides a possible mechanistic model of its synergy with TNF for induction of *PIGR* transcription. TFs binding to the respective sites might interact while bound to the same strand of DNA. A possible cooperation between the STAT-6 site and the NF-κB site was also supported by the finding that NF-κB activation was required for IL-4-mediated increase of pIgR expression (Ackermann and Denning, 2004).

The kinetics of TNF induction of *PIGR* gene transcription is similar to that of IL-4. Like STAT-6 activation in response to IL-4, NF-κB p50/p65 heterodimer activation in response to TNF is rapid and independent of new protein synthesis (Schjerven et al., 2001). However, whereas overexpression of p65 results in activation of pIgR reporter genes in transient transfection assays, overexpression of RelB provides a stronger and more sustained response (Schjerven et al., 2004). The synthesis of this NF-κB family member is induced by TNF in HT-29 cells, suggesting that it could be one *de novo*-synthesized factor required for TNF-mediated induction of the *PIGR* gene. Its crucial role in TNF-mediated activation of pIgR-derived reporter genes was demonstrated by inhibiting RelB production with short interfering (si)RNA, which reduced the level of activation by a similar degree as mutation of the intronic NF-κB site (Schjerven et al., 2004). Nevertheless,

---

FIG. 5.4. Model for the regulation of human *PIGR* gene transcription by cytokines or bacterial and viral products. **Top** (early events): Ligand binding to the TNF and IL-4 receptors activates receptor-associated intracellular kinases, IκB kinase (IKK) and Janus kinase (JAK), respectively. Phosphorylation of signal transducer and activator of transcription (STAT)-6 by JAKs induces dimerization and migration to the nucleus, whereas phosphorylation of IκB results in the release of active NF-κB that also can migrate to the nucleus. Target genes of STAT-6 and NF-κB, which encode factors required for the protein-synthesis-dependent transcriptional activation of the *PIGR* gene, are thus rapidly activated. Activation of TLR4 and TLR3 by LPS and dsRNA, respectively, results in the activation of NF-κB and other signaling pathways. Whereas TLR3 signaling is through the adaptors Trif-TRAM (independent of MyD88), TLR4 can signal both in a MyD88-dependent and MyD88-independent fashion. These TLRs might exist in a preformed state intracellularly or they might be expressed on the surface. **Bottom** (late events): Cycloheximide (CHX) treatment prevents the translation of mRNA for several rapidly activated transcription factors that in turn activate *PIGR* transcription. Putative targets for nuclear "third messengers" in the TNF or TLR activation of pIgR include RelB and interferon regulatory factor (IRF)-1, whereas the newly synthesized and required coactivator for IL-4 induction remains unknown. STAT-6 also acts directly on a pIgR enhancer element.

p50/p65 is also required for this induction, presumably because this heterodimer activates transcription of the *RELB* gene. It is further possible that the p65/p50 heterodimer is important in initiating *PIGR* transcription in response to TNF, because this cytokine exerts a small effect early on that appears to be independent of *de novo* protein synthesis (Nilsen et al., 1999). Chromatin immunoprecipitation assays will be required to determine the time course of binding to the intronic NF-κB site and whether there is an exchange of factors bound to this site as time progresses from the initial stimuli.

5.6.4. Species Conservation of the Intronic Enhancer

The intronic enhancer in the *PIGR* gene was found to be highly conserved in other species (Fig. 5.5). Alignment among human, mouse, dog, and cow revealed that the intronic enhancer is better conserved than the exon 1 and exon 2 encoding the 5′-UTR (untranslated region) and the leader sequence of pIgR. The highest degree of conservation within the intronic enhancer extended 3′ to the region required for IL-4- and TNF-induced transcriptional response. Although no function has been ascribed to this region, the similarity among species suggests that other regulatory elements are located here.

5.6.5. Regulation of pIgR Expression by Microbes and Microbial Products

The pIgR-expressing epithelium forms an interface between the internal body surface and the external environment and is normally colonized by multiple commensal bacterial species, particularly in the large intestine and oropharynx. In the last few years, it has become apparent that the intestinal microbiota

Fig. 5.5. Species conservation of the intronic enhancer in the *PIGR* gene. **Top:** The region encompassing the first and second exon (E1 and E2) and the first intron of the *PIGR* gene from human, cow, mouse, and dog was aligned with Multalin (http://prodes.toulouse.inra.fr/multalin/multalin.html). A score of 100% was given to positions where all four species showed identity, whereas 75% indicated identity between three species and 25% identity between two species. The average identity over a 60-bp window with a 20-bp shift is plotted against nucleotide position in the alignment. Note that the intronic enhancer is better conserved than both exon 1 and exon 2 encoding 5′-UTR and the leader peptide. Because of gaps in all sequences at different positions in the alignment, the alignment is over 8000 bp in spite of the longest intron (mouse) being only 7565 bp. **Bottom:** Sequence alignment of the 250-bp intronic IL-4-responsive enhancer (and about 30 bases downstream). Elements required for IL-4 induction are indicated in bold and underlined, whereas an NF-κB site required for induction by TNF and via TLR3 or TLR4 is marked with a line above the sequence. Sites "B," "C," "D," "E," and "F" are putative IL-4 regulatory elements, although the proteins that bind to these sites have not yet been identified. Numbers in parentheses indicate approximate reduction in fold induction when the sites were mutated individually. Based on data from Schjerven et al. (2000, 2001, 2003) and unpublished data from the authors' laboratory.

exerts a positive influence on the mucosal immune system. Thus, many host cells express receptors for so-called pathogen-associated molecular patterns (PAMPs), which mediate innate responses to microbial components such as LPS, peptidoglycan, double-stranded (ds)RNA or unmethylated DNA (CpG) (Kopp and Medzhitov, 2003). These receptors, known as pattern recognition

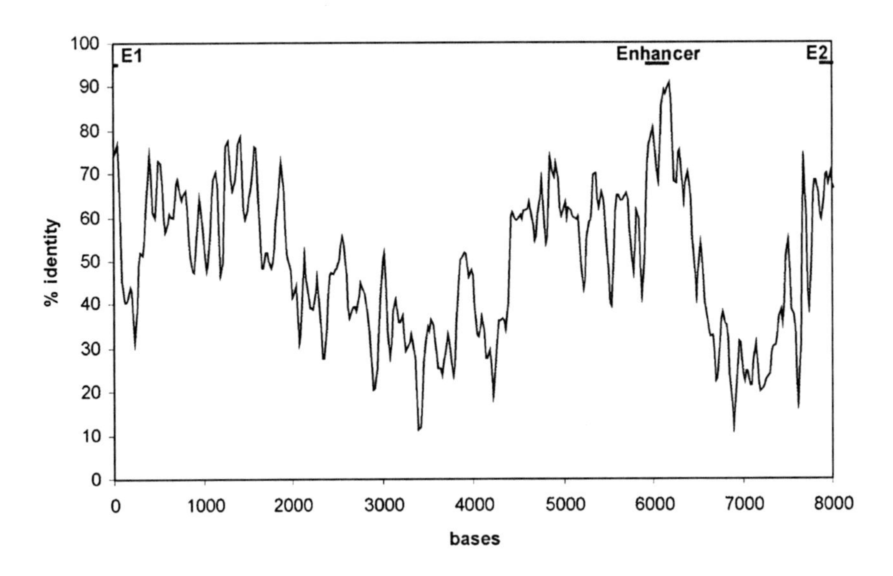


```
Human  CCCTGATTGC TTCCATGTTG GAGCTGA..C TGGTATTTGG CCAAGGCCCC ACTCCAGGCA
  Cow  t-----gca- a--tga-c-- -------cg- ---c------ -t------a- -------agg
  Dog  t--c--g--t a----gt--- -------tg- ---c------ -t---aa--t --c-t--aag
Mouse  a------ca- a----aa--a ag-t.....- a-a...---- -t----.... .--tt---ag

        HNF-1(40%)                "E" (40%)                            120
Human  TTAATCATTG ACCTGGTAAG AAAGGTCTGG GGCAATTAAC CGGACTCAAG AGAGCCTCAA
  Cow  ---------- g-------t- ---a---ca- ---------- ----a----- ----g-----
  Dog  ---------- g---c-a-t- -------cg- ---------- ta--a----- -a-a-g---g
Mouse  --t--t---- --t-a---.- --t-..-at- ---c------ -ccgg-a--a -a-----tt-

                                                  "F" (50%)            180
Human  CGTTGCCATC TGACACCCAT ATTTGT.... ....TCCCAC TCCC..TCCC TTTATTCTTC
  Cow  .-c-----c- c--------- ------ctct acctc----a a---gc---- cca-c-----
  Dog  .-c-----c- ---------- --c---gccc tcctt---ta a---aca-t- c---------
Mouse  .--------- ---------- g-cg--.... .....--tca a-t-..-g-- c-c-c-----

       STAT6 (100%)               "B" (80%)       "C" (60%)            240
Human  CAAAGAACTG CAGAAACAAA ATACCTTTAT ATTCACAAGG AAATAAA.AG GTTCCTTGCT
  Cow  -g-------- g------cg- ---------- c-c------- -------g-- -gg-t-----
  Dog  ---------- g-------g- --g------- c-g------- -------a-- ----------
Mouse  ---------a g--g----g- -a-------- c-ct------ -------ag- ---t-----a

      "D"(70%) NF-κB (60%)                                             300
Human  GGGAAATTCC CCTGCAACAA AGAAATAACC AAGATAAACA ACTCTATTGC CTTTGGCTCA
  Cow  ---------- -tca--ga-- ----gg---- ---------- g----g---- ----------
  Dog  ---g------ -cca-gga-- ---------- ----c----- g--------- ----------
Mouse  ---------- -t-aagg.-- ---------t ---------- ---------- -c-----c--
```

receptors (PRRs), activate intracellular signaling pathways, often converging on NF-κB activation.

The best characterized family of PRRs is the Toll-like receptor (TLR) family and many of its members can be expressed by enterocytes (Cario and Podolsky, 2000). TLR3 is the PRR for dsRNA, a replication intermediate or genome component of many viruses, whereas TLR4 is activated by LPS, a cell wall component of Gram-negative bacteria. It was recently demonstrated that triggering of either TLR3 or TLR4 increased the level of pIgR expression in HT-29 cells, both at the mRNA and the protein levels (Schneeman et al., 2005). Both ligands also increased surface expression of their respective receptors, whereas dsRNA generally resulted in a stronger activation of a proinflammatory gene repertoire than LPS. Activation of pIgR transcription by both ligands appeared to be mediated primarily via the TNF-responsive NF-κB site in intron 1 (Fig. 5.4).

These results are in accordance with the finding that activation via TLRs is not only required for proinflammatory immune defense but might also mediate constitutive homeostatic signaling provided by commensal or mutualistic bacteria (Rakoff-Nahoum et al., 2004). Thus, LPS appears to be a PAMP that can trigger an inflammatory response to invading Gram-negative bacteria, but its presence in the intestinal lumen might be required for maintenance of host cytoprotective genes, including enhanced pIgR expression. In this regard, it is significant that intestinal pIgR mRNA levels were found to be enhanced by colonization of germ-free mice with a prominent Gram-negative component of the commensal gut microbiota *Bacteroides thetaiotaomicron*, (Hooper et al., 2001).

Another bacterial product that has been shown to modulate pIgR expression is butyrate, a fermentation product of dietary fiber and oligosaccharides. Although butyrate had little effect on pIgR expression by itself, it generally enhanced the level induced by proinflammatory cytokines, whereas it reduced the expression mediated by IL-4 (Kvale and Brandtzaeg, 1995). The strongest effect was seen when butyrate was coincubated with IFN-γ and TNF or IFN-γ and IL-1. Whether the effect of butyrate is due to global changes attributable to its activity as an inhibitor of histone deacetylases or to stimulation through butyrate response elements has not been examined.

It would also be interesting to determine what effect butyrate might have on pIgR expression when combined with LPS or other PAMPs. Because signaling through IL-1R, the single cytokine most potentiated by butyrate (Kvale and Brandtzaeg, 1995), and TLRs share many features, it is possible that butyrate could also enhance the effect of LPS on pIgR expression. Indirect support for butyrate regulation of pIgR *in vivo* was provided by a study demonstrating that dietary fructo-oligosaccharides could enhance pIgR expression in mice because these sugars are substrates for bacterial fermentation leading to butyrate production (Nakamura et al., 2004).

Stimulation of HT-29 cells with dsRNA resulted in a stronger proinflammatory response than that induced by LPS, evidenced by higher levels of

IRF-1, IL-8, RANTES, and TNF (Schneeman et al., 2005). Teleologically, this makes sense because whereas Gram-negative bacteria are constituents of the normal microbiota, viruses are not. Thus, although dsRNA and LPS provided a similar degree of pIgR enhancement, the physiological setting of these two PAMPs might be very different: LPS contributing to the "normal" constitutive level of pIgR expression seen in the distal small intestine and colon, and dsRNA regulation possibly resulting in a transient increase in pIgA export during an active intestinal viral infection. Elevated levels of virus-reactive IgA would facilitate clearance of the virus and return to baseline levels of pIgR expression. In support of viral upregulation of pIgR, it was recently shown that reovirus augmented pIgR expression (Pal et al., 2005). Increased pIgR expression was neutralized by preincubating virus particles with specific antiserum, required endosomal acidification, but was achieved with both live virus and inactivated virions.

5.7. Proinflammatory and Anti-Inflammatory Responses in Epithelial Cells

A number of signaling pathways, including TNF, IL-1, and TLRs, converge on the activation of NF-κB (Barton and Medzhitov, 2004; Ben-Neriah and Schmitz, 2004). This transcription factor family is central to host immunity and inflammation (Barton and Medzhitov, 2004; Ben-Neriah and Schmitz, 2004). Transcriptionally active homodimers or heterodimers are retained in the cytosol by an inhibitor (IκB) that is marked for ubiquitin modification and proteasome degradation by IκB kinase (IKK), located downstream of the TNFR, IL-1R, or TLRs in their respective signaling pathways. Interestingly, the IKK complex contains two catalytic subunits: the IKKβ subunit is essential for the activation of NF-κB, whereas the IKKα subunit is required to downregulate the inflammatory response to avoid excessive tissue damage or septic shock (Lawrence et al., 2005). Thus, activation of IKKβ leads to rapid activation and nuclear translocation of p65 (RelA)/p50 heterodimers that activate target genes within minutes or hours. Serine phosphorylation of p65 might be a prerequisite for attracting RNA polymerase II to certain NF-κB-responsive promoters such as the IL-8 promoter (Schmeck et al., 2004). Such phosphorylation can be inhibited by blocking mitogen-activated protein (MAP) kinases or the nuclear kinase MSK1 (Vermeulen et al., 2002). We found that inhibiting MAP kinases greatly diminished the induction of IL-8 expression by TNF while concomitantly increasing the induction of pIgR (our unpublished observations). This suggested that p65 phosphorylation attenuates rather than potentiates activation of *PIGR* transcription by TNF.

An alternative pathway of NF-κB activation has been described where inactive heterodimers of RelB/p100 are processed to the active RelB/p52 form (Senftleben et al., 2001). Several studies have indicated that RelB activates pIgR expression (Schjerven et al., 2001; 2004; Bruno & Kaezel, 2005). Because activation of pIgR expression correlated with *de novo* synthesis of RelB, it is tempting to speculate that a balance in the expression level of proinflammatory and anti-inflammatory genes in epithelial cells is determined by the balance between p65 and RelB.

It will also be interesting to determine if the different roles for IKKα and IKKβ found in macrophages also holds true for enterocytes and to see how the relative activity of these two subunits is regulated by cytokines and PAMPs. A recent study on the effect of long-term exposure of HT-29 cells to TNF found that the levels of RelB and pIgR mRNA remained high for several weeks (Bruno and Kaetzel, 2005). Interestingly, elevated pIgR expression in chronically TNF-stimulated HT-29 cells was also partially due to the increased half-life of pIgR mRNA.

5.8. Concluding Remarks

The intestinal IgA system is essential for mucosal protection and homeostasis. This complex defense system depends on the priming of B-cells in GALT with favored J- chain expression, isotype switching to IgA, and imprinting of chemokine receptors and integrins that will facilitate homing to secretory effector sites. A high constitutive level of epithelial pIgR expression is required to export pIgA from the LP to the lumen, because the pIgR is a sacrificial transport protein that is cleaved to release its cargo. The baseline expression might to some extent rely on PRR-mediated microbial signals from the lumen such as LPS from intestinal Gram-negative bacteria. Furthermore, endogenous proinflammatory or immunoregulatory cytokines as well as signals via PRRs might cause a transient increase in pIgR expression to facilitate enhanced SIgA generation to overcome ongoing infections. Knowledge of how the mucosal IgA system is regulated will benefit vaccination strategies aimed at protecting the vulnerable mucosae. Whereas detailed information on the regulation of pIgR expression has been gathered by a reductionistic approach in *in vitro* model systems, a more difficult holistic approach will be required to understand the mucosal IgA system in the living body. Advances in genomic and proteomic research will greatly facilitate such discoveries.

Acknowledgments. Studies in the authors' laboratories have been supported by the University of Oslo, Rikshospitalet-Radiumhospitalet HF, the Research Council of Norway, the Norwegian Cancer Society, and Anders Jahre's Fund.

References

Ackermann, L. W., and Denning, G. M. (2004). Nuclear factor-κB contributes to interleukin-4- and interferon-dependent polymeric immunoglobulin receptor expression in human intestinal epithelial cells. *Immunology* 111:75–85.

Austin, A. S., Haas, K. M., Naugler, S. M., Bajer, A. A., Garcia-Tapia, D., and Estes, D. M. (2003). Identification and characterization of a novel regulatory factor: IgA-inducing protein. *J. Immunol.* 171:1336–1342.

Bakos, M. A., Kurosky, A., Czerwinski, E. W., and Goldblum, R. M. (1993). A conserved binding site on the receptor for polymeric Ig is homologous to CDR1 of Ig V κ domains. *J. Immunol.* 151: 1346–1352.

Barton, G. M., and Medzhitov, R. (2004). Toll signaling: RIPping off the TNF pathway. *Nat. Immunol.* 5:472–474.

Ben-Neriah, Y., and Schmitz, M. L. (2004). Of mice and men. *EMBO Rep.* 5:668–673.

Bentley, A. M., Durham, S. R., Robinson, D. S., Menz, G., Storz, C., Cromwell, O., Kay, A. B., and Wardlaw, A. J. (1993). Expression of endothelial and leukocyte adhesion molecules interacellular adhesion molecule-1, E-selectin, and vascular cell adhesion molecule-1 in the bronchial mucosa in steady-state and allergen-induced asthma. *J. Allergy Clin. Immunol.* 92+857–868.

Berstad, A. E., and Brandtzaeg, P. (1998). Expression of cell membrane complement regulatory glycoproteins along the normal and diseased human gastrointestinal tract. *Gut* 42:522–529.

Bertrand, F. E., 3rd, Billips, L. G., Gartland, G. L., Kubagawa, H., and Schroeder, H. W., Jr. (1996). The J chain gene is transcribed during B and T lymphopoiesis in humans. *J. Immunol.* 156:4240–4244.

Blanch, V. J., Piskurich, J. F., and Kaetzel, C. S. (1999). Cutting edge: Coordinate regulation of IFN regulatory factor-1 and the polymeric Ig receptor by pro-inflammatory cytokines. *J. Immunol.* 162:1232–1235.

Braathen, R., Sorensen, V., Brandtzaeg, P., Sandlie, I., and Johansen, F. E. (2002). The carboxyl-terminal domains of IgA and IgM direct isotype-specific polymerization and interaction with the polymeric immunoglobulin receptor. *J. Biol. Chem.* 277:42,755–42,762.

Brandtzaeg, P. (1974). Presence of J chain in human immunocytes containing various immunoglobulin classes. *Nature* 252:418–420.

Brandtzaeg, P. (1975). Human secretory immunoglobulin M. An immunochemical and immunohistochemical study. *Immunology* 29:559–570.

Brandtzaeg, P. (1977). Human secretory component—VI. Immunoglobulin-binding properities. *Immunochemistry* 14:179–188.

Brandtzaeg, P. (1985). Role of J chain and secretory component in receptor-mediated glandular and hepatic transport of immunoglobulins in man. *Scand. J. Immunol.* 22:111–146.

Brandtzaeg, P., Baekkevold, E. S., Farstad, I. N., Jahnsen, F. L., Johansen, F. E., Nilsen, E. M., and Yamanaka, T. (1999a). Regional specialization in the mucosal immune system: what happens in the microcompartments? *Immunol. Today* 20:141–151.

Brandtzaeg, P., Carlsen, H. S., and Halstensen, T. S. (2005). The B-cell system in inflammatory bowel disease. In Blumberg, R. S. and Neurath, M. F. (eds.), *Immune Mechanisms in Inflammatory Bowel Disease,* Springer Publishing, New York, pp. 151–170.

Brandtzaeg, P., Farstad, I. N., and Haraldsen, G. (1999b). Regional specialization in the mucosal immune system: Primed cells do not always home along the same track. *Immunol. Today* 20:267–277.

Brandtzaeg, P., Farstad, I. N., Johansen, F. E., Morton, H. C., Norderhaug, I. N., and Yamanaka, T. (1999c). The B-cell system of human mucosae and exocrine glands. *Immunol. Rev.* 171:45–87.

Brandtzaeg, P., Halstensen, T. S., Huitfeldt, H. S., Krajci, P., Kvale, D., Scott, H., and Thrane, P. S. (1992). Epithelial expression of HLA, secretory component (poly-Ig receptor), and adhesion molecules in the human alimentary tract. *Ann. NY Acad. Sci.* 664:157–179.
Brandtzaeg, P., and Johansen, F. E. (2001). Confusion about the polymeric Ig receptor. *Trends Immunol.* 22:545–546.
Brandtzaeg, P., and Johansen, F. E. (2005). Mucosal B cells: Phenotypic characteristics, transcriptional regulation, and homing properties. *Immunol. Rev.* 206:32–63.
Brandtzaeg, P., Nilssen, D. E., Rognum, T. O., and Thrane, P. S. (1991). Ontogeny of the mucosal immune system and IgA deficiency. *Gastroenterol. Clin. North Am.* 20:397–439.
Brandtzaeg, P., and Prydz, H. (1984). Direct evidence for an integrated function of J chain and secretory component in epithelial transport of immunoglobulins. *Nature 311*:71–73.
Brandtzaeg, P., and Tolo, K. (1977). Mucosal penetrability enhanced by serum-derived antibodies. *Nature* 266:262–263.
Bruno, M. E., and Kaetzel, C. S. (2005). Long-term exposure of the HT-29 human intestinal epithelial cell line to TNF causes sustained up-regulation of the polymeric Ig receptor and pro-inflammatory genes through transcriptional and posttranscriptional mechanisms. *J. Immunol.* 174:7278–7284.
Bruno, M. E., West, R. B., Schneeman, T. A., Bresnick, E. H., and Kaetzel, C. S. (2004). Upstream stimulatory factor but not c-Myc enhances transcription of the human polymeric immunoglobulin receptor gene. *Mol. Immunol.* 40:695–708.
Butcher, E. C., and Picker, L. J. (1996). Lymphocyte homing and homeostasis. *Science* 272:60–66.
Buts, J. P., and Delacroix, D. L. (1985). Ontogenic changes in secretory component expression by villous and crypt cells of rat small intestine. *Immunology* 54:181–187.
Campbell, D. J., and Butcher, E. C. (2002). Intestinal attraction: CCL25 functions in effector lymphocyte recruitment to the small intestine. *J. Clin. Invest.* 110:1079–1081.
Cario, E., and Podolsky, D. K. (2000). Differential alteration in intestinal epithelial cell expression of toll-like receptor 3 (TLR3) and TLR4 in inflammatory bowel disease. *Infect. Immun.* 68:7010–7017.
Castigli, E., Wilson, S. A., Garibyan, L., Rachid, R., Bonilla, F., Schneider, L., and Geha, R. S. (2005a). TACI is mutant in common variable immunodeficiency and IgA deficiency. *Nat. Genet.* 37:829–834.
Castigli, E., Wilson, S. A., Scott, S., Dedeoglu, F., Xu, S., Lam, K. P., Bram, R. J., Jabara, H., and Geha, R. S. (2005b). TACI and BAFF-R mediate isotype switching in B cells. *J. Exp. Med.* 201:35–39.
Cerutti, A., Qiao, X., and He, B. (2005). Plasmacytoid dendritic cells and the regulation of immunoglobulin heavy chain class switching. *Immunol. Cell. Biol.* 83:554–562.
Chintalacharuvu, K. R., Tavill, A. S., Louis, L. N., Vaerman, J. P., Lamm, M. E., and Kaetzel, C. S. (1994). Disulfide bond formation between dimeric immunoglobulin A and the polymeric immunoglobulin receptor during hepatic transcytosis. *Hepatology* 19:162–173.
Chintalacharuvu, K. R., Vuong, L. U., Loi, L. A., Larrick, J. W., and Morrison, S. L. (2001). Hybrid IgA2/IgG1 antibodies with tailor-made effector functions. *Clin. Immunol.* 101:21–31.
Chothia, C., and Lesk, A. M. (1987). Canonical structures for the hypervariable regions of immunoglobulins. *J. Mol. Biol.* 196:901–917.

Conley, M. E., and Delacroix, D. L. (1987). Intravascular and mucosal immunoglobulin A: Two separate but related systems of immune defense? *Ann. Intern. Med.* 106:892–899.
Coyne, R. S., Siebrecht, M., Peitsch, M. C., and Casanova, J. E. (1994). Mutational analysis of polymeric immunoglobulin receptor/ligand interactions. Evidence for the involvement of multiple complementarity determining region (CDR)-like loops in receptor domain I. *J. Biol. Chem.* 269:31,620–31,625.
Denning, G. M. (1996). IL-4 and IFN-γ synergistically increase total polymeric IgA receptor levels in human intestinal epithelial cells. Role of protein tyrosine kinases. *J. Immunol.* 156:4807–4814.
Dickinson, B. L., Badizadegan, K., Wu, Z., Ahouse, J. C., Zhu, X., Simister, N. E., Blumberg, R. S., and Lencer, W. I. (1999). Bidirectional FcRn-dependent IgG transport in a polarized human intestinal epithelial cell line. *J. Clin. Invest,* 104:903–911.
Erlandsson, L., Andersson, K., Sigvardsson, M., Lycke, N., and Leanderson, T. (1998). Mice with an inactivated joining chain locus have perturbed IgM secretion. *Eur. J. Immunol.* 28:2355–2365.
Fernandez, M. I., Pedron, T., Tournebize, R., Olivo-Marin, J. C., Sansonetti, P. J., and Phalipon, A. (2003). Anti-inflammatory role for intracellular dimeric immunoglobulin a by neutralization of lipopolysaccharide in epithelial cells. *Immunity* 18:739–749.
Fujioka, H., Emancipator, S. N., Aikawa, M., Huang, D. S., Blatnik, F., Karban, T., DeFife, K., and Mazanec, M. B. (1998). Immunocytochemical colocalization of specific immunoglobulin A with sendai virus protein in infected polarized epithelium. *J. Exp. Med.* 188:1223–1229.
Ghetie, V., and Ward, E. S. (2000). Multiple roles for the major histocompatibility complex class I- related receptor FcRn. *Annu. Rev. Immunol.* 18:739–766.
Giffroy, D., Courtoy, P. J., and Vaerman, J. P. (2001). Polymeric IgA binding to the human pIgR elicits intracellular signalling, but fails to stimulate pIgR-transcytosis. *Scand. J. Immunol.* 53:56–64.
Haelens, A., Verrijdt, G., Schoenmakers, E., Alen, P., Peeters, B., Rombauts, W., and Claessens, F. (1999). The first exon of the human sc gene contains an androgen responsive unit and an interferon regulatory factor element. *Mol. Cell. Endocrinol.* 153:91–102.
Halpern, M. S., and Koshland, M. E. (1970). Noval subunit in secretory IgA. *Nature* 228:1276–1278.
Hamburger, A. E., West, A. P., Jr., and Bjorkman, P. J. (2004). Crystal structure of a polymeric immunoglobulin binding fragment of the human polymeric immunoglobulin receptor. *Structure* 12:1925–1935.
Hamuro, K., Suetake, H., Saha, N. R., Kikuchi, K., and Suzuki, Y., (2007). A Teleost polymeric Ig receptor exhibiting two Ig-like domains transports tetrameric IgM into the skin. *J. Immunonl.* 178:5682–5689.
Hanson, L. A. (1961). Comparative immunological studies of the immune globulins of human milk and of blood serum. *Int. Arch. Allergy Appl. Immunol.* 18:241–267.
Hanson, L. A., and Brandtzaeg, P. (1993). The discovery of secretory IgA and the mucosal immune system. *Immunol. Today* 14:416–417.
Hempen, P. M., Phillips, K. M., Conway, P. S., Sandoval, K. H., Schneeman, T. A., Wu, H. J., and Kaetzel, C. S. (2002). Transcriptional regulation of the human polymeric Ig receptor gene: analysis of basal promoter elements. *J. Immunol.* 169:1912–1921.
Hendrickson, B. A., Conner, D. A., Ladd, D. J., Kendall, D., Casanova, J. E., Corthésy, B., Max, E. E., Neutra, M. R., Seidman, C. E., and Seidman, J. G. (1995). Altered hepatic transport of immunoglobulin A in mice lacking the J chain. *J. Exp. Med.* 182:1905–1911.

Hexham, J. M., White, K. D., Carayannopoulos, L. N., Mandecki, W., Brisette, R., Yang, Y. S., and Capra, J. D. (1999). A human immunoglobulin (Ig)A Cα3 domain motif directs polymeric Ig receptor-mediated secretion. *J. Exp. Med.* 189:747–752.

Hieshima, K., Kawasaki, Y., Hanamoto, H., Nakayama, T., Nagakubo, D., Kanamaru, A., and Yoshie, O. (2004). CC chemokine ligands 25 and 28 play essential roles in intestinal extravasation of IgA antibody-secreting cells. *J. Immunol.* 173:3668–3675.

Hohman, V. S., Stewart, S. E., Rumfelt, L. L., Greenberg, A. S., Avila, D. W., Flajnik, M. F., and Steiner, L. A. (2003). J chain in the nurse shark: implications for function in a lower vertebrate. *J. Immunol.* 170:6016–6023.

Holmgren, J., and Czerkinsky, C. (2005). Mucosal immunity and vaccines. *Nat. Med.* 11:S45–S53.

Hooper, L. V., Wong, M. H., Thelin, A., Hansson, L., Falk, P. G., and Gordon, J. I. (2001). Molecular analysis of commensal host-microbial relationships in the intestine. *Science* 291:881–884.

Iwata, M., Eshima, Y., and Kagechika, H. (2003). Retinoic acids exert direct effects on T cells to suppress Th1 development and enhance Th2 development via retinoic acid receptors. *Int. Immunol.* 15:1017–1025.

Jahnsen, F. L., Haraldsen, G., Aanesen, J. P., Haye, R., and Brandtzaeg, P. (1995). Eosinophil infiltration is related to increased expression of vascular cell adhesion molecule-1 in nasal polyps. *Am. J. Respir. Cell. Mol. Biol.* 12:624–632.

Jenkins, S. L., Wang, J., Vazir, M., Vela, J., Sahagun, O., Gabbay, P., Hoang, L., Diaz, R. L., Aranda, R., and Martâin, M. G. (2003). Role of passive and adaptive immunity in influencing enterocyte-specific gene expression. *Am. J. Physiol.: Gastrointest. Liver Physiol.* 285:G714–G725.

Johansen, F. E., Baekkevold, E. S., Carlsen, H. S., Farstad, I. N., Soler, D., and Brandtzaeg, P. (2005). Regional induction of adhesion molecules and chemokine receptors explains disparate homing of human B cells to systemic and mucosal effector sites: Dispersion from tonsils. *Blood* 106:593–600.

Johansen, F. E., Bosloven, B. A., Krajci, P., and Brandtzaeg, P. (1998). A composite DNA element in the promoter of the polymeric immunoglobulin receptor regulates its constitutive expression. *Eur. J. Immunol.* 28:1161–1171.

Johansen, F. E., Braathen, R., and Brandtzaeg, P. (2000). Role of J chain in secretory immunoglobulin formation. *Scand. J. Immunol.* 52:240–248.

Johansen, F. E., Braathen, R., and Brandtzaeg, P. (2001). The J chain is essential for polymeric Ig receptor-mediated epithelial transport of IgA. *J. Immunol.* 167:5185–5192.

Johansen, F. E., and Brandtzaeg, P. (2004). Transcriptional regulation of the mucosal IgA system. *Trends Immunol.* 25:150-157.

Johansen, F. E., Pekna, M., Norderhaug, I. N., Haneberg, B., Hietala, M. A., Krajci, P., Betsholtz, C., and Brandtzaeg, P. (1999). Absence of epithelial immunoglobulin A transport, with increased mucosal leakiness, in polymeric immunoglobulin receptor/secretory component-deficient mice. *J. Exp. Med.* 190:915–922.

Johansson-Lindbom, B., Svensson, M., Wurbel, M. A., Malissen, B., Marquez, G., and Agace, W. (2003). Selective generation of gut tropic T cells in gut-associated lymphoid tissue (GALT): Requirement for GALT dendritic cells and adjuvant. *J. Exp. Med.* 198:963–969.

Kaetzel, C. S. (2005). The polymeric immunoglobulin receptor: Bridging innate and adaptive immune responses at mucosal surfaces. *Immunol. Rev.* 206:83–99.

Kaetzel, C. S., Blanch, V. J., Hempen, P. M., Phillips, K. M., Piskurich, J. F., and Youngman, K. R. (1997). The polymeric immunoglobulin receptor: structure and synthesis. *Biochem. Soc. Trans.* 25:475–480.

Kaetzel, C. S., Robinson, J. K., Chintalacharuvu, K. R., Vaerman, J. P., and Lamm, M. E. (1991). The polymeric immunoglobulin receptor (secretory component) mediates transport of immune complexes across epithelial cells: A local defense function for IgA. *Proc. Natl. Acad. Sci. USA* 88:8796–8800.

Kang, C. J., Sheridan, C., and Koshland, M. E. (1998). A stage-specific enhancer of immunoglobulin J chain gene is induced by interleukin-2 in a presecretor B cell stage. *Immunity* 8:285–295.

Khattar, N. H., Lele, S. M., and Kaetzel, C. S. (2005). Down-regulation of the polymeric immunoglobulin receptor in non-small cell lung carcinoma: Correlation with dysregulated expression of the transcription factors USF and AP2. *J. Biomed. Sci.* 12:65–77.

Kiyono, H., and Fukuyama, S. (2004). NALT- versus Peyer's-patch-mediated mucosal immunity. *Nat. Rev. Immunol.* 4:699–710.

Kopp, E., and Medzhitov, R. (2003). Recognition of microbial infection by Toll-like receptors. *Curr. Opin. Immunol.* 15:396–401.

Kraehenbuhl, J. P., and Neutra, M. R. (2000). Epithelial M cells: Differentiation and function. *Annu. Rev. Cell. Dev. Biol.* 16:301–332.

Krugmann, S., Pleass, R. J., Atkin, J. D., and Woof, J. M. (1997). Structural requirements for assembly of dimeric IgA probed by site-directed mutagenesis of J chain and a cysteine residue of the α-chain CH2 domain. *J. Immunol.* 159:244–249.

Kunkel, E. J., Campbell, D. J., and Butcher, E. C. (2003). Chemokines in lymphocyte trafficking and intestinal immunity. *Microcirculation* 10:313–323.

Kunkel, E. J., Campbell, J. J., Haraldsen, G., Pan, J., Boisvert, J., Roberts, A. I., Ebert, E. C., Vierra, M. A., Goodman, S. B., Genovese, M. C., et al. (2000). Lymphocyte CC chemokine receptor 9 and epithelial thymus-expressed chemokine (TECK) expression distinguish the small intestinal immune compartment: Epithelial expression of tissue-specific chemokines as an organizing principle in regional immunity. *J. Exp. Med.* 192:761–768.

Kvale, D., and Brandtzaeg, P. (1995). Constitutive and cytokine induced expression of HLA molecules, secretory component, and intercellular adhesion molecule-1 is modulated by butyrate in the colonic epithelial cell line HT-29. *Gut* 36:737–742.

Lansford, R. D., McFadden, H. J., Siu, S. T., Cox, J. S., Cann, G. M., and Koshland, M. E. (1992). A promoter element that exerts positive and negative control of the interleukin 2-responsive J-chain gene. *Proc. Natl. Acad. Sci. USA* 89:5966–5970.

Lawrence, T., Bebien, M., Liu, G. Y., Nizet, V., and Karin, M. (2005). IKKα limits macrophage NF-κB activation and contributes to the resolution of inflammation. *Nature* 434:1138–1143.

Li, T. W., Wang, J., Lam, J. T., Gutierrez, E. M., Solorzano-Vargus, R. S., Tsai, H. V., and Martin, M. G. (1999). Transcriptional control of the murine polymeric IgA receptor promoter by glucocorticoids. *Am. J. Physiol.* 276:G1425–G1434.

Lin, K. I., Angelin-Duclos, C., Kuo, T. C., and Calame, K. (2002). Blimp-1-dependent repression of Pax-5 is required for differentiation of B cells to immunoglobulin M-secreting plasma cells. *Mol. Cell. Biol.* 22:4771–4780.

Lin, Y. C., and Stavnezer, J. (1992). Regulation of transcription of the germ-line Ig α constant region gene by an ATF element and by novel transforming growth factor-β1-responsive elements. *J. Immunol.* 149:2914–2925.

Linderson, Y., Eberhard, D., Malin, S., Johansson, A., Busslinger, M., and Pettersson, S. (2004). Corecruitment of the Grg4 repressor by PU.1 is critical for Pax5-mediated repression of B-cell-specific genes. *EMBO Rep*. 5:291–296.

Litinskiy, M. B., Nardelli, B., Hilbert, D. M., He, B., Schaffer, A., Casali, P., and Cerutti, A. (2002). DCs induce CD40-independent immunoglobulin class switching through BLyS and APRIL. *Nat. Immunol*. 3:822–829.

Lycke, N., Erlandsson, L., Ekman, L., Schèon, K., and Leanderson, T. (1999). Lack of J chain inhibits the transport of gut IgA and abrogates the development of intestinal antitoxic protection. *J. Immunol*. 163:913–919.

Macdonald, T. T., and Monteleone, G. (2005). Immunity, inflammation, and allergy in the gut. *Science* 307:1920–1925.

Mackie, R. I., Sghir, A., and Gaskins, H. R. (1999). Developmental microbial ecology of the neonatal gastrointestinal tract. *Am. J. Clin. Nutr*. 69:1035S–1045S.

Manis, J. P., Tian, M., and Alt, F. W. (2002). Mechanism and control of class-switch recombination. *Trends Immunol*. 23:31–39.

Manz, R. A., Hauser, A. E., Hiepe, F., and Radbruch, A. (2005). Maintenance of serum antibody levels. *Annu. Rev. Immunol*. 23:367–386.

Martin, M. G., Wang, J., Li, T. W., Lam, J. T., Gutierrez, E. M., Solorzano-Vargas, R. S., and Tsai, A. H. (1998). Characterization of the 5'-flanking region of the murine polymeric IgA receptor gene. *Am. J. Physiol*. 275:G778–G788.

Mazanec, M. B., Coudret, C. L., and Fletcher, D. R. (1995). Intracellular neutralization of influenza virus by immunoglobulin A anti-hemagglutinin monoclonal antibodies. *J. Virol*. 69:1339–1343.

Mazanec, M. B., Kaetzel, C. S., Lamm, M. E., Fletcher, D., and Nedrud, J. G. (1992). Intracellular neutralization of virus by immunoglobulin A antibodies. *Proc. Natl. Acad. Sci. USA* 89:6901–6905.

Mazanec, M. B., Nedrud, J. G., Kaetzel, C. S., and Lamm, M. E. (1993). A three-tiered view of the role of IgA in mucosal defense. *Immunol. Today* 14:430–435.

Mestecky, J., and Russell, M. W. (1986). IgA subclasses. *Monogr. Allergy* 19:277–301.

Mestecky, J., Zikan, J., and Butler, W. T. (1971). Immunoglobulin M and secretory immunoglobulin A: Presence of a common polypeptide chain different from light chains. *Science* 171:1163–1165.

Mora, J. R., Bono, M. R., Manjunath, N., Weninger, W., Cavanagh, L. L., Rosemblatt, M., and Von Andrian, U. H. (2003). Selective imprinting of gut-homing T cells by Peyer's patch dendritic cells. *Nature* 424:88–93.

Mosmann, T. R., Gravel, Y., Williamson, A. R., and Baumal, R. (1978). Modification and fate of J chain in myeloma cells in the presence and absence of polymeric immunoglobulin secretion. *Eur. J. Immunol*. 8:94–101.

Mostov, K., and Kaetzel, C. S. (1999). Immunoglobulin transport and the polymeric immunoglobulin receptor. In Ogra, P. L., Mestecky, J., Lamm, M. E. ,Strober, W., Bienenstock, J., and McGhee, J. R. (eds.), *Mucosal Immunology*. Academic Press, San Diego, pp. 181–211.

Nakamura, Y., Nosaka, S., Suzuki, M., Nagafuchi, S., Takahashi, T., Yajima, T., Takenouchi-Ohkubo, N., Iwase, T., and Moro, I. (2004). Dietary fructooligosaccharides up-regulate immunoglobulin A response and polymeric immunoglobulin receptor expression in intestines of infant mice. *Clin. Exp. Immunol*. 137:52–58.

Natvig, I. B., Johansen, F. E., Nordeng, T. W., Haraldsen, G., and Brandtzaeg, P. (1997). Mechanism for enhanced external transfer of dimeric IgA over pentameric IgM: Studies of diffusion, binding to the human polymeric Ig receptor, and epithelial transcytosis. *J. Immunol*. 159:4330–4340.

Nilsen, E. M., Johansen, F. E., Kvale, D., Krajci, P., and Brandtzaeg, P. (1999). Different regulatory pathways employed in cytokine-enhanced expression of secretory component and epithelial HLA class I genes. *Eur. J. Immunol.* 29:168–179.

Norderhaug, I. N., Johansen, F. E., Krajci, P., and Brandtzaeg, P. (1999a). Domain deletions in the human polymeric Ig receptor disclose differences between its dimeric IgA and pentameric IgM interaction. *Eur. J. Immunol.* 29:3401–3409.

Norderhaug, I. N., Johansen, F. E., Schjerven, H., and Brandtzaeg, P. (1999b). Regulation of the formation and external transport of secretory immunoglobulins. *Crit. Rev. Immunol.* 19:481–508.

Odendahl, M., Mei, H., Hoyer, B. F., Jacobi, A. M., Hansen, A., Muehlinghaus, G., Berek, C., Hiepe, F., Manz, R., Radbruch, A., and Dorner, T. (2005). Generation of migratory antigen-specific plasma blasts and mobilization of resident plasma cells in a secondary immune response. *Blood* 105:1614–1621.

Pabst, O., Ohl, L., Wendland, M., Wurbel, M. A., Kremmer, E., Malissen, B., and Forster, R. (2004). Chemokine receptor CCR9 contributes to the localization of plasma cells to the small intestine. *J. Exp. Med.* 199:411–416.

Pal, K., Kaetzel, C. S., Brundage, K., Cunningham, C. A., and Cuff, C. F. (2005). Regulation of polymeric immunoglobulin receptor expression by reovirus. *J. Gen. Virol.* 86:2347–2357.

Persson, C. G., Erjefalt, J. S., Greiff, L., Andersson, M., Erjefalt, I., Godfrey, R. W., Korsgren, M., Linden, M., Sundler, F., and Svensson, C. (1998). Plasma-derived proteins in airway defence, disease and repair of epithelial injury. *Eur. Respir. J.* 11:958–970.

Phalipon, A., and Corthésy, B. (2003). Novel functions of the polymeric Ig receptor: Well beyond transport of immunoglobulins. *Trends Immunol.* 24:55–58.

Piskurich, J. F., Youngman, K. R., Phillips, K. M., Hempen, P. M., Blanchard, M. H., France, J. A., and Kaetzel, C. S. (1997). Transcriptional regulation of the human polymeric immunoglobulin receptor gene by interferon-γ. *Mol. Immunol.* 34:75–91.

Qiu, G., and Stavnezer, J. (1998). Overexpression of BSAP/Pax-5 inhibits switching to IgA and enhances switching to IgE in the I.29 mu B cell line. *J. Immunol.* 161:2906–2918.

Quiding-Jarbrink, M., Nordstrom, I., Granstrom, G., Kilander, A., Jertborn, M., Butcher, E. C., Lazarovits, A. I., Holmgren, J., and Czerkinsky, C. (1997). Differential expression of tissue-specific adhesion molecules on human circulating antibody-forming cells after systemic, enteric, and nasal immunizations. A molecular basis for the compartmentalization of effector B cell responses. *J. Clin. Invest.* 99:1281–1286.

Rakoff-Nahoum, S., Paglino, J., Eslami-Varzaneh, F., Edberg, S., and Medzhitov, R. (2004). Recognition of commensal microflora by toll-like receptors is required for intestinal homeostasis. *Cell* 118:229–241.

Randall, T. D., Heath, A. W., Santos-Argumedo, L., Howard, M. C., Weissman, I. L., and Lund, F. E. (1998). Arrest of B lymphocyte terminal differentiation by CD40 signaling: Mechanism for lack of antibody-secreting cells in germinal centers. *Immunity* 8:733–742.

Rao, S., Karray, S., Gackstetter, E. R., and Koshland, M. E. (1998). Myocyte enhancer factor-related B-MEF2 is developmentally expressed in B cells and regulates the immunoglobulin J chain promoter. *J. Biol. Chem.* 273:26,123–129.

Rinkenberger, J. L., Wallin, J. J., Johnson, K. W., and Koshland, M. E. (1996). An interleukin-2 signal relieves BSAP (Pax5)-mediated repression of the immunoglobulin J chain gene. *Immunity* 5:377–386.

Robinson, J. K., Blanchard, T. G., Levine, A. D., Emancipator, S. N., and Lamm, M. E. (2001). A mucosal IgA-mediated excretory immune system *in vivo*. *J. Immunol.* 166:3688–3692.

Roe, M., Norderhaug, I. N., Brandtzaeg, P., and Johansen, F. E. (1999). Fine specificity of ligand-binding domain 1 in the polymeric Ig receptor: Importance of the CDR2-containing region for IgM interaction. *J. Immunol.* 162:6046–6052.

Salmi, M., and Jalkanen, S. (2005). Lymphocyte homing to the gut: Attraction, adhesion, and commitment. *Immunol. Rev.* 206:100–113.

Schjerven, H., Brandtzaeg, P., and Johansen, F. E. (2000). Mechanism of IL-4-mediated up-regulation of the polymeric Ig receptor: Role of STAT6 in cell type-specific delayed transcriptional response. *J. Immunol.* 165:3898–3906.

Schjerven, H., Brandtzaeg, P., and Johansen, F. E. (2001). A novel NF-κ B/Rel site in intron 1 cooperates with proximal promoter elements to mediate TNF-α-induced transcription of the human polymeric Ig receptor. *J. Immunol.* 167:6412–6420.

Schjerven, H., Brandtzaeg, P., and Johansen, F. E. (2003). Hepatocyte NF-1 and STAT6 cooperate with additional DNA-binding factors to activate transcription of the human polymeric Ig receptor gene in response to IL-4. *J. Immunol.* 170:6048–6056.

Schjerven, H., Tran, T. N., Brandtzaeg, P., and Johansen, F. E. (2004). *De novo* synthesized RelB mediates TNF-induced up-regulation of the human polymeric Ig receptor. *J. Immunol.* 173:1849–1857.

Schmeck, B., Zahlten, J., Moog, K., van Laak, V., Huber, S., Hocke, A. C., Opitz, B., Hoffmann, E., Kracht, M., Zerrahn, J., et al. (2004). *Streptococcus pneumoniae*-induced p38 MAPK-dependent phosphorylation of RelA at the interleukin-8 promotor. *J. Biol. Chem.* 279:53,241–53,247.

Schneeman, T. A., Bruno, M. E., Schjerven, H., Johansen, F. E., Chady, L., and Kaetzel, C. S. (2005). Regulation of the polymeric Ig receptor by signaling through TLRs 3 and 4: Linking innate and adaptive immune responses. *J. Immunol.* 175:376–384.

Senftleben, U., Cao, Y., Xiao, G., Greten, F. R., Krahn, G., Bonizzi, G., Chen, Y., Hu, Y., Fong, A., Sun, S. C., and Karin, M. (2001). Activation by IKKa of a second, evolutionary conserved, NF-κB signaling pathway. *Science* 293:1495–1499.

Shi, M. J., Park, S. R., Kim, P. H., and Stavnezer, J. (2001). Roles of Ets proteins, NF-κB and nocodazole in regulating induction of transcription of mouse germline Ig α RNA by transforming growth factor-β1. *Int. Immunol.* 13:733–746.

Shi, M. J., and Stavnezer, J. (1998). CBF α3 (AML2) is induced by TGF-β1 to bind and activate the mouse germline Ig α promoter. *J. Immunol.* 161:6751–6760.

Shimada, S., Kawaguchi-Miyashita, M., Kushiro, A., Sato, T., Nanno, M., Sako, T., Matsuoka, Y., Sudo, K., Tagawa, Y., Iwakura, Y., and Ohwaki, M. (1999). Generation of polymeric immunoglobulin receptor-deficient mouse with marked reduction of secretory IgA. *J. Immunol.* 163:5367–5373.

Shin, M. K., and Koshland, M. E. (1993). Ets-related protein PU.1 regulates expression of the immunoglobulin J-chain gene through a novel Ets-binding element. *Genes Dev.* 7:2006–2015.

Tarkowski, A., Lue, C., Moldoveanu, Z., Kiyono, H., McGhee, J. R., and Mestecky, J. (1990). Immunization of humans with polysaccharide vaccines induces systemic, predominantly polymeric IgA2-subclass antibody responses. *J. Immunol.* 144:3770–3778.

Tigges, M. A., Casey, L. S., and Koshland, M. E. (1989). Mechanism of interleukin-2 signaling: Mediation of different outcomes by a single receptor and transduction pathway. *Science* 243:781–786.

Tomasi, T. B. (1992). The discovery of secretory IgA and the mucosal immune system. *Immunol. Today* 13:416–418.

Tomasi, T. B., Jr., Tan, E. M., Solomon, A., and Prendergast, R. A. (1965). Characteristics of an immune system common to certain external secretions. *J. Exp. Med.* 121:101–124.

Vermeulen, L., De Wilde, G., Notebaert, S., Vanden Berghe, W., and Haegeman, G. (2002). Regulation of the transcriptional activity of the nuclear factor-κB p65 subunit. *Biochem. Pharmacol.* 64:963–970.

Wallin, J. J., Gackstetter, E. R., and Koshland, M. E. (1998). Dependence of BSAP repressor and activator functions on BSAP concentration. *Science* 279:1961–1964.

Wallin, J. J., Rinkenberger, J. L., Rao, S., Gackstetter, E. R., Koshland, M. E., and Zwollo, P. (1999). B cell-specific activator protein prevents two activator factors from binding to the immunoglobulin J chain promoter until the antigen-driven stages of B cell development. *J. Biol. Chem.* 274:15,959–15,965.

White, K. D., and Capra, J. D. (2002). Targeting mucosal sites by polymeric immunoglobulin receptor-directed peptides. *J. Exp. Med.* 196:551–555.

Woof, J. M., and Mestecky, J. (2005). Mucosal immunoglobulins. *Immunol. Rev.* 206:64–82.

Xu, X., Sun, Y. L., and Hoey, T. (1996). Cooperative DNA binding and sequence-selective recognition conferred by the STAT amino-terminal domain. *Science* 273:794–797.

Yoshida, M., Claypool, S. M., Wagner, J. S., Mizoguchi, E., Mizoguchi, A., Roopenian, D. C., Lencer, W. I., and Blumberg, R. S. (2004). Human neonatal Fc receptor mediates transport of IgG into luminal secretions for delivery of antigens to mucosal dendritic cells. *Immunity* 20:769–783.

Youngman, K. R., Fiocchi, C., and Kaetzel, C. S. (1994). Inhibition of IFN-γ activity in supernatants from stimulated human intestinal mononuclear cells prevents up-regulation of the polymeric Ig receptor in an intestinal epithelial cell line. *J. Immunol.* 153:675–681.

Zhang, Y., and Derynck, R. (2000). Transcriptional regulation of the transforming growth factor-β-inducible mouse germ line Ig α constant region gene by functional cooperation of Smad, CREB, and AML family members. *J. Biol. Chem.* 275:16,979–985.

6
Biological Functions of IgA

Michael W. Russell[1]

6.1. Introduction: The Enigma of IgA

Immunoglobulin A (IgA) is the most enigmatic of immunoglobulins. It is by far the most abundant of human Igs, being present in the blood plasma at concentrations approximating 2–3mg/mL, as well as the dominant isotype in most secretions where its output amounts to some 5–8g/day in adults. Furthermore, its evolutionary origins appear to precede the synapsid–diapsid divergence in tetrapod phylogeny (>300 million years ago) because it is present in both mammals and birds and therefore possibly also in reptiles (reviewed in

[1] Department of Microbiology and Immunology, University at Buffalo, Buffalo, NY 14214, USA.

Peppard et al., 2005); an IgA-like molecule has now been identified in a lizard (Deza et al., 2007). Yet IgA remains inadequately understood, at least with respect to its biological functions. In part this reflects its molecular heterogeneity (Russell et al., 1992) as well as its occurrence in two distinct physiological compartments: the systemic circulation and the mucosal secretions. As detailed in Chapter 1 of this volume, IgA in humans and the anthropoid apes occurs in two subclasses (IgA1 and IgA2) as well as in monomeric, polymeric (dimeric and higher), and secretory forms. Secretory IgA (SIgA) predominates in mucosal secretions, where its activities in the protection of these surfaces against colonization and invasion by pathogens and against injurious toxins are relatively well understood. However, the functions of circulating "serum" IgA remain poorly understood. Despite the metabolic cost involved in synthesizing IgA, which implies that there should be a significant physiological benefit derived from producing it, IgA is also the most readily dispensable isotype, as revealed by IgA deficiency. This is the most common form of primary immunodeficiency occurring in up to 1:400 individuals of Caucasian origin (although less frequently in other populations); yet, affected individuals are usually not severely immunocompromized (Chapter 13). Consideration of all these facts reveals that much about IgA remains perplexing.

Nevertheless, specific IgA antibodies, particularly in mucosal secretions, have been documented to provide significant protection against a variety of toxins, viruses, bacteria, and protozoa in both humans and experimental animal models (Tables 6.1 and 6.2) (reviewed in Russell and Kilian, 2005) (see Chapters 7 and 8). Conventional concepts of SIgA neutralizing toxins and enzymes, inhibiting the adherence of microorganisms to mucosal surfaces, and facilitating their clearance in the mucus layer remain essentially valid. SIgA is well adapted to mucosal protection, because its abundant carbohydrate chains render it hydrophilic and negatively charged. In addition, the secretory component (SC) of SIgA has been shown to protect it from proteolysis (Crottet and Corthésy, 1998), thereby prolonging its survival within enzymatically hostile environments such as the intestinal tract. This is consistent with models showing SC folded around the juxtaposed Fc domains of two IgA monomers that are held together partly by the J-chain polypeptide (Royle et al., 2003), based on computerized predictions derived from other Ig structures and solution studies (Chapter 1). However, a crystallographic model of SIgA, or even of its Fc_2.J.SC segment, is not yet available to confirm this.

6.2. Biological Properties of IgA

6.2.1. Heterogeneity of IgA in Relation to Function

In considering the biological functions of IgA, one should keep in mind that IgA is the most heterogeneous of immunoglobulins, occurring in several molecular forms, subclasses, allotypes, and probably glycoforms, although

TABLE 6.1. Selected examples demonstrating the role of IgA antibody in protective immunity *in vivo*.

Tract	Pathogen	Finding	Ref.
Respiratory	Influenza virus	IgA > IgG protects mice	Renegar and Small (1991)
	Influenza virus	Increased susceptibility in immunized IgA-KO mice	Arulanandam et al. (2001)
	S. pyogenes	SIgA > IgG protects mice	Bessen and Fischetti (1988)
	S. pneumoniae	Human IgA2 > IgA1 protects mice against lethal infection	Janoff et al. (2002)
	Shigella flexneri	SIgA > IgA protects mice	Phalipon et al. (1995, 2002)
Intestinal	Reovirus	IgA mAb protects mice against peroral infection	Kraehenbuhl & Neutra (1992)
	Rotavirus	Reduced protection in orally immunized J-chain–KO mice	Schwartz-Cornil et al. (2002)
	Vibrio cholerae *S. typhimurium*	pIgA protects mice against oral challenge (backpack tumor models)	Winner et al. (1991) Michetti et al. (1992)
	E. coli	Increased frequency of *E. coli* strains in IgA-deficient humans	Friman et al., 2002
	Cholera toxin	Reduced protection in orally immunized J chain-KO mice	Lycke et al. (1999)
	Giardia	IgA protects mice	Langford et al. (2002)
	Cryptosporidium	pIgA protects neonatal mice	Enriquez and Riggs (1998)
Genital	*Chlamydia*	IgA mAb protects mice	Pal et al. (1997)
Ocular	*Acanthamoeba*	pIgA protects hamsters	Leher et al. (1999)

TABLE 6.2. Selected examples of IgA antibody-dependent protective activities *in vitro*.

Pathogen	Finding	Ref.
Influenza virus	pIgA/SIgA > mIgA inhibit hemagglutination inhibition	Renegar et al. (1998)
Sendai and Measles virus	pIgA mAbs inhibit infection in polarized pIgR-expressing cells	Mazanec et al. (1992); Yan et al. (2002)
S. pneumoniae	IgA opsonizes (human PMN)	Janoff et al. (1999), van der Pol et al. (2000)
Neisseria meningitidis	p/mIgA opsonizes (human PMN)	Vidarsson et al. (2001)
Streptococcus mutans	SIgA > mIgA = IgG inhibits adherence to saliva-coated hydroxyapatite	Hajishengallis et al. (1992)
Salmonella enterica	IgA inhibits adhesion/invasion of polarized Hep-2 cells	Iankov et al. (2002)
Clostridium difficile toxin	pIgA > mIgA = IgG neutralizes	Stubbe et al. (2000)
Shigella flexneri LPS	pIgA inhibits NF-κB activation in polarized mouse epithelial cells	Fernandez et al. (2003)

the latter have been little explored. These patterns of heterogeneity also differ between species, implying that its functions might differ in subtle ways also. Most notably, humans along with our close relatives, the great apes, have evolved a novel subclass, IgA1, which has an extended proline-rich and O-glycosylated hinge region (Chapter 1). IgA1 circulates in human blood plasma at a relatively high concentration (2–3mg/mL) and in predominantly monomeric form. In contrast, most other eutherian mammals that have been investigated possess only one IgA isotype that is structurally more akin to IgA2 and that is predominantly dimeric and circulates at concentrations around 0.2 mg/mL, similar to the concentration of human IgA2. Quite what the physiological significance of this difference between humans and other mammals represents is uncertain, but the implication is that the "extra" abundance of monomeric (m) IgA1 fulfills an additional function that remains largely unknown. Steady-state concentrations of Igs in plasma, however, give a misleading impression, because they do not take into account the half-lives of the different isotypes. Whereas IgG has an average circulating half-life of ~21 days, that of IgA1 is 5.9 days and that of IgA2 is 4.5 days (Morell et al., 1973). These authors estimated the synthetic rate of plasma IgA1 as 24 mg/kg/day and that of IgA2 as 4.3 mg/kg/day, whereas for IgG it was ~30 mg/kg/day. Given that IgG consists of four subclasses, it is therefore probable that IgA1 is the most abundantly produced circulating Ig isotype in humans!

Antibody specificities for proteins and polysaccharides have been reported to be differently distributed between the subclasses, such that antibodies to carbohydrates are often preferentially expressed as IgA2; however, the distinction is not absolute (Mestecky and Russell, 1986). Otherwise, few clear functional differences have emerged, as both isotypes bind SC and are represented in SIgA and both bind the IgA Fc receptor on myeloid cells equally. A major difference is that IgA1, but not IgA2, is susceptible to cleavage by bacterial IgA1 proteases (see Sect. 6.2.4), but as these enzymes thereby disrupt its structure and function, the advantage of possessing IgA1 is difficult to grasp and it seems more likely that pathogens have exploited this weakness in human IgA1.

Further heterogeneity arises from the existence of at least two, possibly three (or more), allotypes of human IgA2, which appear to represent constant region domain-swap variants between IgA1 and IgA2 (Chintalacharuvu et al., 1994). Different glycoforms arise from the differential occurrence of N-glycosylation sites between subclasses and allotypes, as well as the presence of O-linked glycans in the hinge region of IgA1 (Mattu et al., 1998). In addition, structural analysis of the N-linked glycans in myeloma proteins has revealed considerable sequence variation, which might also occur in normal IgA (Endo et al., 1994). Because glycans can interact with lectinlike receptors and modulate interactions with Fc receptors and complement components, it is likely that subtle variations in glycosylation affect the functional properties of IgA in ways that have yet to be examined in detail. A particular example, however, is that defective glycosylation of IgA1 might be responsible for IgA nephropathy (Chapter 13).

Some other mammals also possess multiple IgA subclasses. The Lagomorphs (rabbits and their allies) have genes for 13 IgA subclasses, most of which are expressed, although not equally, but their physiological significance remains a mystery (Knight and Rhee, 2005). Limited genetic data also suggest the presence of multiple IgA subclasses in the monotremes and marsupials (Peppard et al., 2005).

Although in humans polymeric (p) IgA constitutes only about 5–10% of total plasma IgA, several studies have shown that the initial serum IgA component of a systemic immune response is pIgA, followed by mIgA (reviewed in Russell et al., 1992). As discussed in Section 6.2.3.2, pIgA is better able to cross-link Fcα receptors on phagocytes and might therefore be advantageous in protection against infection. Otherwise, the functional significance of this, and the maturation of the response toward mIgA, is uncertain. Other studies have revealed that individual IgA-secreting cells first produce pIgA and later mIgA (Moldoveanu et al., 1984). The long-term production of circulating IgA is probably due to plasma cells in bone marrow, which mainly secrete mIgA1 (Hijmans, 1987).

6.2.2. Functions of SIgA at Mucosal Surfaces

6.2.2.1. Inhibition of Adherence

Just as it has become clear that colonizing microorganisms, whether pathogenic or commensal, must adhere to host tissue surfaces, so also has it been recognized that an important function of antibody-mediated defense of the mucosae is the inhibition of microbial adherence. It can readily be envisaged that any isotype of antibody having specificity for microbial adherence epitopes would inhibit their interactions with host receptors. However, SIgA is particularly well suited to this role because of its extensive glycosylation (accounting for 20% by weight), which confers hydrophilicity and negative charge on the molecule. SC, which contains 22% carbohydrate, contributes much of this property to SIgA. The macromolecular bulk of SIgA (400 kDa for dimeric forms) might also be important. Comparative studies of IgG, serum mIgA or pIgA, and SIgA antibodies of the same specificity for antigen have revealed the superiority of the latter in inhibiting adherence of different organisms to host surfaces (Hajishengallis et al., 1992; Phalipon et al., 2002) (see Chapter 8). Even if SIgA antibodies are not specific for adhesin antigens or epitopes, covering the surface of a microbe with a hydrophilic shell might be able to repel attachment of microbes to the surface. Agglutination of microbes is facilitated by the multiple valency (4 in the case of dimers) of SIgA, and this might promote their removal in the mucus stream.

The carbohydrate residues on SIgA might also enable it to inhibit adherence of microorganisms independently of its antibody activity, by binding to carbohydrate-specific adhesins on bacteria. For example, certain strains of *Escherichia coli* possessing mannose-specific type 1 pili can be agglutinated especially by IgA2, which carries mannose-rich glycans, and, as a result, inhibited from adherence to epithelial cells (Wold et al., 1990). In some cases it might be necessary for sialic acid or other terminal residues to be removed to expose the interactive sugar residues (Royle et al., 2003). However, the extent to which these interactions function *in vivo* to inhibit adherence, or conversely to promote it depending on the size of aggregates formed, has been debated (Friman et al., 1996; Liljemark et al., 1979). An interesting if controversial example is represented by *Streptococcus pneumoniae*, which has been proposed to exploit its ability to bind SC to enhance epithelial cell invasion by inducing retrograde reuptake of SC (Brock et al., 2002; Zhang et al., 2000).

6.2.2.2. Interaction with Mucus

An old concept for the biological function of SIgA at mucosal surfaces is that SIgA is arrayed on the surface of the mucus layer to form a kind of immunological "flypaper," allowing entrapped microbes to be swept along with the mucus flow. As appealing as this idea might be, supporting experimental evidence is only tentative. Interactions of SIgA with mucins, possibly involving the mucinlike hinge region of IgA1, or even the formation of disulfide bonds have been proposed (Clamp, 1977), but other more recent studies indicate that SIgA diffuses freely through mucus (Saltzman et al., 1994). Coating of microorganisms with SIgA antibodies reduces their hydrophobicity and facilitates their entrapment in mucus (Edebo et al., 1985; Magnusson and Stjernström, 1982; Phalipon et al., 1995). SIgA is associated with high-molecular-weight fractions of saliva that also contain mucins, and binding of SIgA to mucin MG2 has been described (Biesbrock et al., 1991). Interestingly, when spermatozoa are coated with SIgA, their ability to penetrate cervical mucus is impaired, but treatment with IgA1 protease to remove the Fc and SC regions restores this ability (Bronson et al., 1987). It is likely that cross-linked complexes formed by polyvalent SIgA interfere with sperm mobility, but this is alleviated when IgA1 protease cleaves SIgA1 to monovalent Fab fragments.

6.2.2.3. Neutralization of Toxins, Enzymes, and Viruses

Numerous examples of enzyme and toxin neutralization by SIgA antibodies have been described, including cholera and other enterotoxins (Johnson et al., 1995; Lycke et al., 1987; Stubbe et al., 2000), bacterial neuraminidase, hyaluronidase, or chondroitin sulfatase (Fukui et al., 1973), the glycosyltransferases of *Streptococcus mutans* and *Streptococcus sobrinus*, which are involved in dental caries development (Smith et al., 1985), and bacterial IgA1 proteases (Reinholdt and Kilian, 1995). In some instances, it has been demonstrated that pIgA antibodies, or divalent $F(ab')_2$ fragments of IgA, are more effective than equivalent mIgA or IgG antibodies (Johnson et al., 1995; Norrby-Teglund et al., 2000). This implies that neutralization involves more than simply the blockade of substrate binding or induction of a conformational change that affects enzyme or toxic activity, because this would be independent of isotype, the presence of the Fc region, or molecular conformation. In contrast, the monovalent Fab fragments of IgA1 antibodies to bacterial IgA1 proteases retain inhibitory activity (Gilbert et al., 1983).

Secretory IgA antibodies have been well documented to neutralize a wide variety of viruses. Although in many instances this might be due to inhibition of the binding and uptake of virus by cell receptors, viral replication can be inhibited in various ways, including inhibition of viral uncoating and other intracellular replicative processes depending on the epitope specificity, isotype, and concentration of antibody and the virus and cells involved (Armstrong and Dimmock, 1992; Castilla et al., 1997; Liew et al., 1984). Again, pIgA and SIgA antibodies might be more effective than mIgA—for example, in inhibition of hemagglutination by influenza virus (Renegar et al., 1998). However, pIgA antibodies to gp340 that neutralize the infectivity of Epstein-Barr virus (EBV) for B-cells (via complement receptor CR2) promote infection of colonic carcinoma cells via pIgR, at least in unpolarized cells *in vitro* (Sixbey and Yao, 1992). On the other hand, polarized epithelial cells transport pIgA-complexed EBV from the basal to the apical surface without becoming infected by the virus, both *in vitro* and *in vivo* (Gan et al., 1997). Likewise, IgA antibodies to gp120 might neutralize human immunodeficiency virus (HIV) infection of T-cells (Burnett et al., 1994), whereas IgA antibodies enhance HIV infection of FcαR-expressing monocytes (Janoff et al., 1995; Kozlowski et al., 1995). SIgA or plasma IgA from HIV-1-exposed but uninfected individuals is especially effective in inhibiting the uptake and transcytosis of HIV-1 in epithelial cells (Devito et al., 2000). SIgA antibodies to the ELDKWA epitope of gp41 have been shown to prevent epithelial cell uptake of HIV-1 (Alfsen et al., 2001; Matoba et al., 2004). pIgA antibodies to gp41 can reexport virus to the apical surface of pIgR-expressing epithelial cells (Bomsel et al., 1998), in a process resembling the removal of absorbed antigens (Kaetzel et al., 1991) (see Chapter 7). Furthermore, it is possible for viruses to be neutralized within epithelial cells by pIgA antibodies during their pIgR-mediated transcytosis (Mazanec et al., 1992). For this to occur, vesicles containing replicating virus must interact with the vesicles that carry pIgA across the epithelial cells, and evidence of this has been obtained both *in vitro* and *in vivo* (Feng et al., 2002; Huang et al., 1997; Mazanec et al., 1995; Ruggeri et al., 1998; Yan et al., 2002) (see Chapter 7). In a similar way, the ability of *Shigella* lipopolysaccharide (LPS) to activate nuclear factor (NF-κB) within epithelial cells can be inhibited by pIgA antibody during its pIgR-mediated transcellular transport (Fernandez et al., 2003). The extent to which these mechanisms operate under natural conditions will depend on the presence of IgA antibody-secreting cells of appropriate specificity in the lamina propria adjacent to the site of the viral invasion or LPS uptake. Moreover, as pIgA is transported largely through cells in intestinal crypts, it might not encounter viruses or LPS entering through M-cells or the villi.

6.2.2.4. Inhibition of Antigen Penetration

Uptake of food antigens in the intestine can be inhibited by SIgA antibodies previously developed in response to them (Walker et al., 1972). It has been proposed that this mechanism can be exploited to inhibit the absorption of environmental toxins or carcinogens (Silbart and Keren, 1989). Likewise, absorption of antigen from the airway is inhibited by the simultaneous administration of IgA antibody (Stokes *et al.*, 1975). IgA-deficient subjects show increased absorption of food antigens and formation of circulating immune complexes (Cunningham-Rundles et al., 1981), which might predispose them to greater environmental antigenic challenge as well as increased susceptibility to atopic allergies or autoimmune disease (Stokes et al., 1974). However, more recent studies have shown that allergic patients have increased levels of not only allergen-specific IgE antibodies but also IgA, including SIgA, and IgG antibodies, which are not normally detectable in healthy individuals (Benson et al., 2003; Peebles et al., 2001; Reed et al., 1991). It has been proposed that cleavage of potentially protective IgA1 antibodies by bacterial IgA1 proteases might contribute to this finding (Kilian et al., 1995).

The mechanisms responsible for immune exclusion by SIgA are probably similar to those described earlier, including hydrophilicity, agglutination, and mucus entrapment. It is also possible that the pIgR-mediated transport of pIgA by enterocytes serves to reexport absorbed antigens that become complexed with pIgA antibody in the lamina propria (see Chapter 7). A similar process has been described for the hepatobiliary transport and elimination of antigens complexed to pIgA antibodies (see Sect. 6.2.3.3).

In contrast to the above, it has been proposed that SIgA antibodies can facilitate the uptake of reovirus through the M-cells of Peyer's patches and thereby enhance the mucosal immune response to it (Weltzin et al., 1989) (see Chapter 9). This, however, is difficult to reconcile with another report that describes inhibition of reovirus infection of M-cells by SIgA antibodies (Silvey et al., 2001). Lectinlike IgA receptors on murine M-cells, distinct from pIgR or the asialoglycoprotein receptor, might be responsible, and selective binding of human IgA2 was described (Mantis et al., 2002).

6.2.2.5. Interaction with Innate Defense Factors

Most mucosal secretions contain numerous innate defense factors that are highly effective in killing or inhibiting a broad range of microorganisms (Russell et al., 2005), offering ample opportunity for synergism with SIgA antibodies. Although it has been speculated that SIgA antibodies might target these factors to specific microbes, there is scant molecular evidence for such interactions. The classic example of a SIgA antibody interacting with complement and lysozyme to lyse *E. coli* (Adinolfi et al., 1966) unfortunately proved difficult to reproduce, and it is now thought that undetected contaminants were responsible for the observed effect. The bacteriostatic synergy of lactoferrin and SIgA antibodies (Stephens et al., 1980; Funakoshi et al., 1982) is possibly due to antibody-mediated inhibition of alternative mechanisms of iron acquisition; covalent complexes between lactoferrin and SIgA have been reported (Watanabe et al., 1984). Myeloma IgA1 and IgA2 proteins enhance the ability of lactoperoxidase-H_2O_2-SCN^- to inhibit *S. mutans* metabolism, but this was attributed to stabilization of enzyme activity (Tenovuo et al., 1982). The interaction of SIgA with human secretory leukocyte protease inhibitor has been postulated to have a role in intrauterine defense (Hirano et al., 1999).

6.2.3. Functions of IgA Within the Tissues and Circulation

6.2.3.1. Interactions of IgA with the Complement System

The question of whether IgA activates complement has generated some controversy (reviewed in Russell and Kilian, 2005). It is accepted that IgA does not activate the classical complement pathway (CCP), as IgA molecules do not contain a C1q-binding motif. Statements commonly found in many texts that IgA activates the alternative complement pathway (ACP), however, should be examined by reference to the primary literature and careful consideration of the conditions under which the experiments were performed. Numerous reports describe activation of the ACP by heat-aggregated, chemically cross-linked, or denatured human serum IgA, colostral SIgA, or myeloma proteins or by artificial recombinant IgA antibody constructs produced in transfected cell lines and complexed to haptenated antigen (Boackle et al., 1974; Götze and Müller-Eberhard, 1971; Hiemstra et al., 1988; Valim and Lachmann, 1991). In contrast, human monoclonal and polyclonal IgA antibodies physiologically complexed with antigen do not activate the ACP (Colten and Bienenstock, 1974; Imai et al., 1988; Römer et al., 1980; Russell and Mansa, 1989). However, the same IgA antibodies might activate the ACP when bound to a hydrophobic surface, chemically cross-linked or deglycosylated (Nikolova et al., 1994a; Russell and Mansa, 1989; Zhang and Lachmann, 1994). Interestingly, ACP activation by aggregated IgA depends on the Fc (or Fc.SC) region instead of Fab, which is responsible for ACP activation by IgG (Nikolova et al., 1994a). In heat-aggregated mixtures of human IgG and IgA, C3b fixation by the ACP depends on the proportion of IgG, and C3b becomes covalently coupled to the IgG component (Waldo and Cochran, 1989). Mouse, rat, or rabbit IgA antihapten antibodies complexed with haptenated proteins activate the ACP (Pfaffenbach et al., 1982; Rits et al., 1988; Schneiderman et al., 1990). However, comparison of mouse monoclonal antibodies of different isotypes in studies of complement-mediated solubilization of immune complexes showed that whereas IgM and IgG complexes fix C4 and C3, IgA complexes do not (Stewart et al., 1990). Several factors might contribute to all of these conflicting results. IgA purified by procedures involving exposure to denaturing conditions might be conformationally altered. Recombinant IgA proteins produced in hybridoma or transfectoma cells are often abnormally or incompletely glycosylated. Moreover, heavily haptenated proteins themselves can activate the ACP. Nevertheless, it remains possible that differences in amino acid sequence as well as glycosylation between human and animal IgA result in subtle but important functional differences, including their ability to activate the ACP.

Numerous studies have shown that IgA antibodies can effectively interfere with complement activation mediated by other antibody isotypes. The exacerbation of meningococcal infection in some patients was attributed to the presence of IgA antibody to the capsular polysaccharide which inhibited IgG or IgM antibody-dependent complement-mediated lysis of *Neisseria meningitidis* (Griffiss et al., 1975). Similar findings have been made on the bacteriolysis of *Brucella abortus* (Hall et al., 1971), immune hemolysis of erythrocytes, and the Arthus reaction (Russell-Jones et al., 1980, 1981). Human monoclonal and polyclonal IgA1 antibodies inhibit IgG antibody-dependent CCP activation *in vitro* (Nikolova et al., 1994b; Russell et al., 1989). Interestingly, IgA1 protease-generated Fabα fragments of IgA antibodies also inhibit these IgG- and complement-mediated processes (Jarvis and Griffiss, 1991; Russell et al., 1989). However, the lysis of *N. meningitidis* by IgA antibody to outer membrane proteins (in contrast to antibody to capsular polysaccharide) through a mechanism requiring C1q remains unexplained (Jarvis and Griffiss, 1989; Jarvis and Li, 1997). Some of the most definitive evidence is provided by experiments using recombinant human monoclonal antibodies against meningococcal porin: Whereas IgG antibodies mediated complement-dependent bacteriolysis, IgA with identical antigen-binding domains not only failed to do so but also blocked IgG-dependent bacteriolysis (Vidarsson et al., 2001).

A third pathway of complement activation has been described involving lectins such as the mannose-binding lectin (MBL), which structurally resembles C1q and binds to terminal mannose, fucose, or *N*-acetylglucosamine residues in the presence of calcium. MBL-associated serine proteases, MSP-1 and MSP-2, which are homologous to C1r and C1s, similarly cleave C4 and the remainder of the classical pathway then follows (Møller-Kristensen et al., 2003). pIgA (but not mIgA) can bind MBL and initiate this pathway (Roos et al., 2001). Although the full physiological significance of the lectin pathway has yet to be elucidated, it might explain some of the controversy surrounding complement activation by IgA.

It can be generally concluded that native human IgA antibodies when complexed with antigens have little to no ability to activate complement by either the CCP or ACP. Within the mucosae, where IgA is abundant, the ability to resist complement activation and the consequent inflammatory reactions might help to maintain the integrity of the mucosal barrier. However, some findings remain to be explained, and it is possible that significant differences exist between IgA from humans and other species. In addition, it has been well demonstrated that denatured, conformationally altered, deglycosylated, or chemically modified IgA can activate the ACP. Whether equivalent changes can occur in IgA due to abnormal synthesis or even microbial attack and thereby initiate activation of complement and consequent pathological lesions is an interesting speculation. Support for this notion, however, might be found in IgA nephropathy, in which it is proposed that defective glycosylation of IgA1 leads to its deposition in the renal glomeruli and activation of the ACP (see Chapter 13).

6.2.3.2. Interactions of IgA with Leukocytes (Fig. 6.1)

6.2.3.2.1. Neutrophils and Macrophages

Several early studies, mostly using myeloma IgA proteins or colostral SIgA, indicated that IgA was inhibitory to phagocytosis, bactericidal activity, or chemotaxis by neutrophils or macrophages (reviewed in Kilian et al., 1988).

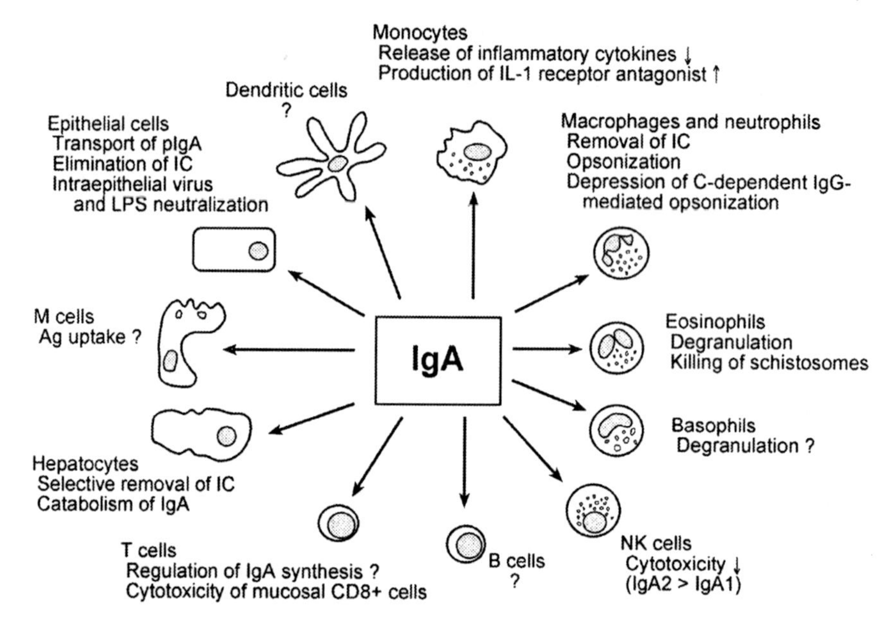

FIG. 6.1. Interactions of IgA with various cell types. Human cells of the myeloid lineage (neutrophils, eosinophils, monocytes, and macrophages) express FcαRI (CD89) through which they can be activated by serum IgA, especially in polymeric form or when aggregated or complexed with antigen. Binding of SIgA (at least by neutrophils) requires Mac-1 as a coreceptor. Expression of FcαRI varies according to the cell type, its state of differentiation or activation, and location. Signal transduction and hence cellular responses depend on association of FcαRI with FcRγ chain. Eosinophils bind and respond especially well to SIgA (or SC), but the nature of the receptor is not clear. Basophils are also reported to degranulate in response to SIgA. Fcα/μR occurs on T- and B-lymphocytes, but its physiological function remains uncertain. The interaction of IgA with NK cells mighty be mediated by lectinlike receptors for carbohydrate determinants. DCs variably express FcαRI or another receptor for IgA, but their response to IgA is controversial. Epithelial cells, including hepatocytes of certain nonprimate animal species, express pIgR, which binds pIgA and thereby transports it to the apical surface where it is released as SIgA. Antigens complexed to pIgA antibodies can be similarly transported by hepatocytes into bile or by intestinal epithelial cells into the gut lumen. pIgA antibodies might also be able to interfere with intracellular viral replication or inhibit responses to LPS within epithelial cells. Enterocytes and M-cells are also reported to bind IgA by other, possibly lectinlike, receptors. Serum IgA is catabolized by hepatocytes, probably after uptake mediated by the asialoglycoprotein receptor. For further details, see text. Reproduced with permission from Russell and Kilian (2005), © Elsevier Inc.

However, it is now known that a receptor for the Fc of IgA, FcαR (CD89), is expressed on myeloid cells and can mediate phagocytosis and other cellular responses to complexed IgA (reviewed in Monteiro and van de Winkel, 2003) (see Chapter 4). The level of expression of FcαR varies between cell types and their activation state. For example, FcαR is upregulated on gingival exudative neutrophils (Fanger et al., 1983; Yuan et al., 2000) but is absent from macrophages isolated from the gut mucosa (Smith et al., 2001). Several activating agents such as phorbol esters, bacterial LPS, or even IgA itself, as well as tumor necrosis factor (TNF)-α, interleukin (IL)-8, and granulocyte monocyte-colony stimulating factor (GM-CSF) enhance the surface expression of FcαR on neutrophils, whereas interferon (IFN)-γ and transforming growth factor (TGF)-β downregulate it (Gessl et al., 1994; Hostoffer et al., 1994; Maliszewski et al., 1985; Nikolova and Russell, 1995; Reterink et al., 1996; Weisbart et al., 1988; Shen et al., 1994). pIgA is more effective than mIgA in cross-linking FcαR (Stewart et al., 1994); indeed, plasma mIgA concentrations are sufficient to saturate FcαR, but in the absence of cross-linking, the cells are not triggered. Association of FcαR with the common FcRγ chain is necessary for signal transduction (Honorio-França et al., 2001; Van Egmond et al., 1999), but its expression varies between different cell types, their state of activation or differentiation, and location (Hamre et al., 2003). In the absence of the signaling FcRγ chain, IgA might be taken up and recycled without inducing inflammatory responses (Launay et al., 1999). Binding of SIgA appears to require Mac-1 (CD11b/CD18) as an accessory receptor (Van Spriel et al., 2002). Thus, numerous factors are involved in determining whether myeloid cells respond to IgA. It is remarkable that mice lack a CD89 homologue, implying that differences exist in the physiological functions of IgA in mice and humans. However, it has been found that galectin-3 can substitute as an IgA receptor, at least in mediating IgA antibody-dependent protection against *Mycobacterium tuberculosis* in a mouse model (Reljic et al., 2004).

Several studies have reported that polyclonal human serum IgA or monoclonal IgA antibodies can promote phagocytic uptake and killing of bacteria such as *S. pneumoniae* or *N. meningitidis* by human neutrophils *in vitro* (Janoff et al., 1999; Van der Pol et al., 2000; Vidarsson et al., 2001). Dependence on complement was variable in these experiments and its precise role is unclear: Nonclassical pathways and complement receptors CR1 or CR3 were implicated, and preactivation of neutrophils by C5a diminished the dependence on complement. IgA-mediated protection against infection has been shown *in vivo* using transgenic mice that express human FcαR, presumably involving opsono-phagocytic mechanisms (Hellwig et al., 2001; Van der Pol et al., 2000; Van Egmond et al., 2000).

In contrast to opsono-phagocytic activation, reports that IgA could downregulate the inflammatory response of LPS-stimulated human monocytes (Wolf et al., 1994, 1996) provoked renewed interest in the concept of IgA as an anti-inflammatory isotype. However, subsequent studies revealed FcαR-dependent signal transduction through Src-family kinases, similar to the pathways induced by other γ-chain-dependent Fc receptors (Gulle et al., 1998). In human alveolar macrophages, pIgA or SIgA downregulates the respiratory burst induced by LPS through inhibition of the ERK1/2 pathway but enhance the response to phorbol ester in association with ERK1/2 phosphorylation and enhance TNF-α release by an ERK1/2-independent mechanism (Ouadrhiri et al., 2002). Studies on the partitioning of ligand-bound FcαRI into membrane lipid rafts with recruitment of tyrosine kinases have suggested that there are temporally regulated signaling events associated with IgA binding (Lang et al., 2002).

6.2.3.2.2. Eosinophils and Basophils

The interaction of IgA with other types of granulocyte (i.e., eosinophils and basophils) and hence its role in defense against parasites and in allergic reactions deserve more attention. A highly glycosylated isoform of FcαRI is found on eosinophils (Decot et al., 2005; Monteiro et al., 1993), and SIgA strongly stimulates the degranulation of these cells (Abu-Ghazaleh et al., 1989). IgA also mediates the killing of schistosomes by eosinophils (Dunne et al., 1993; Grezel et al., 1993). A distinct 15-kDa receptor for SIgA and SC was described on eosinophils (Lamkhioued et al., 1995) although its function and significance remain uncertain. However, an immunoregulatory role has been suggested as SIgA can inhibit IL-2 and IFN-γ secretion and induce that of IL-10 by eosinophils (Woerly et al., 1999). IgA antibodies were long ago shown to inhibit IgE-mediated hypersensitivity (Ishizaka et al., 1963; Russell-Jones et al., 1981). In contrast, the more recent finding that SIgA can induce basophil degranulation (Iikura et al., 1998) suggests a possible role in allergic reactions.

6.2.3.2.3. Lymphocytes

The presence of IgA receptors on lymphocytes has been somewhat controversial, and despite several earlier reports of IgA binding by T- or B-cells, no receptors were defined. However, it is possible that the receptor for IgA and IgM (Fcα/μR) recently found on human and murine lymphocytes accounts for some of those observations (reviewed in Shibuya and Honda, 2006). The transferrin receptor (CD71) also serves as a receptor for IgA1 on B-cells and epithelial cells (Moura et al., 2001). The ability of natural killer (NK) cells to bind IgA, especially IgA2, might be carbohydrate dependent (Komiyama et al., 1986; Mota et al., 2003), but the physiological significance of the cellular inhibition that resulted is uncertain. It currently remains unclear whether IgA has direct functional effects on lymphocytes.

6.2.3.2.4. Dendritic Cells

FcαR has been found on human interstitial dermal and gingival dendritic cells (DCs) as well as monocyte-derived DCs *in vitro*, but not on Langerhans cells (Geissmann et al., 2001). Triggering of monocyte-derived DCs with pIgA complexes induces their functional activation, endocytosis of the complexes, and the production of IL-10, suggesting that interstitial DCs might be able to take up and process IgA-opsonized antigens (Pasquier et al., 2004). Conversely, Heystek et al. (2002) found that FcαR expression was greatly diminished upon differentiation of monocytes into DCs, whereas monocyte-derived DCs bound SIgA independently of FcαR but were not activated as a result. These authors suggested that immature DCs might serve to modulate immune responses to SIgA-complexed antigens at mucosal surfaces. As it has become clear that DCs represent highly variable and plastic types of cell, their expression of receptors for and responses to IgA might also be highly variable and dependent on the precise type of DC, their location, and state of maturity or activation. If intestinal DCs, which protrude into the lumen between the epithelial cells (Mowat, 2005; Rimoldi and Rescigno, 2005), express IgA receptors, then it might be speculated that intestinal IgA antibodies will influence the immune responses initiated by these cells.

6.2.3.3. Interactions of IgA with Epithelial Cells (Fig. 6.1)

Polymeric IgA, along with IgM, interacts with mucosal epithelial cells that express pIgR on their basolateral surfaces. As a result, pIgA is endocytosed and transported apically to be released into the lumen covalently coupled to SC as SIgA (see Chapter 3). In addition to serving as the mechanism for producing SIgA, this process has other functional consequences, some of which have already been discussed when considering the functions of SIgA at mucosal surfaces. The finding that certain animal species, among them rats, mice, and rabbits, have pIgR expressed on hepatocytes that can therefore transport pIgA directly from the blood into bile led to the demonstration that pIgA antibodies can mediate elimination of bound antigens from the circulation by hepatobiliary transport (Peppard et al., 1981; Russell et al., 1981; Socken et al., 1981). This has been proposed as a means of noninflammatory disposal of complex microbial antigens that cannot be broken down in mammalian tissues, or of food antigens absorbed in the intestine (Brown et al., 1984; Russell et al., 1983). However, as human hepatocytes do not express pIgR, this process does not occur in humans (Tomana et al., 1988). Nevertheless, other receptors, such as the asialoglycoprotein receptor, which mediates the uptake of desialylated glycoproteins for catabolism by the liver (Mestecky et al., 1991), and possibly also membrane galactosyltransferase, might contribute to a functionally similar transport process on a smaller scale (Tomana et al., 1993). A portion of desialylated IgA, together with any bound antigen, taken up by these receptors might become missorted into the biliary secretory pathway instead of the lysosomal degradative pathway (Schiff et al., 1984, 1986).

6.2.4. Counteraction of IgA by Bacterial IgA Proteases

One measure of the significance of IgA in protection of the mucosae in humans might be the frequency with which bacterial pathogens have developed countermeasures specific for human IgA. A classic example of this is IgA1 protease, which is expressed by numerous significant human mucosal pathogens but not by closely related nonpathogenic species (Table 6.3). Other species of bacteria produce IgA-binding proteins that have been proposed to interfere with

TABLE 6.3. Bacterial IgA1 proteases.

Homology group	Species	Enzyme type	Cleavage site	
I	*Haemophilus influenzae*	Serine protease	P231-S232	(type 1)
			P235-T236	(type 2)
	H. influenzae biogroup aegyptius		P235-T236	
	H. aegyptius		P231-S232	
	H. parahaemolyticus		P235-T236	
	Neisseria meningitidis		P237-S238	(type 1)
			P235-T236	(type 2)
	N. gonorrhoeae		P237-S238	(type 1)
			P235-T236	(type 2)
II	*Streptococcus pneumoniae*	Metalloprotease	P227-T228	
	S. sanguis		P227-T228	
	S. gordonii		P227-T228	
	S. oralis		P227-T228	
	S. mitis		P227-T228	
	Gemella haemolysans		P227-T228	
?	*Prevotella* sp	Cysteine protease	P223-S224	
?	*Capnocytophaga* sp	Metalloprotease	P223-S224	
?	*Ureaplasma urealyticum*	Serine protease	P235-T236	

functional protective mechanisms exerted by IgA antibodies, although these are not well understood (Kilian and Russell, 2005).

Molecular characterization of the IgA1 proteases and their catalytic mechanisms has revealed three distinct classes of enzyme: serine proteases (*Haemophilus, Neisseria, Ureaplasma*), metalloproteases (*Streptococcus, Capnocytophaga*), and cysteine proteases (*Prevotella*), as well as several different genetic origins (Table 6.3). This means that the same unique enzymatic activity has evolved independently as many as five times in bacterial phylogeny. Yet, all IgA1 proteases show the same effect of cleaving human IgA1 specifically at one or other of the proline-serine or proline-threonine bonds in the hinge region, yielding Fabα and Fcα fragments, which are not further degraded by these enzymes. mIgA, pIgA, and SIgA forms are all susceptible to cleavage, and the Fabα fragments retain antigen-binding activity (Mansa and Kilian, 1986). Apart from the homologous IgA1 proteins of other anthropoid apes (Cole and Hale, 1991; Qiu et al., 1996), no other species of IgA, or human IgA2, is cleaved by these proteases. Investigation of the role of IgA1 proteases as virulence factors is hampered by this exquisite specificity, which precludes the use of conventional animal models. Moreover, many of the bacteria that produce them are exclusively human pathogens. However, indirect inferential evidence suggests that IgA1 proteases contribute to the virulence of the organisms that produce them. One hypothetical mechanism (Kilian and Reinholdt, 1987) concerns the three species implicated in bacterial meningitis, *Haemophilus influenzae, Neisseria meningitidis,* and *S. pneumoniae*, which all produce IgA1 proteases. IgA1 anticapsular antibodies, which might occur in primary infections of children as a result of prior exposure to cross-reacting antigens, are cleaved by IgA1 protease to Fabα fragments, which facilitate instead of preventing invasion of the organisms. Moreover, these Fabα fragments block access of other functionally intact antibodies of the same or different isotype. However, if the IgA1 protease elicits an inhibitory antibody response against itself concomitant with the induction of anticapsular IgA1 antibodies, then protection might be achieved. Several items of evidence lend indirect support to this hypothesis. The IgA1 proteases particularly of *H. influenzae* and to a lesser extent of *N. meningitidis* show extensive antigenic variation, which permits escape from antibody-mediated inhibition (Lomholt et al., 1995). Fabα fragments have been shown to be ineffective in inhibiting adherence, and to inhibit complement activation mediated by IgG antibodies (and resultant bacteriolysis) to the same antigen (Hajishengallis et al., 1992; Janoff et al., 2002; Jarvis and Griffiss, 1991; Reinholdt and Kilian, 1987; Russell et al., 1989; Tyler and Cole, 1998). In addition, adherence of *S. pneumoniae* to epithelial cells is enhanced in the presence of IgA1 antibodies to capsular polysaccharide that have been cleaved by IgA1 protease (Weiser et al., 2003). Analysis of virulence genes in *S. pneumoniae* has also revealed an association of the *iga* gene with pathogenicity in a mouse model, even though murine IgA is not susceptible to cleavage by IgA1 protease, suggesting the possibility of additional activities of IgA1 protease (Polissi et al., 1998). However, it is possible that a paralogous "*iga*" gene was involved in the observed association (Kilian and Russell, 2005).

6.3. IgA and Homeostasis

Immunoglobulin A has long been thought of as a benign form of antibody that lacks the dramatic functional activities commonly associated with other isotypes. Its relatively noninflammatory nature is undoubtedly important at mucosal surfaces where the immune system is continuously exposed to an abundance of microorganisms both pathogenic and harmless, as well as foreign macromolecules. IgA antibodies can form immune complexes with antigens without necessarily eliciting inflammatory reactions that can inflict collateral damage on the host tissues (Brandtzaeg and Tolo, 1977) (see Chapter 10). Yet, it has become clear that IgA can mediate potent responses in cells that possess receptors for it. Key to understanding the physiological role of IgA is the recognition that it is heterogeneous in structure, comprising monomeric, polymeric, and secretory forms, two subclasses (in humans), and possibly several glycoforms. Moreover, these variants are differentially distributed in two distinct compartments (the systemic circulation and mucosal secretions) and are produced with different temporal kinetics. In addition, the expression of cellular IgA receptors is subject to regulation. SIgA probably has little opportunity to interact with either complement or phagocytes at mucosal surfaces, because leukocytes quickly disintegrate in the hypotonic environment of most secretions and a fully functional complement system is not usually present. However, if the mucosal barrier is breached, microorganisms will become exposed to an environment dominated by submucosal pIgA secreted by resident plasma cells as well as SIgA. Infiltrating neutrophils expressing FcαR will be capable of responding to IgA-opsonized organisms and thereby providing immune defense, but the IgA might also afford damage-limiting capability by regulating inflammatory responses. In this connection, it is noteworthy that inflammatory bowel diseases appear to involve increased IgG relative to IgA production in the affected tissues (Baklien et al., 1977).

Observations that preexisting mucosal antibody responses can interfere with the use of live bacterial or viral vectors for the delivery of mucosal vaccines (e.g., Attridge et al., 1997; Roberts et al., 1999; Svennerholm et al., 1981) have suggested that SIgA antibody might regulate the initiation of the immune response. This has been demonstrated in infant mice suckled on immune foster mothers: The neonates' antibody responses to reovirus were suppressed by maternal milk antibodies (Kramer and Cebra, 1995). It is likely that the low level of mucosal antibodies elicited by commensal bacteria prevents overstimulation of the immune system without actually eliminating harmless organisms that are desirable for host survival (Shroff et al., 1995). IgA antibodies to human leukocyte antigen (HLA) class I have been reported to promote the survival of kidney allografts (Koka et al., 1993), but the mechanism underlying this apparent suppression of immune responsiveness is not known. It seems likely that further work on the interaction of SIgA with epithelial cells, M-cells, or DCs will elucidate these observations.

Acknowledgments. I thank numerous colleagues in several institutions around the world for their interactions and discussions, which have been instrumental in developing the concepts expressed in this chapter. Studies in the author's laboratory have been supported by USPHS grants from the National Institute of Dental and Craniofacial Research and the National Institute of Allergy and Infectious Diseases.

References

Abu-Ghazaleh, R. I., Fujisawa, T., Mestecky, J., Kyle, R. A., and Gleich, G. J. (1989). IgA-induced eosinophil degranulation. *J. Immunol.* 142:2393–2400.

Adinolfi, M., Glynn, A. A., Lindsay, M., and Milne, C. M. (1966). Serological properties of γA antibodies to *Escherichia coli* present in human colostrum. *Immunology* 10:517–526.

Alfsen, A., Iniguez, P., Bouguyon, E., and Bomsel, M. (2001). Secretory IgA specific for a conserved epitope on gp41 envelope glycoprotein inhibits epithelial transcytosis of HIV-1. *J. Immunol.* 166:6257–6265.

Armstrong, S. J., and Dimmock, N. J. (1992). Neutralization of influenza virus by low concentrations of hemagglutinin-specific polymeric immunoglobulin A inhibits viral fusion activity, but activation of the ribonucleoprotein is also inhibited. *J. Virol.* 66:3823–3832.

Arulanandam, B. P., Raeder, R. H., Nedrud, J. G., Bucher, D. J., Le, J. H., and Metzger, D. W. (2001). IgA immunodeficiency leads to inadequate Th cell priming and increased susceptibility to influenza virus infection. *J. Immunol.* 166:226–231.

Attridge, S. R., Davies, R., and LaBrooy, J. T. (1997). Oral delivery of foreign antigens by attenuated *Salmonella*: Consequences of prior exposure to the vector strain. *Vaccine* 15:155–162.

Baklien, K., Brandtzaeg, P., and Fausa, O. (1977). Immunoglobulins in jejunal mucosa and serum from patients with adult coeliac disease. *Scand. J. Immunol.* 12:149–159.

Benson, M., Reinholdt, J., and Cardell, L. O. (2003). Allergen-reactive antibodies are found in nasal fluids from patients with birch pollen-induced intermittent allergic rhinitis, but not in healthy controls. *Allergy* 58:386–392.

Bessen, D., and Fischetti, V. A. (1988). Passive acquired mucosal immunity to group A streptococci by secretory immunoglobulin A. *J. Exp. Med.* 167:1945–1950.

Biesbrock, A. R., Reddy, M. S., and Levine, M. J. (1991). Interaction of a salivary mucin-secretory immunoglobulin A complex with mucosal pathogens. *Infect. Immun.* 59:3492–3497.

Boackle, R. J., Pruitt, K. M., and Mestecky, J. (1974). The interactions of human complement with interfacially aggregated preparations of human secretory IgA. *Immunochemistry* 11:543–548.

Bomsel, M., Heyman, M., Hocini, H., Lagaye, S., Belec, L., Dupont, C., and Desgranges, C. (1998). Intracellular neutralization of HIV transcytosis across tight epithelial barriers by anti-HIV envelope protein dIgA or IgM. *Immunity* 9:277–287.

Brandtzaeg, P., and Tolo, K. (1977). Mucosal penetrability enhanced by serum-derived antibodies. *Nature* 266:262–263.

Brock, S. C., McGraw, P. A., Wright, P. F., and Crowe, J. E. (2002). The human polymeric immunoglobulin receptor facilitates invasion of epithelial cells by *Streptococcus pneumoniae* in a strain-specific and cell type-specific manner. *Infect. Immun.* 70:5091–5095.

Bronson, R. A., Cooper, G. W., Rosenfeld, D. L., Gilbert, J. V., and Plaut, A. G. (1987). The effect of an IgA1 protease on immunoglobulins bound to the sperm surface and sperm cervical mucus penetrating ability. *Fertil. Steril.* 47:985–991.

Brown, T. A., Russell, M. W., and Mestecky, J. (1984). Elimination of intestinally absorbed antigen into the bile by IgA. *J. Immunol.* 132:780–782.

Burnett, P. R., VanCott, T. C., Polonis, V. R., Redfield, R. R., and Birx, D. L. (1994). Serum IgA-mediated neutralization of HIV type 1. *J. Immunol.* 152:4642–4648.

Castilla, J., Sola, I., and Enjuanes, L. (1997). Interference of coronavirus infection by expression of immunoglobulin G (IgG) or IgA virus-neutralizing antibodies. *J. Virol.* 71:5251–5258.

Chintalacharuvu, K. R., Raines, M., and Morrison, S. L. (1994). Divergence of human α-chain constant region gene sequences. A novel recombinant α2 gene. *J. Immunol.* 152:5299–5304.

Clamp, J. R. (1977). The relationship between secretory immunoglobulin A and mucus. *Biochem. Soc. Trans.* 5:1579–1581.

Cole, M. F., and Hale, C. A. (1991). Cleavage of chimpanzee secretory immunoglobulin A by *Haemophilus influenzae* IgA1 protease. *Microb. Pathog.* 11:39–46.

Colten, H. R., and Bienenstock, J. (1974). Lack of C3 activation through classical or alternate pathways by human secretory IgA antiblood group A antibody. *Adv. Exp. Med. Biol.* 45:305–308.

Crottet, P., and Corthésy, B. (1998). Secretory component delays the conversion of secretory IgA into antigen-binding competent F(ab'): A possible implication for mucosal defense. *J. Immunol.* 161:5445–5453.

Cunningham-Rundles, C., Brandeis, W. E., Pudifin, D. J., Day, N. K., and Good, R. A. (1981). Autoimmunity in selective IgA deficiency: relationship to anti-bovine protein antibodies, circulating immune complexes and clinical disease. *Clin. Exp. Immunol.* 45:299–304.

Decot, V., Woerly, G., Loyens, M., Loiseau, S., Quatannens, B., Capron, M., and Dombrowicz, D. (2005). Heterogeneity of expression of IgA receptors by human, mouse, and rat eosinophils. *J. Immunol.* 174:628–635.

Devito, C., Broliden, K., Kaul, R., Svensson, L., Johansen, K., Kiama, P., Kimani, J., Lopalco, L., Piconi, S., Bwayo, J. J., Plummer, F., Clerici, M., and Hinkula, J. (2000). Mucosal and plasma IgA from HIV-1-exposed uninfected individuals inhibit HIV-1 transcytosis across human epithelial cells. *J. Immunol.* 165:5170–5176.

Deza, F. G., Espinel, C. S., and Beneitez, J. V. (2007). A novel IgA-like immunoglobulin in the reptile *Eublepharis macularius. Dev. Comp. Immunol.* 31:596–605.

Dunne, D. W., Richardson, B. A., Jones, F. M., Clark, M., Thorne, K. J. I., and Butterworth, A. E. (1993). The use of mouse/human chimaeric antibodies to investigate the roles of different antibody isotypes, including IgA2, in the killing of *Schistosoma mansoni* schistosomula by eosinophils. *Parasite Immunol.* 15:181–185.

Edebo, L., Richardson, N., and Feinstein, A. (1985). The effects of binding mouse IgA to dinitrophenylated *Salmonella typhimurium* on physicochemical properties and interaction with phagocytic cells. *Int. Arch. Allergy Appl. Immunol.* 78:353–357.

Endo, T., Mestecky, J., Kulhavy, R., and Kobata, A. (1994). Carbohydrate heterogeneity of human myeloma proteins of the IgA1 and IgA2 subclasses. *Mol. Immunol.* 31:1415–1422.

Enriquez, F. J., and Riggs, M. W. (1998). Role of immunoglobulin A monoclonal antibodies against P23 in controlling murine *Cryptosporidium parvum* infection. *Infect. Immun.* 66:4469–4473.

Fanger, M. W., Goldstine, S. N., and Shen, L. (1983). Cytofluorographic analysis of receptors for IgA on human polymorphonuclear cells and monocytes and the correlation of receptor expression with phagocytosis. *Mol. Immunol.* 20:1019–1027.

Feng, N. G., Lawton, J. A., Gilbert, J., Kuklin, N., Vo, P., Prasad, B. V. V., and Greenberg, H. B. (2002). Inhibition of rotavirus replication by a non-neutralizing rotavirus VP6-specific IgA mAb. *J. Clin. Invest.* 109:1203–1213.

Fernandez, M. I., Pedron, T., Tournebize, R., Olivo-Marin, J. C., Sansonetti, P. J., and Phalipon, A. (2003). Anti-inflammatory role for intracellular dimeric immunoglobulin A by neutralization of lipopolysaccharide in epithelial cells. *Immunity* 18:739–749.

Friman, V., Adlerberth, I., Connell, H., Svanborg, C., Hanson, L.-Å., and Wold, A. E. (1996). Decreased expression of mannose-specific adhesins by *Escherichia coli* in the colonic microflora of immunoglobulin A-deficient individuals. *Infect. Immun.* 64:2794–2798.

Friman, V., Nowrouzian, F., Adlerberth, I., and Wold, A. E. (2002). Increased frequency of intestinal *Escherichia coli* carrying genes for S fimbriae and haemolysin in IgA-deficient individuals. *Microb. Pathog.* 32:35–42.

Fukui, Y., Fukui, K., and Moriyama, T. (1973). Inhibition of enzymes by human salivary immunoglobulin A. *Infect. Immun.* 8:335–340.

Funakoshi, S., Dot, T., Nakajima, T., Suyama, T., and Tokuda, M. (1982). Antimicrobial effect of human serum IgA. *Microbiol. Immunol.* 26:227–239.

Gan, Y. J., Chodosh, J., Morgan, A., and Sixbey, J. W. (1997). Epithelial cell polarization is a determinant in the infectious outcome of immunoglobulin A-mediated entry by Epstein-Barr virus. *J. Virol.* 71:519–526.

Geissmann, F., Launay, P., Pasquier, B., Lepelletier, Y., Leborgne, M., Lehuen, A., Brousse, N., and Monteiro, R. C. (2001). A subset of human dendritic cells expresses

IgA Fc receptor (CD89), which mediates internalization and activation upon cross-linking by IgA complexes. *J. Immunol.* 166:346–352.
Gessl, A., Willheim, M., Spittler, A., Agis, H., Krugluger, W., and Boltz-Nitulescu, G. (1994). Influence of tumor-necrosis factor-α on the expression of Fc IgG and IgA receptors, and other markers by cultured human blood monocytes and U937 cells. *Scand. J. Immunol.* 39:151–156.
Gilbert, J. V., Plaut, A. G., Longmaid, B., and Lamm, M. E. (1983). Inhibition of microbial IgA proteases by human secretory IgA and serum. *Mol. Immunol.* 20:1039–1049.
Götze, O., and Müller-Eberhard, H. J. (1971). The C3-activator system: An alternative pathway of complement activation. *J. Exp. Med.* 134:90s–108s.
Grezel, D., Capron, M., Grzych, J.-M., Fontaine, J., Lecocq, J.-P., and Capron, A. (1993). Protective immunity induced in rat schistosomiasis by a single dose of the Sm28GST recombinant antigen: Effector mechanisms involving IgE and IgA antibodies. *Eur. J. Immunol.* 23:454–460.
Griffiss, J. M., Broud, D., and Bertram, M. A. (1975). Bactericidal activity of meningococcal antisera. Blocking by IgA of lytic antibody in human convalescent sera. *J. Immunol.* 114:1779–1784.
Gulle, H., Samstag, A., Eibl, M. M., and Wolf, H. M. (1998). Physical and functional association of FcαR with protein tyrosine kinase Lyn. *Blood* 91:383–391.
Hajishengallis, G., Nikolova, E., and Russell, M. W. (1992). Inhibition of *Streptococcus mutans* adherence to saliva-coated hydroxyapatite by human secretory immunoglobulin A (SIgA) antibodies to cell surface protein antigen I/II: reversal by IgA1 protease cleavage. *Infect. Immun.* 60:5057–5064.
Hall, W. M., Manion, R. E., and Zinneman, H. H. (1971). Blocking serum lysis of *Brucella abortus* by hyperimmune rabbit immunoglobulin A. *J. Immunol.* 107:41–46.
Hamre, R., Farstad, I. N., Brandtzaeg, P., and Morton, H. C. (2003). Expression and modulation of the human immunoglobulin A Fc receptor (CD89) and the FcR γ chain on myeloid cells in blood and tissue. *Scand. J. Immunol.* 57:506–516.
Hellwig, S. M. M., Van Spriel, A. B., Schellekens, J. F. P., Mooi, F. R., and Van de Winkel, J. G. J. (2001). Immunoglobulin A-mediated protection against *Bordetella pertussis* infection. *Infect. Immun.* 69:4846–4850.
Heystek, H. C., Moulon, C., Woltman, A. M., Garonne, P., and van Kooten, C. (2002). Human immature dendritic cells efficiently bind and take up secretory IgA without induction of maturation. *J. Immunol.* 168:102–107.
Hiemstra, P. S., Biewenga, J., Gorter, A., Stuurman, M. E., Faber, A., Van Es, L. A., and Daha, M. R. (1988). Activation of complement by human serum IgA, secretory IgA and IgA1 fragments. *Mol. Immunol.* 25:527–533.
Hijmans, W. (1987). Circulating IgA in humans. *Adv. Exp. Med. Biol.* 216B:1169–1174.
Hirano, M., Kamada, M., Maegawa, M., Gima, H., and Aono, T. (1999). Binding of human secretory leukocyte protease inhibitor in uterine cervical mucus to immunoglobulins: Pathophysiology in immunologic infertility and local immune defense. *Fertil. Steril.* 71:1108–1114.
Honorio-França, A. C., Launay, P., Carneiro-Sampaio, M. M. S., and Monteiro, R. C. (2001). Colostral neutrophils express Fcα receptors (CD89) lacking γ chain association and mediate noninflammatory properties of secretory IgA. *J. Leuk. Biol.* 69:289–296.
Hostoffer, R. W., Krukovets, I., and Berger, M. (1994). Enhancement by tumor necrosis factor-α of Fcα receptor expression and IgA-mediated superoxide generation and killing of *Pseudomonas aeruginosa* by polymorphonuclear leukocytes. *J. Infect. Dis.* 170:82–87.

Huang, D. S., Emancipator, S. N., Lamm, M. E., Karban, T. L., Blatnik, F. H., Tsao, H. M., and Mazanec, M. B. (1997). Virus-specific IgA reduces hepatic viral titers *in vivo* on mouse hepatitis virus (MHV) infection. *Immunol. Cell Biol.* 75(Suppl. 1):A12.

Iankov, I. D., Petrov, D. P., Mladenov, I. V., Haralambieva, I. H., and Mitov, I. G. (2002). Lipopolysaccharide-specific but not anti-flagellar immunoglobulin A monoclonal antibodies prevent *Salmonella enterica* serotype enteritidis invasion and replication within Hep-2 cell monolayers. *Infect. Immun.* 70:1615–1618.

Iikura, M., Yamaguchi, M., Fujisawa, T., Miyamasu, M., Takaishi, T., Morita, Y., Iwase, T., Moro, I., Yamamoto, K., and Hirai, K. (1998). Secretory IgA induces degranulation of IL-3-primed basophils. *J. Immunol.* 161:1510–1515.

Imai, H., Chen, A., Wyatt, R. J., and Rifai, A. (1988). Lack of complement activation by human IgA immune complexes. *Clin. Exp. Immunol.* 73:479–483.

Ishizaka, K., Ishizaka, T., and Hornbrook, M. M. (1963). Blocking of Prausnitz-Küstner sensitization with reagin by normal human β_{2A} globulin. *J. Allergy* 34: 395–403.

Janoff, E. N., Fasching, C., Orenstein, J. M., Rubins, J. B., Opstad, N. L., and Dalmasso, A. P. (1999). Killing of *Streptococcus pneumoniae* by capsular polysaccharide-specific polymeric IgA, complement, and phagocytes. *J. Clin. Invest.* 104:1139–1147.

Janoff, E. N., Rubins, J. B., Fasching, C., Plaut, A., and Weiser, J. N. (2002). Inhibition of IgA-mediated killing of *S. pneumoniae* (Spn) by IgA1 protease (IgA1P). *Mucosal Immunol. Update* 10:Abst 2839.

Janoff, E. N., Wahl, S. M., Thomas, K., and Smith, P. D. (1995). Modulation of human immunodeficiency virus type 1 infection of human monocytes by IgA. *J. Infect. Dis.* 172:855–858.

Jarvis, G. A., and Griffiss, J. M. (1989). Human IgA1 initiates complement-mediated killing of *Neisseria meningitidis. J. Immunol.* 143:1703–1709.

Jarvis, G. A., and Griffiss, J. M. (1991). Human IgA1 blockade of IgG-initiated lysis of *Neisseria meningitidis* is a function of antigen-binding fragment binding to the polysaccharide capsule. *J. Immunol.* 147:1962–1967.

Jarvis, G. A., and Li, J. (1997). IgA1-initiated killing of *Neisseria meningitidis*: requirement for C1q and resistance to IgA1 protease. *Immunol. Cell Biol.* 75(Suppl. 1):A12.

Johnson, S., Sypura, W. D., Gerding, D. N., Ewing, S. L., and Janoff, E. N. (1995). Selective neutralization of a bacterial enterotoxin by serum immunoglobulin A in response to mucosal disease. *Infect. Immun.* 63:3166–3173.

Kaetzel, C. S., Robinson, J. K., Chintalacharuvu, K. R., Vaerman, J.-P., and Lamm, M. E. (1991). The polymeric immunoglobulin receptor (secretory component) mediates transport of immune complexes across epithelial cells: A local defense function for IgA. *Proc. Natl. Acad. Sci. USA* 88:8796–8800.

Kilian, M., Husby, S., Høst, A., and Halken, S. (1995). Increased proportions of bacteria capable of cleaving IgA1 in the pharynx of infants with atopic disease. *Pediatr. Res.* 38:182–186.

Kilian, M., Mestecky, J., and Russell, M. W. (1988). Defense mechanisms involving Fc-dependent functions of immunoglobulin A and their subversion by bacterial immunoglobulin A proteases. *Microbiol. Rev.* 52:296–303.

Kilian, M., and Reinholdt, J. (1987). A hypothetical model for the development of invasive infection due to IgA1 protease-producing bacteria. *Adv. Exp. Med. Biol.* 216B:1261–1269.

Kilian, M., and Russell, M. W. (2005). Microbial evasion of IgA functions. In: Mestecky, J., Bienenstock, J., Lamm, M. E., Mayer, L., Strober, W., and McGhee, J. R. (eds.), *Mucosal Immunology*. Elsevier/Academic Press, Amsterdam, pp. 291–303.

Knight, K. L., and Rhee, K.-J. (2005). Organization and expression of genes encoding IgA heavy chain, polymeric Ig receptor, and J chain. In: Mestecky, J., Bienenstock, J., Lamm, M. E., Mayer, L., Strober, W., and McGhee, J. R. (eds.), *Mucosal Immunology,* 3rd ed. Academic Press/Elsevier, San Diego, pp. 183–194.

Koka, P., Chia, D., Terasaki, P. I., Chan, H., Chia, J., Ozawa, M., and Lim, E. (1993). The role of IgA anti-HLA class I antibodies in kidney transplant survival. *Transplantation* 56:207–211.

Komiyama, K., Crago, S. S., Itoh, K., Moro, I., and Mestecky, J. (1986). Inhibition of natural killer cell activity by IgA. *Cell. Immunol.* 101:143–155.

Kozlowski, P. A., Black, K. P., Shen, L., and Jackson, S. (1995). High prevalence of serum IgA HIV-1 infection-enhancing antibodies in HIV-infected persons: masking by IgG. *J. Immunol.* 154:6163–6173.

Kraehenbuhl, J.-P., and Neutra, M. R. (1992). Molecular and cellular basis of immune protection of mucosal surfaces. *Physiol. Rev.* 72:853–879.

Kramer, D. R., and Cebra, J. J. (1995). Role of maternal antibody in the induction of virus specific and bystander IgA responses in Peyer's patches of suckling mice. *Int. Immunol.* 7:911–918.

Lamkhioued, B., Gounni, A. S., Gruart, V., Pierce, A., Capron, A., and Capron, M. (1995). Human eosinophils express a receptor for secretory component. Role in secretory IgA-dependent activation. *Eur. J. Immunol.* 25:117–125.

Lang, M. L., Chen, Y. W., Shen, L., Gao, H., Lang, G. A., Wade, T. K., and Wade, W. F. (2002). IgA Fc receptor (FcαR) cross-linking recruits tyrosine kinases, phosphoinositide kinases and serine/threonine kinases to glycolipid rafts. *Biochem. J.* 364:517–525.

Langford, T. D., Housley, M. P., Boes, M., Chen, J. Z., Kagnoff, M. F., Gillin, F. D., and Eckmann, L. (2002). Central importance of immunoglobulin A in host defense against *Giardia* spp. *Infect. Immun.* 70:11–18.

Launay, P., Patry, C., Lehuen, A., Pasquier, B., Blank, U., and Monteiro, R. C. (1999). Alternative endocytic pathway for immunoglobulin A Fc receptors (CD89) depends on the lack of FcRγ association and protects against degradation of bound ligand. *J. Biol. Chem.* 274:7216–7225.

Leher, H., Zaragoza, F., Taherzadeh, S., Alizadeh, H., and Niederkorn, J. Y. (1999). Monoclonal IgA antibodies protect against *Acanthamoeba* keratitis. *Exp. Eye Res.* 69:75–84.

Liew, F. Y., Russell, S. M., Appleyard, G., Brand, C. M., and Beale, J. (1984). Cross protection in mice infected with influenza A virus by the respiratory route is correlated with local IgA antibody rather than serum antibody or cytotoxic T cell reactivity. *Eur. J. Immunol.* 14:350–356.

Liljemark, W. F., Bloomquist, C. G., and Ofstehage, J. C. (1979). Aggregation and adherence of *Streptococcus sanguis*: role of human salivary immunoglobulin A. *Infect. Immun.* 26:1104–1110.

Lomholt, H., Poulsen, K., and Kilian, M. (1995). Antigenic and genetic heterogeneity among *Haemophilus* and *Neisseria* IgA1 proteases. *Adv. Exp. Med. Biol.* 371A:599–603.

Lycke, N., Eriksen, L., and Holmgren, J. (1987). Protection against cholera toxin after oral immunization is thymus-dependent and associated with intestinal production of neutralizing IgA antitoxin. *Scand. J. Immunol.* 25:413–419.

Lycke, N., Erlandsson, L., Ekman, L., Schön, K., and Leanderson, T. (1999). Lack of J chain inhibits the transport of gut IgA and abrogates the development of intestinal antitoxic protection. *J. Immunol.* 163:913–919.

Magnusson, K.-E., and Stjernström, I. (1982). Mucosal barrier systems. Interplay between secretory IgA (SIgA), IgG and mucins on the surface properties and association of salmonellae with intestine and granulocytes. *Immunology* 45:239–248.

Maliszewski, C. R., Shen, L., and Fanger, M. W. (1985). The expression of receptors for IgA on human monocytes and calcitriol-treated HL-60 cells. *J. Immunol.* 135:3878–3881.

Mansa, B., and Kilian, M. (1986). Retained antigen-binding activity of Fabα fragments of human monoclonal immunoglobulin A1 (IgA1) cleaved by IgA1 protease. *Infect. Immun.* 52:171–174.

Mantis, N. J., Cheung, M. C., Chintalacharuvu, K. R., Rey, J., Corthésy, B., and Neutra, M. R. (2002). Selective adherence of IgA to murine Peyer's patch M cells: Evidence for a novel IgA receptor. *J. Immunol.* 169:1844–1851.

Matoba, N., Magerus, A., Geyer, B. C., Zhang, Y., Muralidharan, M., Alfsen, A., Arntzen, C. J., Bomsel, M., and Mor, T. S. (2004). A mucosally targeted subunit vaccine candidate eliciting HIV-1 transcytosis-blocking Abs. *Proc. Natl. Acad. Sci. USA* 101:13584–13589.

Mattu, T. S., Pleass, R. J., Willis, A. C., Kilian, M., Wormald, M. R., Lellouch, A. C., Rudd, P. M., Woof, J. M., and Dwek, R. A. (1998). The glycosylation and structure of human serum IgA1, Fab, and Fc regions and the role of N-glycosylation on Fcα receptor interactions. *J. Biol. Chem.* 273:2260–2272.

Mazanec, M. B., Coudret, C. L., and Fletcher, D. R. (1995). Intracellular neutralization of influenza virus by immunoglobulin A anti-hemagglutinin monoclonal antibodies. *J. Virol.* 69:1339–1343.

Mazanec, M. B., Kaetzel, C. S., Lamm, M. E., Fletcher, D., and Nedrud, J. G. (1992). Intracellular neutralization of virus by immunoglobulin A antibodies. *Proc. Natl. Acad. Sci. USA* 89:6901–6905.

Mestecky, J., Lue, C., and Russell, M. W. (1991). Selective transport of IgA: cellular and molecular aspects. *Gastroenterol. Clin. North Amr.* 20:441–471.

Mestecky, J., and Russell, M. W. (1986). IgA Subclasses. *Monogr. Allergy* 19:277–301.

Michetti, P., Mahan, M. J., Slauch, J. M., Mekalanos, J. J., and Neutra, M. R. (1992). Monoclonal secretory immunoglobulin A protects mice against oral challenge with the invasive pathogen *Salmonella typhimurium*. *Infect. Immun.* 60:1786–1792.

Moldoveanu, Z., Egan, M. L., and Mestecky, J. (1984). Cellular origins of human polymeric and monomeric IgA: intracellular and secreted forms of IgA. *J. Immunol.* 133:3156–3162.

Møller-Kristensen, M., Thiel, S., Hansen, A. G., and Jensenius, J. C. (2003). On the site of C4 deposition upon complement activation via the mannan-binding lectin pathway or the classical pathway. *Scand. J. Immunol.* 57:556–561.

Monteiro, R. C., Hostoffer, R. W., Cooper, M. D., Bonner, J. R., Gartland, G. L., and Kubagawa, H. (1993). Definition of immunoglobulin A receptors on eosinophils and their enhanced expression in allergic individuals. *J. Clin. Invest.* 92:1681–1685.

Monteiro, R. C., and van de Winkel, J. G. J. (2003). IgA Fc receptors. *Annu. Rev. Immunol.* 21:177–204.

Morell, A., Skvaril, F., Noseda, G., and Barandun, S. (1973). Metabolic properties of human IgA subclasses. *Clin. Exp. Immunol.* 13:521–528.

Mota, G., Manciulea, M., Cosma, E., Popescu, I., Hirt, M., Jensen-Jarolim, E., Calugaru, A., Galatiuc, C., Regalia, T., Tamandl, D., Spittler, A., and Boltz-Nitulescu, G. (2003). Human NK cells express Fc receptors for IgA which mediate signal transduction and target cell killing. *Eur. J. Immunol.* 33:2197–2205.

Moura, I. C., Centelles, M. N., Arcos-Fajardo, M., Malheiros, D. M., Collawn, J. F., Cooper, M. D., and Monteiro, R. C. (2001). Identification of the transferrin receptor as a novel immunoglobulin (Ig)A1 receptor and its enhanced expression on mesangial cells in IgA nephropathy. *J. Exp. Med.* 194:417–425.

Mowat, A. M. (2005). Dendritic cells and immune responses to orally administered antigens. *Vaccine* 23:1797–1799.

Nikolova, E. B., and Russell, M. W. (1995). Dual function of human IgA antibodies: Inhibition of phagocytosis in circulating neutrophils and enhancement of responses in IL-8-stimulated cells. *J. Leuk. Biol.* 57:875–882.

Nikolova, E. B., Tomana, M., and Russell, M. W. (1994a). The role of the carbohydrate chains in complement (C3) fixation by solid-phase-bound human IgA. *Immunology* 82:321–327.

Nikolova, E. B., Tomana, M., and Russell, M. W. (1994b). All forms of human IgA antibodies bound to antigen interfere with complement (C3) fixation induced by IgG or by antigen alone. *Scand. J. Immunol.* 39:275–280.

Norrby-Teglund, A., Ihendyane, N., Kansal, R., Basma, H., Kotb, M., Andersson, J., and Hammarström, L. (2000). Relative neutralizing activity in polyspecific IgM, IgA, and IgG preparations against group A streptococcal superantigens. *Clin. Infect. Dis.* 31:1175–1182.

Ouadrhiri, Y., Pilette, C., Monteiro, R. C., Vaerman, J. P., and Sibille, Y. (2002). Effect of IgA on respiratory burst and cytokine release by human alveolar macrophages: Role of ERK1/2 mitogen-activated protein kinases and NF-κB. *Am. J. Respir. Cell. Mol. Biol.* 26:315–332.

Pal, S., Theodor, I., Peterson, E. M., and De la Maza, L. M. (1997). Monoclonal immunoglobulin A antibody to the major outer membrane protein of the *Chlamydia trachomatis* mouse pneumonitis biovar protects mice against a chlamydial genital challenge. *Vaccine* 15:575–582.

Pasquier, B., Lepelletier, Y., Baude, C., Hermine, O., and Monteiro, R. C. (2004). Differential expression and function of IgA receptors (CD89 and CD71) during maturation of dendritic cells. *J. Leuk. Biol.* 76:1134–1141.

Peebles, R. S., Hamilton, R. G., Lichtenstein, L. M., Schlosberg, M., Liu, M. C., Proud, D., and Togias, A. (2001). Antigen-specific IgE and IgA antibodies in bronchoalveolar lavage fluid are associated with stronger antigen-induced late phase reactions. *Clin. Exp. Allergy* 31:239–248.

Peppard, J., Orlans, E., Payne, A. W., and Andrew, E. (1981). The elimination of circulating complexes containing polymeric IgA by excretion in the bile. *Immunology* 42:83–89.

Peppard, J. V., Kaetzel, C. S., and Russell, M. W. (2005). Phylogeny and comparative physiology of IgA. In: Mestecky, J., Bienenstock, J., Lamm, M. E., Mayer, L., Strober, W., and McGhee, J. R. (eds.), *Mucosal Immunology*. Elsevier/Academic Press, Amsterdam, pp. 195–210.

Pfaffenbach, G., Lamm, M. E., and Gigli, I. (1982). Activation of the guinea pig alternative complement pathway by mouse IgA immune complexes. *J. Exp. Med.* 155:231–247.

Phalipon, A., Cardona, A., Kraehenbuhl, J. P., Edelman, L., Sansonetti, P. J., and Corthésy, B. (2002). Secretory component: A new role in secretory IgA-mediated immune exclusion *in vivo*. *Immunity* 17:107–115.

Phalipon, A., Kaufmann, M., Michetti, P., Cavaillon, J. M., Huerre, M., Sansonetti, P., and Kraehenbuhl, J. P. (1995). Monoclonal immunoglobulin A antibody directed against serotype-specific epitope of *Shigella flexneri* lipopolysaccharide protects against murine experimental shigellosis. *J. Exp. Med.* 182:769–778.

Polissi, A., Pontiggia, A., Feger, G., Altieri, M., Mottl, H., Ferrari, L., and Simon, D. (1998). Large-scale identification of virulence genes from *Streptococcus pneumoniae. Infect. Immun.* 66:5620–5629.

Qiu, J., Brackee, G. P., and Plaut, A. G. (1996). Analysis of the specificity of bacterial immunoglobulin A (IgA) proteases by a comparative study of ape serum IgAs as substrates. *Infect. Immun.* 64:933–937.

Reed, C. E., Bubak, M., Dunnette, S., Blomgren, J., Pfenning, M., Wentz-Murtha, P., Wallen, N., Keating, M., and Gleich, G. J. (1991). Ragweed-specific IgA in nasal lavage fluid of ragweed-sensitive allergic rhinitis patients: increase during the pollen season. *Int. Arch. Allergy Appl. Immunol.* 94:275–277.

Reinholdt, J., and Kilian, M. (1987). Interference of IgA protease with the effect of secretory IgA on adherence of oral streptococci to saliva-coated hydroxyapatite. *J. Dent. Res.* 66:492–497.

Reinholdt, J., and Kilian, M. (1995). Titration of inhibiting antibodies to bacterial IgA1 proteases in human serum and secretions. *Adv. Exp. Med. Biol.* 371A:605–608.

Reljic, R., Crawford, C., Challacombe, S., and Ivanyi, J. (2004). Mouse monoclonal IgA binds to the galectin-3/Mac-2 lectin from mouse macrophage cell lines. *Immunol. Lett.* 93:51–56.

Renegar, K. B., Jackson, G. D. F., and Mestecky, J. (1998). *In vitro* comparison of the biologic activities of monoclonal monomeric IgA, polymeric IgA, and secretory IgA. *J. Immunol.* 160:1219–1223.

Renegar, K. B., and Small, P. A. (1991). Passive transfer of local immunity to influenza virus infection by IgA antibody. *J. Immunol.* 146:1972–1978.

Reterink, T. J. F., Levarht, E. W. N., Klar-Mohamad, N., Van Es, L. A., and Daha, M. R. (1996). Transforming growth factor-beta 1 (TGF-β1) down-regulates IgA Fc-receptor (CD89) expression on human monocytes. *Clin. Exp. Immunol.* 103:161–166.

Rimoldi, M., and Rescigno, M. (2005). Uptake and presentation of orally administered antigens. *Vaccine* 23:1793–1796.

Rits, M., Hiemstra, P. S., Bazin, H., Van Es, L. A., Vaerman, J.-P., and Daha, M. R. (1988). Activation of rat complement by soluble and insoluble rat IgA immune complexes. *Eur. J. Immunol.* 18:1873–1880.

Roberts, M., Bacon, A., Li, J. L., and Chatfield, S. (1999). Prior immunity to homologous and heterologous *Salmonella* serotypes suppresses local and systemic anti-fragment C antibody responses and protection from tetanus toxin in mice immunized with *Salmonella* strains expressing fragment C. *Infect. Immun.* 67:3810–3815.

Römer, W., Rothke, U., and Roelcke, D. (1980). Failure of IgA cold agglutinin to activate C. *Immunobiology* 157:41–46.

Roos, A., Bouwman, L. H., van Gijlswijk-Janssen, D. J., Faber-Krol, M. C., Stahl, G. L., and Daha, M. R. (2001). Human IgA activates the complement system via the mannan-binding lectin pathway. *J. Immunol.* 167:2861–2868.

Royle, L., Roos, A., Harvey, D. J., Wormald, M. R., van Gijlswijk-Janssen, D., Redwan, E.-R. M., Wilson, I. A., Daha, M. R., Dwek, R. A., and Rudd, P. M. (2003). Secretory IgA N- and O-glycans provide a link between the innate and adaptive immune systems. *J. Biol. Chem.* 278:20,140–20,153.

Ruggeri, F. M., Johansen, K., Basile, G., Kraehenbuhl, J.-P., and Svensson, L. (1998). Antirotavirus immunoglobulin A neutralizes virus *in vitro* after transcytosis through epithelial cells and protects infant mice from diarrhea. *J. Virol.* 72:2708–2714.

Russell, M. W., Bobek, L. A., Brock, J. H., Hajishengallis, G., and Tenovuo, J. (2005). Innate humoral defense factors. In: Mestecky, J., Bienenstock, J., Lamm, M. E., Mayer, L., Strober, W., and McGhee, J. R. (eds.), *Mucosal Immunology*. Elsevier/ Academic Press, Amsterdam, pp. 73–93.

Russell, M. W., Brown, T. A., and Mestecky, J. (1981). Role of serum IgA: Hepatobiliary transport of circulating antigen. *J. Exp. Med* 153:968–976.

Russell, M. W., Brown, T. A., Claflin, J. L., Schroer, K., and Mestecky, J. (1983). IgA-mediated hepatobiliary transport constitutes a natural pathway for disposing of bacterial antigens. *Infect. Immun.* 42:1041–1048.

Russell, M. W., and Kilian, M. (2005). Biological activities of IgA. In: Mestecky, J., Bienenstock, J., Lamm, M. E., Mayer, L., Strober, W., and McGhee, J. R. (eds.), *Mucosal Immunology*. Academic Press, San Diego, pp. 267–289.

Russell, M. W., Lue, C., van den Wall Bake, A. W. L., Moldoveanu, Z., and Mestecky, J. (1992). Molecular heterogeneity of human IgA antibodies during an immune response. *Clin. Exp. Immunol.* 87:1–6.

Russell, M. W., and Mansa, B. (1989). Complement-fixing properties of human IgA antibodies. Alternative pathway complement activation by plastic-bound, but not specific antigen-bound, IgA. *Scand. J. Immunol.* 30:175–189.

Russell, M. W., Reinholdt, J., and Kilian, M. (1989). Anti-inflammatory activity of human IgA antibodies and their Fabα fragments: Inhibition of IgG-mediated complement activation. *Eur. J. Immunol.* 19:2243–2249.

Russell-Jones, G. J., Ey, P. L., and Reynolds, B. L. (1980). The ability of IgA to inhibit the complement-mediated lysis of target red blood cells sensitized with IgG antibody. *Mol. Immunol.* 17:1173–1180.

Russell-Jones, G. J., Ey, P. L., and Reynolds, B. L. (1981). Inhibition of cutaneous anaphylaxis and Arthus reactions in the mouse by antigen-specific IgA. *Int. Arch. Allergy Appl. Immunol.* 66:316–325.

Saltzman, W. M., Radomsky, M. L., Whaley, K. J., and Cone, R. A. (1994). Antibody diffusion in human cervical mucus. *Biophys. J.* 66:508–515.

Schiff, J. M., Fisher, M. M., and Underdown, B. J. (1984). Receptor-mediated biliary transport of immunoglobulin A and asialoglycoprotein: Sorting and missorting of ligands revealed by two radiolabeling methods. *J. Cell. Biol.* 98:79–89.

Schiff, J. M., Huling, S. L., and Jones, A. L. (1986). Receptor-mediated uptake of asialoglycoprotein by the primate liver initiates both lysosomal and transcellular pathways. *Hepatology* 6:837–847.

Schneiderman, R. D., Lint, T. F., and Knight, K. L. (1990). Activation of the alternative pathway of complement by twelve different rabbit-mouse chimeric transfectoma IgA isotypes. *J. Immunol.* 145:233–237.

Schwartz-Cornil, I., Benureau, Y., Greenberg, H., Hendrickson, B. A., and Cohen, J. (2002). Heterologous protection induced by the inner capsid proteins of rotavirus requires transcytosis of mucosal immunoglobulins. *J. Virol.* 76:8110–8117.

Shen, L., Collins, J. E., Schoenborn, M. A., and Maliszewski, C. R. (1994). Lipopolysaccharide and cytokine augmentation of human monocyte IgA receptor expression and function. *J. Immunol.* 152:4080–4086.

Shibuya, A., and Honda, S. (2006). Molecular and functional characteristics of the Fcα/μR, a novel Fc receptor for IgM and IgA. *Springer Semin. Immunopathol.* 28:377–382.

Shroff, K. E., Meslin, K., and Cebra, J. J. (1995). Commensal enteric bacteria engender a self-limiting humoral mucosal immune response while permanently colonizing the gut. *Infect. Immun.* 63:3904–3913.

Silbart, L. K., and Keren, D. F. (1989). Reduction of intestinal carcinogen absorption by carcinogen-specific secretory immunity. *Science* 243:1462–1464.

Silvey, K. J., Hutchings, A. B., Vajdy, M., Petzke, M. M., and Neutra, M. R. (2001). Role of immunoglobulin A in protection against reovirus entry into murine Peyer's patches. *J. Virol.* 75:10,870–10,879.

Sixbey, J. W., and Yao, Q. (1992). Immunoglobulin A-induced shift of Epstein-Barr virus tissue tropism. *Science* 255:1578–1580.

Smith, D. J., Taubman, M. A., and Ebersole, J. L. (1985). Salivary IgA antibody to glucosyltransferase in man. *Clin. Exp. Immunol.* 61:416–424.

Smith, P. D., Smythies, L. E., Mosteller-Barnum, M., Sibley, D. A., Russell, M. W., Merger, M., Sellers, M. T., Orenstein, J. M., Shimada, T., Graham, M. F., and Kubagawa, H. (2001). Intestinal macrophages lack CD14 and CD89 and consequently are down-regulated for LPS- and IgA-mediated activities. *J. Immunol.* 167:2651–2656.

Socken, D. J., Simms, E. S., Nagy, B. R., Fisher, M. M., and Underdown, B. J. (1981). Secretory component-dependent hepatic transport of IgA antibody-antigen complexes. *J. Immunol.* 127:316–319.

Stephens, S., Dolby, J. M., Montreuil, J., and Spik, G. (1980). Differences in inhibition of the growth of commensal and enteropathogenic strains of *Escherichia coli* by lactotransferrin and secretory immunoglobulin A isolated from human milk. *Immunology* 41:597–603.

Stewart, W. W., Johnson, A., Steward, M. W., Whaley, K., and Kerr, M. A. (1990). The effect of antibody isotype on the activation of C3 and C4 by immune complexes formed in the presence of serum: Correlation with the prevention of immune precipitation. *Mol. Immunol.* 27:423–428.

Stewart, W. W., Mazengera, R. L., Shen, L., and Kerr, M. A. (1994). Unaggregated serum IgA binds to neutrophil FcαR at physiological concentrations and is endocytosed but cross-linking is necessary to elicit a respiratory burst. *J. Leuk. Biol.* 56:481–487.

Stokes, C. R., Soothill, J. F., and Turner, M. W. (1975). Immune exclusion is a function of IgA. *Nature* 255:745–746.

Stokes, C. R., Taylor, B., and Turner, M. W. (1974). Association of house-dust and grass-pollen allergies with specific IgA antibody deficiency. *Lancet* 2:485-488.

Stubbe, H., Berdoz, J., Kraehenbuhl, J. P., and Corthésy, B. (2000). Polymeric IgA is superior to monomeric IgA and IgG carrying the same variable domain in preventing *Clostridium difficile* toxin A damaging of T84 monolayers. *J. Immunol.* 164:1952–1960.

Svennerholm, A. M., Hanson, L. A., Holmgren, J., Jalil, F., Lindblad, B. S., Khan, S. R., Nilsson, A., and Svennerholm, B. (1981). Antibody responses to live and killed poliovirus vaccines in the milk of Pakistani and Swedish women. *J. Infect. Dis.* 143:707–711.

Tenovuo, J., Moldoveanu, Z., Mestecky, J., Pruitt, K. M., and Mansson-Rahemtulla, B. (1982). Interaction of specific and innate factors of immunity: IgA enhances the antimicrobial effect of the lactoperoxidase system against *Streptococcus mutans*. *J. Immunol.* 128:726–731.

Tomana, M., Kulhavy, R., and Mestecky, J. (1988). Receptor-mediated binding and uptake of immunoglobulin A by human liver. *Gastroenterology* 94:762–770.

Tomana, M., Zikan, J., Moldoveanu, Z., Kulhavy, R., Bennett, J. C., and Mestecky, J. (1993). Interactions of cell-surface galactosyltransferase with immunoglobulins. *Mol. Immunol.* 30:265–275.

Tyler, B. M., and Cole, M. F. (1998). Effect of IgA1 protease on the ability of secretory IgA1 antibodies to inhibit the adherence of *Streptococcus mutans. Microbiol. Immunol.* 42:503–508.

Valim, Y. M. L., and Lachmann, P. J. (1991). The effect of antibody isotype and antigenic epitope density on the complement-fixing activity of immune complexes: A systematic study using chimaeric anti-NIP antibodies with human Fc regions. *Clin. Exp. Immunol.* 84:1–8.

Van der Pol, W. L., Vidarsson, G., Vilé, H. A., Van de Winkel, J. G. J., and Rodriguez, M. E. (2000). Pneumococcal capsular polysaccharide-specific IgA triggers efficient neutrophil effector functions via FcαRI (CD89). *J. Infect. Dis.* 182:1139–1145.

Van Egmond, M., Van Garderen, E., Van Spriel, A. B., Damen, C. A., Van Amersfoort, E. S., Van Zandbergen, G., Van Hattum, J., Kuiper, J., and Van de Winkel, J. G. J. (2000). FcαRI-positive liver Kupffer cells: Reappraisal of the function of immunoglobulin A in immunity. *Nature Med.* 6:680–685.

Van Egmond, M., Van Vuuren, A. J. H., Morton, H. C., Van Spriel, A. B., Shen, L., Hofhuis, F. M. A., Saito, T., Mayadas, T. N., Verbeek, J. S., and Van de Winkel, J. G. J. (1999). Human immunoglobulin A receptor (FcαRI, CD89) function in transgenic mice requires both FcR γ chain and CR3 (CD11b/CD18). *Blood* 93:4387–4394.

Van Spriel, A. B., Leusen, J. H. W., Vilé, H., and Van de Winkel, J. G. J. (2002). Mac-1 (CD11b/CD18) as accessory molecule, for FcαR (CD89) binding of IgA. *J. Immunol.* 169:3831–3836.

Vidarsson, G., van der Pol, W.-L., van den Elsen, J. M. H., Vilé, H., Jansen, M., Duijs, J., Morton, H. C., Boel, E., Daha, M. R., Corthésy, B., and Van de Winkel, J. G. J. (2001). Activity of human IgG and IgA subclasses in immune defense against *Neisseria meningitidis* serogroup B. *J. Immunol.* 166:6250–6256.

Waldo, F. B., and Cochran, A. M. (1989). Mixed IgA-IgG aggregates as a model of immune complexes in IgA nephropathy. *J. Immunol.* 142:3841–3846.

Walker, W. A., Isselbacher, K. J., and Bloch, K. J. (1972). Intestinal uptake of macromolecules: effect of oral immunization. *Science* 177:608–610.

Watanabe, T., Nagura, H., Watanabe, K., and Brown, W. R. (1984). The binding of human milk lactoferrin to immunoglobulin A. *FEBS Lett.* 168:203–207.

Weisbart, R. H., Kacena, A., Schuh, A., and Golde, D. W. (1988). GM-CSF induces human neutrophil IgA-mediated phagocytosis by an IgA Fc receptor activation mechanism. *Nature* 332:647–648.

Weiser, J. N., Bae, D., Fasching, C., Scamurra, R. W., Ratner, A. J., and Janoff, E. N. (2003). Antibody-enhanced pneumococcal adherence requires IgA1 protease. *Proc. Natl. Acad. Sci. USA* 100:4215–4220.

Weltzin, R., Lecia-Jandris, P., Michetti, P., Fields, B. N., Kraehenbuhl, J. P., and Neutra, M. R. (1989). Binding and transepithelial transport of immunoglobulins by intestinal M cells: Demonstration using monoclonal IgA antibodies against enteric viral proteins. *J. Cell Biol.* 108:1673–1685.

Winner, L., Mack, J., Weltzin, R., Mekalanos, J. J., Kraehenbuhl, J.-P., and Neutra, M. R. (1991). New model for analysis of mucosal immunity: intestinal secretion of specific monoclonal immunoglobulin A from hybridoma tumors protects against *Vibrio cholerae* infection. *Infect. Immun.* 59:977–982.

Woerly, G., Roger, N., Loiseau, S., Dombrowicz, D., Capron, A., and Capron, M. (1999). Expression of CD28 and CD86 by human eosinophils and role in the secretion of type 1 cytokines (interleukin 2 and interferon γ): Inhibition by immunoglobulin A complexes. *J. Exp. Med.* 190:487–495.

Wold, A., Mestecky, J., Tomana, M., Kobata, A., Ohbayashi, H., Endo, T., and Svanborg Edén, C. (1990). Secretory immunoglobulin A carries oligosaccharide receptors for *Escherichia coli* type 1 fimbrial lectin. *Infect. Immun.* 58:3073–3077.

Wolf, H. M., Fischer, M. B., Pühringer, H., Samstag, A., Vogel, E., and Eibl, M. M. (1994). Human serum IgA downregulates the release of inflammatory cytokines (tumor necrosis factor-α, interleukin-6) in human monocytes. *Blood* 83:1278–1288.

Wolf, H. M., Hauber, I., Gulle, H., Samstag, A., Fischer, M. B., Ahmad, R. U., and Eibl, M. M. (1996). Anti-inflammatory properties of human serum IgA: Induction of IL-1 receptor antagonist and FcαR (CD89)-mediated down regulation of tumor necrosis factor-α (TNF-α) and IL-6 in human monocytes. *Clin. Exp. Immunol.* 105:537–543.

Yan, H. M., Lamm, M. E., Björling, E., and Huang, Y. T. (2002). Multiple functions of immunoglobulin A in mucosal defense against viruses: an *in vitro* measles virus model. *J. Virol.* 76:10972–10979.

Yuan, Z. N., Gjermo, P., Helgeland, K., and Schenk, K. (2000). Fcα receptor I (CD89) on neutrophils in periodontal lesions. *J. Clin. Periodontol.* 27:489–493.

Zhang, J. R., Mostov, K. E., Lamm, M. E., Nanno, M., Shimida, S., Ohwaki, M., and Tuomanen, E. (2000). The polymeric immunoglobulin receptor translocates pneumococci across human nasopharyngeal epithelial cells. *Cell* 102:827–837.

Zhang, W., and Lachmann, P. J. (1994). Glycosylation of IgA is required for optimal activation of the alternative complement pathway by immune complexes. *Immunology* 81:137–141.

7
Protection of Mucosal Epithelia by IgA: Intracellular Neutralization and Excretion of Antigens

Michael E. Lamm[1]

7.1. Introduction

Mucosal immunoglobulin A (IgA) antibodies are synthesized by local plasma cells in the lamina propria and are largely destined for export through the lining epithelium into the luminal secretions. Here, IgA antibodies can bind antigens and exclude them from the body, as has long been appreciated. It is becoming increasingly apparent, though, that passage through mucosal epithelium creates additional opportunities for IgA antibodies to function in host defense. For example, IgA antibodies against viruses can directly counter infections within mucosal epithelium, and immune complexes formed in the lamina propria containing locally produced IgA antibodies can pass through the epithelium via the same route and mechanism as free IgA. Thus, IgA antibodies might first encounter antigens in three anatomic compartments in relation to mucosal epithelium: in the lumen, in the epithelium itself, or in the lamina propria (Lamm, 1997). The nonclassical defense functions of IgA, in which IgA antibodies initially bind antigens in the lamina propria or inside the lining epithelial cells, are the focus of this chapter.

[1] Department of Pathology, Case Western Reserve University, Cleveland, OH 44106, USA.

7.2. Intraepithelial Cell Neutralization of Viruses by IgA Antibodies

Viruses are obligatory intracellular parasites, and IgA antibodies secreted by lamina propria plasma cells pass through the epithelial cells that cover the mucous membrane by receptor-mediated endocytosis and transcytosis in order to reach the lumen. These facts prompted the following question. What would happen if antivirus IgA antibodies were passing through an epithelial cell that happened to be infected by the same virus? Specifically, would the antibodies actually encounter viral antigens, and, if so, would the virus be inhibited? These issues have been explored in model systems both *in vitro* and *in vivo*.

Studies *in vitro* have taken advantage of the ability to grow polarized monolayers of epithelial cells that express the receptor for IgA, the polymeric Ig receptor (pIgR), on their basolateral surface (see Chapter 3). The cell often employed is the dog kidney epithelial cell line MDCK, transfected so that it expresses pIgR (Mostov and Deitcher, 1986; Tamer et al., 1995). This cell line has long been used by cell biologists for studying epithelial cell traffic because it grows easily *in vitro* and readily polarizes (i.e., forms monolayers in which the cells are attached to their neighbors by tight junctions and have apical and basolateral plasma membrane domains with different compositions, morphologies, and functions). Receptors for internalizing particular viruses may be preferentially expressed at the apical or basolateral side, and release of newly formed virus particles may be similarly polarized.

The original system used to demonstrate intraepithelial cell neutralization of virus by IgA antibody employed polarized monolayers of MDCK cells that expressed pIgR. The cells were grown in two-chambered transwells and infected via the apical surface with Sendai virus, a rodent paramyxovirus. Subsequently, IgA monoclonal antibodies to viral envelope protein were added to the lower chamber, from which they could be internalized by the pIgR. Assays for virus in the apical supernatant and cell lysate, together with appropriate controls, showed that the growth of virus had indeed been inhibited by the ability of the IgA antibody to bind viral protein intracellularly (Mazanec et al., 1992). Analogous model systems have since been used to demonstrate intracellular neutralization of a number of viruses belonging to different classes, including influenza virus (an orthomyxovirus) (Mazanec et al., 1995), measles virus (a paramyxovirus) (Yan et al., 2002), rotavirus (a reovirus) (Corthésy et al., 2006; Feng et al., 2002), and human immunodeficiency virus (HIV: a retrovirus) (Huang et al., 2005; Wright et al., 2006).

From the available studies collectively some tentative general conclusions can be drawn. One, the antiviral effect is mediated by IgA antibodies that are following the normal route of epithelial endocytosis and transcytosis. Thus, to be effective, IgA antibodies must be polymeric, and the epithelial cells must express pIgR at the basolateral surface. Two, IgA antibodies tend to be more effective when directed at antigens displayed on the surface of the virus as opposed to internal viral antigens. This result is thought to reflect that surface antigens like envelope glycoproteins are synthesized in the rough endoplasmic reticulum and then undergo vesicular traffic through the Golgi and trans-Golgi network. Because transcytosing IgA antibodies also undergo vesicular transport, it is presumed that vesicles transporting IgA and vesicles transporting viral envelope proteins meet and fuse at some point in their respective journeys. Indeed, in the above-cited studies together with those of Fujioka et al. (1998), colocalization of viral envelope antigens and IgA antibodies has been demonstrated by immunofluorescence, confocal and electron microscopy. Available evidence suggests apical recycling endosomes as an important locus of intersection of IgA antibodies and viral envelope proteins (Bomsel et al., 1998).

Different classes of viruses may vary in their intracellular life cycles in terms of site and mechanism of initial entry and uncoating, replication, subcellular location where different viral components are made, how the components are assembled, and when and how mature virus particles are released. Thus, it should not be surprising that IgA antibodies to viral antigens might vary considerably in their intracellular antivirus effects, depending on the type of viral component, the reactive epitopes, the characteristics of a particular virus life cycle, and the opportunities for transcytosing IgA antibodies to meet their prospective targets. For example, IgA antibody to an internal component of measles virus that is synthesized on cytoplasmic ribosomes was shown to be capable of a degree of intracellular neutralization (Yan et al., 2002). It has recently been reported that intraepithelial cell neutralization of HIV by IgA antibodies to internal viral proteins can occur during antibody transcytosis from the basolateral to the apical surface (Wright et al., 2006). Polarized epithelial cells expressing pIgR were transfected with HIV proviral DNA, and IgA was added to the basolateral side. Transcytosing IgA antibodies against Gag and RT significantly inhibited HIV replication, as assessed by infection of HeLa cells and analysis of HIV p24 expression. Consistent with intracellular neutralization, colocalization of the internal virus proteins and their IgA antibodies was demonstrated by confocal microscopy. Thus, at least in the context of infections of polarized epithelia, antibody-mediated neutralization might not be restricted to viral surface antigens. Future investigations focused on such questions could lead to new insights into mechanisms by which subcellular organelles and transport systems intersect and interact. In turn, such insights could provide important hints for improved formulation of vaccines designed to stimulate effective mucosal antibodies.

Most of the evidence for the ability of IgA to neutralize viruses inside epithelial cells derives from experiments performed *in vitro*. So, is this phenomenon applicable to living animals? The best evidence that it is comes from studies with rotavirus in which systemically delivered IgA antibodies to the conserved inner core protein VP6 of rotavirus (which would not be accessible to antibody in an intact virus) were able to prevent infection as well as cure an existing infection in mice (Burns et al., 1996; Feng et al., 2002; Schwartz-Cornil et al., 2002). A recent study explored the mechanism by which IgA monoclonal antibodies directed at rotavirus VP6 protein mediate virus neutralization (Corthésy et al., 2006). One such antibody, IgA7D9, appeared to interact directly with purified triple-layer viral particles *in vitro*. However, passively feeding mice with secretory IgA7D9 [complexed to secretory component (SC)] did not prevent viral infection, suggesting that immune exclusion was not the mechanism of protection. In contrast, systemic administration of polymeric IgA7D9, which could be transported across mucosal epithelia, did confer protection. Application of polymeric IgA7D9 to the basolateral surface of polarized Caco-2 intestinal cells significantly reduced replication of rotavirus introduced at the apical surface and prevented rotavirus-induced loss of barrier function. These findings support the conclusion that the IgA antibodies to internal viral epitopes can mediate intracellular neutralization of rotavirus *in vitro* and *in vivo*.

In addition to being able to neutralize viruses inside epithelial cells, IgA antibodies have also been shown to be capable of intracellular neutralization of bacterial LPS and its proinflammatory effects (Fernandez et al., 2003) (see Chapter 8). Intracellular neutralization of microbes and their products could prove to be a general mechanism by which IgA promotes homeostasis at mucosal surfaces.

7.3. Excretion of Antigens Across Epithelial Cells by IgA Antibodies

Mucous membranes provide an effective barrier to exclude foreign substances from the interior of the body. Exclusion is facilitated by mucus, the tight junctions between neighboring epithelial cells in the mucosal lining, and by mucosal antibodies, principally secretory IgA (SIgA). Nevertheless, the barrier is not perfect and small amounts of foreign substances do normally penetrate. Moreover, periodic mucosal infections provide additional opportunities for the body proper to be exposed to foreign antigens, including those of the infecting microorganism as well as other antigens whose entry might be facilitated by infection- and inflammation-induced alterations in local epithelial permeability. Accordingly, there are ample opportunities for mucosal lamina propria, especially in the intestinal tract, to be exposed to foreign antigens. What then is the fate of foreign antigens that access the mucosal lamina propria for whatever reasons? In addition to local phagocytosis or absorption into lymphatics and blood vessels, there is the possibility, even likelihood, that foreign antigens in the lamina propria will be bound by IgA antibodies secreted from nearby plasma cells, leading to the formation of IgA immune complexes. As long as an immune complex contains a molecule of the dimeric form of IgA, which is the majority of the locally synthesized IgA, the complex is potentially able to be endocytosed via epithelial cell pIgR and transcytosed across the epithelium by the same mechanism and route followed by free, noncomplexed dimeric IgA. In this way, IgA would provide a means of "excreting" antigens from mucosal lamina propria into the lumen and in a noninflammatory manner, given the reduced capacity of IgA to excite inflammatory phenomena compared to other classes of antibody like IgG or IgE. Such an excretory immune function for mucosal IgA is envisioned to function both on an ongoing basis as well as during local infections. This excretory function is potentially expandable as a result of increased synthesis of local IgA antibody during antigenic challenge. Furthermore, synthesis of epithelial pIgR can be upregulated by a number of cytokines whose production would be expected to increase during inflammatory reactions, as well as by microbes and their products that interact with Toll-like receptors (TLRs) (Bruno and Kaetzel, 2005; Hayashi et al., 1997; Johansen and Brandtzaeg, 2004; Kvale et al., 1988; Nilsen et al., 1999; Phillips et al., 1990; Piskurich et al., 1997; Schneeman et al., 2005; Sollid et al., 1987; Youngman et al., 1994). Increased synthesis of pIgR could in turn allow epithelial cells to transport increased amounts of IgA and IgA immune complexes.

The initial experiments in support of a local IgA-mediated excretory immune system were performed *in vitro* using the same pIgR-expressing polarized epithelial cell monolayer system described earlier. Soluble IgA-containing immune complexes were placed in the compartment under the monolayer and their transport across the monolayer was followed by sampling the apical supernatant above the monolayer (Kaetzel et al., 1991., 1994). Transport required the presence of polymeric IgA and expression of pIgR by the epithelial cells. The complexes were transported intact with no evidence of intracellular breakdown during transit, consistent with vesicular transport and lack of fusion with lysosomes. Monomeric antibodies like IgG could be transported as well if they were in the same immune complex along with a molecule of dimeric IgA.

Evidence from experiments *in vivo* also supports a mucosal IgA-mediated excretory immune system (Robinson et al., 2001). These experiments utilized mice that were immunized with a protein antigen included in the drinking water. When the IgA antibody response peaked, antigen was injected intravenously. This antigen could be detected immunohistochemically inside the small intestinal crypt, but not villus cells. The experimental design supported the contention that antigen complexed to IgA had entered the crypt cells from the lamina propria.

There are some similarities between these experiments supporting a local IgA-mediated excretory immune system and earlier experiments in rodents demonstrating that circulating polymeric IgA immune complexes can be transported by the liver from blood to bile (Harmatz et al., 1982; Peppard et al., 1981; Russell et al., 1981; Socken et al., 1981), a phenomenon that does not apply to species like humans that do not express pIgR on hepatocytes. Conceptually, however, there are important differences between clearance through the local mucosal epithelium and clearance through the liver. Clearance of IgA immune complexes by hepatic transport in rodents implies that antigen, free or complexed, has already reached the circulation; therefore, this mechanism of clearance should be viewed as part of the systemic defense. In contrast, IgA-mediated excretion from the lamina propria serves to keep antigens in mucosal lamina propria from reaching the circulation in the first place. In this way, the body's systemic exposure to immune complexes is reduced, as well as opportunities for relatively phlogistic isotypes like IgG, which quantitatively exceeds IgA in the blood but not in mucosal lamina propria, to become involved. The local excretory immune system can be envisioned to subject newly arrived antigen to a revolving door: the antigen enters, is bound by relatively noninflammatory IgA antibody, and quickly exits without reaching the systemic circulation.

Not only can IgA excrete soluble antigens but also particles as large as viruses, as demonstrated both *in vivo* and *in vitro* with Epstein-Barr virus (EBV) and measles virus (Gan et al., 1997; Yan et al., 2002). In addition, Bomsel et al. (1998) have shown that IgA antibodies transcytosing epithelial cells from basolateral to apical can intercept HIV that is transcytosing in the opposite direction after endocytosis from the apical side, following which the virus complexed to IgA is diverted back to the apical medium and is prevented from reaching the basolateral medium. This phenomenon can be considered a variant of IgA-mediated excretion, with IgA first meeting the antigen in the epithelial cell—in this case, within apical recycling endosomes—rather than in the lamina propria.

The mucosal excretory immune system could be directly relevant in the context of at least two diseases. IgA nephropathy, the most common form of glomerulonephritis worldwide, often follows bouts of upper respiratory infections. By definition, in IgA nephropathy, affected kidneys contain IgA, presumably in immune complexes. The pathogenesis might involve the induction of an IgA antibody response to a mucosal infection followed by the formation of IgA immune complexes that reach the circulation and deposit in the renal glomeruli (Emancipator et al., 2005). It is thus possible that a deficiency, for whatever reason, in IgA-mediated excretion at the local mucus membrane where the infection occurred could contribute to untoward amounts of IgA immune complexes reaching the circulation. In another disease process, nasopharynageal carcinoma caused by EBV, it has been proposed that introduction of virus complexed to IgA antibody into mucosal epithelial cells via pIgR-mediated endocytosis leads to their ultimate neoplastic transformation. For virus to remain in the epithelial cell and be capable of causing infection or transformation rather than being fully transcytosed, it was further proposed that the affected epithelial cells would have to have their normal polarity disturbed by injury (Gan et al., 1997; Sixby and Yao, 1992). It is significant that a single amino acid polymorphism in pIgR, located within the domain that is cleaved to form SC, has been associated with increased risk for both IgA nephropathy (Obara et al., 2003) and EBV-associated nasopharyngeal cancer (Hirunsatit et al., 2003). It is tempting to speculate that this polymorphism might reduce the rate of cleavage of pIgR to SC and hence the rate of excretion of locally formed IgA immune complexes, posing an increased risk for IgA nephropathy. Similarly, reduced cleavage and release of pIgR–IgA–EBV complexes at the apical surface could increase the opportunity for EBV to remain associated with epithelial cells and promote infection or transformation.

7.4. Concluding Remarks

It is becoming apparent that mucosal IgA can act in host defense in different anatomic compartments in relation to epithelium and by different mechanisms, including those not traditionally associated with antibodies. The anatomic compartments in which IgA antibodies can function evidently include the mucosal luminal secretions, the interior of the epithelial cells that line the mucosa, and the lamina propria beneath the surface. If IgA antibody initially encounters antigen in the lumen, it can prevent it from penetrating the mucosa. If the initial meeting instead takes place in the lining epithelium during the normal transit of IgA from lamina propria to luminal secretions and if the antigen in question belongs to a virus capable of infecting epithelial cells, IgA antibody can directly counter infection. Finally, if IgA antibody binds to antigen, including on an infectious microbe, within the lamina propria, it might be capable of excreting it through the epithelium into the lumen.

Because humoral antibodies have traditionally been viewed as functioning only extracellularly, it should be noted that intracellular virus neutralization by transcytosing IgA antibody is a potentially important exception to the rule. The fact that IgA antibodies can function in mucosae in diverse ways, including intracellularly, would seem to strengthen the rationale for including them when formulating vaccine regimens to protect against microbes that infect mucous membranes and also those that infect internal organs as long as they enter the body through mucosal portals. For the design of such vaccines, it makes sense to attempt to identify the most susceptible antigens and epitopes of individual pathogens. In this context, the ability of IgA antibodies against internal viral proteins to counter infections *in vivo* (Burns et al., 1996) is especially relevant. For example, in the case of influenza A viruses, whose internal proteins tend to be antigenically conserved, such IgA antibodies might be capable of conferring heterosubtypic immunity, which in turn could offer enhanced protection against newly emerging variants with potential to cause pandemics (Takada et al., 2003). It is conceivable that systems employing IgA antibodies and infection of polarized epithelial cells *in vitro* could help identify the most promising antigens (Kumagai, personal communication).

References

Bomsel, M., Heyman, M., Hocini, H., Lagaye, S., Belec, L., Dupont, C., and Desgranges, C. (1998). Intracellular neutralization of HIV transcytosis across tight epithelial barriers by anti-HIV envelope protein dIgA or IgM. *Immunity* 9:277–287.

Bruno, M. E. C., and Kaetzel, C.S. (2005). Long-term exposure of the HT-29 human intestinal epithelial cell line to TNF causes sustained up-regulation of the polymeric Ig receptor and proinflammatory genes through transcriptional and posttranscriptional mechanisms. *J. Immunol.* 174:7278–7284.

Burns, J. W., Siadat-Pajouh, M., Krishnaney, A. A., and Greenberg, H. B. (1996). Protective effect of rotavirus VP6-specific IgA monoclonal antibodies that lack neutralizing activity. *Science* 272:104–107.

Corthésy, B., Benureau, Y., Perrier, C., Fourgeux, C., Parez, N., Greenberg, H., and Schwartz-Cornil, I. (2006). Rotavirus anti-VP6 secretory immunoglobulin A contributes to protection via intracellular neutralization but not via immune exclusion. *J Virol.* 80:10692–10699.

Emancipator, S. N., Mestecky, J., and Lamm, M. E. (2005). IgA nephropathy and related diseases. In: Mestecky J., Bienenstock, J., Lamm, M. E., Mayer, L., McGhee, J. R., and Strober, W. (eds.), *Mucosal Immunology*, 3rd ed. Elsevier, San Diego, pp. 1580–1600.

Feng, N., Lawton, J. A., Gilbert, J., Kuklin, N., Vo, P., Venkataram Prasad, B. V., and Greenberg, H. B. (2002). Inhibition of rotavirus replication by a non-neutralizing, rotavirus VP6-specific IgA mAb. *J. Clin. Invest.* 109:1203–1213.

Fernandez, M. I., Pedron, T., Tournebize, R., Olivo-Marin, J.-C., Sansonetti, P. J., and Phalipon, A. (2003). Anti-inflammatory role for intracellular dimeric immunoglobulin A by neutralization of lipopolysaccharide in epithelial cells. *Immunity* 18:739–749.

Fujioka, H. S., Emancipator, S. N., Aikawa, M., Huang, D. S., Blatnik, F., Karban, T., DeFife, K., and Mazanec, M. B. (1998). Immunocytochemical colocalization of specific immunoglobulin A with sendai virus protein in infected polarized epithelium. *J. Exp. Med.* 188:1223–1229.

Gan, Y. J., Chodosh, J., Morgan, A., and Sixbey, J. W. (1997). Epithelial cell polarization is a determinant in the infectious outcome of immunoglobulin A-mediated entry by Epstein-Barr virus. *J. Virol.* 71:519–526.

Harmatz, P. R., Kleinman, R. E., Bunnel, B. W., Block, K. J., and Walker, W. A. (1982). Hepatobiliary clearance of IgA immune complexes formed in the circulation. *Hepatology* 2:328–333.

Hayashi, M., Takenouchi, N., Asano, M., Kato, M., Tsurumachi, T., Sait, T., and Moro, I. (1997). The polymeric immunoglobulin receptor (secretory component) in a human intestinal epithelial cell line is up-regulated by interleukin-1. *Immunology* 92:220–225.

Hirunsatit, R., Kongruttanachok, N., Shotelersuk, K., Supiyaphun, P., Voravud, N., Sakuntabhai, A., and Mutirangura, A. (2003). Polymeric immunoglobulin receptor polymorphisms and risk of nasopharyngeal cancer. *BMC Genet.* 4:3.

Huang, Y. T., Wright, A., Gao, X., Kulick, L., Yan, H., and Lamm, M. E. (2005). Intraepithelial cell neutralization of HIV-1 replication by IgA. *J. Immunol.* 174:4828–4835.

Johansen, F. E., and Brandtzaeg, P. (2004). Transcriptional regulation of the mucosal IgA system. *Trends Immunol.* 25:150–157.

Kaetzel, C. S., Robinson, J. K., Chintalacharuvu, K. R., Vaerman, J.-P., and Lamm, M. E. (1991). The polymeric immunoglobulin receptor (secretory component) mediates transport of immune complexes across epithelial cells: A local defense function for IgA. *Proc. Natl. Acad. Sci. USA* 88:8796–8800.

Kaetzel, C. S., Robinson, J. K., and Lamm, M. E. (1994). Epithelial transcytosis of monomeric IgA and IgG cross-linked through antigen to polymeric IgA. A role for monomeric antibodies in the mucosal immune system. *J. Immunol.* 152:72–76.

Kvale, D., Brandtzaeg, P., and Lovhaug, D. (1988). Up-regulation of the expression of secretory component and HLA molecules in a human colonic cell line by tumour necrosis factor-α and γ-interferon. *Scand. J. Immunol.* 28:351–357.

Lamm, M. E. (1997). Interaction of antigens and antibodies at mucosal surfaces. *Annu. Rev. Microbiol.* 51:311–340.

Mazanec, M. B., Coudret, C. L., and Fletcher, D.R. (1995). Intracellular neutralization of influenza virus by immunoglobulin A anti-hemagglutinin monoclonal antibodies. *J. Virol.* 69:1339–1343.

Mazanec, M. B., Kaetzel, C. S., Lamm, M. E., Fletcher, D., and Nedrud, J. G. (1992). Intracellular neutralization of virus by immunoglobulin A antibodies. *Proc. Natl. Acad. Sci. USA* 89:6901–6905.

Mostov, K. E., and Deitcher, D. L. (1986). Polymeric immunoglobulin receptor expressed in MDCK cells transcytoses IgA. *Cell* 15:613–621.

Nilsen, E. M., Johansen, F. E., Kvale, D., Krajci, P., and Brandtzaeg, P. (1999). Different regulatory pathways employed in cytokine-enhanced expression of secretory component and epithelial HLA class I genes. *Eur. J. Immunol.* 29:168–179.

Obara, W., Iida, A., Suzuki, Y., Tanaka, T., Akiyama, F., Maeda, S., Ohnishi, Y., Yamada, R., Tsunoda, T., Takei, T., Ito, K., Honda, K., Uchida, K., Tsuchiya, K., Yumura, W., Ujiie, T., Nagane, Y., Nitta, K., Miyano, S., Narita, I., Gejyo, F., Nihei, H., Fujioka, T., and Nakamura, Y. (2003). Association of single-nucleotide polymorphisms in the polymeric immunoglobulin receptor gene with immunoglobulin A nephropathy (IgAN) in Japanese patients. *J. Hum. Genet.* 48:293–299.

Peppard, J., Orlans, E., Payne, A. W. R., and Andrew, E. (1981). The elimination of circulating complexes containing polymeric IgA by excretion in the bile. *Immunology* 42:83–89.

Phillips, J. O., Everson, M. P., Moldoveanu, Z., Lue, C., and Mestecky, J. (1990). Synergistic effect of IL-4 and IFN-γ on the expression of polymeric Ig receptor (secretory component) and IgA binding by human epithelial cells. *J. Immunol.* 145:1740–1744.

Piskurich, J. F., Youngman, K. R., Phillips, K. M., Hempen, P. M., Blanchard, M. H., France, J. A., and Kaetzel, C. S. (1997). Transcriptional regulation of the human polymeric immunoglobulin receptor gene by interferon-γ. *Mol. Immunol.* 34:75–91.

Robinson, J. K., Blanchard, T. G., Levine, A. D., Emancipator, S. N., and Lamm, M. E. (2001). A mucosal IgA-mediated excretory immune system *in vivo*. *J. Immunol.* 166:3688–3692.

Russell, M. W., Brown, T. A., and Mestecky, J. (1981). Role of serum IgA: Hepatobiliary transport of circulating antigen. *J. Exp. Med.* 153:968–976.

Schneeman, T. A., Bruno, M. E. C., Schjerven, H., Johansen, F.-E., Chady, L., and Kaetzel, C. S. (2005). Regulation of the polymeric Ig receptor by signaling through TLRs 3 and 4: Linking innate and adaptive immune responses. *J. Immunol.* 175:376–384.

Schwartz-Cornil, I., Benureau, Y., Greenberg, H., Hendrickson, B. A., and Cohen, J. (2002). Heterologous protection induced by the inner capsid proteins of rotavirus requires transcytosis of mucosal immunoglobulins. *J. Virol.* 76:8110–8117.

Sixbey, J. W., and Yao, Q.-Y. (1992). Immunoglobulin A-induced shift of Epstein-Barr virus tissue tropism. *Science* 255:1578–1580.

Socken, D. J., Simms, E. S., Nagy, B. R., Fisher, M. M., and Underdown, B. J. (1981). Secretory component-dependent hepatic transport of IgA antibody-antigen complexes. *J. Immunol.* 127:316–319.

Sollid, L. M., Kvale, D., Brandtzaeg, P., Markussen, G., and Thorsby, E. (1987). Interferon-γ enhances expression of secretory component, the epithelial receptor for polymeric immunoglobulins. *J. Immunol.* 138:4303–4306.

Takada, A., Matsushita, S., Ninomiya, A., Kawaoka, Y., and Kida, H. (2003). Intranasal immunization with formalin-inactivated virus vaccine induces a broad spectrum of heterosubtypic immunity against influenza A virus infection in mice. *Vaccine* 21:3212–3218.

Tamer, C. M., Lamm, M. E., Robinson, J. K, Piskurich, J. F., and Kaetzel, C. S. (1995). Comparative studies of transcytosis and assembly of secretory IgA in Madin-Darby canine kidney cells expressing human polymeric Ig receptor. *J Immunol.* 155:707–714.

Wright, A., Yan, H., Lamm, M. E., and Huang, Y. T. (2006). Immunoglobulin A antibodies against internal HIV-1 proteins neutralize HIV-1 replication inside epithelial cells. Virology 356:165–170.

Yan, H., Lamm, M. E., Bjorling, E., and Huang, Y. T. (2002). Multiple functions of IgA in mucosal defense against viruses: An *in vitro* measles virus model. *J. Virol.* 76:10972–10979.

Youngman, K. R., Fiocchi, C., and Kaetzel, C. S. (1994). Inhibition of IFN-γ activity in supernatants from stimulated human intestinal mononuclear cells prevents up-regulation of the polymeric Ig receptor in an intestinal epithelial cell-line. *J. Immunol.* 153:675–681.

8
Novel Functions for Mucosal SIgA

Armelle Phalipon[1] and Blaise Corthésy[2]

8.1. Introduction

An important activity of mucosal surfaces is the production of the special type of antibodies referred to as secretory IgA (SIgA) (Lamm, 1997). SIgA is produced predominantly as a dimer complexed with the J-chain and the secretory component (SC). SC is the ectodomain of the polymeric immunoglobulin receptor (pIgR) that either remains bound to polymeric IgA (pIgA) following transcytosis and proteolytic cleavage (Mostov et al., 1980) or is released alone, (i.e., free SC, in mucosal secretions) (Poger and Lamm, 1974) (see Chapter 3). The classical view is that SIgA serves as the first line of defense against microorganisms by agglutinating potential invaders and facilitating their clearance by peristaltic and mucociliary movements, a mechanism called immune exclusion (Mestecky et al., 1999).

In addition to immune exclusion, SIgA, in the course of its pIgR-mediated intracellular transport, contributes to intracellular neutralization of viruses (Bomsel et al., 1998; Corthésy et al., 2006; Fujioka et al., 1998; Huang et al., 2005;

[1] Laboratoire de Pathogénie Microbienne Moléculaire, Institut Pasteur, INSERM U389, 75724 Paris Cedex 15, France.
[2] Laboratoire de Recherche et Développement, Service d'Immunologie et d'Allergie, Centre Hospitalier Universitaire Vaudois, 1011 Lausanne, Switzerland.

Mazanec et al., 1992) and to the removal of stromal antigens that penetrated a breached epithelium (Kaetzel et al., 1991; Robinson et al., 2001) (see Chapter 7). Novel properties of SIgA have been unraveled during the last years, suggesting that this class of antibody possesses multiple associated functions that endow them with the capacity to fine-tune mucosal immune responses. In this chapter, the authors focus their discussion on (1) innate-like properties of SC, including its role in binding of SIgA to mucus and its "scavenger" properties, (2) the transport of SIgA across Peyer's patches (PPs) (also called retrotransport) and its biological consequences, and (3) the role of SIgA in controlling inflammatory responses, both during infection processes and in gut homeostasis.

8.2. Innate-like Properties of SC

The complex nature of the SIgA antibody molecule is instrumental in fulfilling its multiple biological roles. It has long been appreciated that polymerization of the IgA moiety comprising the antigen specificity results in enhanced avidity, which is crucial to efficient cross-linking of antigens. Our understanding of the functions of SIgA-bound or free SC is emerging well beyond its first attributed role of protecting SIgA from protease degradation (Chintalacharuvu and Morrison, 1997; Crottet and Corthésy, 1998; Lindh, 1975). It is becoming much clearer that SC possesses intrinsic "innate-like" properties, including binding to pathogens and preventing their interaction with epithelial cell surface, anchoring of SIgA to mucosal surfaces, and trapping of cytokines (Phalipon and Corthésy, 2003). How these so far overlooked features contribute to the biological functions of SIgA is discussed below.

Free SC is now recognized to participate actively in the protection of mucosal surfaces. For instance, SC isolated from human milk (possibly copurified with lactoferrin) was initially shown to inhibit adhesion of enterotoxigenic *Escherichia coli* colonization factor antigen-1 strains to erythrocytes (Giugliano et al., 1995). Similarly, free SC was reported to bind preferentially to *Clostridium difficile* toxin A, as compared to weak or absent binding of purified light and heavy chains of IgA (Dallas and Rolfe, 1998). Deglycosylation of SC partially reduced the binding of toxin A to hamster brush-border membranes. Likewise, binding of the staphylococcal superantigen-like protein SSL7 to human and bovine SIgA found in milk appears to depend mainly on the SC moiety (Langley et al., 2005). However, that study did not demonstrate that such an interaction is implicated in mucosal protection. To avoid artifactual effects caused by multiple protein and nonprotein glycoconjugates present in milk (Newburg, 1999), recombinant human SC (hSC) produced from transfected Chinese hamster ovary cells was assayed for "scavenger" activity (Perrier et al., 2006). Pure hSC was found to protect Caco-2 and T84 polarized epithelial cell monolayers from the action of toxin A. This protective activity of SC required carbohydrate residues, including galactose and sialic acid, as well as intact disulfide bridges. The involvement of sugar moieties on SC is of interest because of the importance of glycoforms in many functions of the immune system (Rudd et al., 2001). Because toxin A neutralization by free SC occurs at concentrations found in human milk (Underdown et al., 1977), the process identified in *in vitro* models appears to be physiologically relevant.

The role of SC and SIgA in the innate defense of mucosal surfaces has been demonstrated in other infectious models and was shown again to largely rely on the glycosylated nature of SC. Sialyloligosaccharides in SIgA are necessary for preventing epithelial adhesion of *E. coli* through type I fimbrial lectin (Schroten et al., 1998; Wold et al., 1990). Inhibition of adhesion of *Helicobacter pylori* to human gastric surface mucous cells by SIgA purified from human colostrum requires fucose residues on SC (Boren et al., 1993), whereas removal of sialic acid residues from the immunoglobulin moiety is without effect (Falk et al., 1993). The carbohydrate side chains of SIgA serve as a docking site for ricin toxin; human SIgA with no Fab-dependent specificity for ricin reduces attachment to the apical surface of epithelial cell lines in culture and to the luminal surfaces of human intestinal villi via SC and the IgA heavy chain (Mantis et al., 2004). These results are supported by molecular models of human SIgA1 indicating that N- and O-linked glycans protrude extensively from the Fcα and SC domains, allowing accessibility to bacterial and viral adhesins (Mattu et al., 1998; Royle et al., 2003). In addition to preventing contacts between the epithelium and pathogens, such interactions might also promote tethering of antibody–antigen complexes to the associated mucus layer (Biesbrock et al., 1991; Saltzman et al., 1994).

The protective role of free SC has also been reported to take place in the respiratory tract. SC, as well as SIgA, interacts directly with a surface protein of *Streptococcus pneumoniae*, choline binding protein A (CbpA) (Hammerschmidt et al., 1997), which was reported to enhance colonization of the nasopharynx of infant rats (Rosenow et al., 1997) (see Chapter 3). Binding is dependent on stretches present in domains 3 and 4 of SC (Elm et al., 2004; Lu et al., 2003) and on a highly conserved hexapeptide motif within CbpA (Hammerschmidt et al., 2000; Luo et al., 2005). Direct interaction of CbpA with the extracellular domain of pIgR facilitates the invasion of the human nasopharyngeal cell line Detroit 562 (Zhang et al., 2000). However, invasion is observed only when human cell lines are exposed to the unencapsulated, nonpathogenic *S. pneumoniae* strain R6x (Brock et al., 2002). This species and strain limitation favors the interpretation that the excess of free SC and SIgA in secretions will normally compete with cell-associated pIgR for binding to *S. pneumoniae*, thus blocking the attachment of the bacterium to the surface of epithelial cells (Kaetzel, 2001). This would represent another example whereby free SC and SC bound to pIgA contribute to innate immunity.

In addition to its ability to bind to several bacterial molecular patterns, SC has also been shown to potentiate the protective role of pIgA. With the establishment of expression systems producing milligram amounts of specific pIgA and human recombinant SC, it has become possible to compare the biological function of pIgA versus SIgA reconstituted *in vitro*. In a mouse model of respiratory infection by *Shigella flexneri*, the protective capacity of a pIgA specific for this bacterium is enhanced upon its association to SC (Phalipon

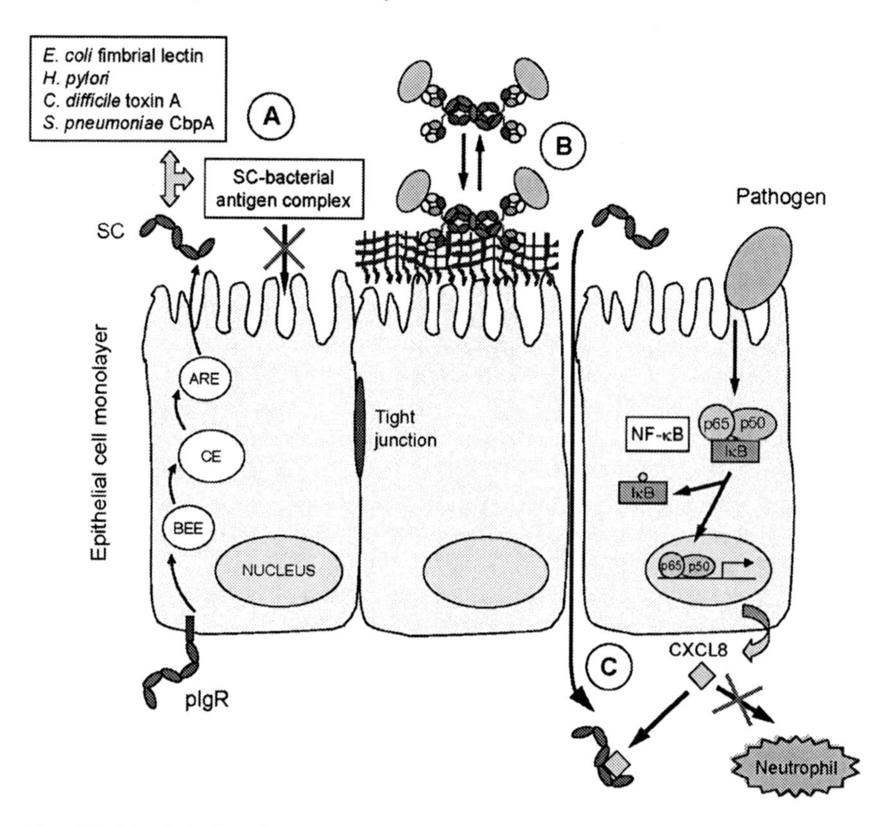

FIG. 8.1. Multiple functions of SC at epithelial surfaces. **(A)** SC alone (or bound to pIgA, not represented) interacts with various microorganism adhesins and toxins through its carbohydrate moieties to prevent attachment to epithelial cell surface, a competitor effect referred to as "scavenger" function in the text. **(B)** SIgA is anchored in the mucus layer overlying the epithelial monolayer through the sugar residues present in SC; this function of SC might favor the establishment of immune complexes between secreted SIgA and microorganisms trapped in the mucus. **(C)** Exposure to pathogens leads to inflammation-induced breaches in the epithelium that permit paracellular entry of macromolecules and microorganisms. Upon binding to proinflammatory CXCL-8, SC contributes to limit recruitment of leukocytes that would otherwise further damage the inflamed epithelium. BEE, basolateral early endosome; CE, common endosome; ARE, apical recycling endosome; boxes p65-p50-IκB represents the cytoplasmic, inactive form of the transcription factor NF-κB, which after IκB phosphorylation acquires the capacity to migrate to the nucleus and activate gene expression.

et al., 2002). SC is involved in establishing local interactions with bronchial mucus that result in the specific localization of the SIgA molecule, compared to the random distribution of pIgA in the lung tissue. This differential distribution of the two forms of the antibody impacts on the number and localization of bacteria within the infected tissue and contributes to the difference observed in the protective capacities of SIgA versus pIgA. *In vitro* deglycosylation of SC with *N*-endoglycosidase H prior to its association to pIgA abrogates the capacity of SC to anchor the reconstituted SIgA molecule to mucus and results in a level of protection comparable to pIgA alone. Thus, SC contributes to efficient SIgA-mediated protection by permitting, through its carbohydrate residues, the appropriate localization of the antibody and, in turn, the exclusion of pathogens from the mucosal surfaces.

Another intriguing issue of the role of SC in innate-like immunity has recently been reported. SC produced by cultures of human primary bronchial cells forms a stable complex with chemokines CXCL-8, epithelial neutrophil-activating peptide-78, growth-related oncogene-α, and RANTES (Marshall et al., 2001). Binding to CXCL-8 requires the carbohydrate moiety of SC. As a consequence of this interaction, neutrophil migration in response to CXCL8 in the micro-Boyden chamber assay, as well as in CXCL-8-mediated neutrophil transendothelial migration, is inhibited by SC in a dose-dependent manner. Whether such a mechanism takes place in the intestinal mucosa remains to be established. Nevertheless, these results suggest that free SC and possibly SIgA can serve as traps for polypeptides involved in proinflammatory processes, thereby limiting subsequent neutrophil responses.

Altogether, this implies that free SC can be seen per se as a factor contributing to the prevention of the interaction of pathogens with mucosal surfaces, as a scavenger of harmful mediators and, when bound to pIgA, as a companion potentializing the antibody molecular function (Fig. 8.1).

8.3. Transport of SIgA Across Peyer's Patches

Although luminal antigens can be trapped by DCs present in the lamina propria and extending their dendrites across the epithelium barrier (Rescigno et al., 2001), most of them gain access to Peyer's patches (PPs) (Owen, 1999) via transfer across M-cells that are scattered among the columnar epithelial cells in the follicle-associated epithelium. SIgA present in the luminal compartment also appears to be translocated (or retrotransported, by opposition to the epithelial secretory pathway ensuring passage from the tissue to the lumen) through M-cells that express a putative SIgA-specific receptor (Roy and Varvayanis, 1987; Weltzin et al., 1989) (see Chapter 9). Irrespective of its antigen-binding specificity, SIgA has the capacity to adhere selectively to M-cells (Mantis et al., 2002), a feature mediated by domains Cα1 and Cα2 of the antibody molecule. The purpose of the IgA-M cell interaction has been addressed in a study based on *in vitro* and *in vivo* observations (Rey et al., 2004) (Fig. 8.2). Incubation of cells recovered from mouse PPs with green fluorescent protein (GFP)-labeled SIgA indicated that DCs, $CD4^+$ T-cells, and B-cells associate selectively with the SIgA. Administration of GFP-labeled SIgA into a murine ligated intestinal loop confirmed *in vivo* targeting of PPs and subsequent transport to the underlying gut-associated lymphoid tissue (GALT). In the GALT, SIgA antibodies are found in association with DCs in the subepithelial dome region and with CD4⁺ T-cells in the interfollicular region rich in T-cells. Such an interaction is dependent on the presence of pIgA in the SIgA molecule. Laser-scanning confocal microscopy indicated that SIgA binds to and is internalized by DCs, whereas only surface association was observed with CD4⁺ T-cells. Targeting of DCs anatomically positioned in the subepithelial dome region underlying the follicle-associated epithelium might provide a mechanistic explanation to the immunogenicity after oral delivery of foreign antigens linked by genetic engineering to SIgA (Corthésy et al., 1996; Zhou et al., 1995). In addition, the interaction of SIgA with CD4⁺ T-cells might modulate local immune responses known to exhibit a cytokine profile dominated by transforming growth factor (TGF)-β, interleukin (IL)-10, and IL-4 (Gonnella et al., 1998). In this respect, it is worth emphasizing that previous investigations of IgA receptors in PPs were focused on the capacity of SIgA to interact specifically with isolated T-cells. (Brière et al., 1988; Kiyono et al., 1982; Sandor et al., 1990). Our observation that SIgA interacts with CD4⁺ T-cells within PPs provides another clue as to the possible regulatory function of this class of antibodies.

8.4. Induction of Immune Responses After SIgA Retrotransport

The possible immunomodulatory effects of the interaction between DC and SIgA have only been investigated in a small number of *in vitro* studies using human monocyte-derived DCs exhibiting a myeloid phenotype (Geissmann et al., 2001; Heystek et al., 2002). These studies gave conflicting results in terms of DC maturation, partly due to differences in the preparation and heterogeneous sources of SIgA. Further, it is highly likely that the properties and functions of these cells differ in their natural tissue environment (Iwasaki and Kelsall, 1999; Williamson et al., 2002). The immunological consequences of the interaction between SIgA and DC have recently been addressed in mice (Favre et al., 2005) (Fig. 8.2). Oral delivery of recombinant SIgA (rSIgA) reconstituted *in vitro* with hSC and mouse pIgA induced anti-hSC-specific antibody responses in mucosal and peripheral tissues. Antibody titers were similar to those obtained when combining hSC with the prototype mucosal adjuvant cholera toxin (CT) used as a control capable to induce well-characterized mucosal immune responses, including Th2 deviation, DC migration, and expression of costimulatory molecules (Anjuère et al., 2004). Because hSC alone leads to very weak antibody production, these results indicate that entry of rSIgA into PPs permits efficient delivery and processing of the associated hSC antigen. Capture of antibody–antigen complexes by DCs is consistent with studies performed with splenic DCs (Mellman and Steinman, 2001) and extends this concept to the GALT compartment. In T-cell proliferation assays, administration of rSIgA alone led to reduced levels of activation

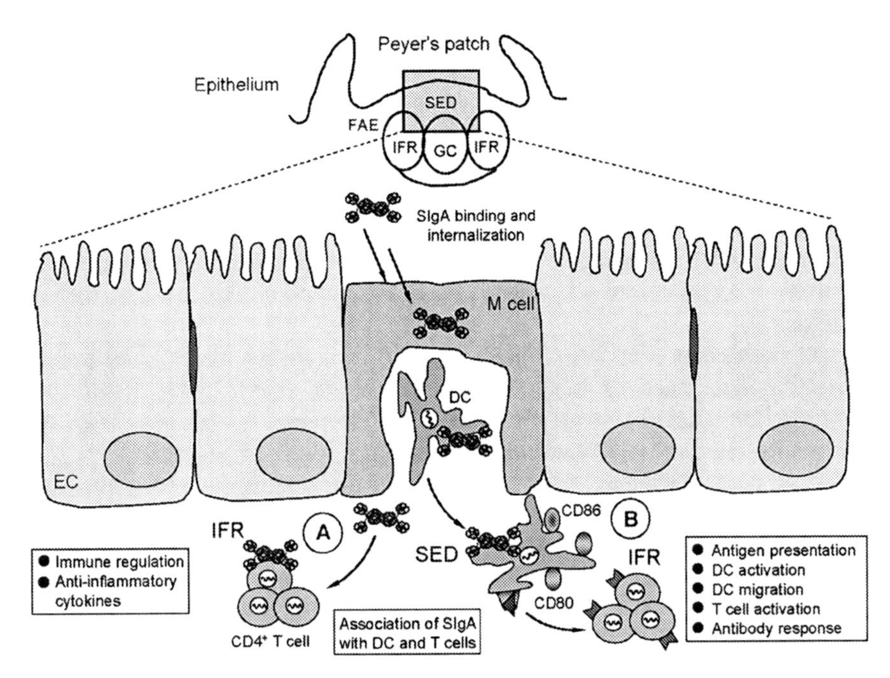

FIG. 8.2. Schematic representation of the trafficking of SIgA *in vivo* after gastrointestinal administration **(A)** and its effect on immune responses **(B)**. SIgA binds to M-cells and is transported across the epithelium to the SED region of PPs. Association with DCs results in the internalization of SIgA, whereas the interaction with $CD4^+$ T-cells is limited to the cell surface. Processing of SIgA by DCs triggers activation within the PP and confers upon DCs the capacity to migrate to the IFR rich in T-cells. When SIgA is used as a foreign antigen carrier, antigen-specific T-cell activation occurs in the draining mesenteric lymph node, resulting in overexpression of Th2 and noninflammatory cytokines, including IL-4, IL-10, and TGF-β. Ultimately, antigen-specific antibodies are recovered in secretions and blood. The upper part of the drawing depicts the overall structure of a PP, and the gray rectangle is the area magnified in the bottom part of the figure. FAE, follicle-associated epithelium; SED, subepithelial dome; IFR, interfollicular region; GC, germinal center; EC, epithelial cell; M cell, microfold cell; DC, dendritic cell. Epitope-loaded MHC class II molecule and costimulation markers are depicted on DCs.

compared to hSC administered with CT. Delivery of hSC in the context of rSIgA caused the cytokine profile to be skewed toward a Th2 phenotype, as reflected by reduced expression of interferon (IFN-γ compared to mice fed hSC in the presence of CT. The mucosal immunomodulatory capacity of rSIgA was further suggested by the increase of TGF-β production by cells isolated from mesenteric lymph nodes. The differences in cytokine profiles between the experimental groups indicate that SIgA triggers mucosal immune responses that defend against luminal antigens while controlling inflammatory processes, an effect not observed with CT. In addition, rSIgA triggers migration of DC to the T-cell-rich regions of PPs, although it is delayed compared to DC migration in mice having received hSC combined with CT. These results provide evidence that mucosal SIgA retrotransport promotes antigen delivery, thus contributing to effector and/or regulatory pathways characteristic of the intestinal mucosal compartment. Neutralization of antigens by SIgA in the lumen, followed by transport through PPs, might also redirect pathogens with the capacity to evade DCs back to cells with antigen-presenting capacity, thus providing the host with a complementary degree of protective immunity.

The expression of CD80, a marker of DC activation, on DCs recovered from PPs and spleen of mice fed with rSIgA was shown to be delayed as compared to mice fed with hSC plus CT. Differences in the intensities of the responses between immunization groups might be explained by the rapid downregulation of CD80 seen after rSIgA administration and the upregulation observed after hSC plus CT administration. These results suggest that DC interacting with SIgA are prone to induce a more regulated, anti-inflammatory response, yet still effective immune response, than that mediated by DCs upon exposure to CT. Provided that DCs have evolved a substantial number of proinflammatory strategies to thwart gastrointestinal pathogens, it is conceivable that counteracting mechanisms exist to guarantee return to steady-state conditions and maintenance of homeostasis. In this respect, cross-talk between DC and epithelial cells producing thymic stromal lymphopoietin conditions the DCs to polarize T-cells toward a Th2 phenotype (IL-6, IL-10) and to promote the production of mucosal IgA (Rimoldi et al., 2005). SIgA present during secondary (recall) stimulation might in addition promote presentation of “neutralized” antigens, thus “educating” targeted DC to respond in a manner that limits the risk of overstimulating the local immune protection system.

8.5. Anti-Inflammatory Properties of (S)IgA

As mentioned earlier, protective mechanisms involving SIgA or free SC tend to limit an uncontrolled induction of inflammation that would lead to extensive damage to the host mucosal barrier and, hence, function. Therefore, it is not surprising that the IgA molecule behaves differently from other classes of antibodies. Indeed, in contrast to circulating IgG, IgM, and IgE antibodies, the Fc portion of IgA does not confer to the molecule effector functions that lead to the release of inflammatory mediators after complement activation and phagocytosis (Fagarasan and Honjo, 2004; Kerr, 1990; Pasquier et al., 2005; Russell et al., 1989) (see Chapter 6). Such properties are consistent with the multitask role of protecting against foreign substances and microbes while not subjecting the mucosa to undue inflammation (Macpherson et al., 2001; Wu and Weiner, 2003).

In the previous paragraphs, we discussed the SIgA- or free-SC-mediated control of inflammation within the lumen and in the PP. Recently, however, another level of control has been reported in epithelial cells. Using an *in vitro* model mimicking pIgR-mediated pIgA transcytosis and lipopolysaccharide (LPS) sampling from the apical surface of polarized epithelial cells, the ability of intracellular pIgA to downregulate proinflammatory responses by epithelial cells was shown to take place (Fernandez et al., 2003). In this model, a protective monoclonal IgA specific for *Shigella flexneri* LPS neutralized this pathogen-associated molecular pattern within the apical recycling endosome compartment, thus precluding LPS-triggered activation of nuclear factor (NF)-κB and subsequent proinflammatory responses (Fig. 8.3). Such a mechanism, by limiting the amount of translocated LPS, might also in turn limit activation of lamina propria leukocytes and reduce the release of associated proinflammatory mediators. This intracellular IgA-mediated neutralization of a proinflammatory response should be envisioned as

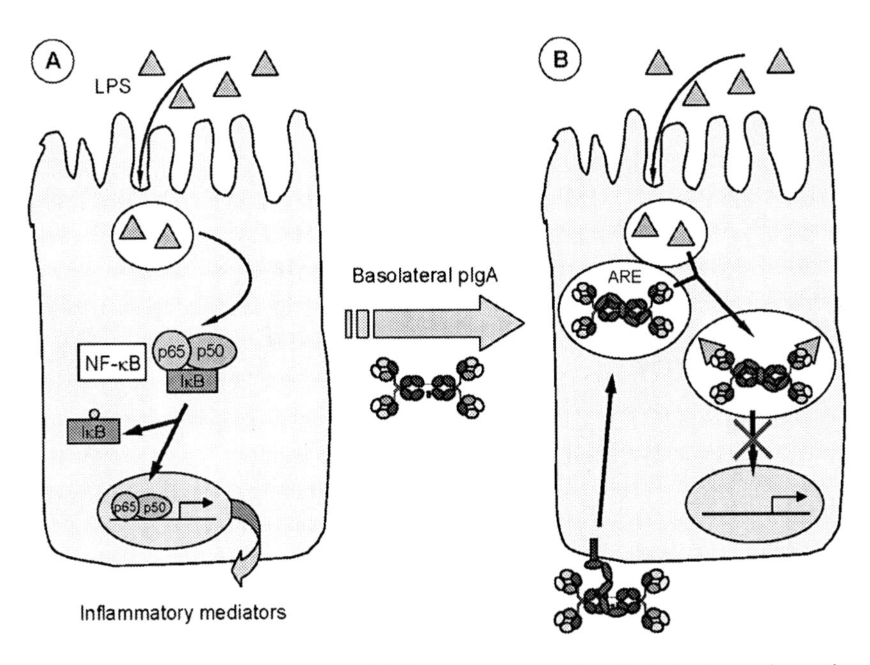

FIG. 8.3. Cellular colocalization of LPS and transcytosing SIgA leads to abrogation of inflammation-mediated processes in epithelial cells (ECs) *in vitro*. LPS taken up by ECs stimulates cellular pathways dependent on Toll-like receptor-4 signaling and kinases that are required for NF-κB activation **(A)**. A model pathogen-associated molecular pattern, LPS, in submembrane vesicles is intercepted by SIgA antibodies on their way to the luminal side of the epithelial cell **(B)**. Such a mechanism promotes mucosal homeostasis by blocking nuclear translocation of NF-κB and subsequent transactivation of proinflammatory gene expression including TNF-α, IL-18, and macrophage inflammatory protein-1α and 2.

an additional mechanism completing the SIgA-mediated immune exclusion occurring in the lumenal compartment and leading to pathogen clearance. It can reasonably be hypothesized that neutralization of LPS might be part of a broader mechanism in the gut to control the huge amount of microbe-associated molecular patterns (lipoproteins, flagellins, DNA and RNA motifs, GPI anchors, outer membrane proteins, lipoteichoic acid, etc.) carried by intestinal bacteria.

Interestingly, another level of control appears to take place in the PP following SIgA-mediated retrotransport of bacteria–SIgA immune complexes. Using a rabbit ileal loop model of infection with *Shigella flexneri*, retrotransport of immune complexes added into the luminal compartment was shown to occur. In contrast to ileal loops receiving bacteria only and exhibiting a massive destruction of the intestinal epithelium, the opsonization of *Shigella* by SIgA within immune complexes guaranteed the absence of tissue alteration even at the level of the PP serving as the preferential site of entry. Modulation of the local immune response within the PP was appraised using real-time reverse transcription–polymerase chain reaction (RT-PCR) analysis of a battery of proinflammatory [IL-6, tumor necrosis factor (TNF)-α, IFN-γ] and noninflammatory (IL-10 and TGF-β) cytokines. Despite the presence of bacteria within the PPs of ileal loops exposed to *Shigella*–SIgA complexes, the level of expression of all four cytokines tested was similar to that measured in PP of mice having received SIgA only. Conversely, upregulation of proinflammatory cytokine expression was obtained in PPs of ileal loops treated with the bacteria alone, whereas the expression of noninflammatory cytokines remained stable (Phalipon and Corthésy, unpublished data). These findings suggest that in the gut environment, noninflammatory effector mechanisms predominate but are overwhelmed in the presence of pathogens affecting the balance of local cytokines. Whatever the mechanism taking place, these results corroborate the data presented in the previous paragraphs, emphasizing the diversity of the processes that have been developed through evolution to maintain the integrity of the mucosal barrier.

8.6. Role of SIgA in Gut Homeostasis

The functions of SIgA that have been discussed thus far mainly concern protection against pathogens. However, the mammalian gastrointestinal tract harbors a complex ecosystem consisting of vast numbers of bacteria in homeostasis with the host immune system. By young adulthood, humans and other mammals are host to about 10^{12} viable bacteria per gram of colonic content consisting of 500–1000 microbial species and outnumbering host cells 100-fold (Hooper and Gordon, 2001). The magnitude of the interaction between commensal bacteria and mammals exerts fundamental influences on the physiology of both. The most impressive feature of this relationship might be that the host not only tolerates but requires colonization by commensal microorganisms for its own development and health. Commensal bacteria act as an important antigenic stimulus for the maturation of GALT, implicated in the induction of local immune responses (Cebra, 1999; Mazmanian et al., 2005; Rhee et al., 2004; Umesaki and Setoyama, 2000). Commensals boost overall SIgA antibody responses in a strain-dependent manner (Cukrowska et al., 2002; Fang et al., 2000; Fukushima et al., 1998; Tejada-Simon et al., 1999). In contrast to mucosal SIgA specific for pathogenic microorganisms, this ensures peaceful control of the endogenous microbiota through a broad spectrum of unselected, low-affinity SIgA (also referred to as "natural" SIgA) to redundant epitopes on gut microorganisms (Stoel et al., 2005). Comparison of wild-type mice to mice deficient in activation-induced cytidine deaminase (AID), whose B-cells cannot switch to IgA, suggests that IgA regulates the microbiota in the gut and limits the activation of local and systemic immune responses, thus maintaining gut homeostasis (Fagarasan et al., 2002; Suzuki et al., 2004).

Introduction of a new microbiota into the intestinal tract of mice lacking T-cells results in induction of some specific SIgA (Macpherson et al., 2000), yet the overall response in the gut was only 20–30% of normal. Deliberate colonization of germ-free mice with a commensal microbe has led to the demonstration that (1) the number of IgA-producing plasma cells reaches a plateau in the lamina propria after 6 weeks, (2) the bacteria in the gut are continuously coated with IgA, and (3) IgA contributes to the maintenance of a steady-state number of bacteria in the feces for up to 1 year (Shroff et al., 1995). Thus, successful establishment of commensal microbiota might rely on the capacity to initially induce and then rapidly downregulate mucosal responses leading to the production of specific SIgA. Although the sampling of low amounts of antigen associated with SIgA capable of entering PPs across M-cells and targeting DCs (Rey et al., 2004) might contribute to the mechanism of downregulating subsequent responses, other inductive sites distributed along the gastrointestinal tract can contribute to the supply of specific regulatory cells in the gut lamina propria (Eberl, 2005; Jang et al., 2004; Lorenz and Newberry, 2004; Suzuki et al., 2000). Sampling of commensals by PPs appears to educate the mucosal immune system to sense the gut content (Macpherson and Uhr, 2004). Capture by DCs in the subepithelial dome region of the PP limits spreading beyond mesenteric lymph nodes, providing a clue as to how local processing can activate local responses, including synthesis of specific IgA. Another mechanism whereby commensal Gram-negative bacteria could promote SIgA-mediated homeostasis is by signaling through Toll-like receptor 4, which upregulates expression of pIgR in epithelial cells while minimizing pro-inflammatory responses (Schneeman et al., 2005). It has been shown that maternal SIgA antibodies in milk influence the pattern of gene expression in intestinal epithelial cells of neonatal mice, including pIgR (Jenkins et al., 2003). Taken together, these results suggest a complex regulatory network in which cross-talk between commensal microbes and host cells regulate the synthesis and transport of SIgA.

The crucial role of SIgA in maintaining bacterial homeostasis is further reflected by the role played by the antibody in microbial biofilm formation *in vitro* (Bollinger et al., 2003, 2006). Indeed, many nonpathogenic bacteria have been observed to be coated with SIgA antibody (van der Waaij et al., 1996), although this intimate contact does not result in the clearance of these bacteria. The potential role of biofilms in complexes of bacteria–bacteria or bacteria–host interactions that take place in the gut remains largely unexplored. Biofilms have been proposed to ensure a mode of steady-state growth of the endogenous microbiota (Costerton et al., 1995). SIgA-mediated biofilm formation might also explain why bacteria that bind SIgA have a selective advantage in the gut (Friman et al., 1996). The demonstration of the involvement of SIgA in biofilm formation in the gut *in vivo* has recently been reported in rats, baboons, and humans (Palestrant et al., 2004). A decapeptide in human SC has also been shown to stimulate the growth of bifidobacteria (Liepke et al., 2002), which might confer to SIgA prebiotic properties after association with nonpathogenic bacteria.

8.7. Concluding Remarks

In addition to conferring increased stability to polymeric IgA and augmenting its biological availability at mucosal surfaces, SC possesses innate-like activities that enhance protection against multiple gastrointestinal infectious agents. Unique properties of SC include the following: (1) its high abundance in milk in the form of free SC (up to 25 μM); (2) its high content of heterogenous glycans, contributing 15–20% of its molecular weight, (3) its specificity for polymeric immunoglobulins (IgA and IgM); (4) its initial synthesis as an epithelial transporter (pIgR), which is "sacrificed" after a single trip across the epithelial cell by cleavage to SC.

Given that SIgA is transported across M-cells in PPs (Rey et al., 2004) and triggers mucosal and systemic immune responses (Favre et al., 2005), it is conceivable that retrotransport of SIgA–antigen complexes contributes significantly to immune regulation by providing a mechanism for luminal antigen to gain access to professional antigen-presenting cells such as DCs in the subepithelial dome region. The modulating effect of SIgA would thus be exerted by masking bacterial and viral epitopes in the local environment where the infection occurs, hence allowing for handling of the invading enemy under neutralizing conditions (Fig. 8.4).

During primary infection, uptake of microbes by M-cells can result in a microbial load too great to be controlled and contained by the cellular immune defenses concentrated within mucosal lymphoid follicles. In the frame of recall responses, association with preexisting neutralizing SIgA will lead to detection of infectious microorganisms by the epithelial sampling machinery in a form limiting the risk of overwhelming the local immune protection system. A major advantage of using the M-cell pathway is that antibodies

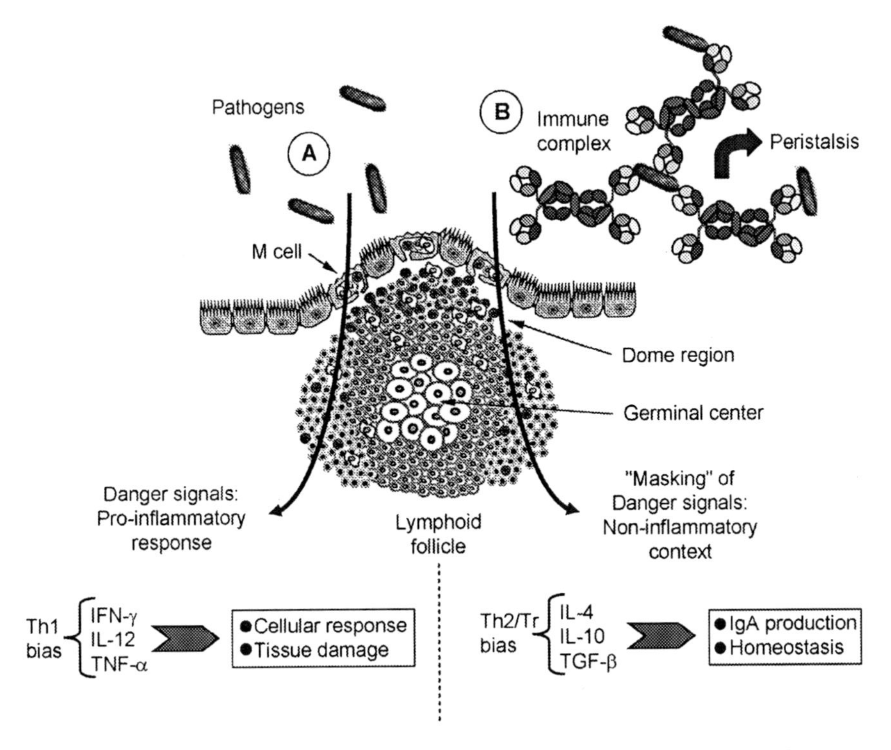

Fig. 8.4. Proposed model to explain the mode of action of SIgA–antigen immune complexes after uptake by the PP in the intestine. **(A)** Pathogens entering across M-cells are processed by underlying antigen-presenting cells, which will trigger neighboring T-cells and production of proinflammatory cytokines. The local cellular responses leading to pathogen neutralization are often accompanied by acute or chronic tissue damage. **(B)** In the scenario involving the entry of immune complexes, Fab- and SC-dependent opsonization by SIgA prevents activation of proinflammatory pathways through masking of pathogen-associated molecular patterns. Resulting cellular responses are skewed toward the production of cytokines governing IgA switch in the mucosal environment as well as induction of tolerance. The conjunction of these events contributes to the maintenance of homeostasis.

or immune complexes are rarely targeted to phagolysosomes, but are carried through the M-cell by pinocytosis and released into the pocket (Owen, 1999). Consequently, the integrity of antigenic structures taken up from the lumen is preserved, allowing selective targeting to underlying antigen-presenting cells that display IgA receptors. Homeostasis is fine-tuned through multiple mechanisms, including classical antigen neutralization mediated by SIgA, binding of antigens to IgG in secretions (Yoshida et al., 2004), and trapping of antigens by lamina propria DC (Niess et al., 2005; Rescigno et al., 2001).

The presence on the SIgA molecule of Fab domains enabling selective association with antigenic structures and multiple glycans serving as nonspecific docking sites for a large battery of pathogens marks this class of antibody with the potential to participate in both adaptive and innate immune responses. Broad-spectrum recognition of markers of inflammation (bacterial and viral infection, toxins, immune mediators) through different patterns associated with SIgA would favor both the "classical" antigen neutralization by the antibody and stimulation of mucosal immune responses under gentle, nondamaging conditions preserving the epithelial barrier. This leads us to hypothesize that in the naturally noninflammatory mucosal environment, SIgA facilitates a return to homeostasis after an inflammatory burst, and under steady-state conditions, it keeps the local environment in balance.

Thus, in terms of humoral immunity at mucosal surfaces, SIgA appears to combine properties of a neutralizing agent (immune exclusion) and of an immunomodulator, underscoring a novel facet of its complex functioning. The information summarized in this chapter illustrates the diversity of biological properties of SIgA, which, given the crucial role of this class of antibody in maintaining the integrity of mucosal barriers, is deserving of further study.

Acknowledgments. Research conducted in the laboratory of B.C. is supported by the Swiss Science Research Foundation (grant Nos 3200–068038 and 3200–109545). A.P. warmly thanks Philippe J. Sansonetti for his unfailing support.

References

Anjuère, F., Luci, C., Lebens, M., Rousseau, D., Hervouet, C., Milon, G., Holmgren, J., Ardavin, C., and Czerkinsky, C. (2004). *In vivo* adjuvant-induced mobilization and maturation of gut dendritic cells after oral administration of cholera toxin. *J. Immunol.* 173:5103–5111.

Biesbrock, A. R., Reddy, M. S., and Levine, M. J. (1991). Interaction of a salivary mucin-secretory immunoglobulin A complex with mucosal pathogens. *Infect. Immun.* 59:3492–3497.

Bollinger, R. R., Everett, M. L., Palestrant, D., Love, S. D., Lin, S. S., and Parker, W. (2003). Human secretory immunoglobulin A may contribute to biofilm formation in the gut. *Immunology* 109:580–587.

Bollinger, R. R., Everett, M. L., Wahl, S. D., Lee, Y. H., Orndorff, P. E., and Parker, W. (2006). Secretory IgA and mucin-mediated biofilm formation by environmental strains of *Escherichia coli*: role of type 1 pili. *Mol. Immunol.* 43:378–387.

Bomsel, M., Heyman, M., Hocini, H., Lagaye, S., Belec, L., Dupont, C., and Desgranges, C. (1998). Intracellular neutralization of HIV transcytosis across tight epithelial barriers by anti-HIV envelope protein dIgA or IgM. *Immunity* 19:277–287.

Boren, T., Falk, P., Roth, K. A., Larson, G., and Normark, S. (1993). Attachment of *Helicobacter pylori* to human gastric epithelium mediated by blood group antigens. *Science* 262:1892–1895.

Brière, F., Paliard, X., and De Vries, J. E. (1988). Induction of the receptor for the Fc portion of IgA by secretory IgA on human T cell lines and T cell clones. *Eur. J. Immunol.* 18:445–450.

Brock, S. C., McGraw, P. A., Wright, P. F., Crowe, J. E., Jr. (2002). The human polymeric immunoglobulin receptor facilitates invasion of epithelial cells by *Streptococcus pneumoniae* in a strain-specific and cell type-specific manner. *Infect. Immun.* 70:5091–5095.

Cebra, J. J. (1999). Influences of microbiota on intestinal immune system development. *Am. J. Clin. Nutr.* 69:1046S–1051S.

Chintalacharuvu, K. R., and Morrison, S. L. (1997). Production of secretory immunoglobulin A by a single mammalian cell. *Proc. Natl. Acad. Sci. USA* 94:6364–6368.

Corthésy, B., Benureau, Y., Perrier, C., Fourgeux, C., Parez, N., Greenberg, H., and Schwartz-Cornil, I. (2006). Rotavirus anti-VP6 secretory IgA contributes to protection via intracellular neutralization but not via immune exclusion. *J. Virol.* 80:10,692–10,699.

Corthésy, B., Kaufmann, M., Phalipon, A., Peitsch, M., Neutra, M. R., and Kraehenbuhl, J. P. (1996). A pathogen-specific epitope inserted into recombinant secretory immunoglobulin A is immunogenic by the oral route. *J. Biol. Chem.* 271:33,670–33,677.

Costerton, J. W., Lewandowski, Z., Caldwell, D. E., Korber, D. R., and Lappin-Scott, H. M. (1995). Microbial biofilms. *Annu. Rev. Microbiol.* 49:711–745.

Crottet, P., and Corthésy, B. (1998). Secretory component delays the conversion of secretory IgA into antigen-binding competent F(ab')2: A possible implication for mucosal defense. *J. Immunol.* 161:5445–5453.

Cukrowska, B., Lodinova-Zadnikova, R., Enders, C., Sonnenborn, U., Schulze, J., and Tlaskalova-Hogenova, H. (2002). Specific proliferative and antibody responses of premature infants to intestinal colonization with nonpathogenic probiotic *E. coli* strain Nissle 1917. *Scand. J. Immunol.* 55:204–209.

Dallas, S. D., and Rolfe, R. D. (1998) Binding of *Clostridium difficile* toxin A to human milk secretory component. *J. Med. Microbiol.* 47:879–888.

Eberl, G. (2005). Inducible lymphoid tissues in the adult gut: recapitulation of a fetal developmental pathway? *Nat. Rev. Immunol.* 5:413–420.

Elm, C., Braathen, R., Bergmann, S., Frank, R., Vaerman, J. P., Kaetzel, C. S., Chhatwal, G. S., Johansen, F. E., and Hammerschmidt, S. (2004). Ectodomains 3 and 4 of human polymeric immunoglobulin receptor (hpIgR) mediate invasion of *Streptococcus pneumoniae* into the epithelium. *J. Biol. Chem.* 279:6296–6304.

Fagarasan, S., Muramatsu, M., Suzuki, K., Nagaoka, H., Hiai, H., and Honjo, T. (2002). Critical roles of activation-induced cytidine deaminase in the homeostasis of gut flora. *Science* 298:1424–1427.

Fagarasan, S., and Honjo, T. (2004). Regulation of IgA synthesis at mucosal surfaces. *Curr. Opin. Immunol.* 16:277–283.

Falk, P., Roth, K. A., Boren, T., Westblom, T. U., Gordon, J. I., and Normark, S. (1993). An *in vitro* adherence assay reveals that *Helicobacter pylori* exhibits cell lineage-specific tropism in the human gastric epithelium. *Proc. Natl. Acad. Sci. USA* 90:2035–2039.

Fang, H., Elina, T., Heikki, A., and Seppo, S. (2000). Modulation of humoral immune response through probiotic intake. *FEMS Immunol. Med. Microbiol.* 29:47–52.

Favre, L., Spertini, F., and Corthésy, B. (2005). Secretory IgA possesse intrinsic modulatory properties stimulating mucosal and systemic immune responses. *J. Immunol.* 175:2793–2800.

Fernandez, M. I., Pedron, T., Tournebize, R., Olivo-Marin, J. C., Sansonetti, P. J., and Phalipon. (2003). A. Anti-inflammatory role for intracellular dimeric immunoglobulin a by neutralization of lipopolysaccharide in epithelial cells. *Immunity* 18:739–749.

Friman, V., Adlerberth, I., Connell, H., Svanborg, C., Hanson, L. A., and Wold, A. E. (1996). Decreased expression of mannose-specific adhesins by *Escherichia coli* in the colonic microflora of immunoglobulin A-deficient individuals. *Infect. Immun.* 64:2794–2798.

Fujioka, H., Emancipator, S. N., Aikawa, M., Huang, D. S., Blatnik, F., Karban, T., DeFife, K., and Mazanec, M. B. (1998). Immunocytochemical colocalization of specific immunoglobulin A with sendai virus protein in infected polarized epithelium. *J. Exp. Med.* 188:1223–1229.

Fukushima, Y., Kawata, Y., Hara, H., Terada, A., and Mitsuoka, T. (1998). Effect of a probiotic formula on intestinal immunoglobulin A production in healthy children. Int. *J. Food Microbiol.* 42:39–44.

Geissmann, F., Launay, P., Pasquier, B., Lepelletier, Y., Leborgne, M., Lehuen, A., Brousse, N., and Monteiro, R. C. (2001). A subset of human dendritic cells expresses IgA Fc receptor (CD89), which mediates internalization and activation upon cross-linking by IgA complexes. *J. Immunol.* 166:346–352.

Giugliano, L. G., Ribeiro, S. T., Vainstein, M. H., and Ulhoa, C. J. (1995). Free secretory component and lactoferrin of human milk inhibit the adhesion of enterotoxigenic *Escherichia coli. J. Med. Microbiol.* 42:3–9.

Gonnella, P. A., Chen, Y., Inobe, J., Komagata, Y., Quartulli, M., and Weiner, H. L. (1998). In situ immune response in gut-associated lymphoid tissue (GALT) following oral antigen in TCR-transgenic mice. *J. Immunol.* 160:4708–4718.

Hammerschmidt, S., Talay, S. R., Brandtzaeg, P., and Chhatwal, G. S. (1997). SpsA, a novel pneumococcal surface protein with specific binding to secretory immunoglobulin A and secretory component. *Mol. Microbiol.* 25:1113–1124.

Hammerschmidt, S., Tillig, M. P., Wolff, S., Vaerman, J. P., and Chhatwal, G. S. (2000). Species-specific binding of human secretory component to SpsA protein of *Streptococcus pneumoniae* via a hexapeptide motif. *Mol. Microbiol.* 36:726–736.

Heystek, H. C., Moulon, C., Woltman, A. M., Garonne, P., and van Kooten, C. (2002). Human immature dendritic cells efficiently bind and take up secretory IgA without the induction of maturation. *J. Immunol.* 168:102–107.

Hooper, L. V., and Gordon, J. I. (2001). Commensal host-bacterial relationships in the gut. *Science* 292:1115–1118.

Huang, Y. T., Wright, A., Gao, X., Kulick, L., Yan, H., and Lamm, M. E. (2005). Intraepithelial cell neutralization of HIV-1 replication by IgA. *J. Immunol.* 174:4828–4835.

Iwasaki, A., and Kelsall, B. L. (1999). Freshly isolated Peyer's patch, but not spleen, dendritic cells produce interleukin 10 and induce the differentiation of T helper type 2 cells. *J. Exp. Med.* 190:229–239.

Jang, M. H., Kweon, M. N., Iwatani, K., Yamamoto, M., Terahara, K., Sasakawa, C., Suzuki, T., Nochi, T., Yokota, Y., Rennert, P. D., Hiroi, T., Tamagawa, H., Iijima, H., Kunisawa, J., Yuki, Y., and Kiyono, H. (2004). Intestinal villous M cells: An antigen entry site in the mucosal epithelium. *Proc. Natl. Acad. Sci. USA* 101:6110–6115.

Jenkins, S. L., Wang, J., Vazir, M., Vela, J., Sahagun, O., Gabbay, P., Hoang, L., Diaz, R. L., Aranda, R., and Martin, M. G. (2003). Role of passive and adaptive

immunity in influencing enterocyte-specific gene expression. *Am. J. Physiol:. Gastrointest. Liver Physiol.* 285:G714–G725.
Kaetzel, C. S. (2001). Polymeric Ig receptor: defender of the fort or Trojan horse? *Curr. Biol.* 11:R35–R38.
Kaetzel, C. S., Robinson, J. K., Chintalacharuvu, K. R., Vaerman, J. P., and Lamm, M. E. (1991). The polymeric immunoglobulin receptor (secretory component) mediates transport of immune complexes across epithelial cells: a local defense function for IgA. *Proc. Natl. Acad. Sci. USA* 88:8796–8800.
Kerr, M. A. (1990). The structure and function of human IgA. *Biochem. J.* 271:285–296.
Kiyono, H., McGhee, J. R., Mosteller, L. M., Eldridge, J. H., Koopman, W. J., Kearney, J. F., and Michalek, S. M. (1982). Murine Peyer's patch T cell clones. Characterization of antigen-specific helper T cells for immunoglobulin A responses. *J. Exp. Med.* 156:1115–1130.
Lamm, M. E. (1997). Interaction of antigens and antibodies at mucosal surfaces. *Annu. Rev. Microbiol.* 51:311–340.
Langley, R., Wines, B., Willoughby, N., Basu, I., Proft, T., and Fraser, J. D. (2005). The staphylococcal superantigen-like protein 7 binds IgA and complement C5 and inhibits IgA-Fc alpha RI binding and serum killing of bacteria. *J. Immunol.* 174:2926–2933.
Liepke, C., Adermann, K., Raida, M., Magert, H. J., Forssmann, W. G., and Zucht, H. D. (2002). Human milk provides peptides highly stimulating the growth of bifidobacteria. *Eur. J. Biochem.* 269:712–718.
Lindh, E. (2005). Increased resistance of immunoglobulin A dimers to proteolytic degradation after binding of secretory component. *J. Immunol.* 114:284–286.
Lorenz, R. G., and Newberry, R. D. (2004). Isolated lymphoid follicles can function as sites for induction of mucosal immune responses. *Ann. NY Acad Sci.* 1029:44–57.
Lu, L., Lamm, M. E., Li, H., Corthésy, B., and Zhang, J. R. (2003). The human polymeric immunoglobulin receptor binds to *Streptococcus pneumoniae* via domains 3 and 4. *J. Biol. Chem.* 278:48,178–48,187.
Luo, R., Mann, B., Lewis, W. S., Rowe, A., Heath, R., Stewart, M. L., Hamburger, A. E., Sivakolundu, S., Lacy, E. R., Bjorkman, P. J., Tuomanen, E., and Kriwacki, R. W. (2005). Solution structure of choline binding protein A, the major adhesin of *Streptococcus pneumoniae. EMBO J.* 24:34–43.
Macpherson, A. J., Gatto, D., Sainsbury, E., Harriman, G. R., Hengartner, H., and Zinkernagel, R. M. (2000). A primitive T cell-independent mechanism of intestinal mucosal IgA responses to commensal bacteria. *Science* 288:2222–2226.
Macpherson, A. J., Hunziker, L., McCoy, K., and Lamarre, A. (2001). IgA responses in the intestinal mucosa against pathogenic and non-pathogenic microorganisms. *Microbes Infect.* 3:1021–1035.
Macpherson, A. J., and Uhr, T. (2004). Induction of protective IgA by intestinal dendritic cells carrying commensal bacteria. *Science* 303:1662–1665.
Mantis, N. J., Cheung, M. C., Chintalacharuvu, K. R., Rey, J., Corthésy, B., and Neutra, M. R. (2002). Selective adherence of IgA to murine Peyer's patch M cells: evidence for a novel IgA receptor. *J. Immunol.* 169:1844–1851.
Mantis, N. J., Farrant, S. A., and Mehta, S. (2004). Oligosaccharide side chains on human secretory IgA serve as receptors for ricin. *J. Immunol.* 172:6838–6845.
Marshall, L. J., Perks, B., Ferkol, T., and Shute, J. K. (2001). IL-8 released constitutively by primary bronchial epithelial cells in culture forms an inactive complex with secretory component. *J. Immunol.* 167:2816–2823.

Mattu, T. S., Pleass, R. J., Willis, A. C., Kilian, M., Wormald, M. R., Lellouch, A. C., Rudd, P. M., Woof, J. M., and Dwek, R. A. (1998). The glycosylation and structure of human serum IgA1, Fab, and Fc regions and the role of N-glycosylation on Fc alpha receptor interactions. *J. Biol. Chem.* 273:2260–2272.

Mazanec, M. B., Kaetzel, C. S., Lamm, M. E., Fletcher, D., and Nedrud, J. G. (1992). Intracellular neutralization of virus by immunoglobulin A antibodies. *Proc. Natl. Acad. Sci. USA* 89:6901–6905.

Mazmanian, S. K., Liu, C. H., Tzianabos, A. O., and Kasper, D. L. (2005). An immunomodulatory molecule of symbiotic bacteria directs maturation of the host immune system. *Cell* 122:107–118.

Mellman, I., and Steinman, R. M. (2001). Dendritic cells: specialized and regulated antigen processing machines. *Cell* 106:255–258.

Mestecky, J., Russell, M. W., and Elson, C. O. (1999). Intestinal IgA: Novel views on its function in the defence of the largest mucosal surface. *Gut* 44:2–5.

Mostov, K. E., Kraehenbuhl, J. P., and Blobel, G. (1980). Receptor-mediated transcellular transport of immunoglobulin: synthesis of secretory component as multiple and larger transmembrane forms. *Proc. Natl. Acad. Sci. USA* 77:7257–7261.

Newburg, D. S. (1999). Human milk glycoconjugates that inhibit pathogens. *Curr. Med. Chem.* 6:117–127.

Niess, J. H., Brand, S., Gu, X., Landsman, L., Jung, S., McCormick, B. A., Vyas, J. M., Boes, M., Ploegh, H. L., Fox, J. G., Littman, D. R., and Reinecker, H. C. (2005). CX3CR1-mediated dendritic cell access to the intestinal lumen and bacterial clearance. *Science* 307:254–258.

Owen, R. L. (1999). Uptake and transport of intestinal macromolecules and microorganisms by M cells in Peyer's patches: A personal and historical perspective. *Semin. Immunol.* 11:157–163.

Palestrant, D., Holzknecht, Z. E., Collins, B. H., Parker, W., Miller, S. E., and Bollinger, R. R. (2004). Microbial biofilms in the gut: visualization by electron microscopy and by acridine orange staining. *Ultrastruct. Pathol.* 28:23–27.

Pasquier, B., Launay, P., Kanamaru, Y., Moura, I. C., Pfirsch, S., Ruffie, C., Henin, D., Benhamou, M., Pretolani, M., Blank, U., and Monteiro, R. C. (2005). Identification of FcalphaRI as an inhibitory receptor that controls inflammation: dual role of FcRgamma ITAM. *Immunity* 22:31–42.

Perrier, C., Sprenger, N., and Corthésy, B. (2006). Glycans on secretory component participate in innate protection against mucosal pathogens. *J. Biol. Chem.* 281:14, 280–14,287.

Phalipon, A., Cardona, A., Kraehenbuhl, J. P., Edelman, L., Sansonetti, P. J., and Corthésy, B. (2002). Secretory component: a new role in secretory IgA-mediated immune exclusion *in vivo*. *Immunity* 17:107–115.

Phalipon, A., and Corthésy, B. (2003). Novel functions of the polymeric Ig receptor: well beyond transport of immunoglobulins. *Trends Immunol.* 24:55–58.

Poger, M. E., and Lamm, M. E. (1974). Localization of free and bound secretory component in human intestinal epithelial cells. A model for the assembly of secretory IgA. *J. Exp. Med.* 139:629–642.

Rescigno, M., Urbano, M., Valzasina, B., Francolini, M., Rotta, G., Bonasio, R., Granucci, F., Kraehenbuhl, J. P., and Ricciardi-Castagnoli, P. (2001). Dendritic cells express tight junction proteins and penetrate gut epithelial monolayers to sample bacteria. *Nat. Immunol.* 2:361–367.

Rey, J., Garin, N., Spertini, F., and Corthésy, B. (2004). Targeting of secretory IgA to Peyer's patch dendritic and T cells after transport by intestinal M cells. *J. Immunol.* 172:3026–3033.

Rhee, K. J., Sethupathi, P., Driks, A., Lanning, D. K., and Knight, K. L. (2004). Role of commensal bacteria in development of gut-associated lymphoid tissues and preimmune antibody repertoire. *J. Immunol.* 172:1118–1124.

Rimoldi, M., Chieppa, M., Salucci, V., Avogadri, F., Sonzogni, A., Sampietro, G. M., Nespoli, A., Viale, G., Allavena, P., and Rescigno, M. (2005). Intestinal immune homeostasis is regulated by the crosstalk between epithelial cells and dendritic cells. *Nat. Immunol.* 6:507–514.

Robinson, J. K., Blanchard, T. G., Levine, A. D., Emancipator, S. N., and Lamm, M. E. (2001). A mucosal IgA-mediated excretory immune system *in vivo. J. Immunol.* 166:3688–3692.

Rosenow, C., Ryan, P., Weiser, J. N., Johnson, S., Fontan, P., Ortqvist, A., and Masure, H. R. (1997). Contribution of novel choline-binding proteins to adherence, colonization and immunogenicity of *Streptococcus pneumoniae. Mol. Microbiol.* 25:819–829.

Roy, M. J., and Varvayanis, M. (1987). Development of dome epithelium in gut-associated lymphoid tissues: association of IgA with M cells. *Cell Tissue Res.* 248: 645–651.

Royle, L., Roos, A., Harvey, D. J., Wormald, M. R., van Gijlswijk-Janssen, D., el-Redwan, R. M., Wilson, I. A., Daha, M. R., Dwek, R. A., and Rudd, P. M. (2003). Secretory IgA N- and O-glycans provide a link between the innate and adaptive immune systems. *J. Biol. Chem.* 278:20,140–20,153.

Rudd, P. M., Elliott, T., Cresswell, P., Wilson, I. A., and Dwek, R. A. (2001). Glycosylation and the immune system. *Science* 291:2370–2376.

Russell, M. W., Reinholdt, J., and Kilian, M. (1989). Anti-inflammatory activity of human IgA antibodies and their Fab alpha fragments: Inhibition of IgG-mediated complement activation. *Eur. J. Immunol.* 19:2243–2249.

Saltzman, W. M., Radomsky, M. L., Whaley, K. J., and Cone, R. A. (1994). Antibody diffusion in human cervical mucus. *Biophys. J.* 66:508–515.

Sandor, M., Waldschmidt, T. J., Williams, K. R., and Lynch, R. G. (1990). IgA-induced avidity maturation of IgA Fc receptors on murine T lymphocytes. *J. Immunol.* 144:4562–4570.

Schneeman, T. A., Bruno, M. E., Schjerven, H., Johansen, F. E., Chady, L., and Kaetzel, C. S. (2005). Regulation of the polymeric Ig receptor by signaling through TLRs 3 and 4: Linking innate and adaptive immune responses. *J. Immunol.* 175:376–384.

Schroten, H., Stapper, C., Plogmann, R., Kohler, H., Hacker, J., and Hanisch, F. G. (1998). Fab-independent antiadhesion effects of secretory immunoglobulin A on S-fimbriated *Escherichia coli* are mediated by sialyloligosaccharides. *Infect. Immun.* 66:3971–3973.

Shroff, K. E., Meslin, K., and Cebra, J. J. (1995). Commensal enteric bacteria engender a self-limiting humoral mucosal immune response while permanently colonizing the gut. *Infect. Immun.* 63:3904–3913.

Stoel, M., Jiang, H. Q., van Diemen, C. C., Bun, J. C., Dammers, P. M., Thurnheer, M. C., Kroese, F. G., Cebra, J. J., and Bos, N. A. (2005). Restricted IgA repertoire in both B-1 and B-2 cell-derived gut plasmablasts. *J. Immunol.* 174:1046–1054.

Suzuki, K., Meek, B., Doi, Y., Muramatsu, M., Chiba, T., Honjo, T., and Fagarasan, S. (2004). Aberrant expansion of segmented filamentous bacteria in IgA-deficient gut. *Proc. Natl. Acad. Sci. USA* 101:1981–1986.

Suzuki, K., Oida, T., Hamada, H., Hitotsumatsu, O., Watanabe, M., Hibi, T., Yamamoto, H., Kubota, E., Kaminogawa, S., and Ishikawa, H. (2000). Gut cryptopatches: direct evidence of extrathymic anatomical sites for intestinal T lymphopoiesis. *Immunity* 13:691–702.

Tejada-Simon, M. V., Lee, J., Ustunol, Z., and Pestka, J. J. (1999). Ingestion of yogurt containing *Lactobacillus acidophilus* and *Bifidobacterium* to potentiate immunoglobulin A responses to cholera toxin in mice. *J. Dairy Sci.* 82:649–660.

Umesaki, Y., and Setoyama, H. (2000). Structure of the intestinal flora responsible for development of the gut immune system in a rodent model. *Microbes Infect.* 2:1343–1351.

Underdown, B. J., DeRose, J., Koczekan, K., Socken, D., and Weicker, J. (1977). Isolation of human secretory component by affinity chromatography on IgM-sepharose. *Immunochemistry* 14:111–118.

van der Waaij, L. A., Limburg, P. C., Mesander, G., and van der Waaij, D. (1996). *In vivo* IgA coating of anaerobic bacteria in human faeces. *Gut* 38:348–354.

Weltzin, R., Lucia-Jandris, P., Michetti, P., Fields, B. N., Kraehenbuhl, J. P., and Neutra, M. R. (1989). Binding and transepithelial transport of immunoglobulins by intestinal M cells: Demonstration using monoclonal IgA antibodies against enteric viral proteins. *J. Cell. Biol.* 108:1673–1685.

Williamson, E., Bilsborough, J. M., and Viney, J. L. (2002). Regulation of mucosal dendritic cell function by receptor activator of NF-kappa B (RANK)/RANK ligand interactions: impact on tolerance induction. *J. Immunol.* 169:3606–3612.

Wold, A. E., Mestecky, J., Tomana, M., Kobata, A., Ohbayashi, H., Endo, T., and Eden, C. S. (1990). Secretory immunoglobulin A carries oligosaccharide receptors for *Escherichia coli* type 1 fimbrial lectin. *Infect. Immun.* 58:3073–3077.

Wu, H. Y., and Weiner, H. L. (2003). Oral tolerance. *Immunol. Res.* 28:265–284.

Yoshida, M., Claypool, S. M., Wagner, J. S., Mizoguchi, E., Mizoguchi, A., Roopenian, D. C., Lencer, W. I., and Blumberg, R. S. (2004). Human neonatal Fc receptor mediates transport of IgG into luminal secretions for delivery of antigens to mucosal dendritic cells. *Immunity* 20:769–783.

Zhang, J. R., Mostov, K. E., Lamm, M. E., Nanno, M., Shimida, S., Ohwaki, M., and Tuomanen, E. (2000). The polymeric immunoglobulin receptor translocates pneumococci across human nasopharyngeal epithelial cells. *Cell* 102:827–837.

Zhou, F, Kraehenbuhl, J. P., and Neutra, M. R. (1995). Mucosal IgA response to rectally administered antigen formulated in IgA-coated liposomes. *Vaccine* 13: 637–644.

9
IgA and Antigen Sampling

Nicholas J. Mantis[1] and Blaise Corthésy[2]

9.1. Introduction

As the primary immunoglobulin class in mucosal secretions, secretory immunoglobulin A (SIgA) antibodies function as the immunological "frontline," protecting the vulnerable surfaces of the intestinal epithelium from pathogenic bacteria, viruses, and toxins encountered in the normal human diet. SIgA also serves as a barrier to commensal microbiota (Johansen et al., 1999; Kelly et al., 2005; Macpherson et al., 2000;), some of which are opportunistic pathogens capable of causing disease if afforded access to the systemic compartment. Protection is achieved primarily by "immune exclusion," a general term referring to the ability of SIgA to coat intestinal antigens, thereby (1) preventing their attachment to epithelial cell receptors and (2) promoting their clearance from the intestinal lumen via peristalsis (Lamm, 1997). SIgA is also of critical importance to neonates whose mucosal immune systems are in the early stages of development. In humans, SIgA is the major immunoglobulin class

[1] Division of Infectious Disease, Wadsworth Center, New York State Department of Health, Albany, NY 12208, USA.
[2] Laboratoire de Recherche et Développement, Service d'Immunologie et d'Allergie, Centre Hospitalier Universitaire Vaudois, 1011 Lausanne, Switzerland.

in colostrum and breast milk, providing passive immunity to a variety of enteric pathogens (Brandtzaeg, 2003).

The gut-associated lymphoid tissues (GALT), consisting of the Peyer's patches (PPs), appendix, and numerous isolated lymphoid aggregates in the colon and rectum, are the principal inductive sites for IgA-committed B-cells (Brandtzaeg et al., 1999; Mowat, 2003; O'Leary and Sweeney, 1986) (see Chapter 2). Antigenic stimuli that drive B-cell somatic mutation within the local germinal centers are derived solely from the intestinal lumen, as lymphoid follicles in the GALT lack afferent lymphatics. Antigen "sampling" and transepithelial transport is accomplished by specialized epithelial cells called M-cells located exclusively within follicle-associated epithelium (FAE) (see sections below). Although the GALT are primarily inductive sites, they might also be important in the maintenance of local immunological memory (Brandtzaeg et al., 1999).

To protect the intestinal epithelium efficiently, the amount and specificity of IgA that is produced and delivered on a daily basis into mucosal secretions must be regulated in response to environmental changes in the intestinal lumen. Although our knowledge of how "immuno-surveillance" is achieved by the GALT is limited, there is increasing evidence to suggest that SIgA itself might function in this process (Favre et al., 2005; Mantis et al., 2002; Neutra et al., 2001). This chapter will summarize our current understanding of how SIgA, in conjunction with M-cells, might mediate the transepithelial transport of antigens from the intestinal lumen to the underlying GALT, possibly influencing primary immune responses and/or assisting in the maintenance of immunological memory.

9.2. M-Cells: The Gateway to the GALT

The GALT comprises B-cell-rich central follicles with germinal centers, flanked by the T-cell-rich interfollicular regions (IFRs). Each follicle is separated from the overlying epithelium by the so-called subepithelial dome region (SED), a dynamic zone rich in B-cells, T-Cells and at least three populations of dendritic cells (DCs) (Brandtzaeg et al., 1999; Iwasaki and Kelsall, 2000). Live and particulate antigens transported across the overlying epithelium, the so-called follicle-associated epithelium (FAE), are sampled by this network of DCs and are postulated to initiate the local immune response (Cheminay et al., 2005; Hopkins and Kraehenbuhl, 1997; Niedergang et al., 2000; Pron et al., 2001; Shreedhar et al., 2003).

Not surprisingly, the FAE has a number of characteristics that distinguish it from the surrounding villus epithelium (VE) and enable it to sample luminal antigens effectively. Whereas the VE is specialized for digestion and absorption of nutrients and is dominated by absorptive enterocytes, mucin-secreting goblet cells, and enteroendocrine cells, the FAE contains few or no goblet or enteroendocrine cells and has reduced levels of certain digestive enzymes (Owen and Bhalla, 1983). Furthermore, the entire FAE is devoid of the polymeric immunoglobulin receptor (pIgR) and is therefore unable to transport IgA from the interstitium to the lumen across this region of the intestinal epithelium (Bjerke and Brandtzaeg, 1988; Brandtzaeg and Bjerke, 1990; Pappo and Owen, 1988; Weltzin et al., 1989). There exist distinct differences between the VE and FAE in accessibility of glycolipid and glycoprotein receptors, which could explain in part why certain viruses and bacteria preferentially attach to the FAE (Frey et al., 1996; Mantis et al., 2000). Although there are biochemical and molecular differences between the FAE and VE (Lelouard et al., 1999; Neutra et al., 2001), these differences have been difficult to analyze in detail because of the challenge of obtaining purified preparations of the FAE. This is likely to change with the advent of novel screening methods and tissue microdissection techniques (Anderle et al., 2005; Higgins et al., 2004; Hooper, 2004; Lo et al., 2003, 2004).

Probably the most distinguishing feature of the FAE is the presence of M-cells, a unique epithelial cell type that is specialized in the uptake and transepithelial transport of particulate antigens, including particles and macromolecules (Neutra et al., 1987; Pappo and Ermak, 1989), viruses (Amerongen et al., 1991; Sicinski et al., 1990; Wolf et al., 1981), bacteria (Jones et al., 1994; Owen et al., 1986; Sansonetti et al., 1996), and parasites (Marcial and Madara, 1986). Indeed, M-cells have been considered the "gateway" to the GALT (Kraehenbuhl and Neutra, 2000; Neutra et al., 1996). The apical and basolateral surfaces of M-cells have distinct features that enable them to deliver mucosal antigens rapidly and efficiently from the lumen to underlying lymphoid follicles. For example, M-cells lack the well-developed brush border and thick glycocalyx present on enterocytes. Consequently, M-cell apical membranes are more accessible to particles, viruses, and bacteria than are the adjacent enterocytes (Frey et al., 1996). In mice and humans, the apical surfaces of M-cells have a pattern of glycosylation that is distinct from the FAE and VE (Clark et al., 1993; Giannasca et al., 1994, 1999). M-cells also express Toll-like receptors on their apical membrane that might facilitate antigen recognition and contribute to signaling in the local environment (Chabot et al., 2006; Tohno et al., 2005).

The M-cell basolateral membrane is deeply invaginated to form a large intraepithelial "pocket" containing specific subpopulations of naive and memory B- and T-cells (Brandtzaeg et al., 1999; Ermak and Owen, 1986; Farstad et al., 1994; Mantis and Wagner, 2004 Yamanaka *et al.*, 1999, 2001), and occasional dendritic cells (Iwasaki and Kelsall, 2001). The pocket brings the M-cell basolateral surface to within a few microns of the apical surface, shortening the distance that transcytotic vesicles must travel to cross the epithelium. Antigens transported by M-cells are sampled by adjacent DCs (Hopkins and Kraehenbuhl, 1997; Niedergang et al., 2000; Pron et al., 2001). The immunological function of the B- and T-cell populations within M-cell pockets, however, is still unclear. The resident T-cells are mostly $CD4^+$, and in humans, the majority display CD45RO, a surface marker typical of memory T-cells (Farstad et al., 1994), which distinguishes them from effector $CD8^+$ intraepithelial lymphocytes. The B-cell population is a mixture of naïve ($SIgD^+$) and memory ($SIgD^-$) cells that are proposed to originate from the underlying B-cell follicles (Brandtzaeg et al., 1999). Based on costimulatory molecule expression and *in vitro* coculture assays, Brandtzaeg and colleagues have proposed that M-cells pockets are extensions of germinal centers (Yamanaka et al., 1999, 2001, 2003) and that memory B-cells in this niche might be engaged in active sampling of luminal antigens and presentation of them to adjacent T-cells (Yamanaka et al., 1999, 2001). Activated T-cells expressing CD40L might in turn induce $CD40^+$ memory B-cell survival and proliferation. A better understanding of the function of lymphocytes within M-cell pockets is critical for understanding the immunological consequences of antigen (and antibody–antigen) sampling by M-cells.

9.3. Transepithelial Transport of SIgA by M-Cells

Antibodies present in colostrum and breast milk provide passive immunity to neonates (Brandtzaeg, 2002). In rodents, for example, IgG antibodies bind selectively to absorptive enterocytes in proximal small intestine and are then transported functionally intact into the circulation (Borel et al., 1996; Fritsche and Borel, 1994; Rodewald and Kraehenbuhl, 1984). Transepithelial transport of IgG is mediated by the neonatal Fcγ receptor FcRn (Simister and Rees, 1985). SIgA antibodies do not bind FcRn; nor are they actively transported into the circulation (Brandtzaeg, 2003). However, SIgA does bind selectively to M-cells. Using immunohistochemistry, Roy and Varvayanis (1987) first described the accumulation of maternally derived SIgA, but not SIgM or IgG, on the luminal face of M-cells in the appendix of 2-, 5-, and 10-day-old suckling rabbits. SIgA was not detected on other epithelial cell types within the FAE or VE. By immuno-electron microscopy, SIgA was evident in M-cell cytoplasmic vesicles, suggesting that antibodies bound to the cell surface are internalized and possibly trafficked to underlying mucosa-associated lymphoid tissues. Moreover, the authors observed IgA-coated bacteria in the proximity of M-cell luminal surfaces, although they were unable to discern whether these bacteria were internalized by M-cells.

We now know that the association of SIgA with the apical surface of M-cells is widespread, having been observed in the appendixes and PPs of neonatal and adult animals, including mice, rats, and rabbits (Kato, 1990; Mantis et al., 2002; Weltzin et al., 1989). There is also evidence to suggest SIgA associates with M-cells in the human ileum (Mantis et al., 2002) (Fig. 9.1). Weltzin et al. (1989) were the first to demonstrate, using antibody–colloidal gold conjugates and transmission electron microscopy, that monoclonal, polyclonal, and SIgA injected into the intestinal lumen of mice adhered to and was internalized by PP M-cells. We have subsequently confirmed this work using biotinylated and fluorophore-labeled SIgA and confocal laser scanning microscopy (Mantis et al., 2002). This highly sensitive and direct method to track and visualize SIgA enabled us to observe the accumulation of SIgA in M-cell cytoplasmic vesicles, along the membrane that defines the M-cell pocket, and within M-cells processes

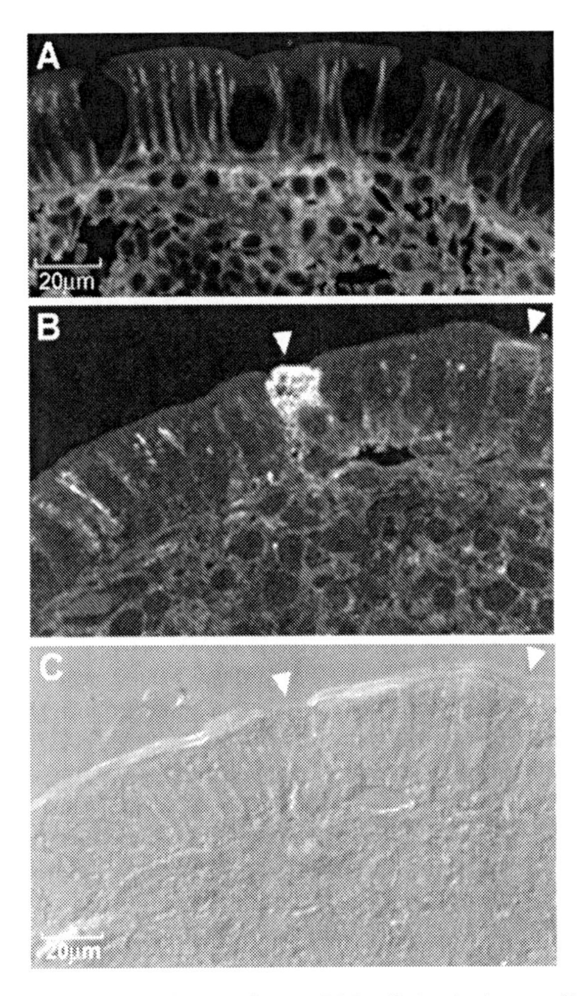

FIG. 9.1. IgA binds to the apical surfaces of M-cells in the human ileum. Paraffin sections of normal ileum from pediatric subjects were stained with biotin-labeled anti-human IgA and streptavidin–FITC, then visualized by confocal laser scanning microscopy (A, B) or differential interference microscopy (C). **(A)** A section of villus epithelium shows IgA-positive cells within the lamina propria, and IgA positive staining along the basolateral surfaces of enterocytes. **(B)** A section of FAE that shows IgA positive staining on two M-cells (arrowheads). **(C)** The same section as (B) visualized by differential interference contrast (DIC) to visualize M-cell morphology (e.g., lack of well-defined apical brush border). From Mantis et al., 2002. Copyright 2002. The American Association of Immunologists, Inc. Reprinted with permission.

that extend through the basal lamina (Fig. 9.2). Interestingly, these basolateral processes have been proposed to make direct contact with subepithelial leukocytes (Giannasca et al., 1994), possibly providing a link between the transport of SIgA across M-cells and communication with underlying target cells.

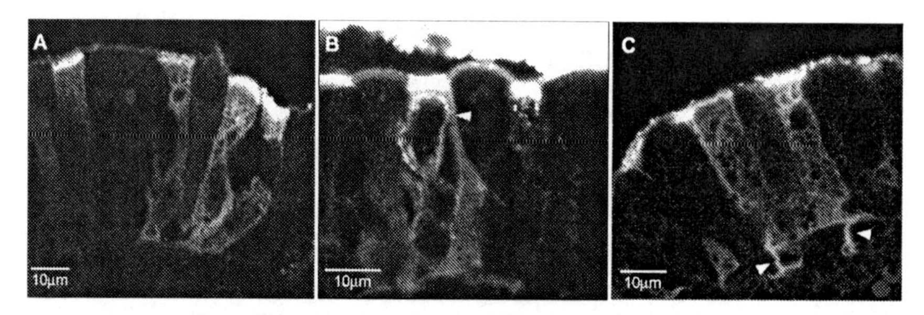

FIG. 9.2. Binding and transepithelial transport of SIgA by PP M-cells. Cy3-labeled mouse monoclonal SIgA was injected into ligated ileal loops from BALB/c mice containing a single PP. Frozen sections (5–7 μm) were viewed by confocal microscopy. **(A)** SIgA was present within M-cells and concentrated on M-cell apical membranes. **(B)** In some M-cells, SIgA was visible in apical vesicles, on lateral membranes (arrowheads), and on the membrane lining the intraepithelial pocket. SIgA did not detectably label cells within the pocket. **(C)** SIgA was also detected in the M-cell processes (arrowhead) that extend through the basal lamina. From Mantis et al., 2002. Copyright 2002. The American Association of Immunologists, Inc. Reprinted with permission.

Recently, Blanco and DiRita (2006) provided evidence that SIgA might enhance the uptake of normally noninvasive pathogens by M-cells. Using M-like cells obtained upon coculture of Caco-2 epithelial cell monolayers with a human B-cell line, the authors demonstrated enhanced uptake of IgA-coated *Vibrio cholerae* by M-cells, as compared to IgG-coated or uncoated bacteria. Although this study leaves many questions unanswered, notably the nature of the IgA receptor on Caco-2 cells and its relevance to the receptor on M-cells *in vivo*, it nonetheless raises the possibility that some aspects of the initial interaction between SIgA and M-cells can be elucidated using a defined *in vitro* model system.

9.4. Fate of SIgA–Antigen Complexes After M-Cell Transport

Secretory IgA–antigen complexes endocytosed by M-cells might be delivered into the M-cell pocket, the SED region, or both. Antibody–antigen complexes delivered into the pocket would be expected to interact with resident memory B- and T-cells (Farstad et al., 1994; Mantis and Wagner, 2004) as well as intraepithelial DCs, whereas SIgA delivered to the SED region would likely interact with the underlying network of DCs (Iwasaki and Kelsall, 2000). By transmission electron microscopy, Weltzin et al. (1989) observed that IgA–colloidal gold complexes were transported into the extracellular space of M-cell pockets within 60 mins. Using Cy3-labeled SIgA and fluorescence microscopy, we observed SIgA primarily in the SED region at 40–60 min, not within the M-cell pocket (Mantis et al., 2002; Rey et al., 2004) (see Chapter 8 for details). Although there are a number of possible explanations for this discrepancy, it is important to note that Weltzin et al. (1989) used particle-associated, aggregated forms of SIgA for their study, whereas we used homogenous, soluble SIgA preparations of well-defined molecular weight and structure. We are intrigued by the possibility that aggregates of SIgA, or SIgA-coated particles, which would mimic SIgA-coated bacteria or viruses, might be trafficked differently than soluble, unaggregated forms of SIgA. At least in macrophages, the intracellular fate of antibody–antigen complexes is determined, in part, by aggregation of Fc receptors induced by bivalent or multivalent antibodies (Mellman and Plutner, 1984; Mellman et al., 1984; Trombetta and Mellman, 2005).

9.5. Formation of SIgA–Antigen Complexes in the Intestinal Lumen

There are at least three mechanisms by which SIgA–antigen complexes can form in the intestinal lumen, which, in turn, could facilitate sampling by M-cells. The first mechanism, and the most apparent, is Fab mediated. For example, SIgA antibodies produced during a primary or secondary immune response following infection can form antibody–antigen complexes upon reexposure to the specific antigen or pathogen. For this to occur, SIgA antibodies must be directed against surface-exposed epitopes such as viral attachment proteins (Helander et al., 2003) or bacterial lipopolysaccharide (LPS) (Apter et al., 1993; Michetti et al., 1994). The formation of Fab-mediated antibody–antigen complexes is not necessarily limited to pathogens, as the commensal microflora stimulate a SIgA response within weeks following colonization (Macpherson et al., 2000), and in human adults, the majority of intestinal bacteria are coated (or "opsonized") with IgA (van der Waaij et al., 1996, 2004). Fab-mediated SIgA–antigen complexes are also expected to form in the lumen of the neonatal intestine in which maternally derived SIgA from milk can interact with pathogenic and toxigenic environmental antigens (Brandtzaeg, 2003). In adults, the uptake of opsonized pathogens (which are presumably rendered noninfectious) and commensals would serve as an effective means of immuno-surveillance capable of boosting and/or maintaining a pool of memory B- and T-cells within the GALT.

The second mechanism by which SIgA–antigen complexes can form is also Fab mediated, but is of lower affinity and less specific than the first situation discussed earlier. SIgA antibodies formed following colonization with a commensal microbiota were shown to be directed against LPS and the outer membrane protein OmpF, which is a porin protein conserved among Gram-negative bacteria (Macpherson et al., 2000). Depending on the specific epitopes, it is possible that SIgA antibodies produced against OmpF or other porins from one strain of commensal bacteria could cross-react with other commensals, albeit with slightly less affinity. There is also evidence for a population of polyreactive SIgA antibodies, possibly the remnants of a primitive immune system, within the intestinal mucosa and secretions that are capable of binding multiple epitopes on commensal and pathogenic bacteria (Quan et al., 1997; Bouvet and Fischetti, 1999; Blanco and DiRita, 2006). These antibodies are predicted to be capable of forming antibody–antigen complexes that could be sampled by M-cells.

The final mechanism by which SIgA–antigen complexes can form in the intestinal lumen occurs through the N- and O-linked oligosaccharide side chains on the Fcα and secretory component (SC) moieties of SIgA and is Fab independent. Both polyvalent toxins and bacterial pathogens can adhere to the exposed carbohydrate side chains on SIgA (Wold et al., 1990; Dallas and Rolfe, 1998; Mantis et al., 2004). Although this interaction is assumed to be defensive in nature (i.e., SIgA serves as a "decoy" that competitively inhibits toxin/pathogen attachment to glycolipid and glycoprotein receptors on the apical surfaces of enterocytes), we speculate that SIgA might also promote toxin/pathogen uptake by M-cells. To our knowledge, this "decoy" function of SIgA has not been examined experimentally.

9.6. Evidence for a Specific IgA Receptor on M-Cells

There is considerable interest in identifying receptors on M-cells that bind and mediate transepithelial transport of IgA (Brayden and Baird, 2004; Neutra et al., 2001). To date, two classes of IgA receptors have been identified: lectinlike receptors, which recognize *N*- and/or *O*-oligosaccharide side chains on IgA (and SC), and Fcα receptors (FcαR), which bind the IgA constant regions. There are three lectinlike receptors with ascribed affinity for IgA: the asialoglycoprotein receptor (ASGP-R), β-1,4-galactosyl transferase (β-1,4-GT), and the transferrin receptor (TfR). ASGP-R is a lectinlike receptor that recognizes oligosaccharide side chains carrying terminal galactose moieties, including those on human IgA (Daniels et al., 1989). Originally considered to be restricted to the sinusoidal membranes of hepatocytes, Mu et al. (1997) identified functional ASGP-R on the apical surfaces of HT-29 human intestinal epithelial cells. *In vivo*, ASGP-R was detected on the luminal surfaces of rat neonatal (but not adult) enterocytes (Hu et al., 1991). β-1,4-GT, which recognizes IgG and IgM in addition to IgA, is found on the surfaces of a wide variety of cell lines, including those of intestinal epithelial origin (Tomana et al., 1993). The TfR binds human IgA1, favoring polymeric forms over monomeric forms (Moura et al., 2001, 2004). With the exception of ASGP-R, which functions in the catabolism of serum IgA, the physiologic significance of these lectinlike receptors in IgA homeostasis remains enigmatic.

The possibility that ASGP-R might be responsible for mediating SIgA attachment to M-cells was investigated, but it proved unsubstantiated (Mantis *et al.*, 2002). Anti-ASGP-R antisera did not label the FAE, and IgA adhered to PP M-cells in ASGP-R-deficient mice. Furthermore, deglycosylated IgA normally associated with M-cells. These data suggest that the Fcα domains, not the side-linked oligosaccharide side chains, are responsible for mediating IgA attachment to M-cells.

There are the three known FcαRs: pIgR, CD89 (FcαRI), and the recently described Fcα/μ receptor (see Chapters 3 and 4). On the basis of existing evidence, none of these Fc receptors are likely to be the IgA receptor on M-cells. Of these candidates, only pIgR is expressed on intestinal epithelial cells, but it is not expressed on the FAE or M-cells (Brandtzaeg, 1978; Pappo and Owen, 1988; Weltzin et al., 1989). CD89 is considered an unlikely candidate because its expression is restricted to cells of the myeloid lineage, including neutrophils, eosinophils, and certain populations of monocyte/macrophages, (Montiero and van de Winkel, 2003; van Egmond et al., 2001). Moreover, a mouse counterpart of CD89 has never been identified. Similarly, Fcα/μ, which shares amino acid similarity with pIgR, is restricted to B-lymphocytes and macrophages (Sakamoto et al., 2001). Kitamura et al. (2000) have described the existence of a putative low-affinity FcαR on HT-29 cells that binds IgA. However, the IgA receptor on HT-29 cells does not bind SIgA, which distinguishes it from the IgA receptor on M cells.

Mantis et al. (2002) took advantage of available natural and recombinant forms of human IgA to identify the regions of IgA necessary for recognition by M-cells,. In a mouse ligated ileal loop assay, human IgA2 but not human IgA1 bound to PP M-cells. The inability of IgA1 to adhere to M-cells was attributed to the extended hinge region between domains Cα1 and Cα2, because a recombinant IgA1 antibody lacking the hinge bound to M-cells in this assay. Domain Cα1 was not sufficient to enable antibody attachment, because a chimeric antibody in which Cα1 of IgA1 was fused to the hinge and γ2–3 constant domains of IgG2 had no detectable affinity for M-cells. From these experiments it was proposed that the IgA receptor on M-cells spans the hinge region and recognizes residues on domains Cα1 and Cα2. If correct, this would clearly distinguish the IgA receptor on M-cells from previously identified Fcα receptors. For example, CD89 recognizes residues within the Cα2–Cα3 interface of human IgA1 and IgA2 (Carayannopoulos et al., 1996; Herr et al., 2003 Pleass et al., 1999). Computer modeling of human SIgA reveals that domains Cα1 and Cα2 are accessible and unobstructed by N-linked oligosaccharides, SC, or the J-chain (Royle et al., 2003). Collectively, these data suggest that M-cells express a novel Fcα-like receptor on their apical surfaces, which appears to recognize its IgA ligand through polypeptide motifs rather than glycan moieties.

9.7. SIgA and Antigen Sampling in GALT Development and Homeostasis

The ability of SIgA to participate in antigen sampling within inductive lymphoid tissues suggests a possible role for SIgA antibodies in GALT development and homeostasis. This unique function of SIgA is discussed in detail in Chapter 8.

9.8. Exploiting SIgA as a Vaccine Delivery Vehicle

The selective adhesion of SIgA to the apical membrane of M-cells promotes the uptake of small amounts of antibody as reflected by the detection of SIgA in the pocket of M-cells (Weltzin et al., 1989). This pioneering work was extended by the demonstration that SIgA can serve as a mucosal vaccine carrier (Corthésy et al., 1996). Insertion of a *Shigella* invasin B epitope into domain 1 of rabbit SC, followed by association with pIgA, yielded recombinant "antigenized" SIgA that was delivered orally to mice in combination with the mucosal adjuvant cholera toxin (CT). Antigenized SIgA elicited the appearance of serum and salivary invasin B-specific antibodies capable of recognizing the antigen in the form of a free peptide or as part of IpaB within *Shigella* lysates. In addition, a specific antibody response was consistently evoked against the rabbit SC carrier. Further studies identified useful sites for introducing T and B epitopes in the sequence of murine SC (Crottet et al., 1999), sites which did not interfere with association with polymeric IgA (Crottet and Corthésy, 1999). Stability of SIgA toward exposure to mouse intestinal washes has been demonstrated in *ex vivo* experiments and represents a strong argument for the use of this class of antibody as delivery vehicle (Crottet and Corthésy, 1998).

However, this experimental design suffered from the limitation that CT was combined with the antigenized SIgA, a combination that could not be given to humans because of unacceptable side effects from the CT (Williams et al., 1999). In order to examine whether SIgA can serve as a substitute for mucosal adjuvant, mice were immunized with recombinant SIgA (rSIgA) molecules containing an SC moiety of human origin (hSC) serving as a source of non-self-epitopes. Significant levels of hSC-specific salivary SIgA and serum IgG were measured after the third immunization of animals received rSIgA orally (Favre et al., 2005). Antibody titers were comparable to those obtained in mice fed hSC + CT or rSIgA + CT, indicating that humoral immune responses can be induced after oral delivery of rSIgA alone. The intrinsic property of SIgA to target DCs within the SED region in PPs confers to this class of antibody the capacity to guarantee efficient mucosal immunotargeting, without the need for extensive molecular engineering (Estrada et al., 1995; Snider et al., 1997). Further, the use of antigenized SIgA complexes affords the flexibility to combine specificity (Fab) and antigenicity (antigenized SC), which is not possible following insertion of antigenic sequences within the heavy-chain CDR3 region, for example (Cook and Barber, 1997).

(Secretory)IgA has been used successfully for passive protection or therapeutic intervention at mucosal surfaces in animal models of infection (respiratory syncytial virus, *Vibrio cholerae*, *Clostridium difficile*, *Shigella flexneri*) and in humans for the prevention of dental caries (reviewed in Corthésy, 2003). In contrast to classical vaccination, passive delivery of specific antibodies enables direct targeting of susceptible mucosal sites where most infections begin. Safety concerns are reduced with mucosally delivered antibody, as it is applied to surfaces normally exposed to foreign materials. Immunity against nonhuman molecules is expected to develop after long-term use only. Upon mucosal administration, problems associated with blood products are reduced, and adverse reactions are unlikely to occur. Another obvious advantage lies in the intrinsic capability of the antibody to neutralize the pathogen at a very early stage of infection or even prior to the initiation of the infection. Together, these observations raise the possibility of combining passive and active immunization for the treatment of, and protection against, mucosal pathogens in the context of a single antibody molecule.

The ability of IgA to selectively target molecules to M-cells was further emphasized in vaccination protocols using either liposomes or microspheres. IgA can enhance the local secretory immune response to antigen in liposomes, apparently by increasing liposome uptake via M-cells (Zhou *et al.*, 1995). Coating of microspheres with human IgA enhanced transport across M-cells and subsequent delivery to mesenteric lymph nodes, compared to bovine serum albumin-coated microspheres (Smith et al., 1995). These data suggest that the surface of M-cells possesses the capacity to determine which kind of antigen will be processed and presented to the gut immune system. A clue to the physiological relevance of such a selective mechanism is given by the observation that SIgA targets DCs in the SED region underlying M-cells in PPs (Rey et al., 2004).

9.9. Concluding Remarks

Without question, the primary function of SIgA is to protect the intestinal epithelium from enteric pathogens and enterotoxins. As summarized in this chapter, however, there is mounting evidence to suggest that SIgA, in both adults and neonates, has a secondary function in facilitating the uptake and delivery of antigens from the intestinal lumen to the GALT (Fig. 9.3). Although the immunological consequence of antibody sampling by M-cells is only beginning to be elucidated (Favre et al., 2005), it is likely that the uptake of SIgA–antigen complexes might be important in intestinal homeostasis and/or immunological memory. The intestinal microbiota, in particular, has a dramatic influence on gastrointestinal physiology, and alterations in the commensal population, which are known to be kept in check in part by SIgA, can be detrimental to human health (Macdonald and Monteleone, 2005; Suzuki et al., 2004). Defining the immunological importance of SIgA–antigen sampling by M-cells to mucosal immunity ultimately awaits the construction of a mouse strain lacking the M-cell IgA receptor. Identification of this novel receptor might now be feasible using precise microdissection techniques combined with sensitive methods for isolating intact nucleic acids (Hooper, 2004). A better understanding of the molecular and immunological mechanisms associated with SIgA–antigen sampling in the intestinal tract will likely have implications for the future development of immunotherapeutics and vaccines to combat inflammatory disorders and infectious diseases.

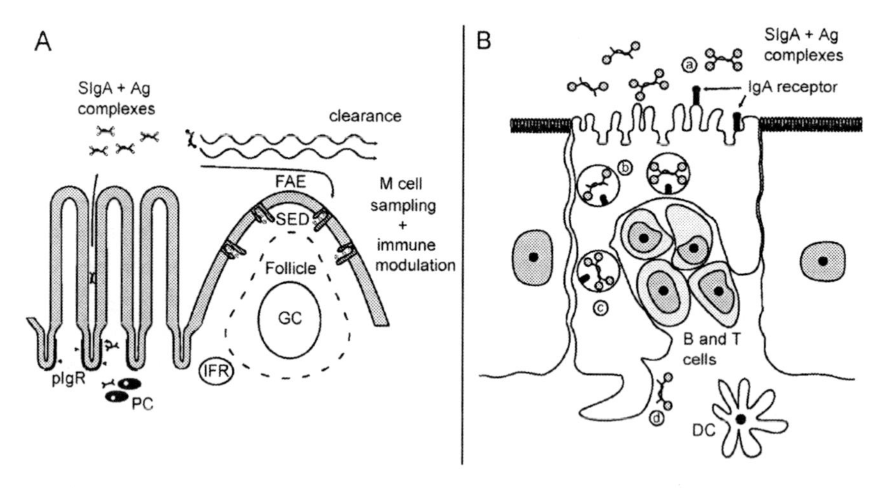

FIG. 9.3. SIgA and antigen sampling by M cells. **(A)** Dimeric and polymeric IgA antibodies produced by plasma cells in the subepithelial connective tissues are transported vectorially across the intestinal epithelium by the pIgR. SIgA–antigen complexes formed within the lumen are cleared from the intestine by peristalsis. However, a fraction of SIgA–antibody complexes are postulated to be "sampled" by M-cells and retrotransported across FAE to underlying lymphocytes. **(B)** Model for the transepithelial transport of SIgA–antigen complexes by M-cells: (a) SIgA and SIgA–antigen complexes adhere to the apical surfaces of M-cells via a novel Fcα-like receptor that recognizes domains Cα1 and Cα2 of IgA; (b, c) SIgA–antigen complexes are internalized by receptor-mediated endocytosis and trafficked by vesicular transport to the basolateral membrane; (d) release of SIgA–antigen complexes into the M-cell intracellular pocket or subepithelial space.

Acknowledgments. The authors would like to thank Dr. Marian Neutra for critically reading this manuscript. N.J.M. gratefully acknowledges financial support from the Wadsworth Center and the National Institutes of Health (USA). B.C. acknowledges the support of the Swiss Science Research Foundation (grants 3200–068038 and 3200–109545).

References

Amerongen, H. M., Weltzin, R., Farnet, C. M., Michetti, P., Haseltine, W. A., and Neutra, M. R. (1991). Transepithelial transport of HIV-1 by intestinal M cells: A mechanism for transmission of AIDS. *J. Acquir. Immun. Def. Synd.* 4:760–765.

Anderle, P., Rumbo, M., Sierro, F., Mansourian, R., Michetti, P., Roberts, M. A., and Kraehenbuhl, J. P. (2005). Novel markers of the human follicle-associated epithelium identified by genomic profiling and microdissection. *Gastroenterology* 129:321–327.

Apter, F. M., Michetti, P., Winner, L. S. d., Mack, J. A., Mekalanos, J. J., and Neutra, M. R. (1993). Analysis of the roles of antilipopolysaccharide and anti-cholera toxin immunoglobulin A (IgA) antibodies in protection against Vibrio cholerae and cholera toxin by use of monoclonal IgA antibodies *in vivo. Infect. Immun.* 61:5279–5285.

Bjerke, K., and Brandtzaeg, P. (1988). Lack of relation between expression of HLA-DR and secretory component (SC) in follicle-associated epithelium of human Peyer's patches. *Clin. Exp. Immunol.* 71:502–507.

Blanco, L. P., and DiRita, V. J. (2006). Antibodies enhance interaction of *Vibrio cholerae* with intestinal M-like cells. *Infect. Immun.* 74:6957–6964.

Borel, Y., Fritsche, R., Borel, H., Dahlgren, U., Dalhman-Hoglund, A., and Hanson, L. A. (1996). Parenteral and oral administration of tolerogens: Protein–IgG conjugates. *Ann. NY Acad. Sci.* 778:80–87.

Bouvet, J. P., and Fischetti, V. A. (1999). Diversity of antibody-mediated immunity at the mucosal barrier. *Infect. Immun.* 67:2687–2691.

Brandtzaeg, P. (1978). Polymeric IgA is complexed with secretory component (SC) on the surface of human intestinal epithelial cells. *Scand. J. Immunol.* 8:39–52.

Brandtzaeg, P. (2002). Role of local immunity and breast feeding in mucosal homeostasis and defense against infections. I. In: Calder, P. Field, C., and Gill, H. (eds.), *Nutrition and Immune Function.,* CABI Publishing, New York.

Brandtzaeg, P. (2003). Mucosal immunity: Integration between mother and the breast-fed infant. *Vaccine* 21:3382–3388.

Brandtzaeg, P., Baekkevold, E. S., Farstad, I. N., Jahnsen, F. L., Johansen, F. E., Nilsen, E. M., and Yamanaka, T. (1999). Regional specialization in the mucosal immune system: what happens in the microcompartments? *Immunol. Today* 20:141–151.

Brandtzaeg, P., and Bjerke, K. (1990). Immunomorphological characteristics of human Peyer's patches. *Digestion* 46:262–273.

Brayden, D. J., and Baird, A. W. (2004). Apical membrane receptors on intestinal M cells: Potential targets for vaccine delivery. *Adv. Drug Deliv. Rev.* 56:721–726.

Carayannopoulos, L., Hexham, J. M., and Capra, J. D. (1996). Localization of the binding site for the monocyte immunoglobulin (Ig) A-Fc receptor (CD89) to the domain boundary between Cα2 and Cα3 in human IgA1. *J. Exp. Med.* 183:1579–1586.

Chabot, S., Wagner, J. S., Farrant, and Neutra, M. R. (2006). TLRs regulate the gatekeeping functions of the intestinal follicle-associated epithelium. *J. Immunol.* 176:4275–4283

Cheminay, C., Mohlenbrink, A., and Hensel, M. (2005). Intracellular Salmonella inhibit antigen presentation by dendritic cells. *J. Immunol.* 174:2892–2899.

Clark, M. A., Jepson, M. A., Simmons, N. L., and Hirst, B. H. (1993). Differential expression of lectin-binding sites defines mouse intestinal M cells. *J. Histochem. Cytochem.* 41:1679–1687.

Cook, J., and Barber, B. H. (1997). Recombinant antibodies with conformationally constrained HIV type 1 epitope inserts elicit glycoprotein 160-specific antibody responses *in vivo. AIDS Res. Hum. Retroviruses* 13:449–460.

Corthésy, B., Kaufmann, M., Phalipon, A., Peitsch, M., Neutra, M. R., and Kraehenbuhl, J. P. (1996). A pathogen-specific epitope inserted into recombinant secretory immunoglobulin A is immunogenic by the oral route. *J. Biol. Chem.* 271:33,670–33,677.

Corthésy, B. (2003). Recombinant secretory immunoglobulin A in passive immunotherapy: linking immunology and biotechnology. *Curr. Pharm. Biotechnol.* 4:51–67.

Crottet, P., and Corthésy, B. (1998). Secretory component delays the conversion of secretory IgA into antigen-binding competent F(ab')2: A possible implication for mucosal defense. *J. Immunol.* 161:5445–5453.

Crottet, P., and Corthésy, B. (1999). Mapping the interaction between murine IgA and murine secretory component carrying epitope substitutions reveals a role of domains II and III in covalent binding to IgA. *J. Biol. Chem.* 274:31,456–31,462.

Crottet, P., Peitsch, M. C., Servis, C., and Corthésy, B. (1999). Covalent homodimers of murine secretory component induced by epitope substitution unravel the capacity of the polymeric Ig receptor to dimerize noncovalently in the absence of IgA ligand. *J. Biol. Chem.* 274:31,445–31,455.

Dallas, S. D., and Rolfe, R. D. (1998). Binding of *Clostridium difficile* toxin A to human milk secretory component. *J. Med. Microbiol.* 47:879–888.

Daniels, C. K., Schmucker, D. L., and Jones, A. L. (1989). Hepatic asialoglycoprotein receptor-mediated binding of human polymeric immunoglobulin A. *Hepatology* 9:229–234.

Ermak, T., and Owen, R. L. (1986). Differential distribution of lymphocytes and accessory cells in mouse Peyer's patches. *Anat. Rec.* 215:144–152.

Estrada, A., McDermott, M. R., Underdown, B. J., and Snider, D. P. (1995). Intestinal immunization of mice with antigen conjugated to anti-MHC class II antibodies. *Vaccine* 13:901–907.

Farstad, I. N., Halstensen, T. S., Fausa, O., and Brandtzaeg, P. (1994). Heterogeneity of M-cell-associated B and T cells in human Peyer's patches. *Immunology* 83:457–464.

Favre, L., Spertini, F., and Corthésy, B. (2005). Secretory IgA possesses intrinsic modulatory properties stimulating mucosal and systemic immune responses. *J. Immunol.* 175:2793–2800.

Frey, A., Giannasca, K. T., Weltsin, R., Giannasca, P. J., Reggio, H., Lencer, W. I., and Neutra, M. R. (1996). Role of the glycocalyx in regulating access of microparticles to apical plasma membranes of intestinal epithelial cells: Implications for microbial attachment and vaccine targeting. *J. Exp. Med.* 184:1045–1059.

Fritsche, R., and Borel, Y. (1994). Prevention of allergic sensitization to β-lactoglobulin with conjugates made of β-lactoglobulin coupled to isologous immunoglobulin G. *J. Allergy Clin. Immunol.* 93:778–786.

Giannasca, P. J., Giannasca, K. T., Falk, P., Gordon, J. I., and Neutra, M. R. (1994). Regional differences in glycoconjugates of intestinal M cells in mice: Potential targets for mucosal vaccines. *Am. J. Physiol.* 267:G1108–G1121.

Giannasca, P. J., Giannasca, K. T., Leichtner, A. M., and Neutra, M. R. (1999). Human intestinal M cells display the sialyl Lewis A antigen. *Infect. Immun.* 67:946–953.

Helander, A., Silvey, K. J., Mantis, N. J., Hutchings, A. B., Chandran, K., Lucas, W. T., Nibert, M. L., and Neutra, M. R. (2003). The viral sigma1 protein and glycoconjugates containing alpha2-3-linked sialic acid are involved in type 1 reovirus adherence to M cell apical surfaces. *J. Virol.* 77:7964–7977.

Herr, A. B., Ballister, E. R., and Bjorkman, P. J. (2003). Insights into IgA-mediated immune responses from the crystal structures of human FcαRI and its complex with IgA1-Fc. *Nature* 423:614–620.

Higgins, L. M., Lambkin, I., Donnelly, G., Byrne, D., Wilson, C., Dee, J., Smith, M., and O'Mahony, D. J. (2004). *In vivo* phage display to identify M cell-targeting ligands. *Pharm. Res.* 21:695–705.

Hooper, L. V. (2004). Laser microdissection: exploring host-bacterial encounters at the front lines. *Curr. Opin. Microbiol.* 7:290–295.

Hopkins, S. A., and Kraehenbuhl, J. P. (1997). Dendritic cells of the murine Peyer's patches colocalize with Salmonella typhimurium avirulent mutants in the subepithelial dome. *Adv. Exp. Med. Biol.* 417:105–109.

Hu, C. B., Lee, E. Y., Hewitt, J. E., Baenziger, J. U., Mu, J. Z., DeSchryver-Kecskemeti, K., and Alpers, D. H. (1991). The minor components of the rat asialoglycoprotein receptor are apically located in neonatal enterocytes. *Gastroenterology* 101:1477–1487.

Iwasaki, A., and Kelsall, B. (2001). Unique functions of CD11b+, CD8a+, and double negative Peyer's patch dendritic cells. *J. Immunol.* 166:4884–4890.

Iwasaki, A., and Kelsall, B. L. (2000). Localization of distinct Peyer's patch dendritic cell subsets and their recruitment by chemokines macrophage inflammatory protein (MIP)-3α, MIP-3β, and secondary lymphoid organ chemokine. *J. Exp. Med.* 191:1381–1394.

Johansen, F. E., Pekna, M., Norderhaug, I. N., Haneberg, B., Hietala, M. A., Krajci, P., Betsholtz, C., and Brandtzaeg, P. (1999). Absence of epithelial immunoglobulin A transport, with increased mucosal leakiness, in polymeric immunoglobulin receptor/secretory component-deficient mice. *J. Exp. Med.* 190:915–922.

Jones, B., Ghori, N., and Falkow, S. (1994). *Salmonella typhimurium* initiated murine infection by penetrating and destroying the specialized epithelial M cells of the Peyer's Patches. *J. Exp. Med.* 180:15–23.

Kato, T. (1990). A study of secretory immunoglobulin A on membranous epithelial cells (M cells) and adjacent absorptive cells of rabbit Peyer's patches. *Gastroenterol. Japon.*,25:15–23.

Kelly, D., Conway, S., and Aminov, R. (2005). Commensal gut bacteria: Mechanisms of immune modulation. *Trends Immunol.* 26:326–333.

Kitamura, T., Garofalo, R. P., Kamijo, A., Hammond, D. K., Oka, J. A., Caflisch, C. R., Shenoy, M., Casola, A., Weigel, P. H., and Goldblum, R. M. (2000). Human intestinal epithelial cells express a novel receptor for IgA. *J. Immunol.* 164:5029–5034.

Kraehenbuhl, J. P., and Neutra, M. R. (2000). Epithelial M cells: Differentiation and function. *Ann. Rev. Cell Develop. Biol.* 16:301–332.

Lamm, M. E. (1997). Interactions of antigens and antibodies at mucosal surfaces. *Ann. Rev. Microbiol.* 51:311–340.

Lelouard, H., Reggio, H., Manget, P., Neutra, M., and Mountcourrier, P. (1999). Mucin related epitopes distinguish M cells and enterocytes in rabbit appendix and Peyer's patches. *Infect. Immun.* 67:357–367.

Lo, D., Tynan, W., Dickerson, J., Mendy, J., Chang, H. W., Scharf, M., Byrne, D., Brayden, D., Higgins, L., Evans, C., and O'Mahony, D. J. (2003). Peptidoglycan recognition protein expression in mouse Peyer's Patch follicle associated epithelium suggests functional specialization. *Cell. Immunol.* 224:8–16.

Lo, D., Tynan, W., Dickerson, J., Scharf, M., Cooper, J., Byrne, D., Brayden, D., Higgins, L., Evans, C., and O'Mahony, D. J. (2004). Cell culture modeling of specialized tissue: Identification of genes expressed specifically by follicle-associated epithelium of Peyer's patch by expression profiling of Caco-2/Raji co-cultures. *Int. Immunol.* 16:91–99.

Macdonald, T. T., and Monteleone, G. (2005). Immunity, inflammation, and allergy in the gut. *Science* 307:1920–1925.

Macpherson, A. J., Gatto, D., Sainsbury, E., Harriman, G. R., Hengartner, H., and Zinkernagel, R. M. (2000). A primitive T cell-independent mechanism of intestinal mucosal IgA responses to commensal bacteria. *Science* 288:2222–2226.

Mantis, N. J., Frey, A. F., and Neutra, M. R. (2000). Accessibility of glycolipid and glycoprotein epitopes on rabbit villus and follicle-associated epithelium. *Am. J. Physiol.* 278:G915–G924.

Mantis, N. J., Cheung, M. C., Chintalacharuvu, K. R., Rey, J., Corthésy, B., and Neutra, M. R. (2002). Selective adherence of IgA to murine Peyer's patch M cells: evidence for a novel IgA receptor. *J. Immunol.* 169:1844–1851.

Mantis, N. J., Farrant, S. A., and Mehta, S. (2004). Oligosaccharide side chains on human secretory IgA serve as receptors for ricin. *J. Immunol.* 172:6838–6845.

Mantis, N. J., and Wagner, J. S. (2004). Analysis of adhesion molecules involved in leukocyte homing into the basolateral pockets of mouse Peyer's patch M cells. *J. Drug Target.* (special issue on Cellular Aspects of Targeting in the Gastrointestinal Tract) 12:79–87.

Marcial, M. A., and Madara, J. L. (1986). Cryptosporidium: Cellular localization, structural analysis of absorptive cell-parasite membrane-membrane interactions in guinea pigs, and suggestion of protozoan transport by M cells. *Gastroenterology* 90:583–594.

Mellman, I., and Plutner, H. (1984). Internalization and degradation of macrophage Fc receptors bound to polyvalent immune complexes. *J. Cell Biol.* 98:1170–1177.

Mellman, I., Plutner, H., and Ukkonen, P. (1984). Internalization and rapid recycling of macrophage Fc receptors tagged with monovalent antireceptor antibody: Possible role of a prelysosomal compartment. *J. Cell Biol.* 98:1163–1169.

Michetti, P., Porta, N., Mahan, M. J., Slauch, J. M., Mekalanos, J. J., Blum, A., Kraehenbuhl, J. P., and Neutra, M. R. (1994). Monoclonal immunoglobulin A prevents adherence and invasion of polarized epithelial cell monolayers by *Salmonella typhimurium. Gastroenterology* 107:915–923.

Monteiro, R. C., and van de Winkel, J. G. (2003). IgA Fc receptors. *Ann. Rev. Immunol.* 21:177–204.

Moura, I. C., Arcos-Fajardo, M., Sadaka, C., Leroy, V., Benhamou, M., Novak, J., Vrtovsnik, F., Haddad, E., Chintalacharuvu, K. R., and Monteiro, R. C. (2004). Glycosylation and size of IgA1 are essential for interaction with mesangial transferrin receptor in IgA nephropathy. *J. Am. Soc. Nephrol.* 15:622–634.

Moura, I. C., Centelles, M. N., Arcos-Fajardo, M., Malheiros, D. M., Collawn, J. F., Cooper, M. D., and Monteiro, R. C. (2001). Identification of the transferrin receptor as a novel immunoglobulin (Ig)A1 receptor and its enhanced expression on mesangial cells in IgA nephropathy. *J. Exp. Med.* 194:417–425.

Mowat, A. M. (2003). Anatomical basis of tolerance and immunity to intestinal antigens. *Nat. Rev. Immunol.* 3:331–341.

Mu, J.-Z., Gordon, M., Shao, J.-S., and Alpers, D. H. (1997). Apical expression of functional asialoglycoprotein receptor in human intestinal epithelial cell line HT-29. *Gastroenterology* 113:1501–1509.

Neutra, M., Mantis, N., and Kraehenbuhl, J. P. (2001). Collaboration of epithelial cells with organized mucosal lymphoid tissue. *Nat. Immunol.* 2:1004–1009.

Neutra, M., Phillips, T., Mayer, E., and Fishkind, D. (1987). Transport of membrane-bound macromolecules by M cells in follicle-associated epithelium of rabbit Peyer's patch. *Cell Tissue Res.* 247:537–546.

Neutra, M. R., Frey, A., and Kraehenbuhl, J. P. (1996). Epithelial M cells: Gateways for mucosal infection and immunization. *Cell* 86:345–348.

Niedergang, F., Sirard, J. C., Blanc, C. T., and Kraehenbuhl, J. P. (2000). Entry and survival of *Salmonella typhimurium* in dendritic cells and presentation of recombinant antigens do not require macrophage-specific virulence factors. *Proc. Natl. Acad. Sci. USA* 97:14,650–14,655.

O'Leary, A. D., and Sweeney, E. C. (1986). Lymphoglandular complexes of the colon: structure and distribution. *Histopathology* 10:267–283.

Owen, R., and Bhalla, D. (1983). Cytochemical analysis of alkaline phosphatase and esterase activities and of lectin-binding and anionic sites in rat and mouse Peyer's Patch M cells. *Am. J. Anat,* 168:199–212.

Owen, R., Pierce, N., Apple, R., and Cray W., Jr. (1986). M Cell transport of *Vibrio cholerae* from the intestinal lumen into Peyer's Patches: A mechanism for antigen sampling and for microbial transepithelial migration. *J. Infect. Dis.* 153:1108–1118.

Pappo, J., and Ermak, T. H. (1989). Uptake and translocation of fluorescent latex particles by rabbit Peyer's patch follicle epithelium: A quantitative model for M cell uptake. *Clin. Exp. Immunol.* 76:144–148.

Pappo, J., and Owen, R. L. (1988). Absence of secretory component expression by epithelial cells overlying rabbit gut-associated lymphoid tissue. *Gastroenterology* 95:1173–1174.

Pleass, R. J., Dunlop, J. I., Anderson, C. M., and Woof, J. M. (1999). Identification of residues in the CH2/CH3 domain interface of IgA essential for interaction with the human Fcα receptor (FcαR) CD89. *J. Biol. Chem.* 274:23,508–23,514.

Pron, B., Boumaila, C., Jaubert, F., Berche, P., Milon, G., Geissmann, F., and Gaillard, J. L. (2001). Dendritic cells are early cellular targets of *Listeria monocytogenes* after intestinal delivery and are involved in bacterial spread in the host. *Cell. Microbiol.* 3:331–340.

Quan, C. P., Berneman, A., Pires, R., Avrameas, S., and Bouvet, J. P. (1997). Natural polyreactive secretory immunoglobulin A autoantibodies as a possible barrier to infection in humans. *Infect. Immun.* 65:3997–4004.

Rey, J., Garin, N., Spertini, F., and Corthésy, B. (2004). Targeting of secretory IgA to Peyer's patch dendritic and T cells after transport by intestinal M cells. *J. Immunol.* 172:3026–3033.

Rodewald, R., and Kraehenbuhl, J. P. (1984). Receptor-mediated transport of IgG. *J. Cell Biol.* 99:159S–164S.

Roy, M. J., and Varvayanis, M. (1987). Development of dome epithelium in gut-associated lymphoid tissues: Association of IgA with M cells. *Cell Tissue Res.* 248:645–651.

Royle, L., Roos, A., Harvey, D. J., Wormald, M. R., van Gijlswijk-Janssen, D., Redwan el, R. M., Wilson, I. A., Daha, M. R., Dwek, R. A., and Rudd, P. M. (2003). Secretory IgA N- and O-glycans provide a link between the innate and adaptive immune systems. *J. Biol. Chem.* 278:20,140–20,153.

Sakamoto, N., Shibuya, K., Shimizu, Y., Yotsumoto, K., Miyabayashi, T., Sakano, S., Tsuji, T., Nakayama, E., Nakauchi, H., and Shibuya, A. (2001). A novel Fc receptor for IgA and IgM is expressed on both hematopoietic and non-hematopoietic tissues. *Eur. J. Immunol.* 31:1310–1316.

Sansonetti, P. J., Arondel, J., Cantey, J. R., Prevost, M. C., and Huerre, M. (1996). Infection of rabbit Peyer's patches by *Shigella flexneri*: Effect of adhesive or invasive bacterial phenotypes on follicle-associated epithelium. *Infect. Immun.* 64:2752–2764.

Shreedhar, V. K., Kelsall, B. L., and Neutra, M. R. (2003). Cholera toxin induces migration of dendritic cells from the subepithelial dome region to T- and B-cell areas of Peyer's patches. *Infect. Immun.* 71:504–509.

Sicinski, P., Rowinski, J., Warchol, J. B., Jarzabek, Z., Gut, W., Szczygiel, B., Bielecki, K., and Koch, G. (1990). Poliovirus type 1 enters the human host through intestinal M cells. *Gastroenterology* 98:56–58.

Simister, N. E., and Rees, A. R. (1985). Isolation and characterization of an Fc receptor from neonatal rat intestine. *Eur. J. Immunol.* 15:733–738.

Smith, M., Thomas, N., Jenkins, P., Miller, N., Cremaschi, D., and Porta, C. (1995). Selective transport of microparticles across Peyer's patch follicle-associated M cells from mice and rats. *Exp. Physiol.* 80:735–743.

Snider, D. P., Underdown, B. J., and McDermott, M. R. (1997). Intranasal antigen targeting to MHC class II molecules primes local IgA and serum IgG antibody responses in mice. *Immunology* 90:323–329.

Suzuki, K., Meek, B., Doi, Y., Muramatsu, M., Chiba, T., Honjo, T., and Fagarasan, S. (2004). Aberrant expansion of segmented filamentous bacteria in IgA-deficient gut. *Proc. Natl. Acad. Sci. USA* 101:1981–1986.

Tohno, M., Shimosato, T., Kitazawa, H., Katoh, S., Iliev, I. D., Kimura, T., Kawai, Y., Watanabe, K., Aso, H., Yamaguchi, T., and Saito, T. (2005). Toll-like receptor 2 is expressed on the intestinal M cells in swine. *Biochem. Biophys. Res. Commun.* 330:547–554.

Tomana, M., Zikan, J., Moldoveanu, Z., Kulhavy, R., Bennett, J. C., and Mestecky, J. (1993). Interactions of cell-surface galactosyltransferase with immunoglobulins. *Mol. Immunol.* 30:265–275.

Trombetta, E. S., and Mellman, I. (2005). Cell biology of antigen processing *in vitro* and *in vivo*. *Annu. Rev. Immunol.* 23:975–1028.

van der Waaij, L. A., Kroese, F. G., Visser, A., Nelis, G. F., Westerveld, B. D., Jansen, P. L., and Hunter, J. O. (2004). Immunoglobulin coating of faecal bacteria in inflammatory bowel disease. *Eur. J. Gastroenterol. Hepatol.* 16:669–674.

van der Waaij, L. A., Limburg, P. C., Mesander, G., and van der Waaij, D. (1996). *In vivo* IgA coating of anaerobic bacteria in human faeces. *Gut* 38:348–354.

van Egmond, M., Damen, C. A., van Spriel, A. B., Vidarsson, G., van Garderen, E., and van de Winkel, J. G. (2001). IgA and the IgA Fc receptor. *Trends Immunol.* 22:205–211.

Weltzin, R., Lucia-Jandris, P., Michetti, P., Fields, B. N., Kraehenbuhl, J. P., and Neutra, M. R. (1989). Binding and transepithelial transport of immunoglobulins by intestinal M cells: Demonstration using monoclonal IgA antibodies against enteric viral proteins. *J. Cell Biol.* 108:1673–1685.

Williams, N. A., Hirst, T. R., and Nashar, T. O. (1999). Immune modulation by the cholera-like enterotoxins: From adjuvant to therapeutic. *Immunol. Today* 20:95–101.

Wold, A. E., Mestecky, J., Tomona, M., Kobata, A., Ohbayashi, H., Endo, T., and Eden, C. S. (1990). Secretory immunoglobulin A carries oligosaccharide receptors for *Escherichia coli* type 1 fimbrial lectin. *Infect. Immun.* 58:3073–3077.

Wolf, J., Rubin, D., Finberg, R., Kauffman, R., Sharpe, A., Trier, J., and Fields, B. (1981). Intestinal M cells: A pathway for entry of reovirus into the host. *Science* 212:471–472.

Yamanaka, T., Helgeland, L., Farstad, I. N., Fukushima, H., Midtvedt, T., and Brandtzaeg, P. (2003). Microbial colonization drives lymphocyte accumulation and differentiation in the follicle-associated epithelium of Peyer's patches. *J. Immunol.* 170:816–822.

Yamanaka, T., Straumfors, A., Morton, H., Fausa, O., Brandtzaeg, P., and Farstad, I. (2001). M cell pockets of human Peyer's patches are specialized extensions of germinal centers. *Eur. J. Immunol.* 31:107–117.

Yamanaka, T., Straumfors, A., Morton, H. C., Rugtveit, J., Fausa, O., Brandtzaeg, P., and Farstad, I. N. (1999). Prominent expression of co-stimulatory molecules B7.1 (CD80) and B7.2 (CD86) by $sIgD^{-}CD20^{lo}$ memory B cells in human Peyer's patch M cell pockets. *Immunol. Lett.* 69:42.

Zhou, F., Kraehenbuhl, J. P., and Neutra, M. R. (1995). Mucosal IgA response to rectally administered antigen formulated liposomes in IgA coated liposomes. *Vaccine* 13:637–644.

10
IgA and Intestinal Homeostasis

Per Brandtzaeg[1] and Finn-Eirik Johansen[1]

[1] Laboratory of Immunohistochemistry and Immunopathology (LIIPAT), Department and Institute of Pathology, Rikshospitalet University Hospital, 0027 Oslo, Norway.

10.1. Introduction

The body is under constant threat of attack by viruses, bacteria, and parasites, and most pathogens use the mucosae as portals of entry. Evolution has therefore provided mammalians with several complex and potent layers of defense. Microorganisms have inhabited Earth for at least 2.5 billion years, and the power of immunity is a result of coevolution in which particularly the commensal bacteria have shaped the body's defense functions in a state of mutualism (Bäckhed et al., 2005; Hooper and Gordon, 2001). In humans, the critical role of the immune system becomes clinically apparent when it is defective. Thus, inherited and acquired immunodeficiency states, or more subtle immunoregulatory defects, are characterized by increased susceptibility to infectious diseases—sometimes caused by the commensal microbiota that is normally considered to be nonpathogenic (Haller and Jobin, 2004; Sansonetti, 2004; Yan and Polk, 2004).

The immune system can be divided into two general arms: innate ("natural" or "nonspecific") immunity and adaptive ("acquired" or "specific") immunity, which work together synergistically (Chaplin, 2003). Notably, the adaptive immune system developed rather late in the phylogeny, and most species survive without it. However, this is not true for mammalians, which have an extremely sophisticated adaptive immune system of both systemic and mucosal type.

There appears to be a great redundancy of mechanisms in both systems—providing robustness to ensure that essential defense functions are preserved.

The success of such a complex overall strategy is evident because most humans have a normal gut despite the fact that its mucosa is covered only by a monolayered and therefore quite vulnerable epithelium. The prevailing mucosal homeostasis is indeed remarkable because the large intestinal surface area—approaching 300 m^2 in an adult—is exposed to an enormous load of commensal bacteria, comprising perhaps up to 800 different species. The human gut microbiota includes ~10^{14} bacteria (i.e., ~10 times the number of body cells), making up a weight of 1–2 kg. In addition, ~1000 kg of food is passing through the gut of an adult every year, and intact food proteins in the amount of 3–10 ng/mL reach the blood circulation after meals (Husby et al., 1985; Paganelli and Levinsky, 1980). In the face of these persistent antigenic challenges, tightly regulated local immune mechanisms are clearly needed to preserve a healthy intestinal mucosa (MacDonald and Monteleone, 2005).

10.2. Nature and Triggering of Innate Immunity

10.2.1. Innate Defense Mechanisms

It is increasingly being appreciated that innate immune mechanisms play a key role for the tuning of adaptive immunity in the gut and maintenance of mucosal homeostasis. Innate responses use preformed or rapidly synthesized effectors and sensors that lead to spontaneous activation or modulation of cellular functions (Chaplin, 2003). In addition to soluble components, the innate defense system comprises surface barriers, professional phagocytes, and dendritic cells (DCs). Together, these functions constitute a primary layer of natural protection against invading microorganisms, with the common goal of restricting their entry into the body by providing (1) physical/structural hindrance and clearance mechanisms (epithelial linings of skin and mucosae, mucus, ciliary function, peristalsis), (2) chemical factors (pH of body fluids, numerous antimicrobial proteins and peptides such as lysozyme, lactoferrin, peroxidase, and defensins), and (3) phagocytic cells (e.g. neutrophils, eosinophils, monocytes/macrophages, and immature DCs). As discussed later in this chapter, challenges of the innate system often lead to activation or modulation of adaptive immunity, including the secretory immune system; such a development might enhance substantially the resistance against, and recovery from, mucosal infections.

10.2.2. Pattern Recognition Receptors

The recognition molecules of the innate immune system are encoded in the germline (Chaplin, 2003). Basically, this system is therefore quite similar among healthy individuals and shows no apparent memory effect; that is, reexposure to the same pathogen will normally elicit more or less the same type of response. The actual cellular receptors sense microbial molecular structures that are conserved and often essential for survival of the microorganisms. Such structures include, for instance, endotoxin or lipopolysaccharide (LPS), teichoic acid, peptidoglycan, and unmethylated CpG motifs of DNA (Beutler and Rietschel, 2003); together, they are traditionally called pathogen-associated molecular patterns (PAMPs), but they also occur in commensal bacteria (Medzhitov, 2001) and are therefore preferably called microbe-associated molecular patterns (MAMPs).

It remains unclear whether the intestinal microbiota through its MAMPs induces a distinct molecular program in the innate immune system, which could explain why commensal bacteria are normally tolerated by the host (Nagler-Anderson, 2001; Philpott et al., 2001). The relative absence of microbial recognition structures from the apical surface of normal gut epithelium (see later) as well as potential attenuation of their signaling by certain bacteria [e.g., modulation of nuclear factor (NF)-κB-mediated responses] also appears of crucial importance to such tolerance (Haller and Jobin, 2004; Kelly et al., 2004, 2005; Kobayashi et al., 2002; Moynagh, 2005; Neish et al., 2000; Rakoff-Nahoum et al., 2004).

The receptors of the innate immune system that recognize PAMPs or MAMPs as cellular triggers are collectively called pattern recognition receptors (PRRs)—many of which belong to the so-called Toll-like receptors (TLRs), which will be discussed later. PRRs are expressed mainly by macrophages and DCs but also by a variety of other cell types, including T- and B-cells as well as epithelial cells (see later). Engagement of PRRs with their intracellular signaling pathways causes controlled activation of the cells; in the case of DCs, this leads to maturation accompanied by production of various cytokines and upregulation or downregulation of surface molecules according to strictly defined kinetics (Ricciardi-Castagnoli and Granucci, 2002). Such cellular modulation will critically influence further development of both innate and adaptive immunity.

Importantly, there are both stereotypical and selective responses of innate host cells to different types of microorganism. In this manner, exogenous triggers can imprint their "signatures" on the immune system. Thus, the plasticity of the innate system prepares the ground for a targeted and enhanced function of adaptive immunity (Liew, 2002).

10.3. Nature and Function of the Intestinal Immune System

10.3.1. Homeostasis-Promoting Mechanisms

During evolutionary modulation, the mucosal immune system has generated two anti-inflammatory layers of defense (Fig. 10.1): (1) immune exclusion performed mainly by secretory immunoglobulin A (SIgA) antibodies to restrict

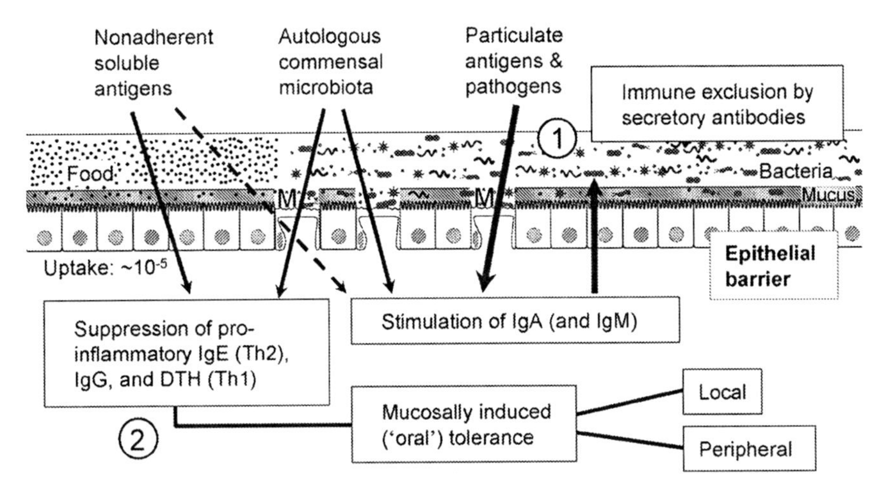

FIG. 10.1. Schematic representation of two major homeostatic immune mechanisms in the gut. **(1)** Immune exclusion limits epithelial colonization of microbes and inhibits penetration of foreign material (magnitude of normal food protein uptake indicated) into the gut mucosa. In cooperation with various innate immune mechanisms (not shown), this first line of defense is principally mediated by pIgR-dependent secretory antibodies of the IgA (and IgM) class. Mucosal production of IgA (and IgM) is strongly stimulated by pathogens and other particulate antigens taken up through thin M-cells (M) in the domes of gut-associated lymphoid tissue (thick solid arrow). The commensal microbiota and penetrating food proteins (magnitude of normal uptake indicated by arrow thickness) are also stimulatory for secretory immunity, but to a lesser extent (thin solid and broken arrows, respectively). **(2)** The latter types of exogenous impact induce, in addition, suppression of proinflammatory humoral immune responses (IgG and Th2 cytokine-dependent IgE antibodies) as well as downregulation of Th1 cytokine-dependent delayed-type hypersensitivity (DTH). Such local and peripheral homeostatic control of Th1 and Th2 responses induced via the gut is collectively termed "oral tolerance."

epithelial penetration and host invasion of microorganisms as well as other potentially dangerous antigens and (2) immunosuppressive mechanisms to avoid local and peripheral hypersensitivity to innocuous luminal antigens. Such mucosally induced inhibitory mechanisms are collectively referred to as "oral tolerance" when stimulated via the gut (Brandtzaeg, 1996; Mowat, 2003) and they probably explain why overt and persistent allergy to food proteins is relatively rare (Bischoff et al., 2000). A similar downregulatory tone of the immune system normally develops against antigenic components of the commensal intestinal microbiota (Duchmann et al., 1997; Helgeland and Brandtzaeg, 2000; Moreau and Gaboriau-Routhiau, 2000).

Secretory immunity can thus be considered as a first line of defense that aims at reducing the need for systemic immunity, because the latter will engage forceful proinflammatory mechanisms to enable elimination of microorganisms when required to save life. Systemic immunity therefore represents a "two-edged sword" prone to cause immunopathology, in contrast to the noninflammatory properties of SIgA antibodies. Oral tolerance appears to be a rather robust backup system for secretory immunity in view of the fact that substantial amounts of intact dietary antigens enter the body after a meal, usually without causing harm (Husby et al., 1985; Paganelli and Levinsky, 1980). However, because both the mucosal barrier function and the immunoregulatory network are poorly developed for a variable period after birth, the neonatal period is particularly critical with regard to infections and priming for allergic disease (Brandtzaeg, 2002a ; Holt, 1995). It is important in this context that the postnatal development of mucosal homeostasis depends on the establishment of a normal microbiota as well as on an adequate timing and the dose of dietary antigens when first introduced (Brandtzaeg, 1998; Moreau and Gaboriau-Routhiau, 2000).

10.3.2. Stimulation of Gut-Associated Lymphoid Tissue

10.3.2.1. Immune-Inductive Tissue Structures

The intestinal IgA system is the best understood component of mucosal immunity. In fact, the gut mucosa contains at least 80% of the body's activated B-cells (Brandtzaeg et al., 1989)—mostly differentiated to IgA-producing blasts and plasma cells (PCs). Their generation depends on complex mechanisms of B-cell induction in gut-associated lymphoid tissue (GALT), which comprises Peyer's patches (PPs), the appendix, and numerous solitary or isolated lymphoid follicles (Brandtzaeg et al., 1987, 1999a; Brandtzaeg and Johansen, 2005 Brandtzaeg and Pabst, 2004). All of these structures are believed to represent inductive sites contributing to intestinal immune responses, whereas the lamina propria and epithelial compartment principally constitute effector sites receiving primed memory/effector B- and T-cells from GALT.

The domes of GALT are covered by a characteristic follicle-associated epithelium, which contains antigen-sampling M-cells. These very thin and bell-shaped specialized epithelial cells transport effectively both living and nonproliferating antigens (especially particles) from the gut lumen into the organized lymphoid tissue (Neutra et al., 2001). Many enteropathogenic infectious agents use the M-cells as portals of entry, so they represent extremely vulnerable parts of the surface epithelium. However, such "gaps" in the epithelial barrier are needed to facilitate efficient induction of mucosal immunity (Fig. 10.1).

Gut-associated lymphoid tissue structures resemble lymph nodes with B-cell follicles, intervening T-cell zones, and a variety of antigen-presenting cells (APCs) such as macrophages and DCs, but there are no afferent lymphatics. Therefore, microbial stimuli must come directly from the epithelial surfaces—mainly via the M-cells but also aided by DCs, which might penetrate the surface epithelium with their processes (Brandtzaeg and Johansen, 2005; Milling et al., 2005; Rescigno et al., 2001). Hence, induction and regulation of mucosal immunity takes place primarily in GALT but also to a lesser extent at the effector sites (Brandtzaeg and Johansen, 2005; Brandtzaeg and Pabst, 2004).

10.3.2.2. Priming and Dispersion of Mucosal B-Cells

Antigens are presented to naïve T-cells in GALT by APCs after intracellular processing to immunogenic peptides. In addition, luminal peptides might be taken up and presented by B-lymphocytes and epithelial cells to subsets of intraepithelial and subepithelial T-lymphocytes (Brandtzaeg et al., 1999a). Both professional mucosal APCs, B-cells, and the small intestinal villous epithelium as well as the follicle-associated epithelium of GALT surrounding the M-cells express antigen-presenting MHC class II molecules— in humans particularly HLA-DR—in addition to classical and nonclassical MHC class I molecules (Brandtzaeg et al., 1987, 1999a; Christ and Blumberg, 1997). Interestingly, MHC class II-expressing naïve and memory B-lymphocytes abound juxtaposed to the M-cells. Such B-cells might present antigens efficiently to T-cells in cognate downregulatory or immunostimulatory interactions (Brandtzaeg et al., 1999a). $CD4^+$ helper T (Th) activated in GALT release cytokines such as transforming growth factor (TGF)-β and interleukin (IL)-10, which might drive the class switch and differentiation of mucosal B-cells to predominantly IgA-committed plasmablasts, although their regulation still remains unclear (Brandtzaeg and Johansen, 2005; Brandtzaeg et al., 2001; Fagarasan and Honjo, 2003).

B-Cells activated in GALT structures migrate rapidly via draining lymphatics to mesenteric lymph nodes, where they might be further stimulated; they next reach thoracic duct lymph and peripheral blood to become seeded by preferential homing mechanisms into distant mucosal effector sites, particularly the intestinal lamina propria, where they finally develop to Ig-producing PCs (Brandtzaeg et al., 1999b; Kunkel and Butcher, 2002). This terminal differentiation is modulated by "second signals" from local antigen-sampling DCs, lamina propria $CD4^+$ T- cells, and available cytokines (Brandtzaeg and Johansen, 2005; Brandtzaeg and Pabst, 2004).

10.3.2.3. Role of SIgA in Exclusion and Inclusion of Antigens

Most B-cells that home to mucosal effector sites apparently belong to clones of an early maturation stage, as indicated by their high level of J- chain regardless of concomitant isotype—although the IgA class normally predominates (Brandtzaeg et al., 1999a, 1999b). Locally produced J-chain-containing dimeric IgA and pentameric IgM are exported to the lumen by the polymeric Ig receptor (pIgR) to provide SIgA and secretory IgM (SIgM) antibodies, as discussed later. In the lumen, SIgA will coat commensal bacteria—demonstrated both in saliva (Brandtzaeg et al., 1968) and in feces (van der Waaij et al., 1996). This coating reduces bacterial access to the epithelial surface and protects against bacterial overgrowth and invasion (Macpherson et al., 2005) and it provides herd defense against the horizontal spread of infectious agents by the fecal-oral route (Wijburg et al., 2006). After the induction of a secretory immune response in the newborn's gut, GALT will temporarily be "shielded" by SIgA—apparently reflecting an antibody-mediated negative feedback mechanism for homeostatic regulation of the mucosal B-cell system (Shroff et al., 1995).

Uptake of SIgA-coated bacteria by M-cells mighty nevertheless be enhanced because an apical receptor specific for IgA/SIgA has been identified on these cells (Mantis *et al.,* 2002; Roy and Varvayanis, 1987). SIgA might exploit this M-cell receptor for targeting of antigens to DCs in the domes of PPs (Rey et al., 2004). This putative positive feedback mechanism might particularly target relevant environmental antigens to GALT of breast-fed infants by means of cognate maternal SIgA antibodies. In this manner, SIgA could act as an instructive immunological enhancer accompanied by a balanced cytokine pattern that promotes homeostasis (Favre et al., 2005). Many reports show that, over time, breast-fed babies develop enhanced secretory immunity (Brandtzaeg, 2002b).

Whether SIgA-coated or not, commensal bacteria sampled by the M-cells mainly become destroyed by subepithelial macrophages, but they are also taken up by DCs in tiny amounts (~0.0001%); this is sufficient to induce an immune response in GALT and mesenteric lymph nodes (Macpherson and Uhr, 2004). In conventional (CVN) clean mice, however, the systemic immune system remains untriggered—a compartmentalization that might contribute to peripheral tolerance or, rather, ignorance of the indigenous microbiota under normal conditions, although this might not be the case in humans (Macpherson et al., 2005).

10.3.3. Stimulation of GALT by Commensal Bacteria

10.3.3.1. Different Ways of Microbial B-Cell Activation

Stimulation of mucosal B-cells might occur by various mechanisms. Some microbial substances are superantigens, which interact with the B-cell receptor (BCR) outside the antigen-binding site—the prototype being *Staphylococcus aureus* protein A. B-Cells exposed to cross-linked protein A show a strong polyclonal response. B-Cells in GALT could be expanded in a similar manner both by exogenous and microbe-induced endogenous superantigens (Lanning et al., 2005).

Other microbial substances called Type 1 T-independent (TI-1) antigens are directly mitogenic for B-cells—including sugars, lipid structures, and certain nucleic acids. Type 2 T-independent (TI-2) antigens, on the other hand, are not by themselves mitogenic but cause extensive cross-linking of BCRc by repeating epitopes; their B-cell activation is exerted through synergy with soluble factors (e.g., cytokines) and interaction with various types of accessory cell. There are notable species differences in the mechanisms involved; for instance, wherease LPS acts as a TI-1 antigen on mouse B-cells, it acts as a TI-2 antigen on human B-cells. Interestingly, microbial polysaccharides might also function as T- cell-dependent antigens and exert profound homeostatic effects on the immune system by stimulating CD4^{+} T-cells after presentation by DCs on MHC class II molecules. Thus, it was recently shown in germ-free (GF) mice monocolonized with the ubiquitous gut commensal *Bacteroides fragilis* that a single bacterial polysaccharide had a striking impact both on lymphoid organogenesis and immune modulation (Mazmanian et al., 2005).

Although PPs containing organized primary follicles with naïve B-cells are present at birth, it takes some days before activated secondary follicles with germinal centers (GCs) appear as a result of stimulation from the environment. Studies in GF and CVN specific pathogen-free rats have demonstrated that bacterial colonization drives intraepithelial T- and B-cell accumulation with formation of M-cell pockets after an initial GC reaction—apparently induced by DC-mediated antigen transport from the lumen (Yamanaka et al., 2003). Whereas TI antibody responses are restricted with regard to affinity and memory, the GC reaction is generally believed to be driven by BCR competition for a limited amount of T-cell-dependent antigens, resulting in memory/effector B-cells that undergo BCR affinity maturation and class switch (Manser, 2004). As reportedly shown for GALT of experimental animals, however, IgA differentiation stimulated by commensal gut bacteria might, under certain conditions, bypass the usual BCR requirement (Macpherson et al., 2001, 2005). Nevertheless, there always appears to be a dependency on some folliclelike aggregates of B-cells that, interestingly, might lack antigen-retaining follicular DCs and GCs.

In line with this observation, recent studies have demonstrated that if there is a sufficient innate drive of the immune system, B- cells might survive with a restricted repertoire and rather low affinity. Thus, in the GF-appendix model in rabbits, it was shown that certain commensal bacteria are quite efficient in promoting GALT development and that such ability depends on certain stress responses in the same bacteria, suggesting a nonspecific (superantigen?) impact on GALT (Rhee et al., 2004). Another study reported independency of BCR engagement for GALT development in mice where the Epstein-Barr virus (EBV) protein LMP2A was transgenically introduced as a constitutive BCR surrogate providing a weak signaling pathway (Casola et al., 2004). It was concluded that commensal bacteria—by interacting with innate immune receptors—can promote the GC reaction in GALT independently of BCR-mediated antigen recognition.

Altogether, it is well established that the gut microbiota is required for activation of GALT with normal intestinal PC development (Crabbé et al., 1970) and that commensal bacteria can shape the BCR repertoire of the host (Lanning et al., 2005). Perhaps the GC reaction during the evolution of GALT to generate a protective antibody repertoire was not antigen-specific but, rather, cross-reactive. The indigenous microbiota might in this context act as polyclonal B-cell activators through several mechanisms, including PRR signaling (Beisner et al., 2005). TLRs expressed by DCs and B-cells are likely molecular candidates to mediate such an innate drive of GALT development (Pulendran and Ahmed, 2006).

10.3.3.2. Cross-reactive SIgA Antibodies as a Homeostatic Mechanism

The peritoneal cavity is recognized as an important source of intestinal B-cells in normal mice, perhaps providing up to 40–50% of the IgA$^+$ PCs in lamina propria (Kroese et al., 1989). The precursors are self-renewing IgM$^+$ B1 (CD5$^+$) cells, which give rise to "natural" cross-reactive SIgA antibodies directed mainly against microbial TI antigens with no clear dependency on a GC reaction (Macpherson et al., 2000). Notably, however, rather than being encoded in germline, murine B1-cells often show hypermutation of Ig heavy-chain V-region (V_H) genes as a sign of selection (Bos et al., 1996). It remains controversial where the B1-cells switch to the IgA phenotype and how they reach the gut mucosa (Brandtzaeg et al., 2001; Brandtzaeg and Pabst, 2004), but their egress from the peritoneal cavity appears to depend on downregulation of their adhesion molecules by MAMP signaling via TLRs (Ha et al., 2006).

To clarify this complex issue, several studies have analyzed the intestinal immune system of GF mice after monoassociation with a variety of noninvasive, commensal bacteria (Shroff et al., 1995; Umesaki et al., 1995). It was generally found that these microbes induce a GC reaction in GALT with generation of IgA$^+$ plasmablasts that accumulate in the lamina propria and produce both "natural" and specific IgA. Individual bacterial species were shown to differ, however, with regard to the maximal amount of total "natural" IgA that they induced and the fraction that could be shown to be specific for antigens of the colonizer (Bos et al., 2001). All tested bacteria apparently elicited a waxing followed by a long-term waning IgA response, which was accompanied by a GC reaction that notably showed a much more rapid development as well as decline. This could be attributed to the "shielding" of GALT from microbial antigens by the production of specific SIgA antibodies, because of the relatively long-term persistence of both specific and "natural" IgA$^+$ PCs in the lamina propria.

Such homeostatic immune modulation has been particularly well documented with segmented filamentous bacteria (SFB, related to *Clostridia*), which become a major gut colonizer of the distal ileum of mice after weaning. Colonization of formerly GF weanlings results in a transient GC reaction in GALT and seeding of the lamina propria with IgA$^+$ plasmablasts producing an SIgA level comprising 50–70% of that seen normally in CVN mice; only about 1% of this IgA has been found to show specificity for the SFB (Jiang et al., 2001; Talham et al., 1999). Interestingly, supercolonization with *Morganella morganii* 100 days following monoassociation with SFB induced little change in production of total intestinal IgA, although the specific response to *M. morganii* increased 20-fold compared to that against SFB (Talham et al., 1999). The chronic GC reaction observed in GALT of CVN mice is therefore likely caused by continuous exposure of the gut to novel microbial antigens. The sustained colonization of commensal gut bacteria might thus provide the necessary chronic stimulation of previously induced "natural" and specific IgA production.

10.3.3.3. Contribution of B1- and B2-Cells to Intestinal IgA Production

The relationship between B1-cells and conventional bone-marrow-derived B2-cells in murine IgA responses induced by the indigenous microbiota remains elusive (Macpherson et al., 2005); when eliminating one of these subsets in genetically manipulated mice, the other subset is likely to occupy the whole intestinal B-cell compartment. Both the B1 and the B2 subset of intestinal IgA$^+$ PCs showed very restricted (oligodisperse) usage of V_H genes and multiple clonally related sequences when the repertoire was analyzed by complementarity-determining region (CDR)3 spectrotyping, cloning, and sequencing (Stoel et al., 2005). Out of 15–20 sequences examined from various types of mouse, there were 2 or more likely clonal relatives based on identical V/D/J junctional sequences. Such restricted V_H-gene usage was seen whether monoassociated immunodeficient recipients of B-cells or CVN immunocompetent mice of several common strains were analyzed. It could have been expected that polyclonal microbial stimuli would rather have induced polydisperse B-cell responses in the gut. Therefore, perhaps both B1 and B2 intestinal cells–stimulated specifically by TI-1 and/or TI-2 bacterial antigens via their BCRs–will generate an oligodisperse population of IgA$^+$ PCs that produce "natural" cross-reactive antibodies. In this process, B-cells with randomly recombined V/D/J segments– but without appreciable N-additions or point mutations in the V_H regions–might be selected by relatively few TI antigens in the gut.

Despite many remaining questions, these murine studies constitute useful information to explain the enormous IgA drive provided by the intestinal microbiota in the absence of a high-affinity BCR development. It appears that the production of large amounts of IgA with a restricted or oligodisperse repertoire provides antibody capacity to bind with low affinity to the numerous redundant epitopes of commensal bacteria to maintain mutualism with the indigenous microbiota; this homeostatic mechanism therefore relies mainly on large quantities of cross-reactive SIgA (Macpherson et al., 2005). Altogether, the induction of IgA by commensal bacteria seems to be a rather primitive system that limits their local colonization and penetration through the epithelial barrier without eliminating them from the gut. Superimposed on this "innate-like" defense, the B2 system has the property to undergo GC-driven high-affinity BCR selection for particular virulence factors of pathogens to clear ongoing infections.

10.3.3.4. Intestinal B-Cell Responses Differ Between Humans and Mice

It remains unclear whether such a two-layered SIgA defense also operates in humans, where there is no evidence to suggest that peritoneal B1-cells contribute to intestinal IgA production (Boursier et al., 2002; Brandtzaeg et al., 2001). In the human gut, both IgA$^+$ and IgM$^+$ PCs have highly mutated V_H-region genes even from childhood—consistent with precursor selection in GCs (Boursier et al., 1999; Dunn-Walters et al., 1997a; Fischer and Kuppers, 1998).

The level of mutations is significantly higher in human intestinal B-cells than that seen in splenic PCs (Dunn-Walters et al., 2000). Moreover, spectrotyping of CDR3 variability shows a rather restricted repertoire for circulating IgA$^+$ cells compared with V_H transcripts from the colon, which are quite diverse, particularly for the V_H1–V_H5 regions (Holtmeier et al., 2000). The IgM V_H-region genes in the human peritoneal cavity likewise exhibit fewer mutations than the corresponding genes from intestinal B-cells (Boursier et al., 2002). Further, the V_H4–34 genes used by IgG and IgA in human peritoneal B-cells show significantly lower numbers of mutations than their mucosal counterparts. Also notably, V_H-gene sequences from human PP B-cells are clonally related to ileal lamina propria PCs, in accordance with a predominant GC derivation from GALT (Dunn-Walters et al., 1997b). Finally, clonally related V_H transcripts are widely distributed along the colon, likewise suggesting seeding from GALT mainly via peripheral blood (Holtmeier et al., 2000).

Altogether, the balance of evidence suggests that human intestinal IgA$^+$ PCs are generated by GC reactions in GALT and that derivation from the peritoneal cavity, or a putative lamina propria switch process, is negligible or absent (Bousier et al., 2005; Brandtzaeg and Johansen, 2005; Brandtzaeg and Pabst, 2004). Nevertheless, considerable levels of cross-reactive "natural" SIgA antibodies directed against self as well as microbial antigens do occur in human external secretions (Bouvet and Fischetti, 1999). The reason for this could be microbial polyclonal activation of GALT independently of BCR-mediated antigen recognition or, alternatively, stimulation via BCR by microbial TI antigens, as discussed earlier.

10.3.4. Antibody-Mediated Mucosal Defense

10.3.4.1. Secretory Immunity and Immune Exclusion

As briefly mentioned earlier, the unique and efficient export of SIgA and SIgM antibodies to the epithelial surface is mediated by pIgR– also known as membrane secretory component (SC)–which is cleaved and sacrificed as bound SC to stabilize the secretory antibodies in the lumen (Brandtzaeg, 2003; Phalipon and Corthésy, 2003). The ligand structure specific for this receptor depends on the J-chain, which is incorporated selectively into dimeric IgA and pentameric IgM (Brandtzaeg, 1974a, 1974b; Brandtzaeg and Prydz, 1984; Johansen et al., 2000, 2001). It has been estimated that such pIgR-mediated transport of dimeric IgA, on average, results in the remarkable daily delivery of ~3 g of SIgA into the intestinal lumen of an adult human (Conley and Delacroix, 1987). Because of this efficient epithelial transport—which basically is constitutive—SIgA can act as a sustained first line of antibody-mediated defense against microorganisms and other antigens; it thereby performs homeostatic modulation of the host response elicited both by commensal bacteria and overt pathogens. Notably, with its high level of cross-reactivity even in humans (Bouvet and Fischetti, 1999), the mucosal IgA system is well designed to provide antibody-mediated immune exclusion of the extremely diverse intestinal microbiota, which expresses a multitude of redundant epitopes, while allowing a persistent host–commensal mutualism. Such production of "natural" SIgA antibodies might at least, in part, reflect the innate drive of the intestinal immune system, as discussed earliere, whereas affinity maturation of B-cells in the GCs of GALT provides a more powerful defense to expel pathogens from the host.

10.3.4.2. Homeostatic Defense Functions of SC and SIgA

Free SC is generated by apical cleavage and epithelial release of unoccupied pIgR (Mostov and Blobel, 1982). Interestingly, free SC has been shown to possess several innate immune properties such as binding enterotoxigenic *Escherichia coli* and reducing the effect of *Clostridium difficile* toxin (Phalipon and Corthésy, 2003). These observations suggest that SC has phylogenetically originated from the innate defense system, like many other proteins involved in adaptive immunity. Importantly, membrane SC has been exploited as the only identifiable epithelial receptor involved in secretory immunity, as documented by the phenotype of pIgR knockout mice, which have no active external transport of dimeric IgA and pentameric IgM (Johansen et al., 1999).

A crucial protective role of secretory antibodies is supported by the fact that such pIgR$^{-/-}$ mice have "leaky" mucosal epithelia and excessive intestinal uptake of antigens from commensal bacteria such as *E. coli,* thereby eliciting a systemic antimicrobial IgG response (Johansen et al., 1999). Also, similarly to J-chain knockout mice that likewise lack both SIgA and SIgM, pIgR$^{-/-}$ mice are less resistant than wild-type mice against cholera toxin and early colonization of the gut epithelium by pathogens (Lycke et al., 1999; Uren et al., 2005). Due to absent secretory immunity, pIgR$^{-/-}$ mice further show reduced protection as well as cross-protection against mucosal challenge in an influenza model with the live A/PR8 strain after intranasal immunization with inactivated influenza vaccines of various types (Asahi et al., 2002).

Altogether, animal and cell culture experiments have suggested that SIgA antibodies promote intestinal homeostasis by neutralizing viruses and bacterial products through noninflammatory mechanisms in various mucosal compartments (Table 10.1 and Fig. 10.2). In addition to traditional luminal neutralization (Davids et al., 2006), it has been demonstrated that dimeric IgA antibodies—when exported by pIgR—can remove antigens from the lamina propria and neutralize viruses within the epithelium or block their transcytosis through polarized epithelial cells (Alfsen et al., 2001; Bomsel et al., 1998; Burns et al., 1996; Feng et al., 2002; Huang et al., 2005; Mazanec et al., 1993, 1995; Robinson et al., 2001). It has also been reported that dimeric IgA can neutralize bacterial LPS within intestinal epithelial cells (Fernandez et al., 2003)—suggesting a novel intracellular, noncytotoxic, and anti-inflammatory role for this antibody class during its transport to the lumen.

TABLE 10.1. Antimicrobial effects of SIgA antibodies.

- Dimeric IgA provides efficient microbial agglutination and virus neutralization.
- Performs noninflammatory extracellular and intracellular immune exclusion by inhibiting epithelial adherence and invasion.
- Exhibits cross-reactive ("innatelike") activity and provides cross-protection.
- SIgA (particularly SIgA2) is quite stable (bound SC stabilizes both isotypes of IgA).
- SIgA is endowed with mucophilic and lectin-binding properties (via bound SC in both isotypes and mannose in IgA2).

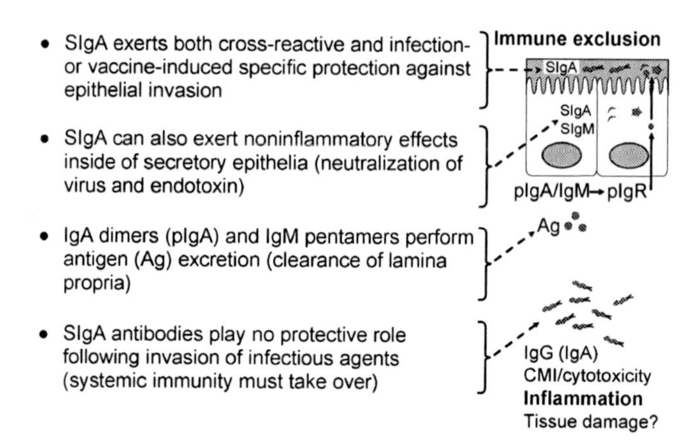

FIG. 10.2. Different principles of SIgA-mediated contribution to mucosal homeostasis. In addition to immune exclusion at the epithelial surface, the pIgR-mediated external transport of dimeric IgA and pentameric IgM (pIgA/IgM) might be exploited for intraepithelial pathogen neutralization and antigen excretion. However, when infection with pathogen invasion takes place, systemic immunity must take over to save life. This involves potent proinflammatory mechanisms such as complement activation by IgG antibodies, cell-mediated immunity (CMI), and cytotoxicity, all of which might cause tissue damage.

10.3.4.3. Role of SIgM and IgG Antibodies

Although SIgA is the chief effector of immune exclusion, SIgM also contributes—particularly in the newborn period and in IgA deficiency (Brandtzaeg et al., 1987, 1991; Brandtzaeg and Nilssen, 1995). In addition, there might be a significant contribution to immune exclusion by serum-derived or locally produced IgG antibodies transferred passively to the lumen by paracellular leakage (Persson et al., 1998)—or perhaps to some extent exported actively by the neonatal Fc receptor (FcRn) expressed on the gut epithelium (Yoshida et al., 2004). Interestingly, monomeric IgA or IgG antibodies, when cross-linked via antigen with dimeric IgA of the same specificity, might contribute to pIgR-mediated epithelial excretion of foreign material from the intestinal lamina propria (Mazanec et al., 1993). Notably, however, because IgG is complement activating, its contribution to surface defense is potentially proinflammatory, which could jeopardize the epithelial barrier function (Brandtzaeg and Tolo, 1977). Such deterioration of local homeostasis is most likely counteracted by a variety of complement regulatory factors produced by mucosal epithelia (Berstad and Brandtzaeg, 1998).

It should finally be noted that when overt infection with microbial invasion occurs, SIgA antibodies will no longer determine the fate of the host, as experimentally documented in pIgR$^{-/-}$ compared with wild-type mice (Sun et al., 2004; Uren et al., 2005). Moreover, studies in mucosally vaccinated wild-type mice challenged with live influenza virus intranasally have suggested that whereas SIgA antibodies are essential to control virus replication locally, serum IgG antibodies protect against clinical illness (Bižanov et al., 2005). However, although systemic immunity might be considered a life-saving layer of defense, it operates at the risk of causing inflammation and tissue damage (Fig. 10.2). Thus, it has been experimentally documented that SIgA antibodies prevent virally induced pathology in the upper airways, whereas IgG antibodies neutralize newly replicated virus after the initiation of infection (Renegar et al., 2004). The lung parenchyma, which lacks a SIgA system, fully depends on serum-derived IgG and monomeric IgA for antibody protection (Daniele, 1990).

10.3.4.4. Interactions Between SIgA and Innate Defense Factors

Several studies have shown that mucosal immunity can be enhanced by cooperation between SIgA and innate defense factors. Immune reactions that take place at epithelial surfaces might stimulate the release of mucus from goblet cells and thereby reinforce the mucosal barrier against penetration of soluble molecules and microorganims (Walker et al., 1982; Walker and Bloch, 1983). Through the affinity of SC to mucus, the mucous barrier is endowed with a "flypaper" effect by topically retaining SIgA antibodies, which can trap antigens (Lim and Rowley, 1982; Phalipon and Corthésy, 2003).

Antigens of immune complexes in the mucous layer are more rapidly degraded by proteolytic enzymes than free antigens (Walker et al., 1975). It has been proposed that IgA1 is especially miscible with mucus, whereas the antibody function of IgA2 might take place mainly in the external secretory fluid (Clamp, 1980). This might be particularly relevant in the distal gut, where there is a predominance of local IgA2 production (Kett et al., 1986); this subclass is also more resistant to degradation by certain proteases other than IgA1 (Kilian et al., 1996). Although the subclass dependency is not determined, it has been proposed that SIgA-coated bacteria retained in mucus form a biofilm on the colonic epithelium (Bollinger et al., 2003); microbial immune exclusion might thus be enhanced by the rapid turnover of the gut epithelium.

In vitro data suggest that SIgA antibodies significantly promote the bacteriostatic effect of lactoferrin— probably by inhibiting bacterial production of iron-chelating agents, which interfere with its function (Rogers and Synge, 1978). SIgA can also enhance the broad antimicrobial spectrum of

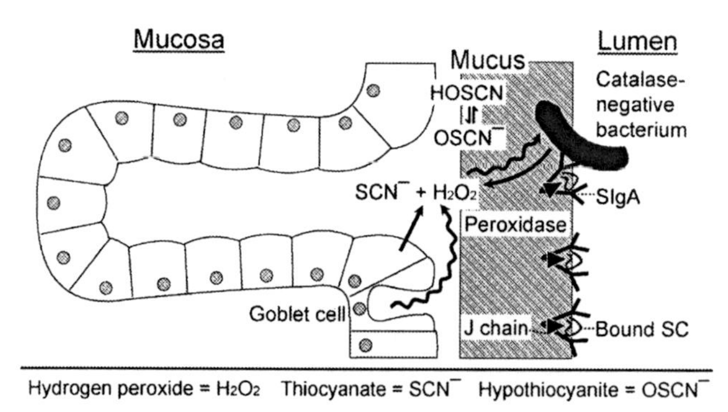

FIG. 10.3. Cooperation between SIgA and the peroxidase defense system. Schematic illustration of postulated microbial immune interactions taking place in the mucous layer of colonic mucosa. Details are discussed in the text.

the peroxidase defense system, apparently with no dependency on antibody specificity (Tenovuo et al., 1982). This effect might be explained by the mucophilic properties of bound SC in SIgA; cross-reactive SIgA antibodies could thus retain bacteria for prolonged and focused action of the enzyme with its biocidal product hypothiocyanate (Fig. 10.3). Human milk is rich in both lactoperoxidase and leukocyte-derived myeloperoxidase (Moldoveanu et al., 1982), which survives enzymatically active in gastric juice (Gothefors and Marklund, 1975). Moreover, peroxidase is produced by colonic goblet cells and is therefore most likely part of the innate gut defense (Venkatachalam et al., 1970).

There is a possibility that catalase-positive bacteria might resist the oxidizing effects derived from H_2O_2 (Fig. 10.3). However, bacterial catalase is restricted to the cytoplasm or periplasm of microorganisms and would therefore not be expected to alter substantially the availability of H_2O_2 as the substrate for peroxidase. Some bacteria might even produce H_2O_2 and, importantly, catalase-positive pathogens have been shown to be cleared efficiently from the airways *in vivo* by the peroxidase defense system (Gerson et al., 2000).

Cooperation between SIgA and the natural antimicrobial actions of gastric acid and intestinal peristalsis was demonstrated in a study of bacterial overgrowth in the jejunum of vagotomized patients; mucosal protection deteriorated when IgA deficiency was combined with a suboptimal function of these two innate defense mechanisms (McLoughlin et al., 1978). As always when IgA is selectively lacking, the study subjects necessarily had compensatory SIgM in their gut fluid (Brandtzaeg and Johansen, 2005). However, although several antimicrobial activities have been identified for this secretory antibody class (Brandtzaeg *et al.*, 1987), it clearly cannot adequately replace SIgA (Table 10.2).

TABLE 10.2. Effect of cooperation between SIgA antibodies and natural defense mechanisms on bacterial overgrowth in the jejunum.

	Clinical conditions without (+) or with (−) defective adaptive or innate immunity				
Variables					
SIgA	+	+	−	−	−
Peristalsis	+	−[a]	+	+	−[a]
Gastric acid	+	−[b]	+	−[b]	−[b]
Effect on jejunal colonization	−	−	−	−	+

[a]Complete vagotomy.
[b]Gastric drainage (gastroenterostomy) or pernicious anemia.
Source: Modified from McLoughlin et al. (1978).

10.3.4.5. Homeostatic Backup Mechanisms in SIgA Deficiency

Despite their compensatory intestinal SIgM, IgA-deficient subjects have raised levels of serum IgG antibodies to dietary antigens and show an increased frequency of infections, allergy, and autoimmune disorders, including celiac disease (Brandtzaeg and Nilssen, 1995; Burrows and Cooper, 1997). The same is claimed to be true for Crohn's disease, with some 20-fold increased incidence associated with IgA deficiency (Hammarström, personal communication). However, observations in IgA knockout mice have questioned the role of SIgA in mucosal defense (Brandtzaeg, 2003); explanations might be underestimation of the impact of compensatory SIgM and the fact that these mice show reduced inflammatory potential at mucosal sites due to decreased APC and Th1-cell functions (Arnaboldi et al., 2005).

As discussed earlier, knockout mice that lack both SIgA and SIgM (pIgR$^{-/-}$ or J chain$^{-/-}$) exhibit decreased resistance to cholera toxin and pathogen colonization in the gut; the leaky epithelial barrier results in a threefold increased intestinal generation of IgA$^+$ PCs (Johansen et al., 1999; Uren et al., 2003). These mice further appear to compensate for their lack of secretory immunity by increasing the number and cytotoxic activity of intraepithelial lymphocytes (IELs) in the gut (Yamazaki et al., 2005). Notably, IELs might not only be involved in antimicrobial defense but also exert a positive effect in the induction of oral tolerance (Grdic et al., 1998; Mennechet et al., 2004). Interestingly, pIgR$^{-/-}$ mice show enhanced ability to mount a tolerogenic response against fed antigen compared with wild-type mice—most likely representing an activated homeostatic mechanism in the face of a decreased mucosal barrier function (Karlsson et al., 2005). Basolateral release of MHC class II-positive exosomelike vesicles ("tolerosomes") from antigen-pulsed gut epithelium induces oral tolerance (Östman et al., 2005); perhaps this mechanism is enhanced in pIgR$^{-/-}$ mice because more antigen is taken up by unshielded epithelial cells.

It has likewise been observed that humans with hypogammaglobulinemia or selective IgA deficiency can have raised numbers of $\alpha\beta^+CD8^+$ and $\gamma\delta^+$ IELs in their gut; possibly, enhanced induction of oral tolerance—in addition to compensatory SIgM—might contribute to the fact that they suffer from relatively little intestinal pathology compared with their frequent clinical airway problems (Brandtzaeg and Nilssen, 1995).

10.4. Mucosal Immunity in Neonatal Defense

10.4.1. Marked Species Differences

In contrast to rodents and ungulates, the newborn human has high levels of circulating maternal IgG because of FcRn-dependent placental transfer in fetal life, and not as a result of breast-feeding. Also unlike ungulates, intestinal uptake of SIgA antibodies from breast milk is of no importance for systemic immunity in infants, except perhaps in the preterm neonate (Brandtzaeg, 2002a). This is so because "gut closure" normally occurs in humans mainly before birth; but the mucosal barrier function might nevertheless be inadequate up to about 2 years of age. Although many variables influencing gut closure remain poorly defined, maturation of the intestinal immune system plays a major role.

Only occasional traces of SIgA and SIgM are exported from the human intestinal mucosa to the gut lumen during the first postnatal period, whereas some IgG is often present—probably reflecting both paracellular and FcRn-mediated transfer from the lamina propria (Harris et al., 2006; Persson et al., 1998), which after 34 weeks of gestation contains readily detectable maternal IgG (Brandtzaeg et al., 1991). IgA^+ PCs are normally undetectable in the mucosa before the infant is 10 days of age, but thereafter a rapid increase takes place—IgM^+ PCs often remaining predominant up to 1 month. On average, little increase of intestinal IgA production usually takes place after 1 year. A much faster establishment of secretory immunity can be seen in developing countries because of a more massive mucosal exposure to microorganisms (Brandtzaeg et al., 1991).

10.4.2. Individual IgA Variations Affecting Immunological Homeostasis

The postnatal mucosal B-cell development shows large individual variations, even within the same population (Brandtzaeg et al., 1991). This disparity might reflect a genetically determined effect on the establishment of the mucosal barrier function—perhaps in part exerted via diversity of the intestinal microbiota among individuals (Zoetendal et al., 2001). It has been proposed on the basis of serum IgA levels that a hereditary risk of atopy is related to a retarded postnatal development of the IgA system (Taylor et al., 1973). This notion was later supported by showing significantly reduced numbers of IgA^+ PCs (with no compensatory increase of IgM^+ PCs) in the jejunal mucosa of atopic children (Sloper et al., 1981). Also, an inverse relationship was found between the serum IgE level and the jejunal IgA$^+$ PC population in children with food-induced atopic eczema (Perkkiö, 1980).

It was subsequently reported that infants born to atopic parents show a significantly higher prevalence of salivary IgA deficiency than age-matched control infants (van Asperen et al., 1985). Interestingly, Kilian et al. (1995) found that 18-month-old infants with presumably IgE-mediated allergic problems had significantly higher proportions of IgA1 protease-producing bacteria in their throats than age-matched healthy controls. This was in keeping with a previous report that showed much less intact SIgA in nasopharyngeal secretions of children with an allergic history than controls with episodes of acute otitis (Sørensen and Kilian, 1984). In this context, it is important to note that it takes up to 3 months after birth before the IgA2$^+$-to-IgA1$^+$ PC ratio in salivary glands has increased to the adult level, with ~33% IgA2$^+$ PCs (Thrane et al., 1991).

Altogether, a poorly developed or enzymatically reduced SIgA-dependent barrier function as part of dysregulated mucosal homeostasis—most likely often combined with hereditary atopic predisposition—might explain the pathogenesis of allergy. This notion accords with a recent study reporting multiple dysregulations of both innate and adaptive immune functions—including reduced IgA production—variably expressed in children with food allergy (Latcham et al., 2003).

Support for such clinical observations has been provided by an experimental model of anaphylaxis in which mice were either sensitized or tolerized to β-lactoglobulin (β-LG) via the gut (Frossard et al., 2004). Compared with anaphylactic mice, the tolerant mice were found to have more β-LG-specific IgA-secreting B-cells in PPs and higher levels of IgA antibodies to β-LG in feces. An increase of β-LG-induced IL-10 and TGF-β production by PP T-cells was also observed in the tolerant mice. This could be a cytokine link to the enhanced IgA–cell differentiation (see earlier). Notably, reduced expression of TGF-β has been found in the gut of food-allergic children, suggesting a deficiency of regulatory T (Treg) cells secreting this suppressive cytokine (Pérez-Machado et al., 2003). A deficiency was recently documented also for the CD4$^+$CD25$^+$ Treg -cell phenotype in peripheral blood of persistently cow's milk-allergic children in contrast to those who outgrew their milk allergy (Karlsson et al., 2004).

10.4.3. Critical Role of Breast-Feeding in Mucosal Defense

The lactating breast is a remarkable production site for SIgA, with a daily output of 0.5-1.0 g (Brandtzaeg, 1983). Experiments in neonatal rabbits have convincingly demonstrated that SIgA is a crucial antimicrobial component of breast milk (Dickinson et al., 1998), in addition to a variety of other factors that might enhance mucosal homeostasis (Brandtzaeg, 2002a).

Mucosal infections are a major killer below the age of 5 years—being responsible for more than 10 million deaths of children annually—mainly in developing countries where infants are highly dependent on SIgA antibodies from breast milk to protect their mucosae; epidemiological data do suggest that the risk of dying from diarrhea is reduced 14–24 times in breast-fed children (Anonymous, 1994). In fact, it appears that breast-feeding is the most efficient feasible intervention measure, with the potential of preventing 13% of all deaths below 5 years of age (Black et al., 2003; Jones et al., 2003).

Although the value of breast-feeding in Westernized countries is clinically most apparent in preterm infants, population studies show that exclusively breast-fed infants are generally better protected against a variety of infections and probably also against allergy, asthma, and celiac disease (Brandtzaeg, 2002a; Ivarsson et al., 2002; Kull et al., 2002; Schoetzau et al., 2002, van Odijk et al., 2003). This strongly suggests that the mucosal barrier function in newborns can be reinforced by breast-feeding, which appears to be particularly important in the face of a parental history of allergic diseases (Benn et al., 2004).

10.4.4. SIgA Antibodies and Induction of Oral Tolerance

Through avoidance of too early intestinal immune activation—for instance, by limiting the upregulation of the costimulatory B7 (CD80/CD86) molecules on APCs (Brandtzaeg, 1998; Chen et al., 2000)—the shielding effect exerted by SIgA from breast milk on the suckling's GALT (see earlier) might contribute to the establishment of oral tolerance, not only against the indigenous microflora but also against dietary antigens such as gluten. Antibodies to gluten peptides are present in breast milk (Juto and Holm, 1992), and breast-feeding has been shown to protect significantly against the development of celiac disease in children (Ivarsson et al., 2002)—an effect that appears to be unrelated to the time of solid food introduction (Brandtzaeg, 1997). Therefore, mixed feeding rather than abrupt weaning seems to promote tolerance to food proteins (Ivarsson et al., 2002). This notion is also supported by reports suggesting that cow's milk allergy is more likely to develop in infants whose mothers have relatively low levels of milk SIgA antibodies to bovine proteins (Savilahti et al., 1991; Järvinen et al., 2000). It is noteworthy in this context that allergic mothers reportedly have relatively low levels of ovalbumin-specific milk SIgA (Casas et al., 2000).

The presence of TGF-β and IL-10 in breast milk might contribute to its tolerogenic properties because these cytokines exert pronounced immunosuppressive effects in the gut (Ishizaka et al., 1994; Steidler et al., 2000) and TGF-β enhances the epithelial barrier function (Planchon et al., 1994). Also, these two cytokines are important switch and differentiation factors in the development of IgA^+ B-cells (Brandtzaeg and Johansen, 2005).

10.5. Microbial and Nutritional Impact on Mucosal Immune Regulation

10.5.1. Effects of Antigen Exposure and Nutrition

The degree of antigenic and mitogenic GALT exposure is decisive for the development of secretory immunity. As discussed earlier, the commensal microbiota is of crucial importance to this end. Thus, the intestinal IgA system of GF mice is normalized after about 4 weeks with CVN microbiota exposure in an ordinary animal facility (Crabbé et al., 1970; Horsfall et al., 1978).

Bacteroides and *E. coli* strains appear to be particularly stimulatory for the development of intestinal IgA^+ PCs (Lodinová et al., 1973; Moreau et al., 1978; Moreau and Gaboriau-Routhiau, 2000). Antigenic constituents of food also exert a significant effect, as suggested by the occurrence of fewer intestinal IgA^+ PCs both in mice fed on hydrolyzed milk proteins (Sagie et al., 1974) and in parenterally fed babies (Knox, 1986). Likewise, mice given total parenteral (intravenous) nutrition have reduced numbers of B- and T-cells in the gut, as well as decreased SIgA levels (Li et al., 1995a, 1995b; Janu et al., 1997), and they show impaired SIgA-dependent influenza-specific immunity (Renegar et al., 2001). The effect of food in the gut lumen could be direct immune stimulation or indirectly mediated by changes in the microbiota or release of gastrointestinal neuropeptides.

In a study of whole-gut lavage obtained from healthy adult volunteers in Dhaka (Bangladesh), the intestinal concentration of IgA was found to be almost 50% higher than in comparable samples obtained from Edinburgh (UK); even more notable, the intestinal IgA antibody titer against LPS core types of *E. coli* was almost seven times higher in the former group of subjects, whereas the levels of ovalbumin antibodies were relatively lower (Hoque et al., 2000). Altogether, therefore, it can be concluded that the large bacterial and dietary antigen load in the gut explains why the greatest density of IgA^+ PCs is seen in the intestinal lamina propria—amounting to some 10^{10} cells per meter of adult intestine, fairly equally distributed in the proximal and distal segments (Brandtzaeg et al., 1989).

In human lactating mammary glands, the PC density is much less—one gland showing an IgA-producing capacity similar to only 1 meter of intestine (Brandtzaeg, 1983). Thus, the daily output of IgA per kilogram wet weight of tissue (minus fat) is not more for lactating mammary glands than for salivary glands. In fact, it remains an enigma how some terminal PC differentiation is accomplished at such concealed secretory effector organs situated at considerable distances from antigen-exposed mucosal surfaces (Brandtzaeg and Johansen, 2005). One possibility is that the response of a subset of mucosal memory B-cells is more dependent on cytokine stimulation than on BCR ligation by antigen (Ehrhardt et al., 2005). Anyhow, the large capacity for storage of dimeric IgA/SIgA in the mammary gland epithelium and duct system—

rather than a high stromal PC density—explains the remarkable output of SIgA during breast-feeding (Brandtzaeg, 1983).

In keeping with the crucial stimulatory effect of luminal antigens on mucosal B-cell differentiation, defunctioning colostomies in children caused a 50% numeric reduction of intestinal lamina propria IgA^+ and IgM^+ PCs after 2–11 months (Wijesinha and Steer, 1982). Prolonged studies of defunctioned ileal segments in lambs revealed an even more striking scarcity of mucosal PCs; this was explained by decreased local proliferation and differentiation of plasmablasts and perhaps reduced B-cell homing from GALT (Reynolds and Morris, 1984). Accordingly, the postnatal establishment of the mucosal IgA system is usually much faster in developing countries than in the industrialized part of the world–a difference that appears to hold true even in undernourished children (Nagao et al., 1993).

Severe vitamin A deficiency, however, reportedly affects adversely mucosal IgA antibody responses in rodents (Wiedermann et al., 1993) but with no consistent disturbance of the epithelial IgA transport (Stephensen et al., 1996); rather, the deficiency might involve lack of retinoic acid-dependent DC-mediated imprinting of gut homing molecules on memory/effector cells in GALT (Iwata et al., 2004; Mora et al., 2006). Nevertheless, undernourished children respond to bacterial overgrowth in the gut with enhanced IgA production and upregulated export of IgA (Beatty et al., 1983); it is of great clinical importance that detrimental effects imposed on the SIgA system by severe malnutrition might be reversed with nutritional rehabilitation (Watson et al., 1985).

10.5.2. Homeostatic Effects of Probiotics

It is possible that a suboptimal development of the SIgA-dependent mucosal barrier function together with inadequate tolerance mechanisms explains the increasing frequency of certain diseases in industrialized countries—particularly allergies and autoimmune/inflammatory disorders (Brandtzaeg, 2002a; Yazdanbakhsh et al., 2002). On the basis of the so-called extended hygiene hypothesis (Rautava et al., 2004), several studies have evaluated the beneficial clinical effect of probiotic preparations derived from commensal gut bacteria (Collins and Gibson, 1999; Isolauri et al., 2001; Kirjavainen and Gibson, 1999). Especially, certain strains of intestinal lactic acid bacteria (LABs)—particularly lactobacilli and bifidobacteria—have been reported to enhance IgA responses, both in humans and experimental animals—apparently in a T-cell-dependent manner (Malin et al., 1996; Prokešová et al., 1998, 1999; Moreau and Gaboriau-Routhiau, 2000; Yasui et al., 1995). A double-blind study of children with a family history of atopy reported the prevalence of atopic dermatitis to be reduced by 50% at the age of 2 years after receiving the probiotic *Lactobacillus* GG strain daily for 6 months after birth (Kalliomäki et al., 2001). It remains unknown whether this beneficial effect was mediated via SIgA enhancement, promotion of oral tolerance, or both.

Similarly, there is some hope that immunization with mycobacterial antigens or bacterial CpG oligonucleotides might skew the cytokine profile toward Th1 and thereby—through cross-regulation and/or Treg-cell induction—dampen Th2-dependent allergic symptoms (Hopkin et al., 1998; von Reyn et al., 1997; Wohlleben and Erb, 2001). Newborns are in fact able to mount a Th1-type immune response when appropriately stimulated (Marchant et al., 1999), similarly to the early mucosal IgA response seen in heavy stimulatory conditions; neonatal expansion of $CD25^+$ Treg-cells can apparently be driven by antigens in the presence of LPS unless counteracted by a strong genetic predisposition for allergy (Haddeland et al., 2005). It is also possible that DNA from probiotic bacteria might induce Treg-cells that enhance mucosal homeostasis because certain strains of LABs appear to act as well by subcutaneous or peritoneal injection as by the oral route (Foligné et al., 2005a, 2005b; Sheil et al., 2004). However, the effect of this approach on the intestinal IgA system has not been studied.

10.5.3. Regulation of pIgR Expression

The postnatal colonization of commensal bacteria is important both to establish and regulate an appropriate barrier function in the gut—including enhanced epithelial expression of pIgR (Table 10.3)—as clearly demonstrated in mouse experiments (Hooper et al., 2001; Neish et al., 2000). Synthesis of pIgR/SC starts in human fetal life around 20 weeks of gestation, but there is a peak of epithelial expression during the first couple of postnatal months—best revealed in salivary glands (Brandtzaeg et al., 1991); this most likely reflects a direct or indirect effect of microbial products following commensal colonization of the mucosae, which starts in the birth canal.

Collectively, the above observations show that the expression of pIgR is both constitutive and subjected to inductive transcriptional upregulation, as first documented at the molecular level in our laboratory (Johansen and Brandtzaeg, 2004). Because one molecule of pIgR is consumed for every ligand

TABLE 10.3. Colonization of germ-free mice with *B. thetaiotaomicron* increases intestinal expression of genes involved in epithelial barrier function and nutritient absorption.

Gene product	Fold Δ over germ-free
Barrier function	
Polymeric Ig receptor (pIgR/SC)	2.6 ± 0.7
Small proline-rich protein 2a (sprr2a)	205 ± 64
Decay-accelerating factor (DAF)	5.7 ± 1.5
Nutrient absorption	
Na^+/glucose cotransporter (SGLT-1)	2.6 ± 0.9
Colipase	6.6 ± 1.9
Liver fatty acid-binding protein (L-FABP)	4.4 ± 1.4

Source: Modified from Hooper et al. (2001).

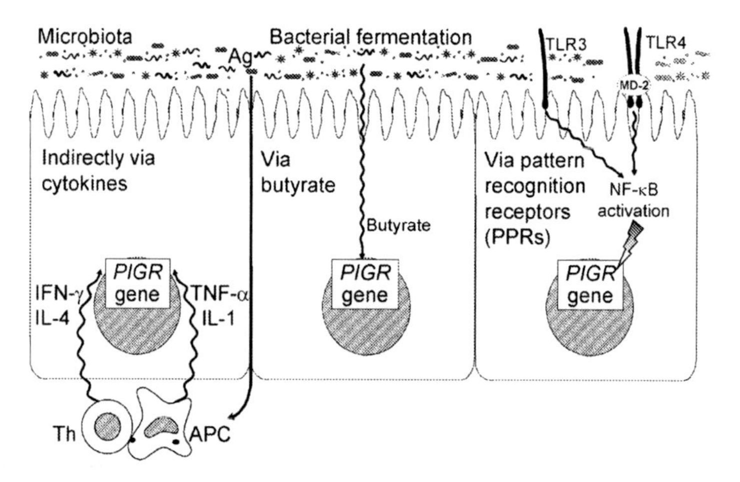

FIG. 10.4. Upregulation of epithelial pIgR expression by microbial activation of the *PIGR* gene locus. Details of the three postulated mechanisms are discussed in the text.

of dimeric IgA or pentameric IgM transported to the lumen, the regulation of pIgR must be crucial for maintenance of intestinal homeostasis.

Celiac disease is a good example of how pIgR expression and SIgA export can be indirectly enhanced via cytokines produced by activated mucosal APCs and T-cells (Fig. 10.4, left panel). There is a remarkable level of interferon (IFN)-γ in the untreated celiac lesion—derived from gluten-specific $CD4^+$ lamina propria T-cells (Nilsen et al., 1995, 1998) and activated intraepithelial $CD8^+$ T-cells (Olaussen et al., 2002). Similar upregulation of pIgR and epithelial transport of dimeric IgA is seen in Sjögren's syndrome and chronic gastritis—supporting the notion that cytokines provide an immunoregulatory link between increased local IgA production and enhanced output of SIgA during infection or low-grade inflammation not causing dysplastic lesions in the secretory epithelium (Brandtzaeg et al., 1992).

10.5.4. Microbial Enhancement of pIgR Expression

Upregulation of pIgR expression directly by microbes or their metabolic products is another intriguing possibility to enhance homeostatic immune functions in the gut. As mentioned earlier, a role for commensal bacteria to this end was suggested by the observation that intestinal pIgR mRNA levels were increased almost threefold when GF mice were colonized with *Bacteroides thetaiotaomicron*, a prominent member of the commensal intestinal microbiota (Table 10.3). Also, it was recently observed that infection with reovirus upregulates expression of pIgR in the human colonic epithelial cell line HT-29 (Pal et al., 2005); this might reflect interaction with epithelial PRRs, as discussed later. Changes in the composition of the indigenous microbiota could thus—either directly by bacterial products or indirectly via cytokines (Fig. 10.4)—explain the reported impact of passive and adaptive immunity on developmental pIgR expression levels in the murine gut (Jenkins et al., 2003).

Altogether, commensal bacteria appear important for maintaining the "tone" of intestinal pIgR expression above its constitutive level. This notion is in keeping with a role for certain strains of commensal gut bacteria to enhance homeostatic immunoregulatory mechanisms—observed directly by administration of probiotics such as LABs (see earlier) or indirectly by adding prebiotics to the diet. Oligosaccharides, for instance, can both promote the growth of LABs and act as a substrate for the intestinal formation of butyrate (C-C-C-COONa), a fermentation product of many anaerobic bacteria in the normal colonic microbiota. Butyrate is an important energy source for colonic epithelial cells and, notably, it can increase gene transcription levels through specific DNA sequences (Glauber et al., 1991).

In this context (Fig. 10.4, middle panel), it is of considerable interest that when we pretreated HT-29 colonic epithelial cells with butyrate, the effect on pIgR/SC expression induced by some cytokines—and particularly various combinations of cytokines—was remarkably enhanced (Fig. 10.5), with the exception of the effect of IL-4 that was reduced (Kvale and Brandtzaeg, 1995). In line with our observations—suggesting an overall positive effect on the pIgR level in the distal gut—Nakamura et al. (2004) reported that mouse pups receiving dietary fructo-oligosaccharides showed significantly enhanced pIgR expression in ileal and colonic epithelium (Fig. 10.6), as well as increased export of SIgA into ileal loops and feces.

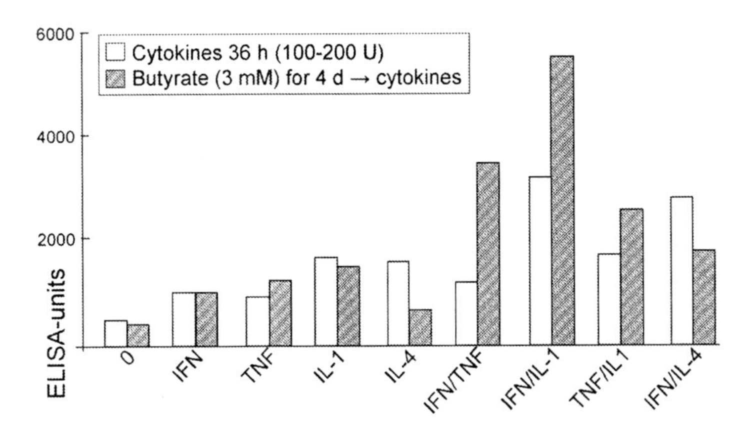

FIG. 10.5. Butyrate enhances cytokine-mediated expression of pIgR in human gut epithelial cells. Preincubation of HT-29 cells for 4 days with butyrate as indicated enhanced the pIgR-inducing effects [shown as enzyme-linked immunosorbent assay (ELISA) units] of TNF-α and IL-1, or these cytokines in combination with IFN-γ, whereas the effect of IL-4 was to reduce pIgR protein expression. Based on data from Kvale and Brandtzaeg (1995).

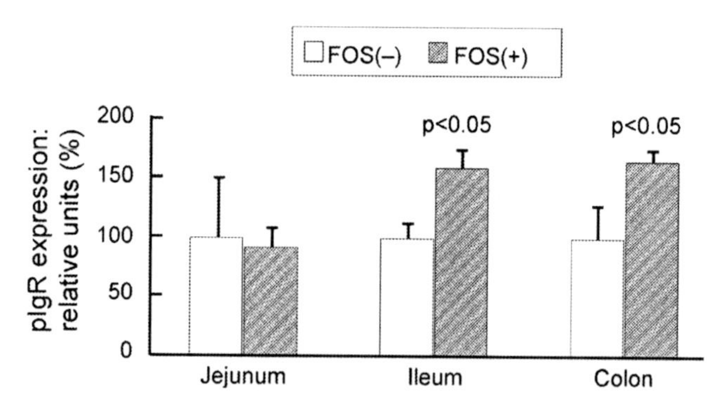

FIG. 10.6. Effect of prebiotics on pIgR expression in murine gut epithelium. Feeding fructooligosaccharides (FOS) to mouse pups enhanced the expression of pIgR in the distal gut (shown as relative units). Based on data from Nakamura et al. (2004).

10.6. Microbial Effects Via Pattern Recognition Receptors

10.6.1. Expression of PRRs in the Human Gut

As previously mentioned, TLRs function as PRRs to sense a variety of microbial constituents or products that trigger innate cellular responses (Medzhitov, 2001). To date, 11 transmembrane TLRs have been identified—acting singly or in combination (Beutler, 2004; Cario, 2005). TLR4 and TLR2 function as the sole conduits for signaling from LPS, which is an integral component of the outer membranes of Gram-negative bacteria. The classical LPS receptor CD14 is anchored in the cell membrane by glycosylphosphatidyl-inositol; in complex with its ligand and TLR4 or TLR2, it represents an important link between innate and adaptive immunity. The same is true for other PRRs that recognize additional PAMPs or MAMPs (Akira, 2003), including TLR9, which binds certain unmethylated CpG motifs of bacterial DNA (Kadowaki et al., 2001; Klinman et al., 1996; Peng et al., 2001). To induce tailored profiles of genes, the TLRs signal partly through "shared" and partly through more "specific" cellular pathways—generally leading to NF-κB activation but with coordinate activation of different transcription factors of the IFN regulatory factor (IRF) family that impose control on the expression pattern (Moynagh, 2005). The result variably includes DC maturation, release of proinflammatory cytokines such as the Th1-inducing IL-12 and IL-18 (Modlin, 2000; Kaisho and Akira, 2001; Medzhitov, 2001), and expression of chemokines and costimulatory molecules (Cario, 2005; Manigold et al., 2000; McInnes et al., 2000).

Although not yet extensively studied in the human gut, subepithelial putative APCs reportedly express certain TLRs. Thus, mRNA for TLR2 and TLR4 has been detected in isolated lamina propria macrophages (Smith et al., 2001), but the proteins were undetectable by immunohistochemistry in the normal state, although a substantial fraction of the subepithelial macrophage-like cells showed positive staining in inflammatory bowel disease (IBD) lesions (Hausmann et al., 2002). Also of note, only negligible levels of CD14 are normally present on these cells, and their cytokine response is usually poor after LPS stimulation (Rugtveit et al., 1997a; Smith et al., 2001). In IBD, however, CD14 expression is strongly elevated on recently recruited, perivascular monocyte-like macrophages (Rugtveit et al., 1997b). Concomitantly, the costimulatory molecules B7.1 (CD80) and B7.2 (CD86) are markedly upregulated on the subepithelial putative APCs, and the LPS-induced proinflammatory cytokine response of macrophage-like cells isolated from IBD lesions is increased (Rugtveit et al., 1997a).

Altogether, lamina propria APCs are normally in a quiescent state *in situ*, both in the human (Brandtzaeg, 2001) and murine (Chirdo et al., 2005) gut. However, in IBD, the mucosal antigen-presenting potential is presumably increased in parallel with the PRR enhancement because monocyte-derived macrophages can be skewed toward differentiation of activated DCs under the influence of TNF-α (Chomarat et al., 2003).

10.6.2. Epithelial Sensing of Microorganisms

The intestinal epithelium appears to have inherent mechanisms to protect itself against activation from the luminal side—unless production of proinflammatory cytokines, chemokines and defensins is needed to protect against pathogens (Philpott et al., 2001; Sansonetti, 2004; Yan and Polk, 2004). Thus, epithelial cells apparently possess sensing systems that allow discrimination between pathogenic and nonpathogenic bacteria in order to initiate an inflammatory reaction only when elimination of invading microorganisms is needed.

Interestingly in this context, nonpathogenic *Salmonella* strains are able to block the NF-κB transcription pathway in human gut epithelial cells *in vitro* and thereby reduce basolateral IL-8 secretion in response to proinflammatory stimuli, including apical infection with wild-type *S. typhimurium* (Neish et al., 2000). In fetal life, the murine gut epithelium is sensitive to MAMPs such as LPS due to intracellular TLR4 expression (Lotz et al., 2006). LPS exposure during vaginal birth induces epithelial activation, leading to subsequent downregulation of TLR signaling in the neonatal gut epithelium and thereby tolerance to MAMPs.

Studies on the expression of CD14 and TLRs on human intestinal epithelial cells have provided inconsistent results, but the HT-29 adenocarcinoma cell line expresses low levels of TLR3 and TLR4 (Schneeman et al., 2005). Double-stranded RNA (dsRNA), a by-product of viral replication, has been identified as a ligand for TLR3, and epithelial cells can apparently be activated by LPS via TLR4 in a CD14-independent manner (Böcker et al., 2003).

The expression of both TLR3 and TLR4 is reportedly upregulated on epithelial cells in IBD lesions— particularly so for TLR4 (Cario and Podolsky, 2000)—although others have failed to confirm this by immunohistochemistry

(Hausmann et al., 2002). Although the signaling cascades for TLR3 and TLR4 are similar, recent studies have shown that differential usage of adaptor molecules might result in disparate biological responses (Hoebe et al., 2003; Oshiumi et al., 2003; Yamamoto et al., 2002, 2003).

10.6.3. Relation Between PRR Signaling and pIgR Upregulation

It was recently reported that pIgR mRNA and protein expression was strongly upregulated in response to both dsRNA/TLR3 and LPS/TLR4 signaling in HT-29 cells (Schneeman et al., 2005). By contrast, dsRNA—but not LPS—increased the mRNA level for TLR3 and TLR4, although the cell surface protein expression of both receptors was enhanced by LPS as well as dsRNA; this suggested that TLR4 could be transported to the cell surface from intracellular stores. A previously characterized binding site for NF-κB in the intron 1 enhancer of the *pIgR* gene (Johansen and Brandtzaeg, 2004; Schjerven et al., 2001) was shown by reporter assays with differently mutated cDNA constructs to be critical for transcriptional activation in response to TLR3 and TLR4 signaling. Analysis of several cytokine/chemokine gene products, including IL-8, demonstrated that TLR3 signaling resulted in a more pronounced proinflammatory response than did TLR4. These data suggested that signaling through TLR4 upregulates pIgR expression while minimizing initiation of inflammation.

Altogether, epithelial TLR engagement by microbes or their products might serve to augment pIgR expression (Fig. 10.4, right panel) and thereby enhance export of SIgA, thus linking the innate and adaptive immune responses to viruses and bacteria. The differential epithelial activation induced via TLR3 and TLR4 implies that particularly the latter receptor might promote SIgA-mediated homeostasis in the presence of commensal Gram-negative bacteria. Experiments in mice have suggested that other epithelial TLRs are also involved in the maintenance of intestinal homeostasis (Bambou et al., 2004; Rakoff-Nahoum *et al.*, 2004). Moreover, evidence is accumulating to suggest that such homeostasis is significantly influenced by cross-talk between epithelial cells and underlying lamina propria cells, particularly macrophages and DCs (Haller et al., 2000; Rimoldi et al., 2005).

10.6.4. Dysregulation of Innate Immunity Might Disturb Mucosal Homeostasis

It follows from the available information that defects in innate immune mechanisms might predispose to abrogated mucosal homeostasis. The best proof of principle in this respect has been provided by the NOD2 family of intracellular sensor molecules carrying a C-terminal leucine-rich repeat domain; its PRR activity recognizes unique muramyl dipeptide (MDP) motifs of peptidoglycans from both Gram-negative and Gram-positive bacteria (Cario, 2005).

The *NOD2 (CARD15)* gene is encoded by the IBD susceptibility locus (*IBD1*) on chromosome 16, and certain mutations of this gene are associated with clinical subsets of Crohn's disease patients—showing a striking dose effect of mutations that affect the PPR function of the molecules (Abreu et al., 2002; Ahmad et al., 2002; Cuthbert et al., 2002; Hampe et al., 2002; Hugot et al., 2001; Mathew and Lewis, 2004; Ogura et al., 2001). These seminal observations provide strong support for the possibility that aberrant sensing of the intestinal microbiota is an early pathogenic event, perhaps involving deficient induction of NF-κB-mediated activation of epithelial Paneth cells leading to insufficient defensin and cryptidin production (Kobayashi et al., 2005; Lala et al., 2003;) and impaired intestinal barrier function (Fig. 10.7). A possible role of SIgA in this context remains to be determined, but membrane targeting of NOD2 is required for NF-κB activation after recognition of MDP in epithelial cells (Barnich et al., 2005). Therefore, one interesting possibility is that membrane-associated pIgR in transcytotic vesicles might provide bacterial products for NOD2 through antigens complexed with receptor-bound dimeric IgA (Fig. 10.2).

Imbalanced triggering of innate immunity could be a common theme for disease initiation on a polygenic susceptibility background in the pathogenesis

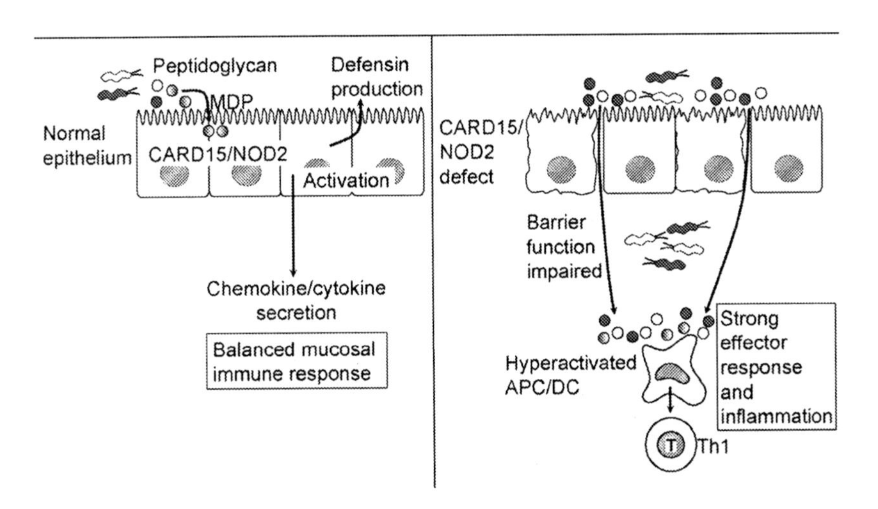

FIG. 10.7. Intestinal homeostasis depends on bacterial reinforcement of the epithelial barrier function. Normal epithelium senses muramyl dipeptide (MDP) motifs from Gram-positive and Gram-negative bacteria via intracellular CARD15/NOD2 molecules. This leads to cellular activation with secretion of protective defensins and cytokines/chemokines, which collaborate with SIgA (not shown) to maintain the barrier function. Mutations of CARD15/NOD2 might result in a leaky epithelium and hyperactivation of antigen-presenting cells (APCs), particularly quiescent lamina propria dendritic cells (DCs). This adverse development might cause strong effector Th1-cell responses with mucosal inflammation.

of a spectrum of clinical IBD entities. In this scenario, it has been suggested that *NOD2/CARD15* mutations might cause deficient signaling for downregulatory mechanisms in the gut, which in the normal state maintain lamina propria APCs/DCs in a quiescent state (Bouma and Strober, 2003; Judge and Lichtenstein, 2002). It has been proposed that one function of NOD2 is to limit the proinflammatory effects of TLR2 stimulation by peptidoglycan at the APC surface; mutant NOD2 is unable to sense MDP, which leads to "gain-of-function" for the TLR2 pathway with enhanced production of IL-12 and chronic inflammation (Watanabe et al., 2004).

However, this theory needs substantiation (Cario, 2005). It has alternatively been suggested that mutated NOD2 itself achieves gain-of-function whereby its N-terminal CARD domains become capable of activating Caspase-1 (Maeda et al., 2005). This enzyme might then cleave off the prodomain of IL-1, leading to secretion of mature IL-1, which promotes the inflammatory process. Admittedly, further studies are needed to discern the precise role of NOD2 in MDP recognition and subsequent intracellular signaling, cytokine/chemokine and defensin production, and perhaps upregulation or downregulation of pIgR. However, the near future will undoubtedly see a body of evidence also for other PRR mutations predisposing to intestinal inflammation.

Secondary alterations resulting in defect transcription of the *pIgR* gene might also result in deterioration of the intestinal barrier function, which could appear as patchy lack of pIgR/SC expression with absent SIgA export (Brandtzaeg et al., 1987). In ulcerative colitis, we have observed a relationship between downregulated pIgR/SC expression and the degree of hyperplastic epithelial lesions—overt dysplasia showing the most reduced immunohistochemical staining (Rognum et al., 1982). Expression of pIgR/SC protein and mRNA correlated—both variables being negatively related to the histological grade of dysplasia (Krajci *et al.*, 1996)—suggesting that this defect is a rather late event in the IBD process.

10.7. Concluding Remarks

Several more or less well-defined factors influence the development of productive IgA-dependent intestinal immunity and oral tolerance. Some of these variables are reciprocally modulated by the immune system to achieve homeostasis in the gut. Increased epithelial permeability for luminal antigens is likely an important primary or secondary event in the pathogenesis of mucosal diseases, including food allergy and IBD. This variable is determined by the individual's age (e.g., preterm versus term infant), interactions among mast cells, nerves, and neuropeptides, concurrent infection, and the epithelium-shielding effect of SIgA provided by breast milk or produced in the infant's gut. The pathological consequences will depend on how fast an intact epithelial barrier function can be attained or reestablished, which is influenced both by the age of the individual and by a successful mounting of intestinal IgA responses, as well as generation of oral tolerance toward innocuous antigens from the diet and commensal bacteria.

Secretory IgA is the best defined effector of the mucosal immune system, and much knowledge has recently been obtained at the molecular level about the constitutive and induced transcriptional regulation of pIgR-mediated secretory antibody export in the gut. Of great importance in infancy is the large capacity for storage of dimeric IgA/SIgA in the mammary gland epithelium and duct system, which explains the remarkable output of SIgA during feeding— serving as passive immunization of the breast-fed baby's gut. Altogether, the secretory immune system is of considerable clinical interest because SIgA not only maintains mutualism with the indigenous microbiota but also forms the first line of immunological defense against infectious agents and other harmful substances. Human studies and characterization of a mouse strain with no secretory immune system (pIgR$^{-/-}$ mice) support the notion that SIgA antibodies are important in reinforcing the intestinal barrier function and promoting mucosal homeostasis.

In summary, it must be emphasized that the vast majority of antigenic challenges confronting the body, including potentially infectious agents, commensal bacteria, and foreign proteins, make contact with mucosal surfaces. Therefore, to maintain homeostasis in the extensive and vulnerable mucosae, they are protected by specialized anti-inflammatory immune defenses in which SIgA antibodies are a prominent effector. The induction of mucosal immunity with its immunoregulatory network is highly dependent on commensal bacteria and the postnatal period is critical. The newborn's adaptation to the environment is significantly modulated by innate ("natural") immune responses, and the SIgA system itself possesses innate features such as production of "natural" cross-reactive antibodies. Through evolution with microbial mutualism, the intestinal immune system has developed two major homeostatic defense layers aiming at control of antigenic challenges and return of the local tissue to a basal state with minimal pathology: immune exclusion provided by noninflammatory SIgA antibodies, which limit epithelial antigen penetration and invasion of microorganisms, and immunosuppressive mechanisms—often referred to as "oral tolerance"—which inhibit overreaction against components of the normal microbiota and other innocuous antigens. Both of these strategies depend on cooperation of adaptive immunity with the innate immune system, including "cross-talk" among commensal bacteria, the intestinal epithelium with its SIgA, and various lamina propria cells. Altogether, it appears that the healthy host not only tolerates the indigenous microbiota but also depends on it for mucosal homeostasis. Dysregulation in one or more of the many homeostatic mechanisms might result in overt intestinal inflammation.

Acknowledgments. Studies in the authors' laboratories are supported by the University of Oslo, the Research Council of Norway, the Norwegian Cancer

Society, Rikshospitalet University Hospital, and Anders Jahre's Fund. Hege Eliassen and Erik K. Hagen provided excellent assistance with the manuscript and figures, respectively.

References

Abreu, M. T., Taylor, K. D., Lin, Y. C., Hang, T., Gaiennie, J., Landers, C. J., Vasiliauskas, E. A., Kam, L. Y., Rojany, M., Papadakis, K. A., Rotter, J. I., Targan, S. R., and Yang, H. (2002). Mutations in NOD2 are associated with fibrostenosing disease in patients with Crohn's disease. *Gastroenterology* 123:679–688.

Ahmad, T., Armuzzi, A., Bunce, M., Mulcahy-Hawes, K., Marshall, S. E., Orchard, T. R., Crawshaw, J., Large, O., de Silva, A., Cook, J. T., Barnardo, M., Cullen, S., Welsh, K. I., and Jewell, D. P. (2002). The molecular classification of the clinical manifestations of Crohn's disease. *Gastroenterology* 122:854–866.

Akira, S. (2003). Mammalian Toll-like receptors. *Curr. Opin. Immunol.* 15:5–11.

Alfsen, A., Iniguez, P., Bouguyon, E., and Bomsel, M. (2001). Secretory IgA specific for a conserved epitope on gp41 envelope glycoprotein inhibits epithelial transcytosis of HIV-1. *J. Immunol.* 166:6257–6265.

Anonymous (1994). A warm chain for breastfeeding (Editorial). *Lancet* 344:1239–1241.

Arnaboldi, P. M., Behr, M. J., and Metzger, D. W. (2005). Mucosal B cell deficiency in $IgA^{-/-}$ mice abrogates the development of allergic lung inflammation. *J. Immunol.* 175:1276–1285.

Asahi, Y., Yoshikawa, T., Watanabe, I., Iwasaki, T., Hasegawa, H., Sato, Y., Shimada, S., Nanno, M., Matsuoka, Y., Ohwaki, M., Iwakura, Y., Suzuki, Y., Aizawa, C., Sata, T., Kurata, T., and Tamura, S. (2002). Protection against influenza virus infection in polymeric Ig receptor knockout mice immunized intranasally with adjuvant-combined vaccines. *J. Immunol.* 168:2930–2938.

Bäckhed, F., Ley, R. E., Sonnenburg, J. L., Peterson, D. A., and Gordon, J. I. (2005). Host-bacterial mutualism in the human intestine. *Science* 307:1915–1920.

Bambou, J. C., Giraud, A., Menard, S., Begue, B., Rakotobe, S., Heyman, M., Taddei, F., Cerf-Bensussan, N., and Gaboriau-Routhiau, V. (2004). *In vitro* and ex vivo activation of the TLR5 signaling pathway in intestinal epithelial cells by a commensal *Escherichia coli* strain. *J. Biol. Chem.* 279:42,984–992.

Barnich, N., Aguirre, J. E., Reinecker, H. C., Xavier, R., and Podolsky, D. K. (2005). Membrane recruitment of NOD2 in intestinal epithelial cells is essential for nuclear factor-κB activation in muramyl dipeptide recognition. *J. Cell. Biol.* 170:21–26.

Beatty, D. W., Napier, B., Sinclair-Smith, C. C., McCabe, K., and Hughes, E. J. (1983). Secretory IgA synthesis in kwashiorkor. *J. Clin. Lab. Immunol.* 12:31–36.

Beisner, D. R., Ch'en, I. L., Kolla, R. V., Hoffmann, A., and Hedrick, S. M. (2005). Cutting edge:innate immunity conferred by B cells is regulated by caspase-8. *J. Immunol.* 175:3469–3473.

Benn, C. S., Wohlfahrt, J., Aaby, P., Westergaard, T., Benfeldt, E., Michaelsen, K. F., Björksten, B., and Melbye, M. (2004). Breastfeeding and risk of atopic dermatitis, by parental history of allergy, during the first 18 months of life. *Am. J. Epidemiol.* 160:217–223.

Berstad, A. E., and Brandtzaeg, P. (1998). Expression of cell-membrane complement regulatory glycoproteins along the normal and diseased human gastrointestinal tract. *Gut* 42:522–529.

Beutler, B. (2004). Inferences, questions and possibilities in Toll-like receptor signalling. *Nature* 430:257–263.

Beutler, B., and Rietschel, E. T. (2003). Timeline: Innate immune sensing and its roots: The story of endotoxin. *Nat. Rev. Immunol.* 3:169–176.

Bischoff, S. C., Mayer, J. H., and Manns, M. P. (2000). Allergy and the gut. *Int. Arch. Allergy Immunol.* 121:270–283.

Bižanov, G., Janakova, L., Knapstad, S. E., Karlstad, T., Bakke, H., Haugen, I. L., Haugan, A., Samdal, H. H., and Haneberg, B. (2005). Immunoglobulin-A antibodies in upper airway secretions may inhibit intranasal influenza virus replication in mice but not protect against clinical illness. *Scand. J. Immunol.* 61:503–510.

Black, R. E., Morris, S. S., and Bryce, J. (2003). Where and why are 10 million children dying every year? *Lancet* 361:2226–2234.

Böcker, U., Yezerskyy, O., Feick, P., Manigold, T., Panja, A., Kalina, U., Herweck, F., Rossol, S., and Singer, M. V. (2003). Responsiveness of intestinal epithelial cell lines to lipopolysaccharide is correlated with Toll-like receptor 4 but not Toll-like receptor 2 or CD14 expression. *Int. J. Colorectal Dis.* 18:25–32.

Bollinger, R. R., Everett, M. L., Palestrant, D., Love, S. D., Lin, S. S., and Parker, W. (2003). Human secretory immunoglobulin A may contribute to biofilm formation in the gut. *Immunology* 109:580–587.

Bomsel, M., Heyman, M., Hocini, H., Lagaye, S., Belec, L., Dupont, C., and Desgranges, C. (1998). Intracellular neutralization of HIV transcytosis across tight epithelial barriers by anti-HIV envelope protein dIgA or IgM. *Immunity* 9:277–287.

Bos, N. A., Bun, J. C., Popma, S. H., Cebra, E. R., Deenen, G. J., van der Cammen, M. J., Kroese, F. G., and Cebra, J. J. (1996). Monoclonal immunoglobulin A derived from peritoneal B cells is encoded by both germ line and somatically mutated VH genes and is reactive with commensal bacteria. *Infect. Immun.* 64:616–623.

Bos, N. A., Jiang, H. Q., and Cebra, J. J. (2001). T cell control of the gut IgA response against commensal bacteria. *Gut* 48:762–764.

Bouma, G., and Strober, W. (2003). The immunological and genetic basis of inflammatory bowel disease. *Nat. Rev. Immunol.* 3:521–533.

Boursier, L., Dunn-Walters, D. K., and Spencer, J. (1999). Characteristics of IgVH genes used by human intestinal plasma cells from childhood. *Immunology* 97:558–564.

Boursier, L., Farstad, I. N., Mellembakken, J. R., Brandtzaeg, P., and Spencer, J. (2002). IgV_H gene analysis suggests that peritoneal B cells do not contribute to the gut immune system in man. *Eur. J. Immunol.* 32:2427–2436.

Boursier, L., Gordon, J. N., Thiagamoorthy, S., Edgeworth, J. D., and Spencer, J. (2005). Human intestinal IgA response is generated in the organized gut-associated lymphoid tissue but not in the lamina propria. *Gasteroenterology* 128:1879–1889.

Bouvet, J. P., and Fischetti, V. A. (1999). Diversity of antibody-mediated immunity at the mucosal barrier. *Infect. Immun.* 67:2687–2691.

Brandtzaeg, P. (1974a). Presence of J chain in human immunocytes containing various immunoglobulin classes. *Nature* 252:418–420.

Brandtzaeg, P. (1974b). Mucosal and glandular distribution of immunoglobulin components: Differential localization of free and bound SC in secretory epithelial cells. *J. Immunol.* 112:1553–1559.

Brandtzaeg, P. (1983). The secretory immune system of lactating human mammary glands compared with other exocrine organs. *Ann. NY Acad. Sci.* 30:353–382.

Brandtzaeg, P. (1996). History of oral tolerance and mucosal immunity. *Ann. NY Acad. Sci.* 778:1–27.

Brandtzaeg, P. (1997). Development of the intestinal immune system and its relation to coeliac disease. In: Mäki, M., Collin, P., and Visakorpi, J. K. (eds.), *Coeliac Disease. Proceedings of the Seventh International Symposium on Coeliac Disease*. Coeliac Disease Study Group, Institute of Medical Technology, University of Tampere, Tampere, Finland (ISBN 951-44-4293-8), pp. 221–244.

Brandtzaeg, P. (1998). Development and basic mechanisms of human gut immunity. *Nutr. Rev.* 56:S5–S18.

Brandtzaeg, P. (2001). Nature and function of gastrointestinal antigen-presenting cells. *Allergy* 56(Suppl. 67):16–20.

Brandtzaeg, P. (2002a). Role of local immunity and breast-feeding in mucosal homeostasis and defence against infections. In: Calder, P. C., Field, C. J., and Gill, H. S. (eds.), *Nutrition and Immune Function,* Frontiers in Nutritional Science, No. 1. CABI Publishing, Oxon, UK, pp. 273–320.

Brandtzaeg, P. (2002b). The secretory immunoglobulin system: regulation and biological significance. Focusing on mammary glands. In: Davis, M. K., Isaacs, C. E., Hanson, L. Å., and Wright, A. L. (eds.), *Integrating Population Outcomes, Biological Mechanisms and Research Methods in the Study of Human Milk and Lactation.* Advances in Experimental Medicine Vol. 503. Kluwer Academic/Plenum, New York, pp. 116.

Brandtzaeg, P. (2003). Role of secretory antibodies in the defence against infections. *Int. J. Med. Microbiol.* 293:3–15.

Brandtzaeg, P., Baekkevold, E. S., Farstad, I. N., Jahnsen, F. L., Johansen, F.-E., Nilsen, E. M., and Yamanaka, T. (1999a). Regional specialization in the mucosal immune system: what happens in the microcompartments? *Immunol. Today* 20:141–151.

Brandtzaeg, P., Baekkevold, E. S., and Morton, H. C. (2001). From B to A the mucosal way. *Nat. Immunol.* 2:1093–1094.

Brandtzaeg, P., Baklien, K., Bjerke, K., Rognum, T.O., Scott, H., and Valnes, K. (1987). Nature and properties of the human gastrointestinal immune system. In: Miller, M., and Nicklin, S. (eds.), *Immunology of the Gastrointestinal Tract*. CRC Press, Boca Raton, FL, pp. 1–85.

Brandtzaeg, P., Farstad, I. N., and Haraldsen, G. (1999b). Regional specialization in the mucosal immune system: primed cells do not always home along the same track. *Immunol. Today* 20:267–277.

Brandtzaeg, P., Fjellanger, I., and Gjeruldsen, S. T. (1968). Adsorption of immunoglobulin A onto oral bacteria *in vivo*. *J. Bacteriol.* 96:242–249.

Brandtzaeg, P., Halstensen, T. S., Huitfeldt, H. S., Krajci, K., Kvale, D., Scott, H., and Thrane, P. S. (1992). Epithelial expression of HLA, secretory component (poly-Ig receptor), and adhesion molecules in the human alimentary tract. *Ann. NY Acad. Sci.* 664:157–179.

Brandtzaeg, P., Halstensen, T. S., Kett, K., Krajci, P., Kvale, D., Rognum, T. O., Scott, H., and Sollid, L. M. (1989). Immunobiology and immunopathology of human gut mucosa: humoral immunity and intraepithelial lymphocytes. *Gastroenterology* 97:1562–1584.

Brandtzaeg, P., and Johansen, F.-E. (2005). Mucosal B cells: phenotypic characteristics, transcriptional regulation, and homing properties. *Immunol. Rev.* 206:32–63.

Brandtzaeg, P., and Nilssen, D. E. (1995). Mucosal aspects of primary B-cell deficiency and gastrointestinal infections. *Curr. Opin. Gastroenterol.* 11:532–540.

Brandtzaeg, P., Nilssen, D. E., Rognum, T. O., and Thrane, P. S. (1991). Ontogeny of the mucosal immune system and IgA deficiency. *Gastroenterol. Clin. North Am.* 20:397–439.

Brandtzaeg, P., and Pabst, R. (2004). Let's go mucosal: communication on slippery ground. *Trends Immunol.* 25:570–577.

Brandtzaeg, P., and Prydz, H. (1984). Direct evidence for an integrated function of J chain and secretory component in epithelial transport of immunoglobulins. *Nature* 311:71–73.

Brandtzaeg, P., and Tolo, K. (1977). Mucosal penetrability enhanced by serum-derived antibodies. *Nature* 266:262–263.

Burns, J. W., Siadat-Pajouh, M., Krishnaney, A. A., and Greenberg, H. B. (1996). Protective effect of rotavirus VP6-specific IgA monoclonal antibodies that lack neutralizing activity. *Science* 272:104–107.

Burrows, P. D., and Cooper, M. D. (1997). IgA deficiency. *Adv. Immunol.* 65:245–276.

Cario, E. (2005). Bacterial interactions with cells of the intestinal mucosa: Toll-like receptors and nod2. *Gut* 54:1182–1193.

Cario, E., and Podolsky, D. K. (2000). Differential alteration in intestinal epithelial cell expression of toll-like receptor 3 (TLR3) and TLR4 in inflammatory bowel disease. *Infect. Immun.* 68:7010–7017.

Casas, R., Böttcher, M. F., Duchén, K., and Björkstén, B. (2000). Detection of IgA antibodies to cat, β-lactoglobulin, and ovalbumin allergens in human milk. *J. Allergy Clin. Immunol.* 105:1236–1240.

Casola, S., Otipoby, K. L., Alimzhanov, M., Humme, S., Uyttersprot, N., Kutok, J. L., Carroll, M. C., and Rajewsky, K. (2004). B cell receptor signal strength determines B cell fate. *Nat. Immunol.* 5:317–327.

Chaplin, D. D. (2003). 1. Overview of the immune response. *J. Allergy Clin. Immunol.* 111:S442–S459.

Chen, Y., Song, K., and Eck, S. L. (2000). An intra-Peyer's patch gene transfer model for studying mucosal tolerance: distinct roles of B7 and IL-12 in mucosal T cell tolerance. *J. Immunol.* 165:3145–3153.

Chirdo, F. G., Millington, O. R., Beacock-Sharp, H., and Mowat, A. M. (2005). Immunomodulatory dendritic cells in intestinal lamina propria. *Eur. J. Immunol.* 35:1831–1840.

Christ, A. D., and Blumberg, R. S. (1997). The intestinal epithelial cell: immunological aspects. *Springer Semin. Immunopathol.* 18:449–461.

Chomarat, P., Dantin, C., Bennett, L., Banchereau, J., and Palucka, A. K. (2003). TNF skews monocyte differentiation from macrophages to dendritic cells. *J. Immunol.* 171:2262–2269.

Clamp, J. R. (1980). Gastrointestinal mucus. In: Wright, R. (ed.), *Recent Advances in Gastrointestinal Pathology*. W.B. Saunders, London, p. 47.

Collins, M. D., and Gibson, G. R. (1999). Probiotics, prebiotics, and synbiotics: approaches for modulating the microbial ecology of the gut. *Am. J. Clin. Nutr.* 69:1052S–1057S.

Conley, M. E., and Delacroix, D. L. (1987). Intravascular and mucosal immunoglobulin A: two separate but related systems of immune defense? *Ann. Intern. Med.* 106:892–899.

Crabbé, P. A., Nash, D. R., Bazin, H., Eyssen, H., and Heremans, J. F. (1970). Immunohistochemical observations on lymphoid tissues from conventional and germ-free mice. *Lab. Invest.* 22:448–457.

Cuthbert, A. P., Fisher, S. A., Mirza, M. M,, King, K., Hampe, J., Croucher, P. J., Mascheretti, S., Sanderson, J., Forbes, A., Mansfield, J., Schreiber, S., Lewis, C. M., and Mathew, C. G. (2002). The contribution of NOD2 gene mutations to the risk and site of disease in inflammatory bowel disease. *Gastroenterology* 122:867–874.

Daniele, R. P. (1990). Immunoglobulin secretion in the airways. *Annu. Rev. Physiol.* 52:177–195.

Davids, B. J., Palm, J. E., Housley, M. P., Smith, J. R., Andersen, Y. S., Martín, M.G., Hendrickson, B. A., Johansen, F.-E., Svard, S. G., Gillin, F. D., and Eckmann, L. (2006). Polymeric immunoglobulin receptor in intestinal immune defense against the lumen-dwelling protozoan parasite *Giardia. J. Immunol.* 177:6821–6990.

Dickinson, E. C., Gorga, J. C., Garrett, M., Tuncer, R., Boyle, P., Watkins, S. C., Alber, S. M., Parizhskaya, M., Trucco, M., Rowe, M. I., and Ford, H. R. (1998). Immunoglobulin A supplementation abrogates bacterial translocation and preserves the architecture of the intestinal epithelium. *Surgery* 124:284–290.

Duchmann, R., Neurath, M., Marker-Hermann, E., and Meyer Zum Buschenfelde, K. H. (1997). Immune responses towards intestinal bacteria: Current concepts and future perspectives. *Z. Gastroenterol.* 35:337–346.

Dunn-Walters, D. K., Boursier, L., and Spencer, J. (1997a). Hypermutation, diversity and dissemination of human intestinal lamina propria plasma cells. *Eur. J. Immunol.* 27:2959–2964.

Dunn-Walters, D. K., Hackett, M., Boursier, L., Ciclitira, P. J., Morgan, P., Challacombe, S. J., and Spencer, J. (2000). Characteristics of human IgA and IgM genes used by plasma cells in the salivary gland resemble those used in duodenum but not those used in the spleen. *J. Immunol.* 164:1595–1601.

Dunn-Walters, D. K., Isaacson, P. G., and Spencer, J. (1997b). Sequence analysis of human IgVH genes indicates that ileal lamina propria plasma cells are derived from Peyer's patches. *Eur. J. Immunol.* 27:463–467.

Ehrhardt, G. R., Hsu, J. T., Gartland, L., Leu, C. M., Zhang, S., Davis, R. S., and Cooper, M. D. (2005). Expression of the immunoregulatory molecule FcRH4 defines a distinctive tissue-based population of memory B cells. *J. Exp. Med.* 202:783–791.

Fagarasan, S., and Honjo, T. (2003). Intestinal IgA synthesis: regulation of front-line body defences. *Nat. Rev. Immunol.* 3:63–72.

Favre, L., Spertini, F., and Corthesy, B. (2005). Secretory IgA possesses intrinsic modulatory properties stimulating mucosal and systemic immune responses. *J. Immunol.* 175:2793–2800.

Feng, N., Lawton, J. A., Gilbert, J., Kuklin, N., Vo, P., Prasad, B. V., and Greenberg, H. B. (2002). Inhibition of rotavirus replication by a non-neutralizing, rotavirus VP6-specific IgA mAb. *J. Clin. Invest.* 109:1203–1213.

Fernandez, M. I., Pedron, T., Tournebize, R., Olivo-Marin, J. C., Sansonetti, P. J., and Phalipon, A. (2003). Anti-inflammatory role for intracellular dimeric immunoglobulin a by neutralization of lipopolysaccharide in epithelial cells. *Immunity* 18:739–749.

Fischer, M., and Kuppers, R. (1998). Human IgA- and IgM-secreting intestinal plasma cells carry heavily mutated VH region genes. *Eur. J. Immunol.* 28:2971–2977.

Foligné, B., Grangette, C., and Pot, B. (2005a). Probiotics in IBD: mucosal and systemic routes of administration may promote similar effects. *Gut* 54:727–728.

Foligné, B., Nutten, S., Steidler, L., et al. (2005b). Potentialities of the TNBS-induced colitis model to evaluate the anti-inflammatory properties of lactic acid bacteria. *Dig. Dis.* 51:390–400.

Frossard, C. P., Hauser, C., and Eigenmann, P. A. (2004). Antigen-specific secretory IgA antibodies in the gut are decreased in a mouse model of food allergy. *J. Allergy Clin. Immunol.* 114:377–382.

Gerson, C., Sabater, J., Scuri, M., Torbati, A., Coffey, R., Abraham, J. W., Lauredo, I., Forteza, R., Wanner, A., Salathe, M., Abraham, W. M., and Conner, G. E. (2000). The lactoperoxidase system functions in bacterial clearance of airways. *Am. J. Respir. Cell Mol. Biol.* 22:665–671.

Glauber, J. G., Wandersee, N. J., Little, J. A., and Ginder, G. D. (1991). 5′-Flanking sequences mediate butyrate stimulation of embryonic globin gene expression in adult erythroid cells. *Mol. Cell. Biol.* 11:4690–4697.

Gothefors, L., and Marklund, S. (1975). Lactoperoxidase activity in human milk and in saliva of newborn infants. *Infect. Immun.* 11:1210–1215.

Grdic, D., Hornquist, E., Kjerrulf, M., and Lycke, N. Y. (1998). Lack of local suppression in orally tolerant CD8-deficient mice reveals a critical regulatory role of $CD8^+$ T cells in the normal gut mucosa. *J. Immunol.* 160:754–762.

Ha, S. A., Tsuji, M., Suzuki, K., Meek, B., Yasuda, N., Kaisho, T., and Fagarasan, S. (2006). Regulation of B1 cell migration by signals through Toll-like receptors. *J. Exp. Med.* 203:2541–2550.

Haddeland, U., Karstensen, A. B., Farkas, L., Bø, K. O., Pirhonen, J., Karlsson, M., Kvåvik, W., Brandtzaeg, P., and Nakstad, B. (2005). Putative regulatory T cells are impaired in cord blood from neonates with hereditary allergy risk. *Pediatr. Allergy Immunol.* 16:104–112.

Haller, D., Bode, C., Hammes, W. P., Pfeifer, A. M., Schiffrin, E. J., and Blum, S. (2000). Non-pathogenic bacteria elicit a differential cytokine response by intestinal epithelial cell/leucocyte co-cultures. *Gut* 47:79–87.

Haller, D., and Jobin, C. (2004). Interaction between resident luminal bacteria and the host: can a healthy relationship turn sour? *J. Pediatr. Gastroenterol. Nutr.* 38:123–136.

Hampe, J., Grebe, J., Nikolaus, S., Solberg, C., Croucher, P. J., Mascheretti, S., Jahnsen, J., Moum, B., Klump, B., Krawczak, M., Mirza, M. M., Foelsch, U. R., Vatn, M., and Schreiber, S. (2002). Association of NOD2 (CARD 15) genotype with clinical course of Crohn's disease: a cohort study. *Lancet* 359:1661–1665.

Harris, N. L., Spoerri, I., Schopfer, J. F., Nembrini, C., Merky, P., Massacand, J., Urban, J. F., Jr., Lamarre, A., Burki, K., Odermatt, B., Zinkernagel, R. M., and Macpherson, A. J. (2006). Mechanisms of neonatal mucosal antibody protection. *J. Immunol.* 177:6256–6262.

Hausmann, M., Kiessling, S., Mestermann, S., Webb, G., Spottl, T., Andus, T., Scholmerich, J., Herfarth, H., Ray, K., Falk, W., and Rogler, G. (2002). Toll-like receptors 2 and 4 are up-regulated during intestinal inflammation. *Gastroenterology* 122:1987–2000.

Helgeland, L., and Brandtzaeg, P. (2000). Development and function of intestinal B and T cells. *Microbiol. Ecol. Health Dis.* 12(Suppl. 2):110–127.

Hoebe, K., Du, X., Georgel, P., Janssen, E., Tabeta, K., Kim, S. O., Goode, J., Lin, P., Mann, N., Mudd, S., Crozat, K., Sovath, S., Han, J., and Beutler, B. (2003).

Identification of Lps2 as a key transducer of MyD88-independent TIR signaling. *Nature* 424:743–748.

Holt, P. G. (1995). Postnatal maturation of immune competence during infancy and childhood. *Pediatr. Allergy Immunol.* 6:59–70.

Holtmeier, W., Hennemann, A., and Caspary, W. F. (2000). IgA and IgM V(H) repertoires in human colon: Evidence for clonally expanded B cells that are widely disseminated. *Gastroenterology* 119:1253–1266.

Hooper, L. V., and Gordon, J. I. (2001). Commensal host-bacterial relationships in the gut. *Science* 292:1115–1118.

Hooper, L. V., Wong, M. H., Thelin, A., Hansson, L., Falk, P. G., and Gordon, J. I. (2001). Molecular analysis of commensal host-microbial relationships in the intestine. *Science* 291:881–884.

Hopkin, J. M., Shaldon, S., Ferry, B., Coull, P., Antrobus, P., Enomoto, T., Yamashita, T., Kurimoto, F., Stanford, J., Shirakawa, T., and Rook, G. (1998). Mycobacterial immunisation in grass pollen asthma and rhinitis. *Thorax* 53(Suppl. 4):S63.

Hoque, S. S., Ghosh, S., and Poxton, I. R. (2000). Differences in intestinal humoral immunity between healthy volunteers from UK and Bangladesh. *Eur.J. Gastroenterol. Hepatol.* 12:1185–1193.

Horsfall, D. J., Cooper, J. M., and Rowley, D. (1978). Changes in the immunoglobulin levels of the mouse gut and serum during conventionalisation and following administration of *Salmonella typhimurium. Aust. J. Exp. Biol. Med. Sci.* 56:727–735.

Huang, Y. T., Wright, A., Gao, X., Kulick, L., Yan, H., and Lamm, M. E. (2005). Intraepithelial cell neutralization of HIV-1 replication by IgA. *J. Immunol.* 174:4828–4835.

Hugot, J. P., Chamaillard, M., Zouali, H., Lesage, S., Cezard, J. P., Belaiche, J., Almer, S., Tysk, C., O'Morain, C. A., Gassull, M., Binder, V., Finkel, Y., Cortot, A., Modigliani, R., Laurent-Puig, P., Gower-Rousseau, C., Macry, J., Colombel, J. F., Sahbatou, M., and Thomas, G. (2001). Association of NOD2 leucine-rich repeat variants with susceptibility to Crohn's disease. *Nature* 411:599–603.

Husby, S., Jensenius, J. C., and Svehag, S.-E. (1985). Passage of undegraded dietary antigen into the blood of healthy adults. Quantification, estimation of size distribution, and relation of uptake to levels of specific antibodies. *Scand. J. Immunol.* 22:83–92.

Ishizaka, S., Kimoto, M., Tsujii, T., and Saito, S. (1994). Antibody production system modulated by oral administration of human milk and TGF-β. *Cell. Immunol.* 159:77–84.

Isolauri, E., Sutas, Y., Kankaanpaa, P., Arvilommi, H., Salminen, S. (2001). Probiotics: Effects on immunity. *Am. J. Clin. Nutr.* 73:444S–450S.

Ivarsson, A., Hernell, O., Stenlund, H., and Persson, L. A. (2002). Breast-feeding protects against celiac disease. *Am. J. Clin. Nutr.* 75:914–921.

Iwata, M., Hirakiyama, A., Eshima, Y., Kagechika, H., Kato, C., and Song, S. Y. (2004). Retinoic acid imprints gut-homing specificity on T cells. *Immunity* 21:527–538.

Janu, P., Li, J., Renegar, K. B., and Kudsk, K. A. (1997). Recovery of gut-associated lymphoid tissue and upper respiratory tract immunity after parenteral nutrition. *Ann. Surg.* 225:707–715.

Järvinen, K. M., Laine, S. T., Jarvenpaa, A. L., and Suomalainen, H. K. (2000). Does low IgA in human milk predispose the infant to development of cow's milk allergy? *Pediatr. Res.* 48:457–462.

Jenkins, S. Wang, L. J., Vazir, M. Vela, J., Sahagun, O., Gabbay, P., Hoang, L., Diaz, R. L., Aranda, R., and Martín, M. G. (2003). Role of passive and adaptive immunity in influencing enterocyte-specific gene expression. *Am. J. Physiol.: Gastrointest. Liver Physiol.* 285:G714–G725.

Jiang, H.-Q., Bos, N. A., and Cebra, J. J. (2001). Timing, localization, and persistence of colonization by segmented filamentous bacteria in the neonatal mouse gut depend on immune status of mothers and pups. *Infect. Immun.* 69:3611–3617.

Johansen, F.-E., Braathen, R., and Brandtzaeg, P. (2000). Role of J chain in secretory immunoglobulin formation. *Scand. J. Immunol.* 52:240–248.

Johansen, F.-E., Braathen, R., and Brandtzaeg, P. (2001). The J chain is essential for polymeric Ig receptor-mediated epithelial transport of IgA. *J. Immunol.* 167:5185–5192.

Johansen, F.-E., and Brandtzaeg, P. (2004). Transcriptional regulation of the mucosal IgA system. *Trends Immunol.* 25:150–157.

Johansen, F.-E., Pekna, M., Norderhaug, I.N., Haneberg, B., Hietala, M.A., Krajci, P., Betsholtz, C., and Brandtzaeg, P. (1999). Absence of epithelial immunoglobulin A transport, with increased mucosal leakiness, in polymeric immunoglobulin receptor/secretory component-deficient mice. *J. Exp. Med.* 190:915–922.

Jones, G., Steketee, R. W., Black, R. E., Bhutta, Z. A., and Morris, S.S., and Bellagio Child Survival Study Group. (2003). How many child deaths can we prevent this year? *Lancet* 362:65–71.

Judge, T., and Lichtenstein, G. R. (2002). The NOD2 gene and Crohn's disease: Another triumph for molecular genetics. *Gastroenterology* 122:826–828.

Juto, P., and Holm, S. (1992). Gliadin-specific and cow's milk protein-specific IgA in human milk. *J. Pediatr. Gastroenterol. Nutr.* 15:159–162.

Kadowaki, N., Antonenko, S., and Liu, Y. J. (2001). Distinct CpG DNA and polyinosinic-polycytidylic acid double-stranded RNA, respectively, stimulate $CD11c^-$ type 2 dendritic cell precursors and $CD11c^+$ dendritic cells to produce type I IFN. *J. Immunol.* 166:2291–2295.

Kaisho, T., and Akira, S. (2001). Dendritic-cell function in Toll-like receptor- and MyD88-knockout mice. *Trends Immunol.* 22:78–83.

Kalliomäki, M., Salminen, S., Arvilommi, H., Kero, P., Koskinen, P., and Isolauri, E. (2001). Probiotics in primary prevention of atopic disease: a randomised placebo-controlled trial. *Lancet* 357:1076–1079.

Karlsson, M. R., Kahu, H., Johansen, F.-E., and Brandtzaeg, P. (2005). Prominent induction of oral tolerance with activation of regulatory $CD4^+CD25^+$ T cells in mice lacking secretory antibodies. In: 12th International Congress of Mucosal Immunology: From Fundamental Biology to Human Disease. June 25–30, Boston, MA, Abstract 53150.

Karlsson, M. R., Rugtveit, J., and Brandtzaeg, P. (2004). Allergen-responsive $CD4^+CD25^+$ regulatory T cells in children who have outgrown cow's milk allergy. *J. Exp. Med.* 199:1679–1688.

Kelly, D., Campbell, J. I., King, T. P., Grant, G., Jansson, E. A., Coutts, A. G., Pettersson, S., and Conway, S. (2004). Commensal anaerobic gut bacteria attenuate inflammation by regulating nuclear-cytoplasmic shuttling of PPAR-γ and RelA. *Nat. Immunol.* 5:104–112.

Kelly, D., Conway, S., and Aminov, R. (2005). Commensal gut bacteria: mechanisms of immune modulation. *Trends Immunol.* 26:326–333.

Kett, K., Brandtzaeg, P., Radl, J., and Haaijman, J. J. (1986). Different subclass distribution of IgA-producing cells in human lymphoid organs and various secretory tissues. *J. Immunol.* 136:3631–3635.

Kilian, M., Husby, S., Høst, A., and Halken, S. (1995). Increased proportions of bacteria capable of cleaving IgA1 in the pharynx of infants with atopic disease. *Pediatr. Res.* 38:182–186.

Kilian, M., Reinholdt, J., Lomholt, H., Poulsen, K., and Frandsen, E. V. (1996). Biological significance of IgA1 proteases in bacterial colonization and pathogenesis: critical evaluation of experimental evidence. *APMIS* 104:321–338.

Kirjavainen, P. V., and Gibson, G. R. (1999). Healthy gut microflora and allergy: factors influencing development of the microbiota. *Ann. Med.* 31:288–292.

Klinman, D. M., Yi, A. K., Beaucage, S. L., Conover, J., and Krieg, A. M. (1996). CpG motifs present in bacteria DNA rapidly induce lymphocytes to secrete interleukin 6, interleukin 12, and interferon γ. *Proc. Natl. Acad. Sci. USA* 93:2879–2883.

Knox, W. F. (1986). Restricted feeding and human intestinal plasma cell development. *Arch. Dis. Child.* 61:744–749.

Kobayashi, K. S., Chamaillard, M., Ogura, Y., Henegariu, O., Inohara, N., Nunez, G., and Flavell, R. A. (2005). Nod2-dependent regulation of innate and adaptive immunity in the intestinal tract. *Science* 307:731–734.

Kobayashi, K., Hernandez, L. D., Galan, J. E., Janeway, C. A., Medzhitov, R., and Flavell, R. A. (2002). IRAK-M is a negative regulator of Toll-like receptor signaling. *Cell* 110:191–202.

Krajci, P., Meling, G. I., Andersen, S. N., Hofstad, B., Vatn, M.H., Rognum, T. O., and Brandtzaeg, P. (1996). Secretory component mRNA and protein expression in colorectal adenomas and carcinomas. *Br. J. Cancer* 73:1503–1510.

Kroese, F. G., Butcher, E. C., Stall, A. M., Lalor, P. A., Adams, S., and Herzenberg, L. A. (1989). Many of the IgA producing plasma cells in murine gut are derived from self-replenishing precursors in the peritoneal cavity. *Int. Immunol.* 1:75–84.

Kull, I., Wickman, M., Lilja, G., Nordvall, S. L., and Pershagen, G. (2002). Breast feeding and allergic diseases in infants-a prospective birth cohort study. *Arch. Dis. Child.* 87:478–481.

Kunkel, E. J., and Butcher, E. C. (2002). Chemokines and the tissue-specific migration of lymphocytes. *Immunity* 16:1–4.

Kvale, D., and Brandtzaeg, P. (1995). Constitutive and cytokine induced expression of HLA molecules, secretory component, and intercellular adhesion molecule-1 is modulated by butyrate in the colonic epithelial cell line HT-29. *Gut* 36:737–742.

Lala, S., Ogura, Y., Osborne, C., Hor, S. Y., Bromfield, A., Davies, S., Ogunbiyi, O., Nunez, G., and Keshav, S. (2003). Crohn's disease and the NOD2 gene: a role for Paneth cells. *Gastroenterology* 125:47–57.

Lanning, D. K., Rhee, K. J., and Knight, K. L. (2005). Intestinal bacteria and development of the B-lymphocyte repertoire. *Trends Immunol.* 26:419–425.

Latcham, F., Merino, F., Lang, A., Garvey, J., Thomson, M. A., Walker-Smith, J. A., Davies, S. E., Phillips, A. D., and Murch, S. H. (2003). A consistent pattern of minor immunodeficiency and subtle enteropathy in children with multiple food allergy. *J. Pediatr.* 143:39–47.

Li, J., Kudsk, K. A., Gocinski, B., Dent, D., Glezer, J., and Langkamp-Henken, B. (1995a). Effects of parenteral and enteral nutrition on gut-associated lymphoid tissue. *J. Trauma* 39:44–51.

Li, J., Kudsk, K. A., Hamidian, M., and Gocinski, B. L. (1995b). Bombesin affects mucosal immunity and gut-associated lymphoid tissue in intravenously fed mice. *Arch. Surg.* 130:1164–1169.

Liew, F. Y. (2002). T_H1 and T_H2 cells: a historical perspective. *Nat. Rev. Immunol.* 2:55–60.

Lim, P. L., and Rowley, D. (1982). The effect of antibody on the intestinal absorption of macromolecules and on intestinal permeability in adult mice. *Int. Arch. Allergy Appl. Immunol.* 68:41–46.

Lodinová, R., Jouja, V., and Wagner, V. (1973). Serum immunoglobulins and coproantibody formation in infants after artificial intestinal colonization with *Escherichia coli* 083 and oral lysozyme administration. *Pediatr. Res.* 7:659–669.

Lotz, M., Gutle, D., Walther, S., Menard, S., Bogdan, C., and Hornef, M. W. (2006). Postnatal acquisition of endotoxin tolerance in intestinal epithelial cells. *J. Exp. Med.* 203:973–984.

Lycke, N., Erlandsson, L., Ekman, L., Schon, K., and Leanderson, T. (1999). Lack of J chain inhibits the transport of gut IgA and abrogates the development of intestinal antitoxic protection. *J. Immunol.* 163:913–919.

MacDonald, T. T., and Monteleone, G. (2005). Immunity, inflammation, and allergy in the gut. *Science* 307:1920–1925.

Macpherson, A. J., Gatto, D., Sainsbury, E., Harriman, G. R., Hengartner, H., and Zinkernagel, R. M. (2000). A primitive T cell-independent mechanism of intestinal mucosal IgA responses to commensal bacteria. *Science* 288:2222–2226.

Macpherson, A. J., Geuking, M. B., and McCoy, K. D. (2005). Immune responses that adapt the intestinal mucosa to commensal intestinal bacteria. *Immunology* 115:153–162.

Macpherson, A. J., Lamarre, A., McCoy, K., Harriman, G. R., Odermatt, B., Dougan, G., Hengartner, H., and Zinkernagel, R. M. (2001). IgA production without μ or δ chain expression in developing B cells. *Nat. Immunol.* 2:625–631.

Macpherson, A. J., and Uhr, T. (2004). Induction of protective IgA by intestinal dendritic cells carrying commensal bacteria. *Science* 303:1662–1665.

Maeda, S., Hsu, L. C., Liu, H., Bankston, L. A., Iimura, M., Kagnoff, M. F., Eckmann, L., and Karin, M. (2005). Nod2 mutation in Crohn's disease potentiates NF-κB activity and IL-1β processing. *Science* 307:734–738.

Malin, M., Suomalainen, H., Saxelin, M., and Isolauri, E. (1996). Promotion of IgA immune response in patients with Crohn's disease by oral bacteriotherapy with Lactobacillus GG. *Ann. Nutr. Metab.* 40:137–145.

Manigold, T., Böcker, U., Traber, P., Dong-Si, T., Kurimoto, M., Hanck, C., Singer, M. V., and Rossol, S. (2000). Lipopolysaccharide/endotoxin induces IL-18 via CD14 in human peripheral blood mononuclear cells *in vitro. Cytokine* 12:1788–1792.

Manser, T. (2004). Textbook germinal centers? *J. Immunol.* 172:3369–3375.

Mantis, N. J., Cheung, M. C., Chintalacharuvu, K. R., Rey, J., Corthesy, B., and Neutra, M. R. (2002). Selective adherence of IgA to murine Peyer's patch M cells: Evidence for a novel IgA receptor. *J. Immunol.* 169:1844–1851.

Marchant, A., Goetghebuer, T., Ota, M. O., Wolfe, I., Ceesay, S. J., De Groote, D., Corrah, T., Bennett, S., Wheeler, J., Huygen, K., Aaby, P., McAdam, K. P., and Newport, M. J. (1999). Newborns develop a Th1-type immune response to *Mycobacterium bovis* bacillus Calmette-Guerin vaccination. *J. Immunol.* 163:2249–2255.

Mathew, C. G., and Lewis, C. M. (2004). Genetics of inflammatory bowel disease: Progress and prospects. *Hum. Mol. Genet.* 13(Spec. No. 1):R161–R168.

Mazanec, M. B., Coudret, C. L., and Fletcher, D. R. (1995). Intracellular neutralization of influenza virus by immunoglobulin A anti-hemagglutinin monoclonal antibodies. *J. Virol.* 69:1339–1343.

Mazmanian, S. K., Liu, C. H., Tzianabos, A. O., and Kasper, D. L. (2005). An immunomodulatory molecule of symbiotic bacteria directs maturation of the host immune system. *Cell* 122:107–118.

Mazanec, M. B., Nedrud, J. G., Kaetzel, C. S., and Lamm, M. E. (1993). A three-tiered view of the role of IgA in mucosal defense. *Immunol. Today*. 14:430–435.

McInnes, I. B., Gracie, J. A., Leung, B. P., Wei, X. Q., and Liew, F. Y. (2000). Interleukin 18: A pleiotropic participant in chronic inflammation. *Immunol Today* 21:312–315.

McLoughlin, G. A., Hede, J. E., Temple, J. G., Bradley, J., Chapman, D. M., and McFarland, J. (1978). The role of IgA in the prevention of bacterial colonization of the jejunum in the vagotomized subject. *Br. J. Surg*. 65:435–437.

Medzhitov, R. (2001). Toll-like receptors and innate immunity. *Nat. Rev. Immunol.* 1:135–145.

Mennechet, F. J., Kasper, L. H., Rachinel, N., Minns, L. A., Luangsay, S., Vandewalle, A., and Buzoni-Gatel, D. (2004). Intestinal intraepithelial lymphocytes prevent pathogen-driven inflammation and regulate the Smad/T-bet pathway of lamina propria $CD4^+$ T cells. *Eur. J. Immunol.* 34:1059–1067.

Milling, S. W., Cousins, L., and MacPherson, G. G. (2005). How do DCs interact with intestinal antigens? *Trends Immunol.* 26:349–352.

Modlin, R. L. (2000). Immunology. A Toll for DNA vaccines. *Nature* 408:659–660.

Moldoveanu, Z., Tenovuo, J., Mestecky, J., and Pruitt, K. M.. (1982). Human milk peroxidase is derived from milk leukocytes. *Biochim. Biophys. Acta* 718:103–108.

Mora, J. R., Iwata, M., Eksteen, B., Song, S. Y., Junt, T., Senman, B., Otipoby, K. L., Yokota, A., Takeuchi, H., Ricciardi-Castagnoli, P., Rajewsky, K., Adams, D. H., and von Andrian, U. H. (2006), Generation of gut-homing IgA-secreting B cells by intestinal dendritic cells. *Science* 314:1157–1160.

Moreau, M. C., Ducluzeau, R., Guy-Grand, D., and Muller, M. C. (1978). Increase in the population of duodenal immunoglobulin A plasmocytes in axenic mice assosiated with different living or dead bacterial strains of intestinal origin. *Infect. Immun.* 121:532–539.

Moreau, M. C., and Gaboriau-Routhiau, V. (2000). Immunomodulation by the gut microflora and probiotics. *Probiotics* 3:69–114.

Mostov, K. E., and Blobel, G. (1982). A transmembrane precursor of secretory component. The receptor for transcellular transport of polymeric immunoglobulins. *J. Biol. Chem.* 257:11,816–11,821.

Mowat, A. M. (2003). Anatomical basis of tolerance and immunity to intestinal antigens. *Nat. Rev. Immunol.* 3:331–341.

Moynagh, P. N. (2005). TLR signalling and activation of IRFs: Revisiting old friends from the NF-κB pathway. *Trends Immunol.* 26:469–476.

Nagao, A. T., Pilagallo, M. I. D. S., and Pereira, A. B. (1993). Quantitation of salivary, urinary and faecal SIgA in children living in different conditions of antigenic exposure. *J. Trop. Pediatr*. 39:278–283.

Nagler-Anderson, C. (2001). Man the barrier! Strategic defences in the intestinal mucosa. *Nat. Rev. Immunol.* 1:59–67.

Nakamura, Y., Nosaka, S., Suzuki, M., Nagafuchi, S., Takahashi, T., Yajima, T., Takenouchi-Ohkubo, N., Iwase, T., and Moro, I. (2004). Dietary fructooligosaccharides up-regulate immunoglobulin A response and polymeric immunoglobulin receptor expression in intestines of infant mice. *Clin. Exp. Immunol.* 137:52–58.

Neish, A. S., Gewirtz, A. T., Zeng, H., Young, A. N., Hobert, M. E., Karmali, V., Rao, A. S., and Madara, J. L. (2000). Prokaryotic regulation of epithelial responses by inhibition of IκB-α ubiquitination. *Science* 289:1560–1563.

Neutra, M. R., Mantis, N. J., and Kraehenbuhl, J. P. (2001). Collaboration of epithelial cells with organized mucosal lymphoid tissues. *Nat. Immunol.* 2:1004–1009.

Nilsen, E. M., Jahnsen, F. L., Lundin, K. E. A., Johansen, F.-E., Fausa, O., Sollid, L. M., Jahnsen, J., Scott, H., and Brandtzaeg, P. (1998). Gluten induces an intestinal cytokine response strongly dominated by interferon-γ in patients with celiac disease. *Gastroenterology* 115:551–563.

Nilsen, E. M., Lundin, K. E. A., Krajcii, P., Scott, H., Sollid, L. M., and Brandtzaeg, P. (1995). Gluten-specific, HLA-DQ restricted T cells from coeliac mucosa produce cytokines with Th1 or Th0 profile dominated by interferon γ. *Gut* 37:766–776.

Ogura, Y., Bonen, D. K., Inohara, N., Nicolae, D. L., Chen, F. F., Ramos, R., Britton, H., Moran, T., Karaliuskas, R., Duerr, R. H., Achkar, J. P., Brant, S. R., Bayless, T. M., Kirschner, B. S., Hanauer, S. B., Nunez, G., and Cho, J. H. (2001). A frameshift mutation in NOD2 associated with susceptibility to Crohn's disease. *Nature* 411:603–606.

Olaussen, R. W., Johansen, F.-E., Lundin, K. E., Jahnsen, J., Brandtzaeg, P., and Farstad, I. N. (2002). Interferon-γ-secreting T cells localize to the epithelium in coeliac disease. *Scand. J. Immunol.* 56:652–664.

Oshiumi, H., Matsumoto, M., Funami, K., Akazawa, T., and Seya, T. (2003). TICAM-1, an adaptor molecule that participates in Toll-like receptor 3-mediated interferon-β induction. *Nat. Immunol.* 4:161–167.

Östman, S., Taube, M., and Telemo, E. (2005). Tolerosome-induced oral tolerance is MHC dependent. *Immunology* 116:464–476.

Paganelli, R., and Levinsky, R. J. (1980). Solid phase radioimmunoassay for detection of circulating food protein antigens in human serum. *J. Immunol. Methods.* 37:333–341.

Pal, K., Kaetzel, C. S., Brundage, K. Cunningham, C., and Cuff, C. F. (2005). Regulation of polymeric immunoglobulin receptor expression by reovirus. *J. Gen. Virol.* 86:2347–2357.

Peng, Z., Wang, H., Mao, X., HayGlass, K. T., and Simons, F. E. (2001). CpG oligodeoxynucleotide vaccination suppresses IgE induction but may fail to down-regulate ongoing IgE responses in mice. *Int. Immunol.* 13:3–11.

Perez-Machado, M. A., Ashwood, P., Thomson, M. A., Latcham, F., Sim, R., Walker-Smith, J. A., and Murch, S. H. (2003). Reduced transforming growth factor-β1-producing T cells in the duodenal mucosa of children with food allergy. *Eur. J. Immunol.* 33:2307–2315.

Perkkiö, M. (1980). Immunohistochemical study of intestinal biopsies from children with atopic eczema due to food allergy. *Allergy* 35:573–580.

Persson, C. G., Erjefalt, J. S., Greiff, L., Erjefalt, I., Korsgren, M., Linden, M., Sundler, F., Andersson, M., and Svensson, C. (1998). Contribution of plasma-derived molecules to mucosal immune defence, disease and repair in the airways. *Scand. J. Immunol.* 47:302–313.

Phalipon, A., and Corthésy, B. (2003). Novel functions of the polymeric Ig receptor: Well beyond transport of immunoglobulins. *Trends Immunol.* 24:55–58.

Philpott, D. J., Girardin, S. E., and Sansonetti, P. J. (2001). Innate immune responses of epithelial cells following infection with bacterial pathogens. *Curr. Opin. Immunol.* 13:410–416.

Planchon, S. M., Martins, C. A., Guerrant, R. L., and Roche, J. K. (1994). Regulation of intestinal epithelial barrier function by TGF-β1. Evidence for its role in abrogating the effect of a T cell cytokine. *J. Immunol.* 153:5730–5739.

Prokešová, L., Ladmanová, P., Cechova, D., Stepánková, R., Kozáková, H., Mlcková, Å., Kuklik, R., and Mára, M. (1999). Stimulatory effects of *Bacillus firmus* on IgA production in human and mice. *Immunol. Lett.* 69:55–56.

Prokešová, L., Mlčková, P., Staňková, I., Chloubová, A., Novotná, V., Ladmanová, P., Chalupná, P., and Mára, M. (1998). Effect of *Bacillus firmus* on antibody formation after mucosal and parenteral immunization in mice. *Immunol. Lett.* 64:161–166.

Pulendran, B., and Ahmed, R. (2006). Translating innate immunity into immunological memory: implications for vaccine development. *Cell* 124:849–863.

Rakoff-Nahoum, S., Paglino, J., Eslami-Varzaneh, F., Edberg, S., and Medzhitov, R. (2004). Recognition of commensal microflora by toll-like receptors is required for intestinal homeostasis. *Cell* 118:229–241.

Rautava, S., Ruuskanen, O., Ouwehand, A., Salminen, S., and Isolauri, E. (2004). The hygiene hypothesis of atopic disease: an extended version. *J. Pediatr. Gastroenterol. Nutr.* 38:378–388.

Renegar, K. B., Johnson, C. D., Dewitt, R. C., King, B. K., Li, J., Fukatsu, K., and Kudsk, K. A. (2001). Impairment of mucosal immunity by total parenteral nutrition: requirement for IgA in murine nasotracheal anti-influenza immunity. *J. Immunol.* 166:819–825.

Renegar, K. B., Small, P. A., Boykins, L. G., and Wright, P. F. (2004). Role of IgA versus IgG in the control of influenza viral infection in the murine respiratory tract. *J. Immunol.* 173:1978–1986.

Rescigno, M., Urbano, M., Valzasina, B., Francolini, M., Rotta, G., Bonasio, R., Granucci, F., Kraehenbuhl, J. P., and Ricciardi-Castagnoli, P. (2001). Dendritic cells express tight junction proteins and penetrate gut epithelial monolayers to sample bacteria. *Nat. Immunol.* 2:361–367.

Rey, J., Garin, N., Spertini, F., and Corthesy, B. (2004). Targeting of secretory IgA to Peyer's patch dendritic and T cells after transport by intestinal M cells. *J. Immunol.* 172:3026–3033.

Reynolds, J. D., and Morris, B. (1984). The effect of antigen on the development of Peyer's patches in sheep. *Eur. J. Immunol.* 14:1–6.

Rhee, K. J., Sethupathi, P., Driks, A., Lanning, D. K., and Knight, K. L. (2004). Role of commensal bacteria in development of gut-associated lymphoid tissues and preimmune antibody repertoire. *J. Immunol.* 172:1118–1124.

Ricciardi-Castagnoli, P., and Granucci, F. (2002). Opinion: Interpretation of the complexity of innate immune responses by functional genomics. *Nat. Rev. Immunol.* 2:881–889.

Rimoldi, M., Chieppa, M., Salucci, V., Avogadri, F., Sonzogni, A., Sampietro, G. M., Nespoli, A., Viale, G., Allavena, P., and Rescigno, M. (2005). Intestinal immune homeostasis is regulated by the crosstalk between epithelial cells and dendritic cells. *Nat. Immunol.* 6:507–514.

Robinson, J. K., Blanchard, T. G., Levine, A. D., Emancipator, S. N., and Lamm, M. E. (2001). A mucosal IgA-mediated excretory immune system *in vivo*. *J. Immunol.* 166:3688–3692.

Rogers, H. J., and Synge, C. (1978). Bacteriostatic effect of human milk on *Escherichia coli*: The role of IgA. *Immunology* 34:19–28.
Rognum, T. O., Elgjo, K., Fausa, O., and Brandtzaeg, P. (1982). Immunohistochemical evaluation of carcinoembryonic antigen, secretory component, and epithelial IgA in ulcerative colitis with dysplasia. *Gut* 23:123–133.
Roy, M. J., and Varvayanis, M. (1987). Development of dome epithelium in gut-associated lymphoid tissues: Association of IgA with M cells. *Cell Tissue Res.* 248:645–651.
Rugtveit, J., Bakka, A., and Brandtzaeg, P. (1997b). Differential distribution of B7.1 (CD80) and B7.2 (CD86) co-stimulatory molecules on mucosal macrophage subsets in human inflammatory bowel disease (IBD). *Clin. Exp. Immunol.* 110:104–113.
Rugtveit, J., Nilsen, E. M., Bakka, A., Carlsen, H., Brandtzaeg, P., and Scott, H. (1997a). Cytokine profiles differ in newly recruited and resident subsets of mucosal macrophages from inflammatory bowel disease. *Gastroenterology* 112:1493–1505.
Sagie, E., Tarabulus, J., Maeir, D. M., and Freier, S. (1974). Diet and development of intestinal IgA in the mouse. *Isr. J. Med. Sci.* 10:532–534.
Sansonetti, P. J. (2004). War and peace at mucosal surfaces. *Nat. Rev. Immunol.* 4:953–964.
Savilahti, E., Tainio, V.-M., Salmenpera, L., Arjomaa, P., Kallio, M., Perheentupa, J., and Siimes, M. A. (1991). Low colostral IgA associated with cow's milk allergy. *Acta Paediatr. Scand.* 80:1207–1213.
Schjerven, H., Brandtzaeg, P., and Johansen, F.-E. (2001). A novel NF-κB/Rel site in intron 1 cooperates with proximal promoter elements to mediate TNF-α-induced transcription of the human polymeric Ig receptor. *J. Immunol.* 167:6412–6420.
Schneeman, T. A., Bruno, M., Schjerven, H., Johansen, F.-E., Chady, L., and Kaetzel, C. S. (2005). Regulation of the polymeric Ig receptor by signaling through TLRs 3 and 4: Linking innate and adaptive immune responses. *J. Immunol.* 175:376–384.
Schoetzau, A., Filipiak-Pittroff, B., Franke, K., Koletzko, S., Von Berg, A., Gruebl, A., Bauer, C. P., Berdel, D., Reinhardt, D., and Wichmann, H. E., and German Infant Nutritional Intervention Study Group. (2002). Effect of exclusive breast-feeding and early solid food avoidance on the incidence of atopic dermatitis in high-risk infants at 1 year of age. *Pediatr. Allergy Immunol.* 13:234–242.
Sheil, B., McCarthy, J., O'Mahony, L., Bennett, M. W., Ryan, P., Fitzgibbon, J. J., Kiely, B., Collins, J. K., and Shanahan, F. (2004). Is the mucosal route of administration essential for probiotic function? Subcutaneous administration is associated with attenuation of murine colitis and arthritis. *Gut* 53:694–700.
Shroff, K. E., Meslin, K., and Cebra, J. J. (1995). Commensal enteric bacteria engender a self-limiting humoral mucosal immune response while permanently colonizing the gut. *Infect. Immun.* 63:3904–3913.
Sloper, K. S., Brook, C. G., Kingston, D., Pearson, J. R., and Shiner, M. (1981). Eczema and atopy in early childhood: low IgA plasma cell counts in the jejunal mucosa. *Arch. Dis. Child.* 56:939–942.
Smith, P. D., Smythies, L. E., Mosteller-Barnum, M., Sibley, D. A., Russell, M. W., Merger, M., Sellers, M. T., Orenstein, J. M., Shimada, T., Graham, M. F., and Kubagawa, H. (2001). Intestinal macrophages lack CD14 and CD89 and consequently are down-regulated for LPS- and IgA-mediated activities. *J. Immunol.* 167:2651–2656.
Sørensen, C. H., and Kilian, M. (1984). Bacterium-induced cleavage of IgA in nasopharyngeal secretions from atopic children. *Acta Pathol. Microbiol. Immunol. Scand. [C]* 92:85–87.

Steidler, L., Hans, W., Schotte, L., Neirynck, S., Obermeier, F., Falk, W., Fiers, W., and Remaut, E. (2000). Treatment of murine colitis by *Lactococcus lactis* secreting interleukin-10. *Science* 289:1352–1355.

Stephensen, C. B., Moldoveanu, Z., and Gangopadhyay, N. N. (1996). Vitamin A deficiency diminishes the salivary immunoglobulin A response and enhances the serum immunoglobulin G response to influenza A virus infection in BALB/c mice. *J. Nutr.* 126:94–102.

Stoel, M., Jiang, H. Q., van Diemen, C. C., Bun, J. C., Dammers, P. M., Thurnheer, M. C., Kroese, F. G., Cebra, J. J., and Bos, N. A. (2005). Restricted IgA repertoire in both B-1 and B-2 cell-derived gut plasmablasts. *J. Immunol.* 174:1046–1054.

Sun, K., Johansen, F.-E., Eckmann, L., and Metzger, D. W. (2004). An important role for polymeric Ig receptor-mediated transport of IgA in protection against *Streptococcus pneumoniae* nasopharyngeal carriage. *J. Immunol.* 173:4576–4581.

Talham, G. L., Jiang, H. Q., Bos, N. A., and Cebra, J. J. (1999). Segmented filamentous bacteria are potent stimuli of a physiologically normal state of the murine gut mucosal immune system. *Infect. Immun.* 67:1992–2000.

Taylor, B., Norman, A. P., Orgel, H. A., Stokes, C. R., Turner, M. W., and Soothill, J. F. (1973). Transient IgA deficiency and pathogenesis of infantile atopy. *Lancet* 2:111–113.

Tenovuo, J., Moldoveanu, Z., Mestecky, J., Pruitt, K. M., and Rahemtulla, B. M. (1982). Interaction of specific and innate factors of immunity: IgA enhances the antimicrobial effect of the lactoperoxidase system against *Streptococcus mutans*. *J. Immunol.* 128:726–731.

Thrane, P. S., Rognum, T. O., and Brandtzaeg, P. (1991). Ontogenesis of the secretory immune system and innate defence factors in human parotid glands. *Clin. Exp. Immunol.* 86:342–348.

Umesaki, Y., Okada, Y., Matsumoto, S., Imaoka, A., and Setoyama, H. (1995). Segmented filamentous bacteria are indigenous intestinal bacteria that activate intraepithelial lymphocytes and induce MHC class II molecules and fucosyl asialo GM1 glycolipids on the small intestinal epithelial cells in the ex-germ-free mouse. *Microbiol. Immunol.* 39:555–562.

Uren, T. K., Johansen, F.-E., Wijburg, O. L., Koentgen, F., Brandtzaeg, P., and Strugnell, R. A. (2003). Role of the polymeric Ig receptor in mucosal B cell homeostasis. *J. Immunol.* 170:2531–2539.

Uren, T. K., Wijburg, O.L. C., Simmons, C., Johansen, F.-E., Brandtzaeg, P., and Strugnell, R. A. (2005). Vaccine-induced protection against gastrointestinal bacterial infections in the absence of secretory antibodies. *Eur. J. Immunol.* 35:180–188.

van Asperen, P. P., Gleeson, M., Kemp, A. S., Cripps, A. W., Geraghty, S. B., Mellis, C. M., and Clancy, R. L. (1985). The relationship between atopy and salivary IgA deficiency in infancy. *Clin. Exp. Immunol.* 62:753–757.

van der Waaij, L. A., Limburg, P. C., Mesander, G., and van der Waaij, D. (1996). *In vivo* IgA coating of anaerobic bacteria in human faeces. *Gut* 38:348–354.

van Odijk, J., Kull, I., Borres, M. P., Brandtzaeg, P., Edberg, U., Hanson, L. Å., Host, A., Kuitunen, M., Olsen, S. F., Skerfving, S., Sundell, J., and Wille, S. (2003). Breastfeeding and allergic disease: A multidisciplinary review of the literature (1966–2001) on the mode of early feeding in infancy and its impact on later atopic manifestations. *Allergy* 58:833–843.

Venkatachalam, M. A., Soltani, M. H., and Fahimi, H. D. (1970). Fine structural localization of peroxidase activity in the epithelium of large intestine of rat. *J. Cell. Biol.* 46:168–173.

von Reyn, C. F., Arbeit, R. D., Yeaman, G., Waddell, R. D., Marsh, B. J., Morin, P., Modlin, J. F., and Remold, H. G. (1997). Immunization of healthy adult subjects in the United States with inactivated *Mycobacterium vaccae* administered in a three-dose series. *Clin. Infect. Dis.* 24:843–848.

Walker, W. A., and Bloch, K. J. (1983). Gastrointestinal transport of macromolecules in the pathogenesis of food allergy. *Ann. Allergy* 51:240–245.

Walker, W. A., Lake, A. M., and Bloch, K. J. (1982). Immunologic mechanisms for goblet cell mucous release: possible role in mucosal defense. In: Strober, W., Hanson, L. Å., and Sell, K. W. (eds.), *Recent Advances in Mucosal Immunity*. Raven Press, New York.

Walker, W. A., Wu, M., Isselbacher, K. J., and Bloch, K. J. (1975). Intestinal uptake of macromolecules. III. Studies on the mechanism by which immunization interferes with antigen uptake. *J. Immunol.* 115:854–861.

Watanabe, T., Kitani, A., Murray, P. J., and Strober, W. (2004). NOD2 is a negative regulator of Toll-like receptor 2-mediated T helper type 1 responses. *Nat. Immunol.* 5:800–808.

Watson, R. R., McMurray, D. N., Martin, P., and Reyes, M. A. (1985). Effect of age, malnutrition and renutrition on free secretory component and IgA in secretion. *Am. J. Clin. Nutr.* 42:281–288.

Wiedermann, U., Hanson, L. A., Holmgren, J., Kahu, H., and Dahlgren, U. I. (1993). Impaired mucosal antibody response to cholera toxin in vitamin A-deficient rats immunized with oral cholera vaccine [published erratum appears in Infection and Immunity 61:5431]. *Infect. Immun.* 61:3952–3957.

Wijburg, O. L., Uren, T. K., Simpfendorfer, K., Johansen, F.-E., Brandtzaeg, P., and Strugnell, R. A. (2006). Innate secretory antibodies protect against natural *Salmonella typhimurium* infection. *J. Exp. Med.* 203:21–26.

Wijesinha, S. S., and Steer, H. W. (1982). Studies of the immunoglobulin-producing cells of the human intestine: The defunctioned bowel. *Gut* 23:211–214.

Wohlleben, G., and Erb, K. J. (2001). Atopic disorders: a vaccine around the corner? *Trends Immunol.* 22:618–626.

Yamanaka, T., Helgeland, L., Farstad, I. N., Midtvedt, T., Fukushima, H., and Brandtzaeg, P. (2003). Microbial colonization drives lymphocyte accumulation and differentiation in the follicle-associated epithelium of Peyer's patches. *J. Immunol.* 170:816–822.

Yamamoto, M., Sato, S., Hemmi, H., Hoshino, K., Kaisho, T., Sanjo, H., Takeuchi, O., Sugiyama, M., Okabe, M., Takeda, K., and Akira, S. (2003). Role of adaptor TRIF in the MyD88-independent toll-like receptor signaling pathway. *Science* 301:640–643.

Yamamoto, M., Sato, S., Mori, K., Hoshino, K., Takeuchi, O., Takeda, K., and Akira, S. (2002). Cutting edge: a novel Toll/IL-1 receptor domain-containing adapter that preferentially activates the IFN-β promoter in the Toll-like receptor signaling. *J. Immunol.* 169:6668–6672.

Yamazaki, K., Shimada, S., Kato-Nagaoka, N., Soga, H., Itoh, T., and Nanno, M. (2005). Accumulation of intestinal intraepithelial lymphocytes in association with lack of polymeric immunoglobulin receptor. *Eur. J. Immunol.* 35:1211–1219.

Yan, F., and Polk, D. B. (2004). Commensal bacteria in the gut: learning who our friends are. *Curr. Opin. Gastroenterol.* 20:565–571.

Yasui, H., Kiyoshima, J., and Ushijima, H. (1995). Passive protection against rotavirus-induced diarrhea of mouse pups born to and nursed by dams fed *Bifidobacterium breve* YIT4064. *J. Infect. Dis.* 172:403–409.

Yazdanbakhsh, M., Kremsner, P. G., and van Ree, R. (2002). Allergy, parasites, and the hygiene hypothesis. *Science* 296:490–494.
Yoshida, M., Claypool, S. M., Wagner, J. S., Mizoguchi, E., Mizoguchi, A., Roopenian, D. C., Lencer, W. I., and Blumberg, R. S. (2004). Human neonatal Fc receptor mediates transport of IgG into luminal secretions for delivery of antigens to mucosal dendritic cells. *Immunity* 20:769–783.
Zoetendal, E. G., Akkermans, A. D. L., Akkermans-van Vliet, W. M., Wisser, J. A. G. M., and de Vos, W. M. (2001). The host genotype affects the bacterial community in the human gastrointestinal tract. *Microb. Ecol. Health Dis.* 13:129–134.

11 IgA and Respiratory Immunity

Dennis W. Metzger[1]

11.1. General Overview of the Respiratory Immune System

Large numbers of microbes and microparticles enter the airways with every breath and the respiratory tract thus represents a major portal of entry for various viral and bacterial pathogens. In addition to mechanical defenses such as coughing, sneezing, and the action of ciliated epithelia, mucosal-associated lymphoid tissue (MALT) plays a critical role in protection of the upper and lower respiratory tracts against microbial challenge. Although the respiratory tract represents about 25% of the total 400 m^2 of mucosal tissue in the adult human, little is known about immune function in the airways. Much of our

[1] Center for Immunology and Microbial Disease, Albany Medical College, Albany, New York 12208, USA.

current understanding is actually based on information obtained from studies of the gastrointestinal system, despite the fact that many differences exist between lymphoid tissues in these two areas.

Organized nasal-associated lymphoid tissue (NALT) and bronchus-associated lymphoid tissue (BALT) are present in the upper respiratory tract and the importance of these tissues in protection against infectious disease is well accepted. However, NALT and BALT generally tend to have few germinal centers and thus resemble the isolated lymphoid tissues of the intestine rather than Peyer's patches (see Chapter 2). The lung contains little, if any, organized mucosal tissue and the dominant cellular population involved in defense in this organ appears to include alveolar macrophages rather than lymphocytes. Indeed, normal bronchoalveolar lavage fluid (BALF) consists of 90% or more macrophages. In addition, M-cells are virtually absent in the normal lung (Pabst, 1992). Lymphocytes in the lung are generally sparse and not organized, although foci of inflammation can occur during asthma and, in some cases, organized lymphoid tissues have been reported to develop in the lungs of both humans and mice during inflammation (Chvatchko et al., 1996; Delventhal et al., 1992; Tschernig and Pabst, 2000).

Antigen-specific B-cells are induced to isotype switch and undergo somatic mutation in the germinal centers of MALT inductive areas (NALT and BALT for the respiratory tract) under the influence of cytokines and microenvironmental influences (Shimoda et al., 2001; Zuercher et al., 2002). It is believed that transforming growth factor (TGF)-β drives switching to IgA, although why this is the major isotype produced by mucosal B-cells remains unknown. It is also unknown whether this switching is driven by T-cells or mucosal epithelial cells because both cell types can produce a multitude of cytokines, including TGF-β (Salvi and Holgate, 1999). In the bronchi and lungs, immunoglobulin G (IgG)-secreting B-cells are also highly represented, but it is not known whether these B-cells are activated in the MALT or draining cervical lymph nodes. After activation, B-cells home to effector mucosal tissues such as the lung under the influence of selected chemokines and adhesion molecules. Although $\alpha_4\beta_7$ and MADCAM-1 interactions appear to be critical for homing of B-cells to gut mucosal tissue, only low levels of MADCAM-1 are present on bronchial endothelial cells. Homing to the respiratory tract appears to involve $\alpha_4\beta_1$–VCAM1 interactions, the same interactions that are involved in recruiting systemic lymphocytes to sites of inflammation (Kunkel and Butcher, 2003). Nevertheless, CCR10 is upregulated on cells destined to home to the respiratory tract but is absent on systemic lymphocytes (Kunkel et al., 2003). Similarly, CCL28, the chemokine ligand for CCR10, is expressed preferentially by mucosal epithelial cells (Pan et al., 2000). Thus, CCR10–CCL28 interactions appear to direct trafficking of lymphocytes to mucosal tissues, and $\alpha_4\beta_1$–VCAM1 interactions ensure homing to the respiratory (and urogenital) tracts. This differential usage of homing receptors would explain seminal findings showing that adoptively transferred, IgA-secreting B-cells that are obtained from the upper respiratory tract preferentially traffic to recipient airways and show only low levels of trafficking to other organs, including the gut (Husband and Gowans, 1978; McDermott and Bienenstock, 1979; Rudzik et al., 1975; Weisz-Carrington et al., 1979). However, there are likely other regulatory factors involved because IgA-secreting B-cells, but not IgM-secreting or IgG-secreting B-cells, show preferential homing to mucosal tissues.

As stated earlier, the lung has traditionally been considered to be an effector mucosal B-cell site rather than an inductive site. Jones and Ada (1986, 1987) showed that antibody-secreting cells can be expressed in the lung after influenza virus infection or immunization with inactivated virus. Furthermore, protection was correlated with the magnitude of the regional lung antibody response. That such antibody-secreting cells might actually form directly in the lung is suggested by recent studies by Randall and colleagues, who found that organized BALT-like structures are induced in the lung by influenza virus infection (Moyron-Quiroz et al., 2004). These BALT-like structures contain distinct B-cell follicles and T-cell areas and, unlike the organized lymphoid tissues in the periphery and gut (De Togni et al., 1994; Fagarasan et al., 2002), can develop in the absence of lymphotoxin-α. In fact, mice lacking lymphotoxin-α clear influenza infection more rapidly and survive higher doses of virus challenge than do their normal counterparts, suggesting that peripheral lymphoid organs actually interfere with effective mucosal immunity (Moyron-Quiroz et al., 2004). Although these BALT-like structures are only induced after infection and therefore cannot be considered constitutive, like NALT, it should be noted that even NALT is found much less frequently in humans compared to rodents. Furthermore, in the normal respiratory tract, NALT is relatively disorganized, with M-cells virtually absent. Thus, many believe that NALT also fully develops only in response to antigenic stimulation (Bienenstock, 2005). Similarly, it is known that commensal gut bacteria influence expression of intestinal IgA-producing cells and that such cells are nearly absent in germ-free animals (Hooper, 2004; Mazmanian et al., 2005; Rhee et al., 2004; Stepankova et al., 1980). This suggests that, in general, the development of mucosal lymphoid tissue requires antigenic stimulation. Finally, it should be noted that $CD28^{-/-}$ and lymphotoxin-$\alpha^{-/-}$ mice, which lack peripheral and Peyer's patch germinal centers, still can develop functional IgA antibodies in the gut (Fagarasan et 1., 2001; Gardby et al., 2003; Kang et al., 2002) (see Chapter 2).

11.2. Immunodeficiency Models to Understand the Role of IgA in Respiratory Immunity

Important clues about the role of IgA in immune protection have come from patients with congenital IgA immunodeficiency, the most common human immunodeficiency disease (see Chapter 13). Although it is believed that IgA is critical for protection of mucosal surfaces against microbial infections, IgA-deficient patients are generally not immunocompromised and usually can clear infections as effectively as normal individuals (Ballow, 2002; Burks and Steele, 1986; Burrows and Cooper, 1997). Protection in the absence of IgA is typically attributed to compensatory increases in expression of secretory IgM (SIgM) and IgG at mucosal surfaces (Brandtzaeg et al., 1987a, 1987b; Burks and Steele, 1986). Although most IgA-deficient patients are healthy, there is an increased incidence of several disorders in this population and these disorders tend to be localized to the respiratory tract. For example, the most common infections associated with IgA immune deficiency include recurrent ear infections, sinusitis, bronchitis, and pneumonia. IgA immunodeficiency also has been found to be associated with an increased incidence of allergy (Burks and Steele, 1986).

In an effort to understand further the importance of IgA in mucosal immune responses, several strains of immune-deficient mice have been developed. These include mice with genetic disruptions in the *Igh-2* gene locus encoding the α H-chain constant region and mice with disruptions in genes required for proper assembly of polymeric IgA and/or its transport across mucosal epithelia. Although human IgA deficiency appears to be caused by defects in regulatory mechanisms rather than in IgA constant region genes or genes involved in IgA transport, the murine models that have been developed do provide potentially important clues into the role of IgA in protection against microbial pathogens.

11.2.1. J-chain$^{-/-}$ Mice

The J-chain is involved in polymerization of Ig as well as the interaction of polymeric IgA and IgM with the polymeric Ig receptor (pIgR) (Hendrickson et al., 1995, 1996; Lycke et al., 1999) (see Chapter 1). Thus, one would expect decreased expression of IgA at mucosal surfaces in these mice due to altered transport across the epithelial barrier. However, it was found that the J-chain$^{-/-}$ mice created by Hendrickson et al. (1996) had levels of IgA in nasal washes and BAL (as well as breast milk and intestinal mucosal surfaces) that were similar to those in J-chain$^{+/+}$ mice. This IgA was monomeric rather than polymeric and was not bound to the secretory component. Whereas the results with these mice suggested the existence of a pIgR-independent mechanism for mucosal IgA transport, another set of J chain$^{-/-}$ mice created by Erlandsson et al. (1998) yielded contrasting results. In the latter case, the J-chain$^{-/-}$ mice had greatly reduced levels of mucosal IgA, normal numbers of IgA antibody-forming cells in the lamina propria and increased levels of serum IgA, all results that would be expected from a defect in pIgR-mediated transport (Erlandsson et al., 1998; Johansen et al., 2000). Curiously, the animals also had significantly decreased levels of serum IgM as well as IgM-producing cells, a finding that has yet to be explained. Utilizing these mice, Lycke et al. (1999) demonstrated the importance of SIgA in protection from oral cholera toxin challenge. The potential susceptibility of these animals to respiratory infections has not been investigated to date.

11.2.2. pIgR$^{-/-}$ Mice

Polymeric IgR$^{-/-}$ mice lack the receptor necessary for transcytosis of pIg across the mucosal epithelium and, in rodents and rabbits (but not humans), hepatobiliary transport of serum pIgA (Johansen et al., 1999; Shimada et al., 1999) (see Chapter 3). Consequently, these mice completely lack SIg and concomitantly contain 100-fold greater levels of serum IgA compared to pIgR$^{+/+}$ mice. Lack of SIg in these mice does not affect development of oral tolerance following antigen feeding, nor does it influence development of cytotoxic T-cell responses after intranasal immunization (Uren et al., 2003). Furthermore, as would be predicted, systemic immunization is unaffected by the lack of pIgR, with the exception of increased serum IgA levels (Uren et al., 2005). Although these animals have reduced, but detectable, levels of IgA in the small intestine and fecal extracts, salivary IgA levels are unchanged. This latter finding indicates that pIgR function is not completely required for external IgA transport, in keeping with the results in J-chain$^{-/-}$ mice by Hendrickson et al. (1995). Nevertheless, albumin levels in saliva and IgG levels in saliva, small intestine, and feces are increased in pIgR$^{-/-}$ mice, suggesting increased mucosal permeability that presumably results from increased irritation of the mucosal epithelium by commensal organisms and ingested antigen, which, in turn, is likely caused by a lack of protective SIgA (Johansen et al., 1999) (see Chapter 10). Recent studies have implicated pIgR-mediated transport of IgA in protection against respiratory infections (see Sects, 11.4 and 11.5).

11.2.3. IgA$^{-/-}$ Mice

The development of IgA-deficient mice (IgA$^{-/-}$) has provided an extremely useful model for the further characterization of the biological activities of IgA. Disruption of the Iα exon, the α-switch region and 5′ α-heavy-chain genes cause these mice to be completely incapable of switching to and expressing IgA (Harriman et al., 1999). This deficiency was also found to cause increased levels of IgM, IgG1, and IgG2b in both the serum and gut, possibly in order to compensate for the lack of IgA. Interestingly, IgG3 and IgE levels are markedly reduced in IgA$^{-/-}$ mice, although it has not been determined if this is the result of decreased isotype switching or reduced Ig production from postswitched B-cells (Arnaboldi et al., 2005; Harriman et al., 1999). It is unlikely that disruption of the Cα gene locus would affect transcription of both Cε and Cγ_3 gene regions, because whereas Cε is in close proximity to the Cα gene region, Cγ3 is far upstream. Thus, an explanation for this decrease has yet to be found. T-cells from IgA$^{-/-}$ mice also show aberrant activity, a finding that might be related to the apparent influence of IgA on B-cell function and homeostasis, a topic that will be discussed in detail below (see Sect. 11.6).

11.3. The Role of IgA in Protection Against Viral Respiratory Infections

Intranasal infection of mice with influenza virus is a common model to examine the contributions of various immune components to respiratory tract protection. It is generally accepted that antibodies are pivotal for protection against lethal virus infection, whereas T-cells are required for recovery, although antibodies are likely involved in viral clearance as well (Huber et al., 2001; Nguyen et al., 2001a). It has been shown by Renegar and Small (1991b) that intravenous injection of pIgA anti-influenza virus hemagglutinin antibody results in transport of the antibody into nasal secretions and protection against homotypic virus challenge. IgG antibody was found not to be protective, although this might have been due to failure of the injected IgG to reach the appropriate mucosal tissues. It was further shown by these investigators that in convalescent mice, which had recovered from a sublethal infection and then were rechallenged, intranasal inoculation of anti-IgA abrogated protection induced by the priming virus, but anti-IgG or anti-IgM had no effect (Renegar and Small, 1991a). Protection in these studies was measured by quantitating levels of virus in nasal washes. This was followed by studies showing that nasal IgA depletion through chemically defined parenteral nutrition similarly led to loss of protection in immune mice, as determined by nasal virus shedding even though serum IgG antibody levels were unaffected as were numbers of IgG-secreting plasma cells in the spleens and nasal cavities of the treated mice (Renegar et al., 2001).

The advent of $IgA^{-/-}$ mice allowed the role of IgA in influenza immunity to be readdressed (Arulanandam et al., 2001; Benton et al., 2001; Harriman et al., 1999; Mbawuike et al., 1999). Naive $IgA^{-/-}$ mice are clearly not more susceptible to virus infection than wild-type animals (Arulanandam et al., 2001; Mbawuike et al., 1999), but in immunized animals the situation appears to be more complicated. Initial studies by Mbawuike et al. (1999) found that after intranasal immunization with influenza subunit vaccine plus cholera toxin as adjuvant, $IgA^{+/+}$ and $IgA^{-/-}$ mice were equally protected against lethal virus challenge. Arulanandam et al. (2001) reported identical results when mice were immunized intranasally with subunit vaccine in the presence of interleukin (IL)-12 as adjuvant. An important distinction was observed, however, when immunization was performed in the absence of adjuvant. In this case, 75% of vaccinated $IgA^{+/+}$ mice survived challenge, whereas only 13% of $IgA^{-/-}$ mice survived (Arulanandam et al., 2001). Thus, it appears that the presence of adjuvant can overcome the lack of protective IgA expression. In apparent contrast to this observation is the finding that $IgA^{-/-}$ mice are fully protected from lethal virus challenge after recovery from a sublethal dose (Benton et al., 2001). This latter protection was seen not only in relation to survival but also in viral titers in nasal washes. Curiously, $IgA^{-/-}$ mice display altered T-cell function, especially in relation to phytohemagglutinin (PHA) responsiveness (Arulanandam et al., 2001; Mbawuike et al., 1999) and development of Th1 function (Zhang *et al.*, 2002b) (see Chapter 10). These mice also have defects in expression of other Ig isotypes, including IgG3 and IgE (Arnaboldi et al., 2005; Mbawuike et al., 1999). Such defects are likely related to altered B-cell antigen-presentation function in these mice, a topic that will be discussed in Sect. 11.6).

Using $pIgR^{-/-}$ mice immunized with inactivated influenza virus, it was found that protection against live virus challenge restricted to the upper respiratory tract was highly dependent on SIgA expression (Asahi et al., 2002). Protection in the nasal cavity could not be substituted by serum IgG as might be the case with infection in the lung. Similarly, Renegar et al. (2004) found that only pIgA, but not IgG, could prevent virally induced pathology in the nasal cavity, although IgG did neutralize newly replicated virus after infection had been initiated and was able to prevent pathology in the lung.

Heterosubtypic immunity to influenza virus is of immense interest for potential vaccination approaches because such immunity, if successfully induced, would allow protection against a variety of type A viruses regardless of the virus subtype used for vaccination. It is believed that subtypic immunity would best be conferred by using internal core proteins that are highly conserved among different isolates and expressed only in infected cells. As such, CD8 T-cells recognizing such internal proteins expressed by infected cells would likely be critical for subtypic protection (Nguyen et al., 1999). Indeed, after sublethal virus infection, Epstein and colleagues found that $IgA^{-/-}$ mice as well as mice lacking all Ig could control replication of heterosubtypic virus in the lungs, but depletion of CD4 or CD8 T-cells abrogated this protection (Benton et al., 2001). However, others have found that heterosubtypic immunity is dependent on B-cells. Nguyen et al. (2001b) reported that $Ig^{-/-}$ or $CD4^{-/-}$ mice primed with a sublethal dose of virus did not survive a lethal heterosubtypic or homotypic challenge, but $CD8^{-/-}$ mice did survive. Tumpey et al. (2001) found similar results using subunit vaccination followed by heterosubtypic challenge and measurement of viral lung titers. The reason for these differing conclusions regarding the role of B-cells in heterosubtypic protection is unknown but might relate to the types of B-cell-deficient mice used for the studies; Benton et al. (2001) employed specific $IgA^{-/-}$ mice and $J_H^{-/-}$ mice, whereas Nguyen et al. (1999, 2001a, 2001b) and Tumpey et al. (2001) employed *Igh-6*$^{-/-}$ (μMT) mice. The μMT strain is known to have defects not only in B-cell expression but also in dendritic antigen-presenting cell (APC) function (Moulin et al., 2000). Nevertheless, immunized $pIgR^{-/-}$ mice showed a decreased ability to clear influenza virus from the upper respiratory tract following subsequent challenge with homologous or heterologous influenza viruses (Asahi et al., 2002; Asahi-Ozaki et al., 2004). Thus, mucosal IgA does appear to play an important function in controlling heterosubtypic influenza virus infection.

The potential role of IgA in protection against other respiratory viruses such as Sendai virus (Mazanec et al., 1987) and respiratory syncytial virus

(Weltzin et al., *1994*, 1996) has been examined by passive transfer of antibody, and it has been found that both IgA and IgG are equally capable of mediating protection. Mice specifically disrupted in IgA expression or transport, however, have yet to be examined for immunity to these pathogens.

11.4. The Role of IgA in Protection Against Bacterial Respiratory Infections

The most common bacterial infections in the respiratory tract and the middle ear are caused by the encapsulated, pyogenic bacteria including *Streptococcus pneumoniae, Hemophilus influenzae*, and *Moraxella catarrhalis* (Murphy and Sethi, 1992). A mouse model of *S. pneumoniae* infection has been used extensively to determine the role of IgA in protection and it has been demonstrated that such protection can be transferred to mice with human IgA antibody (Steinitz et al., 1986). Furthermore, it has been found that human pIgA can mediate *S. pneumoniae* killing through complement receptors on phagocytes (Janoff et al., 1999).

A comparison of respiratory protection in wild-type and $IgA^{-/-}$ mice against *Shigella flexneri* was performed by Way and colleagues (1999). An attenuated form of this pathogen that is defective in oxidase expression is 100-fold less active in mediating lethal disease but can induce mucosal immunity against the fully virulent bacterial strain. However, it was found that protection was not dependent on IgA expression. A similar result was published by Murthy et al. (2004), who examined pulmonary protection against *Chlamydia trachomatis*. Infection of naïve, unimmunized mice was studied and it was found that $IgA^{+/+}$ and $IgA^{-/-}$ mice contained equivalent amounts of pulmonary bacteria 10 days after intranasal infection. In both of the above studies, protection correlated with histological inflammation in the peribronchiolar areas of the lung and, in fact, $IgA^{-/-}$ mice demonstrated more extensive inflammatory changes in response to *C. trachomatis* infection than wild-type mice. $IgA^{-/-}$ animals also expressed significantly higher levels of serum IgG antibodies, suggesting that these antibodies played an important role in protection.

In contrast to the above studies, other groups have reported a pivotal role for mucosal IgA in protection from bacterial lung infection. Lynch and colleagues (2003) found that intranasal vaccination of $IgA^{+/+}$ mice with pneumococcal polysaccharide conjugated to diphtheria toxoid induced protective immunity in adults against subsequent nasal carriage with *S. pneumoniae* type 14. The same vaccination regimen can protect neonatal mice against otitis media (Sabirov and Metzger, 2006). Such protection was not observed in $IgA^{-/-}$ mice. Dependence on mucosal IgA for protection against *S. pneumoniae* carriage was also seen using $pIgR^{-/-}$ mice (Sun et al., 2004). Lack of protection in immunized $IgA^{-/-}$ animals was directly correlated with the absence of antibody in nasal secretions. Furthermore, protection could be observed in $IgA^{-/-}$ mice if the bacteria were first opsonized with serum antibody before intranasal challenge. It is important to note that the type 14 pneumococcal strain used in these studies does not induce inflammation in the respiratory tract nor does it cause systemic infection when instilled intranasally (Sun *et al.*, 2004). On the other hand, protection against intranasal challenge with *S. pneumoniae* type 3, a strain that does induce significant inflammation and systemic infection, does not require SIgA antibody (Sun et al., 2004). A potential caveat to the interpretation of these results is that the secretory component (the cleaved extracellular domain of pIgR) can inhibit adherence of some strains of *S. pneumoniae* to respiratory epithelium by direct binding to proteins on the bacterial cell surface, thus providing "innate" protection that is unrelated to IgA antibody specificity (see Chapters 3 and 8).

The influence of IgA in protection against *Mycobacterium bovis* BCG has also been examined using IgA$^{-/-}$ and pIgR$^{-/-}$ mice (Rodriguez et al., 2005; Tjarnlund et al., 2006). Mice were intranasally immunized with mycobacterium surface antigen PstS-1 using cholera toxin as adjuvant and intranasally challenged 2 weeks later with BCG. After immunization, no significant differences were observed between wild-type and IgA$^{-/-}$ or pIgR$^{-/-}$ mice with regard to levels of IgM or IgG antibodies in serum, saliva, or BALF (except for the absence of IgA in IgA$^{-/-}$ mice and increased serum IgA levels in pIgR$^{-/-}$ mice, as expected). Numbers of lung cells secreting tumor necrosis factor (TNF)-α and interferon (IFN)-γ were much lower in IgA$^{-/-}$ or pIgR$^{-/-}$ mice compared to wild-type controls after immunization, a finding that was distinct from the observations of Murthy et al. (2004), who quantified IFN-γ and TGF-β lung mRNA levels after *C. trachomatis* infection. Following challenge with BCG, unimmunized IgA$^{+/+}$ and IgA$^{-/-}$ mice showed essentially no differences in lung or BALF bacterial load. However, the immunized wild-type mice were able to control infection, whereas IgA$^{-/-}$ or pIgR$^{-/-}$ mice were severely compromised in this regard. Again, the IgA$^{-/-}$ and pIgR$^{-/-}$ mice produced significantly less IFN-γ and TNF-α in the lungs after infection compared to wild-type mice. Although it might be considered surprising that IgA is important for protection against an intracellular bacterium such as BCG, the results are in agreement with passive transfer studies by another group demonstrating protection against this pathogen by IgA monoclonal antibody (Williams et al., 2004). Interestingly, Reljic et al. (2006) recently reported that the protective efficacy of passively transferred IgA antibody against *M. tuberculosis* is significantly increased by coinoculation of IFN-γ.

11.5. What Can We Conclude About the Importance of IgA in Protection Against Respiratory Pathogens?

From the above discussion regarding the use of IgA-deficient animals, it is clear that different laboratories have obtained disparate results and made contrasting conclusions regarding the importance of IgA in protection of

TABLE 11.1. Protection against respiratory influenza virus infection in IgA-deficient mice.

Mouse model	Intranasal immunization	Associated inflammation?	IgA required for protection?	Comments	References
$IgA^{-/-}$	Subunit vaccine + cholera toxin	Yes	No	Cholera toxin likely induced lung inflammation	Mbawuike et al. (1999a)
$IgA^{-/-}$	Subunit vaccine alone	No	Yes		Arulanandam et al. (2001)
$IgA^{-/-}$	Subunit vaccine + IL-12	Yes	No		Arulanandam et al. (2001); unpublished observations
IgA deficient (parenteral nutrition)	Live virus to awake mice	No	Yes	Virus used for immunization likely restricted to upper respiratory tract	Renegar et al. (2001)
$J_H^{-/-}$	Live virus	Yes	No	Heterosubtypic immunity	Benton et al. (2001)
$IgA^{-/-}$	Live virus	Yes	No	Heterosubtypic immunity	Benton et al. (2001)
$pIgR^{-/-}$	Subunit vaccine + cholera toxin	Yes	Yes	Heterosubtypic immunity Measured nasal virus titers only	Asahi et al. (2002, 2004)

TABLE 11.2. Protection against respiratory bacterial infections in IgA-deficient mice.

Mouse model	Bacterial infection	Intranasal immunization	Associated inflammation?	IgA required for protection?	References
$IgA^{-/-}$	*Shigella flexneri* 2a cydC	Attenuated bacteria	Yes	No	Way et al. (1999)
$IgA^{-/-}$	*Chlamydia trachomatis*	None	Yes	No	Murthy et al. (2004)
$IgA^{-/-}$	*Streptococcus pneumoniae* type 14	Polysaccharide conjugate vaccine + IL-12	No	Yes	Lynch et al. (2003)
$pIgR^{-/-}$	*Streptococcus pneumoniae* type 14	Polysaccharide conjugate vaccine + IL-12	No	Yes	Sun et al. (2004)
$pIgR^{-/-}$	*Streptococcus pneumoniae* type 3	Polysaccharide conjugate vaccine + IL-12	Yes	No	Sun et al. (2004)
$IgA^{-/-}$	*Mycobacterium tuberculosis* BCG	Subunit vaccine + cholera toxin	?	Yes	Rodriguez et al. (2005)
$pIgR^{-/-}$	*Mycobacterium tuberculosis* BCG	Subunit vaccine + cholera toxin	?	Yes	Tjarnlund et al. (2006)

the respiratory tract from infectious agents. How can this conundrum be resolved?

The salient features of various studies analyzing protection against influenza virus in IgA-deficient mice are summarized in Table 11.1, and those with bacterial infection models are summarized in Table 11.2. Many of these models have included induction of inflammatory states, either through use of cholera toxin or live pathogen priming for immunization (sublethal doses of influenza virus) or during actual challenge (Shigella, Chlamydia, *S. pneumoniae* type 3). In general, it is the presence of inflammation that appears to determine whether IgA is necessary for any observed protection; that is, in the presence of inflammation, IgA-deficient mice are protected from infection, and in the absence of inflammation, IgA-deficient mice are not protected. This could best be explained by the fact that inflammation leads to (1) increased dendritic cell/lymphocyte activation and (2) damage to the mucosal epithelial barrier, increased blood vessel permeability, and enhanced transudation of IgG antibodies from the bloodstream. Thus, inflammation could enhance the efficacy of vaccination as well as protection upon challenge. Another variable involves the site of pathogen challenge. In the case of the lung, which has a high level of blood vessel penetration and thus IgG transudation, inflammation would tend to obscure the requirement for IgA antibody. However, in the upper respiratory tract, in which few blood vessels are present and little IgG is transudated, a requirement for IgA would become more apparent. Thus, in the studies by Asahi and colleagues (2002; Asahi-Ozaki et al., 2004), which included immunization in the presence of cholera toxin, a treatment that would be expected to induce significant inflammation, IgA was still found to be critical for protection because only virus titers in the upper respiratory tract were measured. However, in studies in which virus titers in the lung and/or survival were measured, it is likely that an amount of inflammation adequate to allow IgG transudation will be sufficient for protection. In the experiments utilizing BCG subunit immunization (Rodriguez et al., 2005; Tjarnlund et al., 2006), although inflammation was likely induced during vaccination due to the presence of cholera toxin, later bacterial challenge led to a more chronic infection in which significant numbers of bacteria were only observed 4 weeks after initial infection. This latter case contrasts with respiratory pathogens such as type 3 pneumococci, *S. flexneri*, or *C. trachomatis*, in which significant lung inflammation is typically observed within days after challenge (Murthy et al., 2004; Way et al., 1999).

From the above discussion, it appears likely that the experimental model used and the amount of inflammation induced influences the apparent need for protective IgA. The ability of IgG to substitute for IgA likely relates at least partially, to the site of infection; that is, IgA is required for protection of the upper respiratory tract, whereas both IgA and IgG can be involved in protecting the lungs. Nevertheless, the human body expends a considerable amount of energy in producing 3 g of IgA per day (see Chapter 2). It has been suggested that IgA is perfectly suited for protecting mucosal surfaces in a noninflammatory manner brcause it doe not activate complement nor induce inflammatory reactions, and the studies in mice would tend to confirm this concept. Clearly, it would be preferable to contain infections in the lung (and other mucosal sites) before painful and potentially serious inflammation develops. However, if the endothelial and epithelial barriers are breached, IgG will transudate from the serum and serve as a backup system to prevent potential blood-borne infection. What about IgA immunodeficient humans who generally do not show any significant effects of their immunodeficiency? (See Chapter 13 for a detailed discussion of IgA deficiency in humans.) First, it must be recognized that clinical IgA immunodeficiency is defined by the presence of <50 μg/mL of serum IgA. Because we do not understand the basis for this immunodeficiency in humans but do know that α H-chains are not disrupted, it is possible that IgA-deficient individuals do contain low levels of IgA that are sufficient to protect mucosal surfaces. In addition, considering the potential for compensation by SIgM and transudated serum IgG, these individuals might have subtle defects in protection that are not recognized. In fact, it is the subset of patients with defects in both IgA and IgG expression that fare the worst clinically, and IgA-deficient mice tend to resemble this subset in showing an associated defect in isotype switching to Ig isotypes other than IgA.

11.6. IgA Expression and Mucosal B-Cell Homeostasis

Whereas it appears that IgA in the respiratory tract is important for protection against infections, its role during lung inflammation, such as that observed in asthma, has been less clear and only recently investigated in detail. Although it is widely believed that IgA can mediate protection without inducing inflammation, some reports have demonstrated a negative correlation between IgA levels and allergic sensitization (Burks and Steele, 1986; Ostergaard and Eriksen, 1979). Allergen-specific IgA can be isolated from the BALF of asthmatic patients (Nahm *et al.*, 1998) and is increased during periods of high allergen exposure (Reed et al., 1991). IgA can mediate degranulation of human eosinophils, the major cell type present in the inflammatory infiltrate of asthmatics (Abu-Ghazaleh *et al.*, 1989). Levels of CD89 (FcαR1) on eosinophils are increased in allergic patients, and IgA levels correlate with the levels of eosinophil products in BALF from these patients (Monteiro et al., 1993; Nahm and Park, 1997). Recently, the role of IgA in asthma was investigated using IgA$^{-/-}$ mice and a murine model of allergic lung inflammation (Arnaboldi et al., 2005). Induction of allergic lung inflammation in mice involves intraperitoneal sensitization with ovalbumin emulsified in alum, followed 3 weeks later by daily intranasal challenge with soluble ovalbumin. This regimen induces Th2 activation, lung eosinophil recruitment, and inflammatory changes characteristic of the early stages of human asthma, although the tissue remodeling typically seen in later stages of human disease is not mimicked in most mouse strains. Curiously, upon immunization and challenge of IgA$^{-/-}$ mice, there were significantly reduced levels of total and IgG1 ovalbumin-specific antibodies and decreased IL-4 and IL-5 in BALF compared to IgA$^{+/+}$ controls (Arnaboldi et al., 2005). The IgA$^{-/-}$ mice also had reduced pulmonary inflammation with fewer inflammatory cells in lung tissue and BALF. Nonspecific inflammation induced by bleomycin was not affected by the lack of IgA, although the mechanism responsible for this form of inflammation is likely different from allergic lung inflammation and primarily involves infiltration of neutrophils rather than eosinophils. A comparison of pIgR$^{-/-}$ and pIgR$^{+/+}$ mice demonstrated no differences in levels of inflammation, suggesting that IgA bound to secretory component is not necessary for ovalbumin-induced lung inflammation, although a role for transudated IgA in lung secretions due to "mucosal leakiness" in pIgR$^{-/-}$ mice could not be ruled out.

Further studies in IgA-deficient mice demonstrated an important role for IgA expression in B-cell homeostasis and apparent trafficking to mucosal tissues, including the lung. First, although pIgR$^{-/-}$ mice are unable to produce SIgA due to the defect in Ig transport, they do express highly elevated levels of serum IgA. Uren et al. (2003) found that these mice had threefold greater numbers of IgA^{+} B220^{-} plasmablasts in the gut lamina propria, a mucosal effector site, compared to pIgR$^{+/+}$ mice. Numbers of IgA^{+} cells in the Peyer's patches, a mucosal inductive site, were within normal range. In the converse situation, a deficiency of serum IgA was found to cause dramatically reduced B-cell numbers in mucosal effector tissues (Arnaboldi et al., 2005). Analysis of pulmonary leukocyte populations in IgA$^{+/+}$ and IgA$^{-/-}$ mice revealed significantly fewer B-cells in the lungs of IgA$^{-/-}$ mice. This difference was observed in the presence or absence of allergic lung inflammation. The decrease in B-cells was not restricted to IgA^{+} cells; there was also a complete absence of IgG-staining cells in lung sections from IgA$^{-/-}$ mice following the induction of allergic lung inflammation, a condition that generates substantial numbers of IgG^{+} cells in the lungs of IgA$^{+/+}$ mice.

What are the implications of altered B-cell homeostasis in IgA-deficient mice and why do IgA$^{-/-}$ mice show the absence of ovalbumin-induced allergic lung inflammation? The answer might lie in the known functions of B-cells other than antibody secretion, particularly antigen presentation. Indeed, it has been reported that protection against infectious diseases of the mucosa, including the lung, can be dependent on B-cells, but not Ig (Lund et al., 2003; Maaser et al., 2004). In addition, whereas transfer of whole splenocyte populations from ovalbumin-primed mice to naïve mice results in profound inflammatory responses after challenge of the recipients with allergen, removal of antigen-specific B-cells from the transferred population, either by isolating CD4^{+} T-cells or by depleting CD19^{+} B-cells, results in loss of responsiveness. It has also been found that both B-cells and dendritic cells are required for optimal CD4 T-cell activation, with B-cells primarily responsible for expanding the stimulated T-cell population following initial activation by dendritic cells (Kurt-Jones *et al.*, 1988; Constant, 1999; Linton *et al.*, 2000; Bradley *et al.*, 2002).

In the lung, antigen presentation by B cells is likely to be of paramount importance since alveolar macrophages are generally poor antigen presenting cells (Kradin et al., 1987; Leemans et al., 2005; Lipscomb et al., 1986; Lyons et al., 1986; Toews et al., 1984). $IgA^{-/-}$ mice, containing few B-cells in the lung, would thus show a defect in antigen presentation. This would explain why intranasal immunization of $IgA^{-/-}$ mice with influenza subunit vaccine plus adjuvant (to activate dendritic APC/lymphocyte function) induces protective immunity, whereas vaccination of these mice in the absence of an adjuvant results in a loss of protection (Arulanandam et al., 2001). $IgA^{-/-}$ mice show a defect in Th1 activity (Zhang et al., 2002a) as well as defective switching to IgG3 and IgE (Arnaboldi et al., 2005; Harriman et al., 1999). B-cell-deficient μMT mice as well as $IgA^{-/-}$ mice have alterations in the cytokine microenvironment compared to wild-type mice (Arnaboldi et al., 2005; Gonnella et al., 2001), including T-cell-derived cytokines. All of these findings might relate to defective B-cell expression in the lungs of $IgA^{-/-}$ mice and, consequently, loss of adequate antigen presentation and immune responsiveness. In fact, it has been found that $IgA^{-/-}$ mice lack adequate antigen presentation function for Th cell priming (Arulanandam et al., 2001).

11.7. Concluding Remarks

In this chapter, an attempt has been made to provide an overview of recent studies that have exploited murine models of IgA immunodeficiency to examine the role of IgA in respiratory immune function. In addition, an effort has been made to present a unifying concept to explain apparently conflicting results obtained from the various laboratories using these animal models. In essence, it is likely that IgA antibody provides an important first line of defense against infections of the respiratory tract. If these infections progress to a point such that significant inflammation occurs, especially in the lung, epithelial and endothelial barriers will become compromised and transudation of IgG antibody will occur and this IgG will provide a second line of defense to prevent systemic, lethal spread of the infection. Thus, the major role of respiratory IgA will be to provide protection against both morbidity and mortality, whereas IgG provides a backup mechanism to ensure survival of the host. This model predicts that IgA will be found to be necessary in experiments in which only upper respiratory tract pathogen titers are measured or protection is observed in the absence of inflammation, but IgG will appear to be sufficient for survival under highly inflammatory (i.e., lethal conditions). Simultaneously, IgA can influence B-cell expression at mucosal effector sites because overexpression of serum IgA in $pIgR^{-/-}$ mice results in increased B-cell expression in these tissues and absence of IgA causes a lack of B-cell expression. Considering the apparent importance of B-cells for antigen presentation in the lung, their absence in IgA$^{-/-}$ mice prevents adequate pulmonary T-cell priming, which might be the reason for associated T-cell deficiencies observed in these mice. The mechanisms responsible for IgA's influence on B-cell expression and function in mucosal tissues and the relevance of the findings made in murine IgA immunodeficiency models to our understanding of human disease remain to be determined.

Acknowledgments. The work in the author's laboratory is supported by NIH grants RO1 AI041715 and PO1 AI056320 and by Philip Morris USA Inc. and Philip Morris International.

References

Abu-Ghazaleh, R. I., Fujisawa, T., Mestecky, J., Kyle, R. A., and Gleich, G. J. (1989). IgA-induced eosinophil degranulation. *J. Immunol.* 142:2393–2400.

Arnaboldi, P. M., Behr, M. J., and Metzger, D. W. (2005). Mucosal B cell deficiency in IgA$^{-/-}$ mice abrogates the development of allergic lung inflammation. *J. Immunol.* 175:1276–1285.

Arulanandam, B. P., Raeder, R. H., Nedrud, J. G., Bucher, D. J., Le, J., and Metzger, D. W. (2001). IgA immunodeficiency leads to inadequate Th cell priming and increased susceptibility to influenza virus infection. *J. Immunol.* 166:226–231.

Asahi, Y., Yoshikawa, T., Watanabe, I., Iwasaki, T., Hasegawa, H., Sato, Y., Shimada, S., Nanno, M., Matsuoka, Y., Ohwaki, M., Iwakura, Y., Suzuki, Y., Aizawa, C., Sata, T., Kurata, T., and Tamura, S. (2002). Protection against influenza virus infection in polymeric Ig receptor knockout mice immunized intranasally with adjuvant-combined vaccines. *J. Immunol.* 168:2930–2938.

Asahi-Ozaki, Y., Yoshikawa, T., Iwakura, Y., Suzuki, Y., Tamura, S., Kurata, T., and Sata, T. (2004). Secretory IgA antibodies provide cross-protection against infection with different strains of influenza B virus. *J. Med. Virol.* 74:328–335.

Ballow, M. (2002). Primary immunodeficiency disorders: Antibody deficiency. *J. Allergy Clin. Immunol.* 109: 581–591.

Benton, K. A., Misplon, J. A., Lo, C. Y., Brutkiewicz, R. R., Prasad, S. A., and Epstein, S. L. (2001). Heterosubtypic immunity to influenza A virus in mice lacking IgA, all Ig, NKT cells, or gamma delta T cells. *J. Immunol.* 166:7437–7445.

Bienenstock, J. (2005). Mucosal and other mechanisms of resistance in the respiratory tract: An overview. In: Mestecky, J., Lamm, M. E., McGhee, J. R., Bienenstock, J., Mayer, L., and Strober, W. (eds.), *Mucosal Immunology,* 3rd. ed. Elsevier Academic Press, Boston, pp. 1401–1402.

Bradley, L. M., Harbertson, J., Biederman, E., Zhang, Y., Bradley, S. M., and Linton, P. J. (2002). Availability of antigen-presenting cells can determine the extent of CD4 effector expansion and priming for secretion of Th2 cytokines *in vivo. Eur. J. Immunol.* 32:2338–2346.

Brandtzaeg, P., Karlsson, G., Hansson, G., Petruson, B., Bjorkander, J., and Hanson, L. A. (1987a). Immunohistochemical study of nasal mucosa in patients with selective IgA deficiency. *Int. Arch. Allergy Appl. Immunol.* 82:483–484.

Brandtzaeg, P., Karlsson, G., Hansson, G., Petruson, B., Bjorkander, J., and Hanson, L. A. (1987b). The clinical condition of IgA-deficient patients is related to the proportion of IgD- and IgM-producing cells in their nasal mucosa. *Clin. Exp. Immunol.* 67:626–636.

Burks, A. W., and Steele, R. W. (1986). Selective IgA deficiency. *Ann. Allergy* 57:3–13.

Burrows, P. D., and Cooper, M. D. (1997). IgA deficiency. *Adv. Immunol.* 65:245–276.

Chvatchko, Y., Kosco-Vilbois, M. H., Herren, S., Lefort, J., and Bonnefoy, J. Y. (1996). Germinal center formation and local immunoglobulin E (IgE) production in the lung after an airway antigenic challenge. *J. Exp. Med.* 184:2353–2360.

Constant, S. L. (1999). B lymphocytes as antigen-presenting cells for CD4+ T cell priming *in vivo*. *J. Immunol.* 162:5695–5703.

Delventhal, S., Hensel, A., Petzoldt, K., and Pabst, R. (1992). Effects of microbial stimulation on the number, size and activity of bronchus-associated lymphoid tissue (BALT) structures in the pig. *Int. J. Exp. Pathol.* 73:351–357.

De Togni, P., Goellner, J., Ruddle, N. H., Streeter, P. R., Fick, A., Mariathasan, S., Smith, S. C., Carlson, R., Shornick, L. P., Strauss-Schoenberger, J., Russel, J. H., Karr, R, and Chaplin, D. D. (1994). Abnormal development of peripheral lymphoid organs in mice deficient in lymphotoxin. *Science* 264:703–707.

Erlandsson, L., Andersson, K., Sigvardsson, M., Lycke, N., and Leanderson, T. (1998). Mice with an inactivated joining chain locus have perturbed IgM secretion. *Eur. J. Immunol.* 28:2355–2365.

Fagarasan, S., Kinoshita, K., Muramatsu, M., Ikuta, K., and Honjo, T. (2001). In situ class switching and differentiation to IgA-producing cells in the gut lamina propria. *Nature* 413:639–643.

Fagarasan, S., Muramatsu, M., Suzuki, K., Nagaoka, H., Hiai, H., and Honjo, T. (2002). Critical roles of activation-induced cytidine deaminase in the homeostasis of gut flora. *Science* 298:1424–1427.

Gardby, E., Wrammert, J., Schon, K., Ekman, L., Leanderson, T., and Lycke, N. (2003). Strong differential regulation of serum and mucosal IgA responses as revealed in CD28-deficient mice using cholera toxin adjuvant. *J. Immunol.* 170:55–63.

Gonnella, P. A., Waldner, H. P., and Weiner, H. L. (2001). B cell-deficient (μMT) mice have alterations in the cytokine microenvironment of the gut-associated lymphoid tissue (GALT) and a defect in the low dose mechanism of oral tolerance. *J. Immunol.* 166:4456–4464.

Harriman, G. R., Bogue, M., Rogers, P., Finegold, M., Pacheco, S., Bradley, A., Zhang, Y., and Mbawuike, I. N. (1999). Targeted deletion of the IgA constant region in mice leads to IgA deficiency with alterations in expression of other Ig isotypes. *J. Immunol.* 162:2521–2529.

Hendrickson, B. A., Conner, D. A., Ladd, D. J., Kendall, D., Casanova, J. E., Corthesy, B., Max, E. E., Neutra, M. R., Seidman, C. E., and Seidman, J. G. (1995). Altered hepatic transport of immunoglobulin A in mice lacking the J chain. *J. Exp. Med.* 182:1905–1911.

Hendrickson, B. A., Rindisbacher, L., Corthesy, B., Kendall, D., Waltz, D. A., Neutra, M. R., and Seidman, J. G. (1996). Lack of association of secretory component with IgA in J chain-deficient mice. *J. Immunol.* 157:750–754.

Hooper, L. V. (2004). Bacterial contributions to mammalian gut development. *Trends Microbiol.* 12:129–134.

Huber, V. C., Lynch, J. M., Bucher, D. J., Le, J., and Metzger, D. W. (2001). Fc receptor-mediated phagocytosis makes a significant contribution to clearance of influenza virus infections. *J. Immunol.* 166:7381–7388.

Husband, A. J., and Gowans, J. L. (1978). The origin and antigen-dependent distribution of IgA-containing cells in the intestine. *J. Exp. Med.* 148:1146–1160.

Janoff, E. N., Fasching, C., Orenstein, J. M., Rubins, J. B., Opstad, N. L., and Dalmasso, A. P. (1999). Killing of *Streptococcus pneumoniae* by capsular polysaccharide-specific polymeric IgA, complement, and phagocytes. *J. Clin. Invest.* 104:1139–1147.

Johansen, F. E., Braathen, R., and Brandtzaeg, P. (2000). Role of J chain in secretory immunoglobulin formation. *Scand. J. Immunol.* 52:240–248.

Johansen, F. E., Pekna, M., Norderhaug, I. N., Haneberg, B., Hietala, M. A., Krajci, P., Betsholtz, C., and Brandtzaeg, P. (1999). Absence of epithelial immunoglobulin A transport, with increased mucosal leakiness, in polymeric immunoglobulin receptor/secretory component-deficient mice. *J. Exp. Med.* 190:915–921.

Jones, P. D., and Ada, G. L. (1986). Influenza virus-specific antibody-secreting cells in the murine lung during primary influenza virus infection. *J. Virol.* 60:614–619.

Jones, P. D., and Ada, G. L. (1987). Influenza-specific antibody-secreting cells and B cell memory in the murine lung after immunization with wild-type, cold-adapted variant and inactivated influenza viruses. *Vaccine* 5:244–248.

Kang, H. S., Chin, R. K., Wang, Y., Yu, P., Wang, J., Newell, K. A., and Fu, Y. X. (2002). Signaling via LTbetaR on the lamina propria stromal cells of the gut is required for IgA production. *Nat. Immunol.* 3:576–582.

Kradin, R. L., McCarthy, K. M., Dailey, C. I., Burdeshaw, A., Kurnick, J. T., and Schneeberger, E. E. (1987). The poor accessory cell function of macrophages in the rat may reflect their inability to form clusters with T cells. *Clin. Immunol. Immunopathol.* 44:348–363.

Kunkel, E. J., and Butcher, E. C. (2003). Plasma-cell homing. *Nat. Rev. Immunol.* 3:822–829.

Kunkel, E. J., Kim, C. H., Lazarus, N. H., Vierra, M. A., Soler, D., Bowman, E. P., and Butcher, E. C. (2003). CCR10 expression is a common feature of circulating and mucosal epithelial tissue IgA Ab-secreting cells. *J. Clin. Invest.* 111: 1001–1010.

Kurt-Jones, E. A., Liano, D., HayGlass, K. A., Benacerraf, B., Sy, M. S., and Abbas, A. K. (1988). The role of antigen-presenting B cells in T cell priming *in vivo*. Studies of B cell-deficient mice. *J. Immunol.* 140:3773–3778.

Leemans, J. C., Thepen, T., Weijer, S., Florquin, S., van, R. N., van de Winkel, J. G., and van der, P. T. (2005). Macrophages play a dual role during pulmonary tuberculosis in mice. *J. Infect. Dis.* 191:65–74.

Linton, P. J., Harbertson, J., and Bradley, L. M. (2000). A critical role for B cells in the development of memory CD4 cells. *J. Immunol.* 165: 5558–5565.

Lipscomb, M. F., Lyons, C. R., Nunez, G., Ball, E. J., Stastny, P., Vial, W., Lem, V., Weissler, J., and Miller, L. M. (1986). Human alveolar macrophages: HLA-DR-positive macrophages that are poor stimulators of a primary mixed leukocyte reaction. *J. Immunol.* 136:497–504.

Lund, F. E., Schuer, K., Hollifield, M., Randall, T. D., and Garvy, B. A. (2003). Clearance of *Pneumocystis carinii* in mice is dependent on B cells but not on P carinii-specific antibody. *J. Immunol.* 171:1423–1430.

Lycke, N., Erlandsson, L., Ekman, L., Schon, K., and Leanderson, T. (1999). Lack of J chain inhibits the transport of gut IgA and abrogates the development of intestinal antitoxic protection. *J. Immunol.* 163:913–919.

Lynch, J. M., Briles, D. E., and Metzger, D. W. (2003). Increased protection against pneumococcal disease by mucosal administration of conjugate vaccine plus interleukin-12. *Infect. Immun.* 71: 4780–4788.

Lyons, C. R., Ball, E. J., Toews, G. B., Weissler, J. C., Stastny, P., and Lipscomb, M. F. (1986). Inability of human alveolar macrophages to stimulate resting T cells correlates with decreased antigen-specific T cell-macrophage binding. *J. Immunol.* 137:1173–1180.

Maaser, C., Housley, M. P., Iimura, M., Smith, J. R., Vallance, B. A., Finlay, B. B., Schreiber, J. R., Varki, N. M., Kagnoff, M. F., and Eckmann, L. (2004). Clearance of *Citrobacter rodentium* requires B cells but not secretory immunoglobulin A (IgA) or IgM antibodies. *Infect. Immun.* 72:3315–3324.

Mazanec, M. B., Nedrud, J. G., and Lamm, M. E. (1987). Immunoglobulin A monoclonal antibodies protect against Sendai virus. *J. Virol.* 61:2624–2626.

Mazmanian, S. K., Liu, C. H., Tzianabos, A. O., and Kasper, D. L. (2005). An immunomodulatory molecule of symbiotic bacteria directs maturation of the host immune system. *Cell* 122:107–118.

Mbawuike, I. N., Pacheco, S., Acuna, C. L., Switzer, K. C., Zhang, Y., and Harriman, G. R. (1999). Mucosal immunity to influenza without IgA: An IgA knockout mouse model. *J. Immunol.* 162:2530–2537.

McDermott, M. R., and Bienenstock, J. (1979). Evidence for a common mucosal immunologic system. I. Migration of B immunoblasts into intestinal, respiratory, and genital tissues. *J. Immunol.* 122:1892–1898.

Monteiro, R. C., Hostoffer, R. W., Cooper, M. D., Bonner, J. R., Gartland, G. L., and Kubagawa, H. (1993). Definition of immunoglobulin A receptors on eosinophils and their enhanced expression in allergic individuals. *J. Clin. Invest.* 92:1681–1685.

Moulin, V., Andris, F., Thielemans, K., Maliszewski, C., Urbain, J., and Moser, M. (2000). B lymphocytes regulate dendritic cell (DC) function *in vivo*: increased interleukin 12 production by DCs from B cell-deficient mice results in T helper cell type 1 deviation. *J. Exp. Med.* 192:475–482.

Moyron-Quiroz, J. E., Rangel-Moreno, J., Kusser, K., Hartson, L., Sprague, F., Goodrich, S., Woodland, D. L., Lund, F. E., and Randall, T. D. (2004). Role of inducible bronchus associated lymphoid tissue (iBALT) in respiratory immunity. *Nat. Med.* 10:927–934.

Murphy, T. F., and Sethi, S. (1992). Bacterial infection in chronic obstructive pulmonary disease. *Am. Rev. Respir. Dis.* 146:1067–1083.

Murthy, A. K., Sharma, J., Coalson, J. J., Zhong, G., and Arulanandam, B. P. (2004). Chlamydia trachomatis pulmonary infection induces greater inflammatory pathology in immunoglobulin A deficient mice. *Cell. Immunol.* 230:56–64.

Nahm, D. H., Kim, H. Y., and Park, H. S. (1998). Elevation of specific immunoglobulin A antibodies to both allergen and bacterial antigen in induced sputum from asthmatics. *Eur. Respir. J.* 12:540–545.

Nahm, D. H., and Park, H. S. (1997). Correlation between IgA antibody and eosinophil cationic protein levels in induced sputum from asthmatic patients. *Clin. Exp. Allergy* 27:676-–681.

Nguyen, H. H., Moldoveanu, Z., Novak, M. J., Van Ginkel, F. W., Ban, E., Kiyono, H., McGhee, J. R., and Mestecky, J. (1999). Heterosubtypic immunity to lethal influenza A virus infection is associated with virus-specific CD8(+) cytotoxic T lymphocyte responses induced in mucosa-associated tissues. *Virology* 254:50–60.

Nguyen, H. H., Van Ginkel, F. W., Vu, H. L., McGhee, J. R., and Mestecky, J. (2001a). Heterosubtypic immunity to influenza A virus infection requires B cells but not CD8+ cytotoxic T lymphocytes. *J. Infect. Dis.* 183:368–376.

Nguyen, H. H., Van Ginkel, F. W., Vu, H. L., McGhee, J. R., and Mestecky, J. (2001b). Heterosubtypic immunity to influenza A virus infection requires B cells but not CD8+ cytotoxic T lymphocytes. *J. Infect. Dis.* 183:368–376.

Ostergaard, P. A., and Eriksen, J. (1979). Association between HLA-A1,B8 in children with extrinsic asthma and IgA deficiency. *Eur. J. Pediatr.* 131:263–270.

Pabst, R. (1992). Is BALT a major component of the human lung immune system? *Immunol. Today* 13:119–122.

Pan, J., Kunkel, E. J., Gosslar, U., Lazarus, N., Langdon, P., Broadwell, K., Vierra, M. A., Genovese, M. C., Butcher, E. C., and Soler, D. (2000). A novel chemokine ligand for CCR10 and CCR3 expressed by epithelial cells in mucosal tissues. *J. Immunol.* 165:2943–2949.

Reed, C. E., Bubak, M., Dunnette, S., Blomgren, J., Pfenning, M., Wentzmurtha, P., Wallen, N., Keating, M., and Gleich, G. J. (1991). Ragweed-specific IgA in nasal lavage fluid of ragweed-sensitive allergic rhinitis patients: Increase during the pollen season. *Int. Arch. Allergy Appl. Immunol.* 94:275–277.

Reljic, R., Clark, S.O., Williams, A. Falero-Diaz, G., Singh, M., Challacombe, S., Marsh, P.D., and Ivanyi, J. (2006). Intranasal IFNγ extends passive IgA antibody protection of mice against *Mycobacterium tuberculosis* lung infection. *Clin. Exp. Immunol.* 143:467–473.

Renegar, K. B., Johnson, C. D., Dewitt, R. C., King, B. K., Li, J., Fukatsu, K., and Kudsk, K. A. (2001). Impairment of mucosal immunity by total parenteral nutrition: requirement for IgA in murine nasotracheal anti-influenza immunity. *J. Immunol.* 166:819–825.

Renegar, K. B., and Small, P. A. (1991a). Passive transfer of local immunity to influenza virus infection by IgA antibody. *J. Immunol.* 146:–1978.

Renegar, K. B., and Small, P. A. (1991b). Immunoglobulin A mediation of murine nasal anti-influenza virus immunity. *J. Virol.* 65:2146–2148.

Renegar, K. B., Small, P. A., Boykins, L. G., and Wright, P. F. (2004). Role of IgA versus IgG in the control of influenza viral infection in the murine respiratory tract. *J. Immunol.* 173:1978–1986.

Rhee, K. J., Sethupathi, P., Driks, A., Lanning, D. K., and Knight, K. L. (2004). Role of commensal bacteria in development of gut-associated lymphoid tissues and preimmune antibody repertoire. *J. Immunol.* 172:1118–1124.

Rodriguez, A., Tjarnlund, A., Ivanji, J., Singh, M., Garcia, I., Williams, A., Marsh, P. D., Troye-Blomberg, M., and Fernandez, C. (2005). Role of IgA in the defense against respiratory infections IgA deficient mice exhibited increased susceptibility to intranasal infection with Mycobacterium bovis BCG. *Vaccine* 23:2565–2572.

Rudzik, R., Clancy, R. L., Perey, D. Y., Day, R. P., and Bienenstock, J. (1975). Repopulation with IgA-containing cells of bronchial and intestinal lamina propria after transfer of homologous Peyer's patch and bronchial lymphocytes. *J. Immunol.* 114:1599–1604.

Sabirov, A., and Metzger, D. W. (2006). Intranasal vaccination of neonatal mice with polysaccharide conjugate vaccine for protection against pneumococcal otitis media. *Vaccine* 24:5584–5592.

Salvi, S., and Holgate, S. T. (1999). Could the airway epithelium play an important role in mucosal immunoglobulin A production? *Clin. Exp. Allergy* 29:1597–1605.

Shimada, S., Kawaguchi-Miyashita, M., Kushiro, A., Sato, T., Nanno, M., Sako, T., Matsuoka, Y., Sudo, K., Tagawa, Y., Iwakura, Y., and Ohwaki, M. (1999). Generation of polymeric immunoglobulin receptor-deficient mouse with marked reduction of secretory IgA. *J. Immunol.* 163:5367–5373.

Shimoda, M., Nakamura, T., Takahashi, Y., Asanuma, H., Tamura, S., Kurata, T., Mizuochi, T., Azuma, N., Kanno, C., and Takemori, T. (2001). Isotype-specific selection of high affinity memory B cells in nasal-associated lymphoid tissue. *J. Exp. Med.* 194:1597–1607.

Steinitz, M., Tamir, S., Ferne, M., and Goldfarb, A. (1986). A protective human monoclonal IgA antibody produced *in vitro*: anti-pneumococcal antibody engendered by Epstein-Barr virus-immortalized cell line. *Eur. J. Immunol.* 16:187–193.

Stepankova, R., Kovaru, F., and Kruml, J. (1980). Lymphatic tissue of the intestinal tract of germfree and conventional rabbits. *Folia Microbiol. (Praha)* 25:491–495.

Sun, K., Johansen, F. E., Eckmann, L., and Metzger, D. W. (2004). An important role for polymeric Ig receptor-mediated transport of IgA in protection against *Streptococcus pneumoniae* nasopharyngeal carriage. *J. Immunol.* 173:4576–4581.

Tjarnlund, A., Rodriguez, A., Cardoona, P. J., Guirado, E., Ivanyi, J., Singh, M., Troye-Blomberg, M., and Fernandez, C. (2006). Polymeric IgR knockout mice are more susceptible to mycobacterial infections in the respiratory tract than wild-type mice. *Int. Immunol.* 5:807–816.

Toews, G. B., Vial, W. C., Dunn, M. M., Guzzetta, P., Nunez, G., Stastny, P., and Lipscomb, M. F. (1984). The accessory cell function of human alveolar macrophages in specific T cell proliferation. *J. Immunol.* 132:181–186.

Tschernig, T., and Pabst, R. (2000). Bronchus-associated lymphoid tissue (BALT) is not present in the normal adult lung but in different diseases. *Pathobiology* 68:1–8.

Tumpey, T. M., Renshaw, M., Clements, J. D., and Katz, J. M. (2001). Mucosal delivery of inactivated influenza vaccine induces B-cell-dependent heterosubtypic cross-protection against lethal influenza A H5N1 virus infection. *J. Virol.* 75: 5141–5150.

Uren, T. K., Johansen, F. E., Wijburg, O. L., Koentgen, F., Brandtzaeg, P., and Strugnell, R. A. (2003). Role of the polymeric Ig receptor in mucosal B cell homeostasis. *J. Immunol.* 170:2531–2539.

Uren, T. K., Wijburg, O. L., Simmons, C., Johansen, F. E., Brandtzaeg, P., and Strugnell, R. A. (2005). Vaccine-induced protection against gastrointestinal bacterial infections in the absence of secretory antibodies. *Eur. J. Immunol.* 35:180–188.

Way, S. S., Borczuk, A. C., and Goldberg, M. B. (1999). Adaptive immune response to *Shigella flexneri* 2a cydC in immunocompetent mice and mice lacking immunoglobulin A. *Infect. Immun.* 67:2001–2004.

Weisz-Carrington, P., Roux, M. E., McWilliams, M., Phillips-Quagliata, J. M., and Lamm, M. E. (1979). Organ and isotype distribution of plasma cells producing specific antibody after oral immunization: Evidence for a generalized secretory immune system. *J. Immunol.* 123:1705–1708.

Weltzin, R., Hsu, S. A., Mittler, E. S., Georgakopoulos, K., and Monath, T. P. (1994). Intranasal monoclonal immunoglobulin A against respiratory syncytial virus protects against upper and lower respiratory tract infections in mice. *Antimicrob. Agents Chemother.* 38:2785–2791.

Weltzin, R., Traina-Dorge, V., Soike, K., Zhang, J. Y., Mack, P., Soman, G., Drabik, G., and Monath, T. P. (1996). Intranasal monoclonal IgA antibody to respiratory syncytial virus protects rhesus monkeys against upper and lower respiratory tract infection. *J. Infect. Dis.* 174:256–261.

Williams, A., Reljic, R., Naylor, I., Clark, S. O., Falero-Diaz, G., Singh, M., Challacombe, S., Marsh, P. D., and Ivanyi, J. (2004). Passive protection with immunoglobulin A antibodies against tuberculous early infection of the lungs. *Immunology* 111:328–333.

Zhang, Y., Pacheco, S., Acuna, C. L., Switzer, K. C., Wang, Y., Gilmore, X., Harriman, G. R., and Mbawuike, I. N. (2002a). Immunoglobulin A-deficient mice exhibit altered T helper 1-type immune responses but retain mucosal immunity to influenza virus. *Immunology* 105:286–294.

Zhang, Y. X., Pacheco, S., Acuna, C. L., Switzer, K. C., Wang, Y., Gilmore, X., Harriman, G. R., and Mbawuike, I. N. (2002b). Immunoglobulin A-deficient mice exhibit altered T helper 1-type immune responses but retain mucosal immunity to influenza virus. *Immunology* 105:286–294.

Zuercher, A. W., Coffin, S. E., Thurnheer, M. C., Fundova, P., and Cebra, J. J. (2002). Nasal-associated lymphoid tissue is a mucosal inductive site for virus-specific humoral and cellular immune responses. *J. Immunol.* 168:1796–1803.

12 IgA and Reproductive Tract Immunity

Charu Kaushic[1] and Charles R. Wira[2]

12.1. Introduction

Mucosal epithelia line the interior surfaces of the body and account for roughly 75% of the human body's interface with the environment. The gastrointestinal, respiratory, urinary, and reproductive tracts are all components of the mucosal system. Although each of these tracts serves unique physiological functions, they are also part of a distinct network comprising the common mucosal immune system (Mestecky, 2005a). The combination of physiological functions and external environment drive the immune system at each mucosal surface to make unique and distinct adaptations. For example, the repertoire of commensal organisms typically present on a mucosal surface and pathogenic organisms capable of invading that mucosal barrier are quite

[1] McMaster University, Department of Pathology and Molecular Medicine, Hamilton, Ontario, Canada.
[2] Dartmouth Medical School, Department of Physiology, Lebanon, NH-03756, USA.

distinct for each mucosal surface, and this drives each mucosal surface to develop unique features (Sansonetti, 2004; Wira et al., 2005b). Similarly, the physiological function of each mucosal surface is quite distinct and that requires additional adjustment of immune responses (Mestecky et al., 2005a; Pitman and Blumberg, 2000).

Among the mucosal surfaces, the female reproductive tract is unique in the diverse, even contrary, nature of immunological demands it has evolved to meet. Physiologically, it is required to provide a supportive immunologic environment for allogeneic spermatozoa and to facilitate the growth of a semiallogeneic fetus. At the same time, it is required to provide protection against microorganisms, including large numbers of pathogenic bacteria and viruses that can be transmitted during sexual intercourse. In response to these requirements, reproductive tracts of different species have evolved to protect the genital mucosa whenever it is likely to encounter "danger" (reviewed in Brandtzaeg, 1997). For example, in species such as mouse and rat, immune cells and antibodies are present in the uterus predominantly at estrus, the receptive phase of the cycle in females when pathogenic organisms might be introduced along with seminal plasma (Head and Gaede, 1986; Kaushic et al., 1998; Parr and Parr, 1985, 1991). In contrast, in humans, sexually transmitted organisms can be introduced into the female genital tract at any time in the reproductive cycle. Consequently, organized lymphoid structures are present in the endometrium of women throughout the menstrual cycle, although the number and type of immune cell present in these structures vary (Bjercke and Brandtzaeg, 1993; Kamat and Isaacson, 1987; Laguens et al., 1990; Pudney et al., 2005; Yeaman et al., 1997, 2001).

Humoral responses mediated by antibodies are key components of immune protection on mucosal surfaces (Russell and Mestecky, 2002; Woof and Mestecky, 2005). IgA and, to a lesser extent, IgG are recognized as the predominant antibody isotypes in mucosal secretions (Mestecky et al., 2005b; Parr and Parr, 1996; Woof and Mestecky, 2005). Most of this immunoglobulin A (IgA), including that present on reproductive tract, is considered "natural" in that these are polyreactive antibodies to common surface antigens of microorganisms that colonize the mucosal surface, allowing for binding and neutralizing of a broad range of pathogens (MacPherson et al., 2000; Ochsenbein et al., 1999; Ochsenbein and Zinkernagel, 2000). Antigen-specific humoral responses in the gastrointestinal and respiratory tracts are also predominantly mediated through IgA (Brandtzaeg, 2003; Mestecky et al., 2005a). However, the reproductive tract might be unique in that the antigen-specific humoral response is constituted of both IgG and IgA antibodies (Brandtzaeg, 1997; Mestecky et al., 2005c; Parr and Parr, 1996; Russell and Mestecky, 2002). Further, this response is under the regulation of reproductive hormones in the female, allowing for exquisite control required for procreation and pathogen protection (Wira et al., 2005a). Information regarding the humoral response in the male reproductive tract is sketchy at best (reviewed in Anderson and Pudney, 2005). This chapter will review current knowledge regarding the presence and origin of IgA and its function in male and female reproductive tracts, especially as it pertains to protection against genital tract pathogens, and highlight some unique aspects of IgA in the reproductive tract.

12.2. IgA in the Female Reproductive Tract

A number of studies have examined the type and amount of immunoglobulins present in the reproductive tract of females from different species (reviewed in Parr and Parr, 1996). The absolute amount of IgG and IgA in cervical mucus and vaginal secretions has been reported to be quite variable (Bjercke and Brandtzaeg, 1993; Kutteh and Mestecky, 1994; Rebello et al., 1975; Schumacher et al., 1977; Tourville et al., 1970). Sampling methods and changes in Ig levels at different times in the reproductive cycle could account for much of this variability. Despite this, there seems to be consensus that, unlike other mucosal surfaces such as small intestines and lungs, where IgA is the predominant antibody isotype found in the secretions, IgG is found in significant quantities in the female reproductive tract (Mestecky *et al.*, 2005c; Parr and Parr, 1997). Reports vary on whether IgG is the dominant isotype in genital secretions or if both IgG and IgA are present in comparable amounts (Parr and Parr, 1996; Waldmann et al., 1971). Most studies have reported that the level of IgG is twofold to sixfold higher than that of IgA in cervical secretions.

There is also a distinct difference in the subtype of IgA found in the female genital tract compared to other mucosal surfaces. Whereas IgA1 is the predominant subtype found in secretions of small intestines, saliva, and lungs, IgA2 dominates in rectal, large intestine, and cervicovaginal secretions (Crago et al., 1984; Kett et al., 1986; Russell and Mestecky, 2002). One of the functional differences in subtypes of IgA is the type of antigen they recognize, with IgA1 recognizing mainly protein antigens and IgA2 recognizing mainly carbohydrates (Russell et al., 1992). It is interesting to note that the three surfaces predominated by IgA2 are colonized by large numbers of commensal bacteria. These surfaces are also more likely to come into contact with pathogenic organisms than are the upper and more sterile regions of the intestinal and reproductive tracts (small intestine and uterus, respectively). IgA2 is also resistant to proteases secreted by pathogenic bacteria that can easily cleave IgA1 (Senior et al., 2000). Thus, IgA subclass bias might be related to the type of antigen recognition that is useful in that location.

The presence of large amounts of IgA in the genital secretions raises the issue of the source of this antibody. Opinions vary regarding the relative contributions of locally synthesized versus serum IgA. Over the last three decades, a number of studies have examined the distribution of cells producing IgA, IgG, and other antibody isotypes as well as the expression of the J-chain and polymeric immunoglobulin receptor (pIgR) in different parts of the female reproductive tract (Bjercke and Brandtzaeg, 1993; Kutteh et al., 1988; Rebello et al., 1975; Tourville et al., 1970; Vaerman and Ferin, 1975). These studies found that ovarian and uterine tissue had barely detectable numbers of Ig-producing cells. Endocervix and ectocervix had a greater preponderance of Ig-producing cells, most of them secreting IgA (reviewed in Parr and Parr, 1996). The majority of these cells were J-chain-positive, indicating that IgA was secreted in its polymeric form. The vaginal tissue and fallopian tubes were also found to contain Ig-producing cells, although fewer in number than in the cervix. The transport of IgA into mucosal secretions is dependent on the ability of epithelial cells to produce pIgR, which transports pIgA (and pIgM) from the basolateral to the apical surface and releases secretory IgA (SIgA) (or SIgM) into the lumen (see Chapter 3). SIgA comprises pIgA bound to the secretory component (SC), the extracellular domain of pIgR. Immunohistochemical studies have utilized antibodies to SC as a marker of pIgR expression. SC was detected primarily in the columnar epithelial cells lining the fallopian tubes, uterus, and endocervix. The pseudostratified squamous epithelium lining the ectocervix and vaginal lumen in humans was reported to be devoid of SC. This difference in the type of epithelium found at different sites might account for the different molecular forms of IgA (monomeric, polymeric, and secretory IgA) found in different locations of the reproductive tract (Kutteh and Mestecky, 1994; Waldmann et al., 1971). Secretions from the vaginal mucosa, which is lined by pseudostratified epithelium that does not produce pIgR, contain large amounts of monomeric IgA (45%) that likely crosses the mucosa by passive transudation. On the other hand, cervical secretions contain as much as 80% of IgA in polymeric form, which is transported locally by pIgR-expressing columnar epithelial cells lining the endocervix.

The cervix therefore seems to be the most immunologically active tissue in the female reproductive tract with respect to IgA levels in secretions, numbers of IgA plasma cells, and abundant pIgR expression in epithelial cells (Schumacher, 1988; Vaerman and Ferin, 1975). The immunological function of the cervix is significant, given its role as a transitional tissue between the more exposed environment of the vaginal mucosa and the relatively protected environment of the upper reproductive tract (uterus and fallopian tubes).

A number of studies have compared the presence of IgA in genital secretions of various animal species (reviewed in Parr and Parr, 1996). There are distinct species differences in expression of pIgR, presence of IgA plasma cells, and relative ratios of IgG to IgA in genital secretions. In the mouse, IgA plasma cells are abundant in the uterus but not in the cervix or vagina. SC was detected in both the upper and lower reproductive tracts (Canning and Billington, 1983; McDermott et al., 1980; Parr and Parr, 1985; Rachman et al., 1983). In contrast, in the rat, plasma cells have not been detected in the reproductive tract. Serum IgA might transudate directly into secretions and/or be transported by pIgR-expressing epithelial cells (Parr and Parr, 1989; Sullivan et al., 1983; Sullivan and Wira, 1981). IgA plasma cells and pIgR/SC have also been examined in female genital tracts of monkeys, horses, cows, and pigs. Whereas species differences are present, the significance of these dissimilarities are not clear, although they might relate to physiological differences in the reproductive processes. For example, in rats as well as humans, semen is deposited in the vagina, but in mice, it is deposited directly into the uterus. This might explain the presence of abundant plasma cells found in the uterus of the mouse, but not in humans or rats.

A unique aspect of IgA in the female reproductive tract is its regulation of by reproductive hormones (see Sect. 12.5).

12.3. IgA in the Male Reproductive Tract

Relatively few studies have examined the levels of Igs and the number and type of plasma cell in the male reproductive tract. Although human seminal plasma is reported to contain as much as 40 mg/mL of protein, the bulk of it comes from secretions of the prostate, seminal vesicles, and related glands (Alexander and Anderson, 1987Anderson and Pudney, 2005; Russell and Mestecky, 2002). IgG, IgA, and IgM have been detected in human seminal plasma and preejaculate; however, there is considerable variation in the relative levels reported in different studies. Although preejaculate has been shown to contain more IgA than IgG, there is no clear consensus regarding the relative levels of IgG and IgA in seminal plasma (Mestecky and Fultz, 1999). As is the case with sample collection from most mucosal surfaces, variability in measurements is likely due to collection procedures, difficulty in obtaining samples free from degrading enzymes, and assay variability. IgA plasma cells have been observed in the prostate and the penile urethra, and pIgA has been detected in the epithelial cells of testis, epididymis, vas deferens, seminal vesicles, prostate, and penile urethra, suggesting pIgR-mediated transport. (Anderson and Pudney, 2005).

The penile urethra appears to be the most immunologically active site in the male reproductive tract. Immunohistochemical studies demonstrated the presence of IgA plasma cells in the lamina propria of the urethra and in the glands of Littre, and they showed that the penile epithelium expressed abundant pIgR. (Pudney and Anderson, 1995). Located in the penile urethra, the glands of Littre secrete the highly viscous component of preejaculatory fluid. IgA and IgM plasma cells were also found to contain the J-chain, indicating their ability to form pIgs. Information on the subtypes and molecular forms of IgA in the male reproductive tract is limited. One study found that the distribution of IgA subclasses in seminal plasma resembles that of serum, with a predominance of IgA1, unlike the enrichment of IgA2 in female reproductive tract secretions (Moldoveanu et al., 2005). All three molecular forms of IgA were found in the seminal plasma: monomeric IgA, pIgA, and SIgA. The penile foreskin is devoid of any IgA plasma cells and does not express pIgR.

12.4. Origin of IgA Plasma Cells in the Male and Female Reproductive Tracts

In the late 1970s, the concept of a common mucosal immune system was proposed (McDermott and Bienenstock 1979; McDermott et al., 1980), and subsequent studies have confirmed the existence of a network of immune cells, including but not limited to B- and T-cells, that circulate systemically but home preferentially into mucosal tissues (Brandtzaeg and Johansen, 2005; Kunkel and Butcher, 2003) (see Chapter 2). Earlier studies had shown that Peyer's patches in the intestines are a source of IgA precursor cells that, when transferred, populate the recipient's intestines and other mucosal tissues (Craig and Cebra, 1971; McDermott and Bienenstock, 1979). However, there is increasing evidence of additional preferential homing between various mucosal tissues (Brandtzaeg et al., 1999). Detailed analysis of homing receptors on T- and B-cells in the circulation indicate that the majority of mucosal lymphocytes homing to intestinal and respiratory mucosa express $\alpha4\beta7$ integrin that binds to the mucosal addressin MADCAM-1 (Brandtzaeg and Johansen, 2005). However, these particular homing receptors and mucosal addressins do not appear to be expressed by lymphocytes homing into the reproductive tract or by the genital tissues (Johansson et al., 1999; Mestecky et al., 2005c). The precise homing mechanism for the reproductive tract is not clear. Several studies have implicated $\alpha e\beta7$ (CD103) and/or $\alpha4\beta1$ or $\alpha L\beta2$ (LFA-1) as homing receptors for lymphocytes targeted to the reproductive tract (Johansson et al., 1999; Kaul et al., 2003). These receptors recognize VCAM-1 and ICAM-1, respectively, which have been shown to be present in genital tract tissues. Locally produced chemokines might also contribute to mucosal homing. The chemokine receptor CCR10 has been shown to be expressed by all mucosal lymphocytes, including IgA-secreting plasmablasts (Lazarus et al., 2003). Whether genital tract tissues express the corresponding mucosal chemokine CCL28 remains to be determined.

Although some studies have clearly demonstrated the presence of IgA plasma cells in the human cervix, it is still unclear whether all or only part of IgA in the genital secretions is of local origin. IgA plasma cells have been localized in the female reproductive tract of some species, including humans and mice, but not in others, such as rats (Parr and Parr, 1996) (see Sect. 12.2). IgA plasma cells have been identified in the oviduct, endocervix, and vaginal tissue of women, but not in the endometrium (Kelly and Fox, 1979). Because SIgA is found in uterine secretions, it is possible that serum-derived IgA is transported across pIgR-expressing endometrial cells. Similarly in rats, the absence of plasma cells in the uterus suggests that IgA and IgG in rat uterine secretions are derived at least in part from the serum immunoglobulins, consistent with *in vivo* transcytosis studies using intravenously administered radiolabeled IgA (Sullivan and Wira, 1984).

12.5. Regulation of IgA in the Reproductive Tract by Sex Hormones

The IgA and IgG levels in uterine secretions vary markedly in intact animals during the reproductive cycle, with higher levels of Igs measured at the time of ovulation than at any other stage of the cycle (Wira and Merritt, 1977).

These observations led to the conclusion that estradiol and progesterone are the principal hormones responsible for regulating IgA and IgG in uterine secretions. When ovariectomized rats were treated with estradiol, IgA and IgG levels in uterine secretions were elevated relative to saline-treated controls (Wira and Sandoe, 1980). IgA and IgG levels in cervicovaginal secretions are also hormonally controlled, but, in contrast to the uterus, levels are lowered in response to hormone treatment (Wira and Sullivan, 1985). Progesterone was found to have an inhibitory effect, blocking estradiol-stimulated increases in uterine IgA and IgG secretion and reducing levels of IgA and IgG in cervicovaginal secretions in the presence or absence of estradiol (Wira and Sandoe, 1980; Wira and Sullivan, 1985). Uterine ligation, which prevents leakage of secretions from the uterus into the vagina, had no effect on the response of IgA and IgG to estradiol treatment, demonstrating that uterine and vaginal responses are separate and distinct.

Because IgA is transported into secretions at mucosal surfaces by pIgR, studies were undertaken to determine whether hormones regulate the level of SC (the cleaved extracellular domain of pIgR) in secretions of the reproductive tract. When estradiol was given to ovariectomized rats for 3 days, SC levels increased in uterine secretions and decreased in cervicovaginal secretions (Wira and Sullivan, 1985). Subsequent studies showed that whereas IgA of blood origin enters uterine tissues within 2–4 h after estradiol treatment, IgA transport from the tissues into the lumen requires uterine epithelial cell production of pIgR (Sullivan and Wira, 1983; 1984).

To examine the mechanisms whereby estradiol exerts its effect on uterine pIgR, organ cultures of uteri from saline- and estradiol-treated rats were incubated with actinomycin D and cycloheximide (Wira et al., 1984). SC accumulation in the incubation media was inhibited, suggesting that hormonal regulation of uterine pIgR is mediated through *de novo* mRNA and protein synthesis (Wira et al., 1984). In subsequent studies, a pIgR cDNA probe was used to measure rat uterine pIgR mRNA levels in the presence or absence of estradiol. When ovariectomized rats were treated with estradiol for 3 days, levels of pIgR mRNA increased in the uterus and decreased in the vagina (Kaushic et al., 1995, 1997). These findings indicate that estradiol regulates uterine and vaginal epithelial pIgR levels by controlling pIgR mRNA expression in the female reproductive tract. Furthermore, these studies indicate that the actions of estradiol on pIgR expression are separate and distinct and vary within specific regions of the reproductive tract.

In the human female reproductive tract, uterine intraepithelial content of IgA is also known to rise during the secretory phase of the menstrual cycle (Hurlimann et al., 1978; Kelly and Fox, 1979; Rebello et al., 1975; Tourville et al., 1970). This IgA accumulation is most likely regulated by the level of pIgR expression (Brandtzaeg, 1981). However, uterine secretion of IgA appears to peak at around the time of ovulation (Schumacher, 1980), whereas SC levels in uterine secretions remain elevated throughout the secretory phase. One possible reason for this apparent discrepancy involves the availability of uterine IgA for transfer. As pointed out earlier, in normal endometria, the numbers of IgA-containing plasma cells are low (Hurlimann et al., 1978; Kelly and Fox, 1979; Rebello et al., 1975; Tourville et al., 1970). During the time of ovulation, though, levels of stromal IgA increase (Kelly and Fox, 1979). This process appears to be due to the estrogen-induced transudation of serum IgA into the uterus (Sullivan and Wira, 1983, 1984). Thus, with increased levels of both IgA and SC at ovulation, IgA secretion would be expected to ensue. Data from studies of transport of IgA into uterine secretions of rats support this hypothesis (Sullivan and Wira 1983; Sullivan et al., 1983).

In the human, the amount of IgA and IgG in the uterus varies with the stage of the menstrual cycle (Tauber et al., 1985). IgG levels in secretions from the uterine mucosa are highest during the periovulatory phase, whereas levels in the fallopian tube are lowest at that time. In contrast, IgA and IgG levels in cervical secretions also vary with the stage of the menstrual cycle, with lowest levels measured at mid-cycle (Schumacher, 1980). Suppression of IgA and IgG in cervical secretions throughout the menstrual cycle was observed when women were treated with oral contraceptives.

To examine the influence of the menstrual cycle on pIgR in the human uterus, uterine washings were collected in the operating room from women prior to gynecological surgery. Samples were collected under sterile conditions using the Gravlee jet wash device (Gravlee, 1969; Sylvan et al., 1981). Levels of SC in human uterine secretions varied considerably with the stage of the menstrual cycle (Sullivan et al., 1984). Amounts of SC, when expressed as the percentage of total wash protein, were highest during the secretory phase, significantly reduced during the proliferative phase, and lowest during menstruation. Total levels of SC were also greatest during the secretory phase, averaging approximately twofold higher than SC content in proliferative and menstrual samples. These variations could not be accounted for by blood contamination of uterine wash samples. In other studies, Bjercke and Brandtzaeg (1993) demonstrated by two-color immunofluorescence that pIgR expression on uterine epithelial cells was low during the proliferative phase and high during the secretory phase of the cycle. Menge and Mestecky (1993) showed that the surface expression of SC is increased by estradiol treatment of freshly isolated human endometrial epithelial cells and endometrial adenocarcinoma cell lines.

Taken together, these studies demonstrate that sex hormones regulate IgA, IgG, and the IgA transporter pIgR in both the human and rodent female reproductive tract. These changes occur normally as part of the adaptive and innate systems that protect the reproductive tract from potential pathogens. These biological changes must be taken into account when evaluating the efficiency of vaccine approaches. Without taking the stage of the cycle into account, the possibility exists that false-negative values will be obtained.

12.6. Role of IgA in Protection Against Genital Tract Infections

Sexually transmitted genital tract infections are fairly common in both men and women and account for significant morbidity and mortality. Globally, 40 million new cases of sexually transmitted infections (STIs) are reported annually, leading to a significant burden on health care systems worldwide.

Reproductive tract secretions from men and women contain large amounts of Igs, the so-called "natural antibodies" most of which are directed against microorganisms, including *Escherichia coli*, *Trichomonas*, *Candida*, and herpes simplex virus (HSV) (Parr and Parr, 1996). Cervicovaginal antibodies against *Candida* are mainly SIgA (Waldmann et al., 1971). It is not clear how humoral responses to these microorganisms are induced, although it is likely that they are initiated by exposure at mucosal surfaces such as oral, intestinal, and vaginal mucosae. Also unclear is whether infection with these microorganisms is required for induction of humoral immunity or if chronic exposure is sufficient. Unlike Peyer's patches in intestines, the female and male reproductive tracts do not appear to contain organized inductive sites (Kutteh et al., 2005). More recent studies, however, have demonstrated the presence of lymphoid aggregates (LAs) in human endometrium, comprising macrophages, B-cells, and T-cells, which might serve as local induction sites (Yeaman et al., 1997, 2001). However, these LAs change in size and composition under the influence of reproductive hormones and their ability to induce functional immune responses *in vivo* remains to be determined. Fahey et al. (2006) have demonstrated that human uterine epithelial cells as well as macrophages/dendritic cells in the uterine stroma are able to process and present antigen to autologous T-cells. Using a highly purified preparation of freshly isolated epithelial cells, they found that the epithelial cells process and present tetanus toxoid recall antigen to drive autologous T-cell proliferation. These studies included epithelial cells from 11 patients, each of which was able to process and present antigen. To rule out the possibility that antigen presentation was due to a contaminating antigen-presenting cell (APC), epithelial cell preparations were analyzed for leukocyte markers; less than 2% of the cells in the epithelial cell preparations expressed a leukocyte phenotype. In experiments in which myeloid cells were added to the APC assay, more than 10-fold higher APC contamination was needed to account for the observed proliferation of T-cells. These studies suggest that uterine epithelial cells are active in antigen presentation. Other sites such as the transition zone in the cervix are also very rich in dendritic cells, T-cells, and B-cells (Pudney et al., 2005; Kaushic et al., unpublished data). Vaginal epithelium contains large numbers of Langerhans cells and subepithelial dendritic cells that have been shown to express MHC class II and to be quite efficient at antigen presentation (Parr and Parr, 1991; Zhao et al., 2003). The natural IgA antibodies found in the genital secretions have been shown to be polyreactive with broad specificity, possibly to be able to bind a large repertoire of microorganisms.

As for the functions of IgA in the reproductive tract, possibly the most important is the prevention of attachment and/or neutralization of pathogenic bacteria and viruses, before they can colonize and/or infect the genital mucosa (reviewed in Woof and Mestecky, 2005; Woof and Kerr, 2006). Prevention of attachment of microorganisms to epithelial cells provides the first line of defense on mucosal surfaces. This type of "immune exclusion" is accomplished by non-neutralizing IgA and IgG antibodies (Burns et al., 1996; Michetti et al., 1992). In addition to non-neutralizing antibodies, polyreactive antibodies with broad specificities bathe the mucosal epithelium. There is some evidence to show that pIgA is more efficient at neutralizing antigen than are monomeric IgA and IgG (Stubbe et al., 2000). One explanation could be that multivalent antibodies promote more efficient agglutination of microorganisms. Interestingly, there is also evidence that the IgA antibodies function to clear sperm and bacteria from the female reproductive tract following coitus (Parr and Parr, 1988, 1996). Studies in mice show that following mating, large numbers of bacteria are found in the uterus, possibly due to introduction of these along with semen. Many of the bacteria had IgG and IgA bound to them and were agglutinated. The uterus was cleared and returned to the normal sterile state within 2–3 days of mating, prior to the arrival of embryos for implantation on day 5.

In addition to preventing adherence, IgA has also been shown to neutralize viruses within epithelial cells (Mazanec et al., 1992, 1995). During the transport of dimeric IgA through the epithelial cell by pIgR, IgA can intercept and neutralize microorganisms such as viruses and intracellular bacteria in the vesicular compartments of epithelial cells (see Chapter 7). Acting through immune exclusion, IgA binds to microorganisms on the basolateral surface of the epithelium and transport them back across the epithelium via the pIgR transport system in an exocytosis type of mechanism. This would also provide an effective means of eliminating excess immune complexes (Mazanec et al., 1993). Renegar et al.. (2006) reported that intravenous administration of polymeric monoclonal IgA antibody protected mice against experimental influenza infection of the uterus. Anti-influenza SIgA was detected in uterine secretions, but it was not clear whether the clearance of virus involved immune exclusion and/or intracellular neutralization. Finally, there is evidence that IgA can also destroy infected cells by antibody-mediated cellular cytotoxicity (ADCC) (van Egmond et al., 2001).

Secretory IgA at mucosal surfaces has been associated with anti-inflammatory properties, because binding of SIgA to antigens does not trigger inflammatory reactions or activate complement (see Chapter 5). Monomeric IgA, which is also present in the genital tract, can induce inflammation by interacting with myeloid IgA receptor FcαRI (CD89) (Morton et al., 1996) (see Chapter 4). Human neutrophils express FcαRI, which, when engaged, induces phagocytosis, respiratory burst activity, and release of inflammatory mediators and cytokines. This IgA-induced inflammation might serve as a second line of defense under pathological conditions when neutrophils are recruited into the genital tract tissues (van Egmond et al., 2001).

12.6.1. Common STIs and IgA

Although humoral immune responses are frequently generated in humans following infection with sexually transmitted microorganisms, the role of mucosal antibodies in clearing pathogens in a natural infection and conferring protection is unclear. In animal models, there is abundant evidence that vaccination strategies resulting in induction of mucosal antibodies, especially IgA against microbial agents, provide protection against subsequent challenge (see Sect. 12.7). A detailed examination of all STIs is beyond the scope of this chapter (for a review, see Russell et al., 2005); however, in the following subsections, we will examine the role of IgA in selected examples of both viral and bacterial STIs.

12.6.1.1. HSV-2

Herpes simplex virus type 2 (HSV-2) is one of the most common sexually transmitted viruses. It is estimated that in the United States alone, 45 million adults are infected by this virus and approximately 1 in every 4 sexually active adults is seropositive for HSV-2 (Fleming et al., 1997). Once infected, individuals carry the virus in a latent state in the peripheral nervous system. From time to time the virus is reactivated and is shed into genital secretions, sometimes accompanied by genital lesions, but frequently asymptomatically. Clinical and epidemiological studies indicate that women are more susceptible to infection than men (Bolan et al., 1999). Although HSV-2 rarely leads to mortality, it remains a significant cause of morbidity and an enormous drain on health systems worldwide.

Herpes simplex virus type 2 is a cytopathic virus that primarily infects the genital epithelium, leading to ulceration and lesions and subsequent migration of the virus into peripheral nerves (Corey and Wald, 1999). Primary infection of HSV-2 leads to robust T-cell-mediated and humoral immune responses. Immune responses to primary infection protect against reinfection with HSV-2, and subsequent HSV-2 recurrences are usually shorter and less severe than primary infections (Cunningham and Mikloska, 2001). Specific antibodies have been detected in serum and genital secretions of both males and females after primary and recurrent infection. Ashley et al. (1994) showed that following primary genital infection, cervical IgM responses to HSV-2 were seen within 6–10 days, followed by specific IgG and IgA responses that persisted for weeks. Other studies have demonstrated the presence of HSV-1- and HSV-2-specific IgG and IgA antibodies in cervicovaginal secretions of seropositive African women that were clinically asymptomatic (Mbopi-Keou et al., 2003). The presence of the antibodies was inversely associated with HSV-2 DNA shedding. Neutralization activity in cervicovaginal secretions was also correlated with the presence of IgG and IgA antibodies. Whether humoral response and mucosal antibodies play a critical role in protection against HSV-2 is difficult to determine from these clinical studies.

Whereas human studies are limited to clinical observations that document serum and/or mucosal antibodies, extensive experimentation has been conducted in mouse models of genital HSV-2 infection to understand the role played by humoral immunity in natural infection and vaccination strategies (reviewed in Parr and Parr, 2003). These studies demonstrate that both cell-mediated and humoral immunity play a role in clearance of HSV-2 infection and protection against subsequent exposure. Parr and Parr (2000) showed that B-cell-deficient mice had diminished immunity to intravaginal challenge with increased viral shedding. However, in these and other studies, genital IgG appeared to be more important than IgA in mediating protection (Hendrickson et al., 2000; Parr and Parr, 2000). Passive intravaginal immunization with monoclonal or polyclonal IgG or IgA antibodies provided protection against infection.

Vaccination studies in mice have demonstrated the effectiveness of mucosal antibodies in conferring protection against genital challenge with HSV-2. Intranasal immunization, in particular, leads to long-term humoral responses that protect against intravaginal challenge (Gallichan and Rosenthal 1995; Parr and Parr 1999). Similar protection was seen following induction of local antibodies by other immunization routes (intramuscular, intravaginal) with different immunogens (attenuated HSV-2, HSV-2 DNA, recombinant adenovirus expressing HSV-2 antigens) and adjuvants (calcium phosphate nanoparticles, CpG oligonucleotides) (Gallichan et al., 1993, 2001; He et al., 2002; Kuklin et al., 1997; Morrison et al., 1998; Ashkar et al., 2003).

12.6.1.2. *Chlamydia*

Chlamydia trachomatis, an obligate Gram-negative intracellular bacterium, is one of the most common sexually transmitted bacterial infections and is a leading cause of infertility in women (Schachter and Grayston 1998). In the United States alone, 4 million new cases are reported every year. As with many other STIs, the majority of chlamydial infections are asymptomatic and disproportionately affect women (Morell, 1995). The pathologic sequelae of this infection in women can be severe, including pelvic inflammatory disease, salpingitis, ectopic pregnancy, and infertility (Morrison et al., 1992; Schachter 1989).

Human studies of *Chlamydia* infection are limited to clinical documentation of antibodies in response to infection or vaccination. Natural infection leads to protection against subsequent challenge, although this protection appears to be serovar-specific and of limited duration (Stagg, 1998). The presence of antichlamydial antibodies in endocervical secretions was found to be associated with decreased bacterial shedding (Brunham et al., 1983). It has also been reported that antichlamydial IgA in the genital tract expedites the clearance of infection in conjunction with antibiotic therapy (Cunningham, 1995). Studies of chlamydial infection in mice have generally supported the concept of a protective role for antibodies (reviewed in Patton and

Lichtenwalner, 1998). However, local IgA antibody responses failed to protect interferon (IFN)-receptor knockout mice against genital tract infection with *Chlamydia,* indicating that cellular immunity also plays a critical role in protection during natural infection (Johansson et al., 1997a). Surprisingly, B-cell-deficient mice were found to resolve primary genital infections with *Chlamydia* in a manner indistinguishable from immunocompetent mice (Johansson et al., 1997b; Su et al., 1997). In contrast to primary infection, antichlamydial antibodies were shown to play an important role in adaptive immunity to secondary challenge in mice (Morrison et al., 2000; Morrison and Morrison, 2005a), although this antibody-mediated defense did not appear to require IgA (Morrison and Morrison, 2005b).

The examples of viral and bacterial STIs discussed earlier suggest that mucosal antibody responses are involved in protecting and/or resolving natural genital tract infections. Whether induction of local antibody responses by immunization is sufficient for protection against these infections remains to be determined.

12.6.2. IgA and HIV

Twenty-five years into the human immunodeficiency virus (HIV) pandemic, it has finally been recognized that HIV is primarily a mucosal infection that is sexually transmitted (Brenchley et al., 2006; Hasse 2005; Shattock and Moore, 2003). Globally, 80–90% of new HIV infections are initiated through sexual contact with the intestinal or genital tract. In 2004, the Joint United Nations Programme on HIV/AIDS reported for the first time that women now account for almost 50% of HIV infections (UNAIDS, 2004). Unfortunately, much of the research in last two decades has been focused on HIV pathogenesis and immune responses in the blood. Hence, much needs to be learned about the immune correlates for protection at the mucosal surfaces where infection is initiated.

Data have been accumulating slowly over the last decade indicating that antibody and cytotoxic T-lymphocyte responses in the mucosa might be critical for protection against HIV (Bhardwaj and Walker, 2003; Burton 2002; Kozlowski and Neutra, 2003). Results from two groups of investigators demonstrated that resistance to HIV infection correlated with high levels of IgA against gp160 envelope protein in the genital tract. One of the groups examined the "highly exposed persistently sero-negative" (HEPS) Kenyan sex worker cohort and found that purified IgA from saliva and cervicovaginal secretions blocked infection of peripheral blood mononuclear cells by HIV-1 Clade A, B, C, and D viral isolates (Devito et al., 2000; Kaul et al., 1999). The second study, from Italy, also found gp160-specific IgA in serum, saliva, and cervicovaginal secretions of exposed, noninfected male and female partners of HIV patients (Mazzoli et al., 1997, 1999). More recently, it was shown that the antienvelope IgA responses were against a conserved gp41 neutralizing epitope that is rarely recognized by IgA in infected individuals (Clerici et al., 2002). These studies suggest that in resistant individuals, robust IgA responses against neutralizing epitopes are induced following persistent exposure to subinfectious doses of virus. These individuals might be protected against infection following subsequent exposure to high doses of HIV. This concept was supported by a study showing increased susceptibility in some of the women in the HEPS cohort following a few months break in sex work (Kaul et al., 2001). Although these studies are promising, they are not without controversy, because HIV-specific IgA antibodies were not detected in studies on other HEPS cohorts (Belec et al., 1989Buchacz et al., 2001; Dorrell et al., 2000). Whether these apparent discrepancies were due to inherent differences in cohort sampling or technical issues in detection of HIV-specific IgA remains to be determined.

Another line of support for the role of antibodies in blocking HIV transmission comes from studies of simian immunodeficiency virus (SIV) infection in macaques. Passive immunization of neonatal macaques with hyperimmune serum against SIV completely blocked oral transmission of SIVmac251, a nonpathogenic molecular clone of SIV that can replicate in primate cells (van Rompay et al., 1998). These results were subsequently confirmed in a macaque model of infection with SHIV, a hybrid SIV–HIV virus. Passive intravenous immunization with human IgG monoclonal antibodies (mAbs) against HIV envelope proteins was highly effective in preventing oral and vaginal transmission of both CXCR4- and CCR5- tropic SHIV (Mascola et al., 1999, 2000). In another study, a triple combination of anti-HIV IgG mAbs [2F5 anti-gp41 ELDKWA peptide, 2G12 anti-gp120 C3-V4, and F105 anti-gp 120 CD4 binding (CD4bs)] prevented oral transmission of SHIV to newborn macaques (Ferrantelli and Ruprecht, 2002). Other investigators have blocked vaginal SIV transmission in macaques with a combination of intravenous 2F5, 2G12, and HIV immune globulin, or human b12 anti-CD4bs IgG alone (Mascola et al., 2000). It is not clear whether these IgG antibodies prevented viral entry at the mucosal surface or eliminated the virus in the mucosa postentry through neutralization or ADCC. The latter mechanism is suggested because IgG was not detected in high amounts in the vaginal secretions of the protected macaques.

Two groups have described the generation of HIV-specific human IgA mAbs by *in vitro* class switching of the IgG mAbs described above (Liu et al., 2003; Wolbank et al., 2003). The IgA mAbs were shown to be superior to the matched IgG mAbs in *in vitro* assays of HIV neutralization, but, as yet, there have been no reports of their use for passive immunization against HIV infection *in vivo*. It will be important to compare the relative effectiveness of monomeric versus pIgA mAbs in passive immunization studies, because only the latter can be transported by pIgR across mucosal epithelia into secretions of the genital and intestinal tracts. In addition, SIgA antibodies might confer superior protection due to the innate immune activities of SC (see

Chapter 8). Antiviral SIgA has been detected in secretions from vaccinated monkeys protected against mucosal SIV and SHIV transmission, but whether these antibodies are critical in preventing transmission is unknown (Devito et al., 2004; Israel et al., 1999). Overall, the studies outlined earlier using HEPS cohorts and vaccinated primates clearly indicate that mucosal antibodies, especially anti-HIV IgA, might play an important role in a successful vaccination strategy.

In contrast to the studies in primates and HIV-resistant humans, in whom a potent IgA response might play an important role in the protection against mucosal HIV infection, a number of studies have found that in HIV-infected individuals, there might be selective defects in the production of HIV-specific IgA (Mestecky et al., 2004; Wright et al., 2002). One of these studies compared HIV-specific antibodies in mucosal samples from HIV-1-infected and uninfected individuals in six different laboratories (Wright et al., 2002). Irrespective of the assay used in different laboratories, HIV-1-specific IgA was absent in most samples. Interestingly, all but one laboratory identified HIV-1-specific IgG in rectal washes of all infected individuals, despite the fact that these secretions contained 10-fold lower levels of total IgG than IgA. There did not appear to be a problem with the assays, as all of the laboratories found comparable levels of influenza-specific IgA in nasal, salivary, and rectal secretions from HIV-1-infected and uninfected individuals. In another study, HIV-1-specific IgG and IgA antibodies were measured in external secretions, including tears, nasal, rectal, and vaginal washes, saliva, semen, urine, and sera from 50 HIV-1-infected individuals and 20 controls (Mestecky et al., 2004). Antibody levels were analyzed by enzyme-linked immunosorbent asssay and confirmed by Western blot. HIV-1-specific IgA antibodies were either absent or present at low concentrations, even in secretions with characteristically high IgA levels (saliva, tears, and rectal and nasal washes). These investigators concluded that HIV-1 does not induce vigorous specific IgA responses. A more complex picture emerged from a study of 75 HIV-1-infected individuals by Moja et al. (2000). Significant levels of anti-HIV-1 IgA were detected in parotid saliva from 57% of these individuals, but only about half of the salivary SIgA isolates were found to neutralize *in vitro* infectivity of HIV-1. Most of the IgA antibodies were specific for the gp160 envelope glycoprotein, but the neutralizing activity was not directed against the third variable loop or the ELDKWA epitope. The specificity of mucosal IgA antibodies in HIV-1-infected individuals and their effectiveness in preventing transmission remain to be determined.

The presence of high levels of HIV-1-specific IgA in HEPS cohorts, in contrast to the weak and variable production of IgA antibodies in HIV-1-infected patients, has resulted in an ongoing controversy regarding the role of IgA in protection against HIV infection. To address this apparent contradiction, it will be important to compare numbers of IgA antibody secreting cells between HIV-resistant and HIV-infected individuals. It is possible that resistance to HIV-1 infection correlates with the ability to mount robust IgA responses, whereas those who have defects in HIV-1-specific IgA responses become infected.

There is considerable evidence from *in vitro* cell culture studies that IgA antibodies specific for HIV envelope glycoproteins can block both the uptake and transcytosis of HIV through epithelial cells (Alfsen et al., 2001; Bomsel et al., 1998; Huang et al., 2005; Wright et al., 2006). Such studies have used epithelial cell lines from a variety of sources, or primary intestinal epithelial cells grown to confluence on permeable filter supports (see Chapter 7). Such culture systems provide an opportunity to distinguish between productive infection of epithelial cells and transcytosis of virus with shedding from the basolateral surface. They can also be used to evaluate intracellular neutralization of virus by IgA versus prevention of viral attachment or entry. Using such a system, a series of studies have shown that most epithelial cells do not become productively infected and amplify HIV-1, even though they can sequester and transmit virus to susceptible cells. Some epithelial cells express CCR5 and/or CXCR4, as well as the alternate HIV-1 receptor GalCer, which might be exploited by HIV for entry into the epithelium, especially in the intestinal mucosa (Fotopoulos et al., 2002; Meng et al., 2002). Whether similar or other mechanisms are used by HIV-1 for invasion of genital epithelium remains to be seen. Coincubation of HIV-1-infected cells with monoclonal Abs or colostral IgA prior to application on the apical surface of epithelial monolayers prevented infection of target cells present in the basolateral compartment (Hocini and Bomsel, 1999). Further experiments showed that these antibodies prevented gp120- or gp41-mediated attachment to the epithelial cell and thereby disrupted attachment and subsequent HIV transport through the cell. These investigators have recently reported that IgA antibodies specific for CCR5, purified from serum, saliva, or cervicovaginal secretions of HIV-1-exposed seronegative or long-term nonprogressor HIV-1-seropositive individuals, inhibited transport of HIV-1 across monolayers of human endometrial or intestinal epithelial cells. In other studies, IgM mAbs and purified dimeric IgA against HIV-1 applied to the basolateral surface of epithelial monolayers was shown to neutralize intracellular HIV (Bomsel et al., 1998; Wright et al., 2006).

The results of these cell culture studies suggest that SIgA antibodies might block entry and/or neutralize HIV-1 within the epithelium, thus preventing access of the virus to target cells in the lamina propria. It is difficult, however, to examine such mechanisms at mucosal surfaces *in vivo*. In addition, non-neutralizing antibodies (especially SIgA) could coat cell-free or cell-associated HIV-1 and promote virus entrapment in the mucus layer of the intestinal and genital tracts (see Chapter 8). Identification of the HIV epitopes recognized by both neutralizing and non-neutralizing SIgA antibodies will be important for the development of rational vaccination studies.

12.7. Induction of IgA in the Reproductive Tract by Vaccination

12.7.1. Vaccination Strategies

Although the effectiveness of local IgA (and IgG) antibodies in genital tract in natural infections is debatable, there is clear evidence that induction of antibody responses could lead to a highly effective vaccination strategy for STIs. A number of studies in animal models have shown that antibodies can be induced in the genital tract following either systemic immunization or mucosal immunization in the genital tract or at other mucosal surfaces (Mestecky and Russell, 2000; ; Parr and Parr, 2005; Russell and Mestecky, 2002). Local immunization in the genital tract typically leads to robust antibody responses when particulate, cellular, or live attenuated agents are used as immunogens. In contrast, soluble antigens elicit relatively poor responses. Immunization with live attenuated HSV-2 has been shown in a number of studies to protect against genital infection with wild-type HSV-2 (reviewed in Parr and Parr, 2003).

Earlier studies had shown that both oral and reproductive tract immunization result in the presence of specific antibodies in uterine and vaginal secretions. Ogra and Ogra (1973) demonstrated that inactivated polio virus, given either orally or deposited locally into the uterus or vagina of women, resulted in antipolio antibodies in uterine and cervicovaginal secretions. In contrast, when soluble horseradish peroxidase was placed in the uterus, no antibody response was detected (Vaerman and Férin, 1974). Several studies focusing on vaginal and cervical deposition of antigen have demonstrated local antibody production in the lower genital tract (Bell and Wolf, 1967; Kerr, 1955; Parr *et al.*, 1988; Yang and Schumacher, 1979). The observation that gastrointestinal as well as intrauterine and vaginal immunization lead to the accumulation of antibodies in genital tract secretions (Wira and Prabhala, 1993; Wira and Sandoe, 1987, 1989) shows that specific IgA and IgG antibodies in uterine and vaginal secretions are derived from both distal and local exposure to antigen. Despite this progress, much remains to be learned about the mechanisms of immune protection in the reproductive tract. Studies of the effectiveness of other mucosal routes for inducing antibodies in the reproductive tract suggested that the rectal and oral routes might have some promise. Rectal immunization of women with inactivated influenza virus resulted in induction of low levels of influenza-specific IgA in cervical secretions 28 days postimmunization (Crowley-Nowick *et al.*, 1997). Both IgA and IgG antibodies were detected in vaginal and cervical secretions 6 months postimmunization. Oral immunization with attenuated *Salmonella typhi* vaccine also led to humoral responses, particularly in saliva and vaginal washes (Kantele et al., 1998). In the same studies, rectal immunization appeared to target rectal and nasal secretions and tears. Oral immunization followed by rectal boosting appeared to be the most successful strategy for inducing antibody responses in the genital tract (Kutteh et al., 1993; Mestecky et al., 2005c).

An important development in strategies for antibody induction in the genital tract comes from the demonstration that intranasal immunization induces antibodies in the reproductive tract. Experimental models of STIs have reproducibly shown that intranasal immunization leads to long-term antibody responses that are protective against subsequent challenges (Gallichan and Rosenthal, 1995; Gallichan et al., 2001; Pal et al., 1996). Studies in monkeys and humans have also demonstrated that genital IgA and IgG responses following intranasal immunization are of higher magnitude than responses induced by other routes (Kozlowski et al., 2002; Marx et al., 1993; Russell et al., 1996).

12.7.2. Effects of Reproductive Hormones

An important aspect of understanding and designing vaccines to STIs is the consideration that both susceptibility and immune responses in females to STIs are regulated by sex hormones. Many animal models of STIs rely on altering the hormonal environment of the female genital tract to induce infection (Parr and Parr, 2003; Patton and Rank, 1992). In mouse studies of HSV-2 and *Chlamydia*, Depo-Provera, a long-acting synthetic progestin, is commonly used to facilitate infection. The stage of the menstrual cycle and use of oral contraceptives influence the susceptibility of women to infection with HSV-2, *Chlamydia,* and *Candida.* Hormonal contraception, including oral contraceptives and Depo-Provera, have been shown in various studies to be a biologic factor linked to HIV-1 acquisition (Lavreys et al., 2004; Mostad, 1998; Wang et al., 2004). The impact of hormonal environment was clearly demonstrated in studies in monkeys, in which subcutaneous progesterone implants enhanced susceptibility to vaginal transmission of SIV and conferred protection (Marx et al., 1996; Smith et al., 2000).

Using an animal model of genital *Chlamydia* infection, we found that normally resistant female rats are rendered susceptible following progesterone treatment (Kaushic et al., 1998, 2000). In mouse models of genital herpes, viral infection is influenced by the stage of the estrus cycle (reviewed in Parr and Parr, 2003). Female mice are susceptible at diestrus but not at estrus, the phase of the reproductive cycle with the highest level of circulating estrogen. More recently, a series of studies examined the influence of sex hormones on both susceptibility and immune responses to genital HSV-2 infection (Kaushic et al., 2003). Treatment of mice with the long-acting progestational formulation Depo-Provera was found to increase their susceptibility to infection 100-fold. Longer treatments of immunized mice with Depo-Provera resulted in poor HSV-2-specific mucosal immune responses, including production of IgG and IgA (Gillgrass et al., 2003). Consequently, these mice were not protected from subsequent intravaginal challenges with HSV-2.

Studies with ovariectomized mice treated with exogenous sex hormones have provided further evidence that sex hormones regulate susceptibility and immune responses to HSV-2 (Gillgrass et al., 2005a, 2005b). Estradiol-treated mice were protected from primary infection in these studies. In the absence of hormones, or in the presence of progesterone, mice were highly susceptible to genital HSV-2 infection. Interestingly, the opposite effect was observed after immunization: Progesterone-treated mice exhibited strong IgG and IgA anti-HSV2 responses and were protected against infection, whereas estradiol-treated mice were not protected. These studies indicate that vaccination strategies that seek to induce sterilizing immunity in the genital tract against STIs need to take into consideration the influence that sex hormones have on susceptibility to sexually transmitted agents, as well as their effects on mucosal immune responses.

References

Alexander, N. J., and Anderson, D. J. (1987). Immunology of semen. *Fertil. Steril.* 47:192–205.

Alfsen, A., Iniguez, P., Bouguyon, E., and Bomsel, M. (2001). Secretory IgA specific for a conserved epitope on gp41 envelope glycoprotein inhibits epithelial transcytosis of HIV-1. *J. Immunol.* 166:6257–6265.

Anderson, D. J., and Pudney, J. (2005). Human male genital tract immunity and experimental models. In: Mestecky, J., Bienenstock, J., Lamm, M., Strober, W., McGhee, J. and Mayer, L. (eds.), *Mucosal Immunology*, 3rd ed., Academic Press, San Diego, pp. 1647–1659.

Ashkar, A. A., Bauer, S. Mitchell, W. J., Vieira, J., and Rosenthal, K. L. (2003). Local delivery of CpG oligodeoxynucleotides induces rapid changes in the genital mucosa and inhibits replication, but not entry, of herpes simplex virus type 2. *J. Virol.* 77:8948–8956.

Ashley, R. L., Corey, L., Dalessio, J., Wilson, P., Remington, M., Barnum, G., and Trethewey, P. (1994). Protein-specific cervical antibody responses to primary genital herpes simplex virus type 2 infections. *J. Infect. Dis.* 170:20–26.

Belec, L., Georges, A. J., Steenman, G., and Martin, P. M. (1989). Antibodies to human immunodeficiency virus in vaginal secretions of heterosexual women..*J Infect. Dis.* 160:385–391.

Bell, E. B., and Wolf, B. (1967). Antibody synthesis *in vitro* by the rabbit vagina against diphtheria toxoid. *Nature* 214:423–424.

Bhardwaj, N., and Walker, B. D. (2003). Immunotherapy for AIDS virus infections: cautious optimism for cell-based vaccine. *Nat. Med.* 9:13–14.

Bjercke, S., and Brandtzaeg, P. (1993). Glandular distribution of immunoglobulins, J chain, secretory component and HLA-DR in the human endometrium throughout the menstrual cycle. *Hum. Reprod.* 8:1420–1425.

Bolan, G., Ehrhardt, A. A., and Wasserheit, J. N. (1999). Gender perspectives and STDs. In: Holmes, K. K., Sparling, P. F., Mardh, P. A., Stamm, W. E., Piot, P., and Wasserheit, J. N. (eds.), *Sexually Transmitted Diseases*, 3rd ed., McGraw-Hill, New York, pp. 117–127.

Bomsel, M., Heyman, M., Hocini, H., Lagaye, S., Belec, L., Dupont, C., and Desgranges, C. (1998). Intracellular neutralization of HIV transcytosis across tight epithelial barriers by anti-HIV envelope protein dIgA or IgM. *Immunity* 9:277–287.

Bomsel, M., Pastori, C., Tudor, D., Alberti, C., Garcia, S., Ferrari, D., Lazzarin, A., and Lopalco, L. (2007). Natural mucosal antibodies reactive with first extracellular loop of CCR5 inhibit HIV-1 transport across human epithelial cells. *AIDS* 21:13–22.

Brandtzaeg, P. (1981). Transport models for secretory IgA and IgM. *Clin. Exp. Immun.* 44:221–232.

Brandtzaeg, P. (1997). Mucosal immunity in the female genital tract. *J. Reprod. Immunol.* 36:23–50.

Brandtzaeg, P. (2003). Role of secretory antibodies in defence against infections. *Int. J. Med. Microbiol.* 293:3–15.

Brandtzaeg, P., Farstad, I., and Haraldsen, G. (1999). Regional specialization in the mucosal immune system: Primed cells do not always home along the same tract. *Immunol. Today* 20:267–277.

Brandtzaeg, P., and Johansen, F.-E. (2005). Mucosal B cells: Phenotypic characteristics, transcriptional regulation, and homing properties. *Immunol. Rev.* 206:32–63.

Brenchley, J., Price, D, and Douek, D. (2006). HIV disease: Fallout from a mucosal catastrophe? *Nat. Immunol.* 7:235–239.

Brunham, R. C., Kuo, C. C., Cles, L., and Holmes, K. K. (1983). Correlation of host immune response with quantitative recovery of *Chlamydia trachomatis* from the human endocervix. *Infect. Immun.* 39:1491–1494.

Buchacz, K., Parekh, B. S., Padian, N. S., van der Straten, A., Phillips, S., Jonte, J., and Holmberg, S. D. (2001). HIV-specific IgG in cervicovaginal secretions of exposed HIV-uninfected female sexual partners of HIV-infected men. *AIDS Res. Hum. Retroviruses* 17:1689–1693.

Burns, J. W., Siadat-Pajouh, M., Krishnaney, A. A., and Greenberg, H. B. (1996). Protective effect of rotavirus VP6-specific IgA monoclonal antibodies that lack neutralizing activity. *Science* 272:104–107.

Burton, D. R. (2002). Antibodies, viruses and vaccines. *Nat. Rev. Immunol.* 2:706–713.

Canning, M., and Billington, W. (1983). Hormonal regulation of immunoglobulin and plasma cells in the mouse uterus. *J. Endocrinol.* 97:419–424.

Clerici, M., Barassi, C., Devito, C., Pastori, C., Piconi, S., Trabattoni, D., Longhi, R., Hinkula, J., Broliden, K., and Lopalco, L. (2002). Serum IgA of HIV-exposed uninfected individuals inhibits HIV through recognition of a region within the α-helix of gp41. *AIDS* 16:1731–1741.

Corey, L., and Wald, A. (1999). Genital herpes. In: Holmes, K. K., Sparling, P. F., Mardh, P. A., Stamm, W. E., Piot, P., and Wasserheit, J. N. (eds.), *Sexually Transmitted Diseases*, 3rd ed., McGraw-Hill, New York, pp. 285–312.

Crago, S. S., Kutteh, W. H., Moro, I., Allansmith, M. R., Radl, J., Haaijman, J. J., and Mestecky, J. (1984). Distribution of IgA1-, IgA2-, and J chain-containing cells in human tissues. *J. Immunol.* 132:16–18.

Craig, S. W., and Cebra, J. (1971). Peyer's patches: An enriched source of precursors for IgA producing immunocytes in the rabbit. *J. Exp. Med.* 134:188–200.

Crowley-Nowick, P. A., Bell, M. C., Brockwell, R., Edwards, R. P., Chen, C., Partridge, E. E., and Mestecky, J. (1997). Rectal immunization for induction of specific antibody in the genital tract of women. *J. Clin. Immunol.* 17:370–379.

Cunningham, A. L., and Mikloska, Z. (2001). The holy grail: Immune control of human herpes simplex virus infection and disease. *Herpes* 8(Suppl. 1):6A–10A.

Cunningham, D. S. (1995). Immune response characteristics in women with chlamydial genital tract infection. *Gynecol. Obstet. Invest.* 39:54–59.

Devito, C., Hinkula, J., Kaul, R., Lopalco, L., Bwayo, J. J., Plummer, F., Clerici, M., and Broliden, K. (2000). Mucosal and plasma IgA from HIV-exposed seronegative individuals neutralize a primary HIV-1 isolate. *AIDS* 14:1917–1920.

Devito, C., Zuber, B., Schroder, U., Benthin, R., Okuda, K., Broliden, K., Wahren, B., and Hinkula, J. (2004). Intranasal HIV-1–gp160–DNA/gp41 peptide prime-boost immunization regimen in mice results in long-term HIV-1 neutralizing humoral mucosal and systemic immunity. *J. Immunol.* 173:7078–7089.

Dorrell, L., Hessell, A. J., Wang, M., Whittle, H., Sabally, S., Rowland-Jones, S., Burton, D. R., and Parren, P. W. (2000). Absence of specific mucosal antibody responses in HIV-exposed uninfected sex workers from the Gambia. *AIDS* 14:1117–1122.

Fahey, J. V., Wallace, P. K., Johnson, K., Guyre, P. M., and Wira, C. R. (2006). Antigen presentation by human uterine epithelial cells to autologous T cells. *Am. J. Reprod. Immunol.* 55:1–11.

Ferrantelli, F., and Ruprecht, R. M. (2002). Neutralizing antibodies against HIV: Back in the major leagues? *Curr. Opin. Immunol.* 14:495–502.

Fleming, D. T., McQuillan, G. M., Johnson, R. E., Nahmias, A. J., Aral, S. O., Lee, F. K., and St. Louis, M. E. (1997). Herpes simplex virus type 2 in the United States, 1976 to 1994. *N. Engl. J. Med.* 337:1105–1111.

Fotopoulos, G., Harari, A., Michetti, P., Trono, D., Pantaleo, G., and Kraehenbuhl, J. P. (2002). Transepithelial transport of HIV-1 by M cells is receptor-mediated. *Proc. Natl. Acad. Sci. USA* 99:9410–9414.

Gallichan, W. S., Johnson, D. C., Graham, F. L., and Rosenthal, K. L. (1993). Mucosal immunity and protection after intranasal immunization with recombinant adenovirus expressing herpes simplex virus glycoprotein B. *J. Infect. Dis.* 168:622–629.

Gallichan, W. S., and Rosenthal, K. L. (1995). Specific secretory immune responses in the female genital tract following intranasal immunization with a recombinant adenovirus expressing glycoprotein B of herpes simplex virus. *Vaccine* 13:1589–1595.

Gallichan, W. S., Woolstencroft, R. N., Guarasci, T., McCluskie, M. J., Davis, H. L., and Rosenthal, K. L. (2001). Intranasal immunization with CpG oligodeoxynucleotides as an adjuvant dramatically increases IgA and protection against herpes simplex virus-2 in the genital tract. *J. Immunol.* 166:3451–3457.

Gillgrass, A. E., Ashkar, A. A., Rosenthal, K. L., and Kaushic, C. (2003). Prolonged exposure to progesterone prevents induction of protective mucosal responses following intravaginal immunization with attenuated herpes simplex virus type 2. *J. Virol.* 77:9845–9851.

Gillgrass, A. E., Fernandez, S. A., Rosenthal, K. L., and Kaushic, C. (2005a). Estradiol regulates susceptibility following primary exposure to genital herpes simplex virus type 2, while progesterone induces inflammation. *J. Virol.* 79:3107–3116.

Gillgrass, A. E., Tang, V. A., Towarnicki, K. M., Rosenthal, K. L., and Kaushic, C. (2005b). Protection against genital herpes infection in mice immunized under different hormonal conditions correlates with induction of vagina-associated lymphoid tissue. *J. Virol.* 79:3117–3126.

Gravlee, L. C. (1969). Jet irrigation method for the diagnosis of endometrial carcinoma. *Obstet. Gynecol.* 34:168–172.

Hasse, A. (2005). Perils at the mucosal front lines for HIV and SIV and their hosts. *Nat. Rev. Immunol.* 5:783–792.

He, Q., Mitchell, A., Morcol, T., and Bell, S. J. (2002). Calcium phosphate nanoparticles induce mucosal immunity and protection against herpes simplex virus type 2. *Clin. Diagn. Lab. Immunol.* 9:1021–1024.

Head, J. R., and Gaede, S. D. (1986). Ia antigen expression in the rat uterus. *J. Reprod. Immunol.* 9:137–153.

Hendrickson, B. A., Guo, J., Brown, I., Dennis, K., Marcellino, D., Hetzel, J., and Herold, B. C. (2000). Decreased vaginal disease in J-chain-deficient mice following herpes simplex type 2 genital infection. *Virology* 271:155–162.

Hocini, H., and Bomsel, M. (1999). Infectious human immunodeficiency virus can rapidly penetrate a tight human epithelial barrier by transcytosis in a process impaired by mucosal immunoglobulins. *J. Infect. Dis.* 179(Suppl. 3):S448–S453.

Huang, Y. T., Wright, A., Gao, X., Kulick, L., Yan, H., and Lamm, M. E. (2005). Intraepithelial cell neutralization of HIV-1 replication by IgA. *J. Immunol.* 174:4828–4835.

Hurlimann, J., Dayal, R., and Gloor, E. (1978). Immunoglobulins and secretory component in endometrium and cervix: Influence of inflammation and carcinoma. *Virch. Arch. Path. Anat. Histol.* 377:211–223.

Israel, Z. R., Gettie, A., Ishizaka, S. T., Mishkin, E. M., Staas, J., Gilley, R., Montefiori, D., Marx, P. A., and Eldridge, J. H. (1999). Combined systemic and mucosal immunization with microsphere-encapsulated inactivated simian immunodeficiency virus elicits serum, vaginal, and tracheal antibody responses in female rhesus macaques. *AIDS Res. Hum. Retroviruses* 15:1121–1136.

Johansson, E. L., Rudin, A., Wassen, L., and Holmgren, J. (1999). Distribution of lymphocytes and adhesion molecules in human cerrvix and vagina. *Immunology* 96:272–277.

Johansson, M., Schon, K., Ward, M., and Lycke, N. (1997a). Genital tract infection with *Chlamydia trachomatis* fails to induce protective immunity in γ-interferon receptor-deficient mice despite a strong local immunoglobulin A response. *Infect. Immun.* 65:1032–1044.

Johansson, M., Ward, M., and Lycke, N. (1997b). B-cell-deficient mice develop complete immune protection against genital tract infection with *Chlamydia trachomatis. Immunology* 92:422–428.

Kamat, B. R., and Isaacson, P. G. (1987). The immunocytochemical distribution of leukocytic subpopulations in human endometrium. *Am. J. Pathol.* 127:66–73.

Kantele, A., Hakkinen, M., Moldoveanu, Z., Lu, A., Savilahti, E., Alvarez, R. D., Michalek, S., and Mestecky, J. (1998). Differences in immune responses induced by oral and rectal immunizations with *Salmonella typhi* Ty21a: Evidence for compartmentalization within the common mucosal immune system in humans. *Infect. Immun.* 66:5630–5635.

Kaul, R., Rowland-Jones, S. L., Kimani, J., Fowke, K., Dong, T., Kiama, P., Rutherford, J., Njagi, E., Mwangi, F., Rostron, T., Onyango, J., Oyugi, J., MacDonald, K. S., Bwayo, J. J., and Plummer, F. A. (2001). New insights into HIV-1 specific cytotoxic T-lymphocyte responses in exposed, persistently seronegative Kenyan sex workers. *Immunol. Lett.* 79:3–13.

Kaul, R., Thottingal, P., Kimani, J., Kiama, P., Waigwa, C. W., Bwayo, J. J., Plummer, F. A., and Rowland-Jones, S. L. (2003). Quantitative *ex vivo* analysis of functional virus-specific CD8 T lymphocytes in the blood and genital tract of HIV-infected women. *AIDS* 17:1139–1144.

Kaul, R., Trabattoni, D., Bwayo, J. J., Arienti, D., Zagliani, A., Mwangi, F. M., Kariuki, C., Ngugi, E. N., MacDonald, K. S., Ball, T. B., Clerici, M., and Plummer, F. A. (1999). HIV-1-specific mucosal IgA in a cohort of HIV-1-resistant Kenyan sex workers. *AIDS* 13:23–29.

Kaushic, C., Ashkar, A. A., Reid, L. A., and Rosenthal, K. L. (2003). Progesterone increases susceptibility and decreases immune responses to genital herpes infection. *J. Virol.* 77:4558–4565.

Kaushic, C., Frauendorf, E., Rossoll, R. M., Richardson, J. M., and Wira, C. R. (1998). Influence of the estrous cycle on the presence and distribution of immune cells in the rat reproductive tract. *Am. J. Reprod. Immunol.* 39:209–216.

Kaushic, C., Frauendorf, E., and Wira, C. R. (1997). Polymeric immunoglobulin A receptor in the rodent female reproductive tract: expression in vagina and tissue specific mRNA regulation by sex hormones. *Biol. Reprod.* 57:958–966.

Kaushic, C., Murdin, A. D., Underdown, B. J., and Wira, C. R. (1998). *Chlamydia trachomatis* infection in the female reproductive tract of the rat: Influence of progesterone on infectivity and immune response. *Infect. Immun.* 66:893–898.

Kaushic, C., Richardson, J. M., and Wira, C. R. (1995). Regulation of polymeric immunoglobulin A receptor messenger ribonucleic acid expression in rodent uteri: effect of sex hormones. *Endocrinology* 136:2836–2844.

Kaushic, C., Zhou, F., Murdin, A. D., and Wira, C. R. (2000). Effects of estradiol and progesterone on susceptibility and early immune responses to *Chlamydia trachomatis* infection in the female reproductive tract. *Infect. Immun.* 68:4207–4216.

Kelly, J., and Fox, H. (1979). The local immunological defense system in the human endometrium. *J. Reprod. Immunol.* 1:39–45.

Kerr, W. R. (1955). Vaginal and uterine antibodies in cattle with particular reference to *Brucella abortus. Br. Vet. J.* 111:169–178.

Kett, K., Brandtzaeg, P., Radl, J., and Haaijman, J. J. (1986). Different subclass distribution of IgA-producing cells in human lymphoid organs and various secretory tissues. *J. Immunol.* 136:3631–3635.

Kozlowski, P. A., and Neutra, M. R. (2003). The role of mucosal immunity in prevention of HIV transmission. *Curr. Mol. Med.* 3:217–228.

Kozlowski, P. A., Williams, S. B., Lynch, R. M., Flanigan, T. P., Patterson, R. R., Cu-Uvin, S. and Neutra, M. R. (2002). Differential induction of mucosal and systemic antibody responses in women after nasal, rectal, or vaginal immunization: influence of the menstrual cycle. *J. Immunol.* 169:566–574.

Kuklin, N., Daheshia, M., Karem, K., Manickan E., and Rouse, B. T. (1997). Induction of mucosal immunity against herpes simplex virus by plasmid DNA immunization. *J. Virol.* 71:3138–3145.

Kunkel, E., and Butcher, E. (2003). Plasma cell homing. *Nat. Rev. Immunol.* 3:822–829.

Kutteh, W. H., Edwards, R. P., Menge, A. C., and Mestecky, J. (1993). IgA immunity in female reproductive tract secretions. In: Griffin, P. D., and Johnson, P. (eds.), *Local Immunity in Reproductive Tract Tissues,* Oxford University Press, New Delhi, pp. 229–243.

Kutteh, W. H., Hatch, K. D., Blackwell, R. E., and Mestecky, J. (1988). Secretory immune system of the female reproductive tract. I. Immunoglobulin and secretory component-containing cells. *Obstet. Gynecol.* 71:56–60.

Kutteh, W. H., and Mestecky, J. (1994). Secretory immunity in the female reproductive tract. *Am. J. Reprod. Immunol.* 31:40–46.

Kutteh, W. H., Mestecky, J., and Wira, C. R. (2005). Mucosal immune system in the human female reproductive tract. In: Mestecky, J., Bienenstock, J., Lamm, M., Strober, W., McGhee, J., and Mayer, L. (eds.), *Mucosal Immunology*, 3rd ed., Academic Press, San Diego, pp. 1631–1646.

Laguens, G., Goni, J. J., laguens, M.,. Goni, J. M., and Laguens, R. (1990). Demonstration and characterization of HLA-DR positive cells in the stroma of human endometrium. *J. Reprod. Immunol.* 18:179–186.

Lavreys, L., Baeten, J. M., Martin, H. L., Jr., Overbaugh, J., Mandaliya, K., Ndinya-Achola, J., and Kreiss, J. K. (2004). Hormonal contraception and risk of HIV-1 acquisition: results of a 10-year prospective study. *AIDS* 18:695–697.

Lazarus, N. H., Kunkel, E. J., Johnston, B., Wilson, E., Youngman, K. R., and Butcher, E. C. (2003). A common mucosal chemokine (mucosae-associated epithelial chemokine/CCL28) selectively attracts IgA plasmablasts. *J. Immunol.* 170:3799–3805.

Liu, F., Bergami, P. L., Duval, M., Kuhrt, D., Posner, M., and Cavacini, L. (2003). Expression and functional activity of isotype and subclass switched human monoclonal antibody reactive with the base of the V3 loop of HIV-1 gp120. AIDS R*es. Hum. Retroviruses* 19:597–607. MacPherson, A., Gatto, D., Sainsbury, E., Harriman, G., Hengartner, H., and Zinkernagel, R. M. (2000). A primitive T cell independent mechanism of intestinal mucosal IgA responses to commensal bacteria. *Science* 288:2222–2226.

Marx, P. A., Compans, R. W., Gettie, A., Staas, J. K., Gilley, R. M., Mulligan, M. J., Yamshchikov, G. V., Chen, D., and Eldridge, J. H. (1993). Protection against vaginal SIV transmission with microencapsulated vaccine. *Science* 260:1323–1327.

Marx, P. A, Spira, A. I., Gettie, A., Dailey, P. J., Veazey, R. S., Lackner, A. A., Mahoney, C. J., Miller, C. J., Claypool, L. E., Ho, D. D., and Alexander, N. J. (1996). Progesterone implants enhance SIV vaginal transmission and early virus load. *Nat. Med.* 2:1084–1089.

Mascola, J. R., Lewis, M. G., Stiegler, G., Harris, D., VanCott, T. C., Hayes, D., Louder, M. K., Brown, C. R., Sapan, C. V., Frankel, S. S., Lu, Y., Robb, M. L., Katinger H., and Birx, D. L. (1999). Protection of macaques against pathogenic simian/human immunodeficiency virus 89.6PD by passive transfer of neutralizing antibodies. *J. Virol.* 73:4009–4018.

Mascola, J. R., Stiegler, G., VanCott, T. C., Katinger, H., Carpenter, C. B., Hanson, C. E., Beary, H., Hayes, D., Frankel, S. S., Birx D. L., and Lewis, M. G. (2000). Protection of macaques against vaginal transmission of a pathogenic HIV-1/SIV chimeric virus by passive infusion of neutralizing antibodies. *Nat. Med.* 6:207–210.

Mazanec, M. B., Coudret, C. L., and Fletcher, D. R. (1995). Intracellular neutralization of influenza virus by immunoglobulin A anti-hemagglutinin monoclonal antibodies. *J. Virol.* 69:1339–1343.

Mazanec, M. B., Kaetzel, C. S., Lamm, M. E., Fletcher, D., and Nedrud, J. G. (1992). Intracellular neutralization of virus by immunoglobulin A antibodies. *Proc. Natl. Acad. Sci. USA* 89:6901–6905.

Mazanec, M. B., Nedrud, J. G., Kaetzel, C. S., and Lamm, M. E. (1993). A three-tiered view of the role of IgA in mucosal defense. *Immunol. Today* 14:430–435.

Mazzoli, S., Lopalco, L. Salvi, A., Trabattoni, D., Lo Caputo, S., Semplici, F., Biasin, M., Blé, C., Cosma, A., Pastori, C., Meacci, F., Mazzotta, F., Villa, M. L., Siccardi, A. G., and Clerici, M. (1999). Human immunodeficiency virus (HIV)-specific IgA and HIV neutralizing activity in the serum of exposed seronegative partners of HIV-seropositive persons. *J. Infect. Dis.* 180:871–875.

Mazzoli, S., Trabattoni, D., Lo Caputo, S., Piconi, S., Blé, C., Meacci, F., Ruzzante, S., Salvi, A., Semplici, F., Longhi, R., Fusi, M. L., Tofani, N., Biasin, M., Villa, M. L., Mazzotta, F., and Clerici, M. (1997). HIV-specific mucosal and cellular immunity in HIV-seronegative partners of HIV-seropositive individuals. *Nat. Med.* 3:1250–1257.

Mbopi-Keou, F. X., Belec, L., Dalessio, J., Legoff, J., Gresenguet, G., Mayaud, P., Brown D. W., and Morrow, R. A. (2003). Cervicovaginal neutralizing antibodies to herpes simplex virus (HSV) in women seropositive for HSV Types 1 and 2. *Clin. Diagn. Lab. Immunol.* 10:388–393.

McDermott, M. R., and Bienenstock, J. (1979). Evidence for a common mucosal immunologic system. I. Migration of B immunoblasts into intestinal, respiratory, and genital tissues. *J. Immunol.* 122:1892–1898.

McDermott, M. R., Clark, D. A., and Bienenstock, J. (1980). Evidence for a common mucosal immunologic system. II. Influence of the estrous cycle on B immunoblast migration into genital and intestinal tissues. *J. Immunol.* 124:2536–2539.

Meng, G., Wei, X., Wu, X., Sellers, M. T., Decker, J. M., Moldoveanu, Z., Orenstein, J. M., Graham, M. F., Kappes, J. C., Mestecky, J., Shaw, G. M., and Smith, P. D. (2002). Primary intestinal epithelial cells selectively transfer R5 HIV-1 to CCR5+ cells. *Nat. Med.* 8:150–156.

Menge, A. C., and Mestecky, J. (1993). Surface expression of secretory component and HLA class II DR antigen on glandular epithelial cells from human endometrium and two endometrial adenocarcinoma cell lines. *J. Clin. Immunol.* 13:259–264.

Mestecky, J., and Fultz, P. (1999). Mucosal immune system of the human genital tract. *J. Infect. Dis.* 179(Suppl. 3):S470–S474.

Mestecky, J., Jackson, S., Moldoveanu, Z., Nesbit, L. R., Kulhavy, R., Prince, S. J., Sabbaj, S., Mulligan, M. J., and Goepfert, P. A. (2004). Paucity of antigen-specific IgA responses in sera and external secretions of HIV-type 1-infected individuals. *AIDS Res. Hum. Retroviruses* 20:972–988.

Mestecky, J., Bienenstock, J., Lamm, M., Strober, W., McGhee, J., and Mayer, L. (eds.). (2005a). *Mucosal Immunology*, 3rd ed., Academic Press, San Diego.

Mestecky, J., Moldoveanu, Z., and Russell, M. W. (2005c). Immunologic uniqueness of the genital tract: challenge for vaccine development. *Am. J. Reprod. Immunol.* 53:208–214.

Mestecky, J., Moro, I., Kerr, M. A., and Woof, J. M. (2005b). Mucosal immunoglobulins. In: Mestecky, J., Bienenstock, J., Lamm, M., Strober, W., McGhee, J., and Mayer, L. (eds.), *Mucosal Immunology*, 3rd ed., Academic Press, San Diego, pp. 153–181.

Mestecky, J., and Russell, M. W. (2000). Induction of mucosal immune responses in the human genital tract. *FEMS Immunol. Med. Microbiol.* 27:351–355.

Michetti, P., Mahan, M. J., Slauch, J. M., Mekalanos J. J., and Neutra, M. R. (1992). Monoclonal secretory immunoglobulin A protects mice against oral challenge with the invasive pathogen *Salmonella typhimurium*. *Infect. Immun.* 60:1786–1792.

Moldoveanu, Z., Huang, W. Q., Kulhavy, R., Pate, M. S., and Mestecky, J. (2005). Human male genital tract secretions: Both mucosal and systemic immune compartments contribute to the humoral immunity. *J. Immunol.* 175:4127–4136.

Moja, P., Tranchat, C., Tchou, Pozzetto, B., Lucht, F., Dewsgranges, C., and Genin, C. (2000). Neutralization of human immunodeficiency virus type 1 (HIV-1) mediated by parotid IgA of HIV-1-infected patients. *J. Infect. Dis.* 181:1607–1613.

Morell, V. (1995). Attacking the causes of "silent" infertility. *Science* 269:775–777.

Morrison, L. A., Da Costa, X. J., and Knipe, D. M. (1998). Influence of mucosal and parenteral immunization with a replication-defective mutant of HSV-2 on immune responses and protection from genital challenge. *Virology* 243:178–187.

Morrison, R. P., Manning, D. S., and Caldwell, H. D. (1992). Immunology of Chlamydia trachomatis infections. In: Quinn, T. C. (ed.), Sexually Transmitted Diseases ,Raven Press, New York, pp. 57–84.

Morrison, S. G., and Morrison, R. P. (2005a). A predominant role for antibody in acquired immunity to chlamydial genital tract reinfection. *J. Immunol.* 175:7536–7542.

Morrison, S. G., and Morrison, R. P. (2005b). The protective effect of antibody in immunity to murine chlamydial genital tract reinfection is independent of immunoglobulin A. *Infect. Immun.* 73:6183–6186.

Morrison, S. G., Su, H., Caldwell, H. D., and Morrison, R. P. (2000). Immunity to murine *Chlamydia trachomatis* genital tract reinfection involves B cells and CD4(+) T cells but not CD8(+) T cells. *Infect. Immun.* 68:6979–6987.

Morton, H. C., van Egmond, M., and van de Winkel, J. G. (1996). Structure and function of human IgA Fc receptors (FcaR). *Crit. Rev. Immunol.* 16:423–440.

Mostad, S. B. (1998). Prevalence and correlates of HIV-1 shedding in female genital tract. *AIDS Res. Hum. Retroviruses* 14(Suppl. 1):S11–S15.

Ochsenbein, A. F., Fehr, T., Lutz, C., Suter, M., Brombacher, F., Hengartner, H., and Zinkernagel, R. M. (1999). Control of early viral and bacterial distribution and disease by natural antibodies. *Science* 286:2156–2159.

Ochsenbein, A. F., and Zinkernagel, R. M. (2000). Natural antibodies and complement link innate and acquired immunity. *Immunol. Today* 21:624–630.

Ogra, P. L., and Ogra, S. S. (1973). Local antibody response to poliovaccine in the human female genital tract. *J. Immunol.* 110:1307–1311.

Pal, S., Peterson, E. M., and de la Maza, L. M. (1996). Intranasal immunization induces long-term protection in mice against a *Chlamydia trachomatis* genital challenge. *Infect. Immun.* 64:5341–5348.

Parr, E. L., and Parr, M. B. (1988). Anti-bacterial IgA and IgG in mouse uterine luminal fluid, vaginal washings and serum. *J. Reprod. Immunol.* 13:65–72.

Parr, E. L., and Parr, M. B. (1997). Immunoglobulin G is the main protective antibody in mouse vaginal secretions after vaginal immunization with attenuated herpes simplex virus type 2. *J Virol* 71:8109–8115.

Parr, E. L., Parr, M. B., and Thapar, M. (1988). A comparison of specific antibody responses in mouse vaginal fluid after immunization by several routes. *J. Reprod. Immunol.* 14:165–176.

Parr, E. L., and Parr, M. B. (2005). Female genital tract infections and immunity in animal models. In: Mestecky, J., Bienenstock, J., Lamm, M., Strober, W., McGhee, J., and Mayer, L. (eds.). *Mucosal Immunology*, 3rd ed., Academic Press, San Diego, pp. 1613–1630.

Parr, M. B., and Parr, E. L. (1989). Immunohistochemical investigation of secretory component and immunoglobulin A in the genital tract of the female rat. *J. Reprod. Fertil.* 85:105–113.

Parr, M. B., and Parr, E. L. (1985). Immunohistochemical localization of immunoglobulins A, G and M in the mouse female genital tract. *J. Reprod. Fertil.* 74:361–370.

Parr, M. B., and Parr, E. L. (1991). Langerhans cells and T lymphocyte subsets in the murine vagina and cervix. *Biol. Reprod.* 44:491–498.

Parr, M. B., and Parr, E. L. (1996). Immunoglobulins in the female genital tract. In: Bronson, R. D., Alexander, N. J., Anderson, D. J., Branch, D. W., and Kutteh, W. H. (eds.), *Reproductive Immunology*, Blackwell Science, Cambridge, MA, pp. 275–308.

Parr, M. B., and Parr, E. L. (1999). Immune responses and protection against vaginal infection after nasal or vaginal immunization with attenuated HSV-2. *Immunol. Today* 98:639–645.

Parr, M. B., and Parr, E. L. (2000). Immunity to vaginal herpes simplex virus-2 infection in B-cell knockout mice. *Immunology* 101:126–131.

Parr, M. B., and Parr, E. L. (2003). Vaginal immunity in the HSV-2 mouse model. *Int. Rev. Immunol.* 22:43–63.

Patton, D. L., and Lichtenwalner, A. B. (1998). Animal models for the study of chlamydial infections. In: Stephens, R. S., Byrne, G. I., Christiansen, G., Clarke, I. N., Groyston, J. T., Rank, R. G., Ridgway, G. L., Saikku, P., Schachter, J., and Stamm, W. E. (eds.), *Chlamydial Infections, Proceedings of the 9th International Symposium on Human Chlamydial Infection*, International Chlamydia Symposium, San Francisco, pp. 641–650.

Patton, D. L., and Rank, R. G. (1992). Animal models for the study of pelvic inflammatory disease. In: Quinn, T. C. (ed.), *Sexually Transmitted Diseases*, Raven Press, New York, pp. 85–111.

Pitman, R. S., and Blumberg, R. S. (2000). First line of defense: The role of the intestinal epithelium as an active component of the mucosal immune system. *J. Gastroenterol.* 35:805–814.

Pudney, J., and Anderson, D. J. (1995). Immunobiology of the human penile urethra. *Am. J. Pathol.* 147:155–165.

Pudney, J., Quayle, A. J., and Anderson, D. J. (2005). Immunological microenvironments in the human vagina and cervix: mediators of cellular immunity are concentrated in the cervical transformation zone. *Biol. Reprod.* 73:1253–1263.

Rachman, F., Casimiri, V., Psychoyos, A., and Bernard, O. (1983). Immunoglobulins in the mouse uterus during the oestrous cycle. *J. Reprod. Fertil.* 69:17–21.

Rebello, R., Green, F. H., and Fox, H. (1975). A study of the secretory immune system of the female reproductive tract. *Br. J. Obstet. Gynecol.* 82:812–816.

Renegar, K. B., Menge, A., and Mestecky, J. (2006). Influenza virus infection of the murine uterus: A new model for antiviral immunity in the female reproductive tract. *Viral Immunol.* 19:613–622.

Russell, M. W., Lue, C., van den Wall Bake, A. W., Moldoveanu, Z., and Mestecky, J. (1992). Molecular heterogeneity of human IgA antibodies during an immune response. *Clin. Exp. Immunol.* 87:1–6.

Russell, M. W., and Mestecky, J. (2002). Humoral immune responses to microbial infections in the genital tract. *Microbes Infect.* 4:667–677.

Russell, M. W., Moldoveanu, Z., White, P. L., Sibert, G. J., Mestecky, J., and Michalek, S. M. (1996). Salivary, nasal, genital, and systemic antibody responses in monkeys immunized intranasally with a bacterial protein antigen and the Cholera toxin B subunit. *Infect. Immun.* 64:1272–1283.

Russell, M. W., Sparling, P., Morrison, R. P., Cauci, S., Fidel, P., Martin, D., Hook E., and Mestecky, J. (2005). Mucosal immunology of sexually transmitted diseases. In: Mestecky, J., Bienenstock, J., Lamm, M., Strober, W., McGhee, J., and Mayer, L. (eds.). *Mucosal Immunology*, 3rd ed., Academic Press, San Diego, pp. 1693–1720.

Sansonetti, P. J. (2004). War and peace at mucosal surfaces. *Nat. Rev. Immunol.* 4:953–964.

Schachter, J. (1989). Pathogenesis of chlamydial infections. 8:206–220.

Schachter, J., and Grayston, J. T. (1998). Epidemiology of human chlamydial infections. In: Stephens, R. S., Byrne, G. I., Christiansen, G., Clarke, I. N., Groyston, J. T., Rank,

R. G., Ridgway, G. L., Saikku, P., Schachter, J., and Stamm, W. E. (eds.), *Chlamydial Infections, Proceedings of the 9th International Symposium on Human Chlamydial Infection*, International Chlamydia Symposium, San Francisco, pp. 3–10.

Schumacher, G. (1988). Immunology of spermatozoa and cervical mucus. *Hum. Reprod.* 3:289–300.

Schumacher, G., Kim, M. H., Hossenianan, A., and Dupon, C. (1977). Immunoglobulins, proteinase inhibitors, albumin, and lysozyme in human cervical mucus. *Am. J. Obstet. Gynecol.* 129:629–636.

Schumacher, G. F. B. (1980). Humoral immune factors in the female reproductive tract and their changes during the cycle. In: Dinsda, D., and Schumacher, G. (eds.), *Immunological Aspects of Infertility and Fertility Control*, Elsevier, New York, pp. 93–141.

Senior, B. W., Dunlop, J. I., Batten, M. R., Kilian, M., and Woof, J. M. (2000). Cleavage of a recombinant human immunoglobulin A2 (IgA2) –IgA1 hybrid antibody by certain bacterial IgA1 proteases. *Infect. Immun.* 68:463–469.

Shattock, R., and Moore, J. (2003). Inhibiting sexual transmission of HIV-1 infection. *Nat. Rev. Microbiol.* 1:25–34.

Smith, S. M., Baskin, G. B., and Marx, P. A. (2000). Estrogen protects against vaginal transmission of simian immunodeficiency virus. *J. Infect. Dis.* 182:708–715.

Stagg, A. J. (1998). Vaccines against *Chlamydia*: approaches and progress. *Mol. Med. Today* 4:166–173.

Stubbe, H., Berdoz, J., Kraehenbuhl, J. P., and B. Corthésy (2000). Polymeric IgA is superior to monomeric IgA and IgG carrying the same variable domain in preventing *Clostridium difficile* toxin A damaging of T84 monolayers. *J. Immunol.* 164:1952–1960.

Su, H., Feilzer, K., Caldwell, H. D., and Morrison, R. P. (1997). *Chlamydia trachomatis* genital tract infection of antibody-deficient gene knockout mice. *Infect. Immun.* 65:1993–1999.

Sullivan, D. A., Richardson, G. S., MacLaughlin, D. T., and Wira, C. R. (1984). Variations in the levels of secretory component in human uterine fluid during the menstrual cycle. *J. Steroid Biochem.* 20:509–513.

Sullivan, D. A., Underdown, B. J., and Wira, C. R. (1983). Steroid hormone regulation of free secretory component in the rat uterus. *Immunology* 49:379–386.

Sullivan, D. A., and Wira, C. R. (1981). Estradiol regulation of secretory component in the female reproductive tract. *J. Steroid Biochem.* 15:439–444.

Sullivan, D. A., and Wira, C. R. (1983). Hormonal regulation of immunoglobulins in the rat uterus: Uterine response to a single estradiol treatment. *Endocrinology* 112:260–268.

Sullivan, D. A., and Wira, C. R. (1984). Hormonal regulation of immunoglobulins in the rat uterus: Uterine response to multiple estradiol treatments. *Endocrinology* 114:650–658.

Sylvan, P. E., MacLaughlin, D. T., Richardson, G. S., Scully, R. E., and Nikrui, N. (1981). Human uterine fluid proteins associated with secretory phase endometrium: Progesterone-induced proteins? *Biol. Reprod.* 24:423–429.

Tauber, P. F., Wettich, W., Nohlen, M., and Zaneveld, L. J. D. (1985). Diffusible proteins of the mucosa of the human cervix, uterus, and fallopian tubes: Distribution and variations during the menstrual cycle. *Am. J. Obstet. Gynecol.* 15:1115–1125.

Tourville, D. R., Ogra, S., Lippes, J., and Tomsi, T. (1970). The human female reproductive tract: Immunohistochemical localization of γA, γG,γM, secretory piece and lactoferrin. *Am. J. Obstet. Gynecol.* 108:1102–1108.

UNAIDS 2004 Report on HIV/AIDS Epidemic-July (2004). UNAIDS, Geneva, Switzerland.

Vaerman, J.-P., and Férin, J. (1974). Local immunological response in the vagina, cervix and endometrium. *Acta Endocrinol.* 194:281–305.

Vaerman, J.-P., and Férin, J. (1975). Local immunological response in vagina, cervix and endometrium. *Acta Endocrinol.*. 194(Suppl.):281–305.

van Egmond, M., Damen, C. A., van Spriel, A. B., Vidarsson, G., van Garderen, E., and van de Winkel, J. G. (2001). IgA and the IgA Fc receptor. *Trends Immunol.* 22:205–211.

van Rompay, K. K., Berardi, C. J., Dillard-Telm, S., Tarara, R. P., Canfield, D. R., Valverde, C. R., Montefiori, D. C., Cole, K. S., Montelaro, R. C., Miller, C. J., and Marthas, M. L. (1998). Passive immunization of newborn rhesus macaques prevents oral simian immunodeficiency virus infection. *J. Infect. Dis.* 177:1247–1259.

Waldmann, R., Cruz, J., and Rowe, D. (1971). Immunoglobulin levels to *Candida albicans* in human cervicovaginal secretions. *Clin. Exp. Immunol.* 9:427–434.

Wang, C. C., McClelland, R. S., Overbaugh, J., Reilly, M., Panteleeff, D. D., Mandaliya, K., Chohan, B., Lavreys, L., Ndinya-Achola, J., and Kreiss, J. K. (2004). The effect of hormonal contraception on genital tract shedding of HIV-1. *AIDS* 18:205–209.

Wira, C. R., Crane-Godreau, M. A., and Grant-Tschudy, K. S. (2005a). Endocrine regulation of the mucosal immune system in the female reproductive tract. In: Mestecky, J., Bienenstock, J., Lamm, M., Strober, W., McGhee, J., and Mayer, L. (eds.), *Mucosal Immunology*, 3rd ed., Academic Press, San Diego, pp. 1661–1678

Wira, C. R., Grant-Tschudy, K. S., and Crane-Godreau, M. A. (2005b). Epithelial cells in the female reproductive tract: A central role as sentinels of immune protection. *Am. J. Reprod. Immunol.* 53:65–76.

Wira, C. R., and Merritt, K. (1977). Effect of the estrous cycle, castration and pseudopregnancy on *E. coli* in the uterus and uterine secretions of the rat. *Biol. Reprod.* 17:519–522.

Wira, C. R., and Prabhala, R. H. (1993). The female reproductive tract is an inductive site for immune responses: Effect of estradiol and antigen on antibody and secretory component levels in uterine and cervico-vaginal secretions following various routes of immunization. In: Griffin, P. D., and Johnson, P. M. (eds.), *Scientific Basis of Fertility Regulation. Local Immunity in Reproductive Tract Tissues,* Oxford University Press, New York, pp. 271–293.

Wira, C. R., and Sandoe, C. P. (1980). Hormone regulation of immunoglobulins: Influence of estradiol on IgA and IgG in the rat uterus. *Endocrinology* 106:1020–1026.

Wira, C. R., and Sandoe, C. P. (1987). Specific IgA and IgG antibodies in the secretions of the female reproductive tract: Effects of immunization and estradiol on expression of this response *in vivo. J. Immunol.* 138:159–164.

Wira, C. R., and Sandoe, C. P. (1989). Effect of uterine immunization and oestradiol on specific IgA and IgG antibodies in uterine, vaginal and salivary secretions. *Immunology* 68:24–30.

Wira, C. R., Stern, J. E., and Colby, E. (1984). Estradiol regulation of secretory component in the uterus of the rat: evidence for involvement of RNA synthesis. *J. Immunol.* 133:2624–2628.

Wira, C. R., and Sullivan, D. A. (1985). Estradiol and progesterone regulation of IgA, IgG and secretory component in cervico-vaginal secretions of the rat. *Biol. Reprod.* 32:90–95.

Wolbank, S., Kunert, R., Stiegler, G., and Katinger, H. (2003). Characterization of human class-switched polymeric (immunoglobulin M [IgM] and IgA) anti-human immunodeficiency virus type 1 antibodies 2F5 and 2G12. *J. Virol.* 77:4095–4103.

Woof, J. M., and Kerr, M. A. (2006). The function of immunoglobulin A in immunity. *J. Pathol.* 208:270–82.

Woof, J. M., and Mestecky, J. (2005). Mucosal immunoglobulins. *Immunol. Rev.* 206:64–82.

Wright, P. F., Kozlowski, P. A., Rybczyk, G. K., Goepfert, P., Staats, H. F., VanCott, T. C., Trabattoni, D., Sannella, E., and Mestecky, J. (2002). Detection of mucosal antibodies in HIV type 1-infected individuals. *AIDS Res. Hum. Retroviruses* 18:1291–1300.

Wright, A., Yan, H., Lamm, M. E., and Huang, Y. T. (2006). Immunoglobulin A antibodies against internal HIV-1 proteins neutralize HIV-1 replication inside epithelial cells. *Virology* 356:165–170.

Yang, S. L,. and Schumacher, G. F. B. (1979). Immune response after vaginal application of antigens in the rhesus monkey. *Fertil. Steril.* 32:588–598.

Yeaman, G. R., Collins, J. E., Fanger, M. W., Wira, C. R., and Lydyard, P. M. (2001). CD8+ T cells in human uterine endometrial lymphoid aggregates: Evidence for accumulation of cells by trafficking. *Immunology* 102:434–440.

Yeaman, G. R., Guyre, P. M., Fanger, M. W., Collins, J. E., White, H. D., Rathbun, W., Orndorff, K. A., Gonzalez, J., Stern, J. E., and Wira, C. R. (1997). Unique CD8+ T cell-rich lymphoid aggregates in human uterine endometrium. *J. Leuk. Biol.* 61:427–435.

Zhao, X., Deak, E., Soderberg, K., Linehan, M., Spezzano, D., Zhu, J., Knipe, D. M., and Iwasaki, A. (2003). Vaginal submucosal dendritic cells, but not Langerhans cells, induce protective Th1 responses to herpes simplex virus-2. *J. Exp. Med.* 197:153–162.

13
IgA-Associated Diseases

Jiri Mestecky[1] and Lennart Hammarström[2]

13.1. Introduction

In many species, including humans, IgA is produced in quantities that far exceed the combined synthesis of all other immunoglobulin (Ig) isotypes (Conley and Delacrox, 1987; Mestecky et al., 2005). It is estimated that in humans, ~66 mg/kg body weight (i.e., 4–5 g in a 70-kg adult) is produced daily, in contrast with ~25 mg/kg/day for IgG and much less for other isotypes. IgA displays diverse

[1] University of Alabama at Birmingham, Departments of Microbiology and Medicine, Birmingham, AL 35294, USA.
[2] Huddinge University Hospital, Karolinska Institute, Department of Clinical Immunology, S-141 86 Huddinge, Sweden.

metabolic properties due to its heterogeneity of molecular forms (two subclasses, both of which can form monomers and polymers) and two distinct major sites of production (bone marrow and mucosal tissues) (Mestecky et al., 2005). Approximately two-thirds of the IgA in the body is produced as J-chain containing dimers and tetramers by mucosal plasma cells and is selectively transported into external secretions of the digestive, respiratory, and genitourinary tracts to form secretory IgA (SIgA) (see Chapter 3). There is a marked preponderance of the IgA1 subclass in most secretions, with the exception of the large intestine and female genital tract, where IgA2 equals or slightly exceeds IgA1 (see Chapters 1 and 2). Greater than 95% of the IgA in plasma is in the form of monomers of mainly bone marrow origin, about 85% of which is of the IgA1 subclass (Alley et al., 1982; Kutteh et al., 1982a). In humans, only a very small amount of plasma IgA reaches the external secretions. Almost all of the plasma IgA is catabolized in the liver by hepatocytes, which interact with the glycan moieties of IgA through the membrane-expressed asialoglycoprotein receptor (Mestecky et al., 2005) (see Chapter 3).

Pathologically high levels of IgA and IgA-containing immune complexes in plasma and/or external secretions can result from increased production or decreased catabolism of IgA. Elevated IgA levels occur in several human diseases, including bone marrow and intestinal malignancies (multiple myeloma and α-chain disease), diseases of the liver (cirrhosis), kidneys (IgA nephropathy), skin (dermatitis herpetiformis, Henoch-Schoenlein purpura), joints (rheumatoid arthritis), and many infections (Heremans, 1974). Relative increases in levels of antigen-specific IgA can be induced in sera and external secretions by mucosal or systemic immunization with various types of protein, glycoprotein, or polysaccharide antigen (Russell et al., 1992) (see Chapter 14).

Circulating levels of IgA in the plasma of newborns and children is physiologically low, due to the fact that IgA, unlike IgG, is not transported across the placental barrier. Maternal SIgA antibodies are delivered to the newborn in colostrum and milk, but they are not effectively absorbed from the neonatal gut and remain in the lumen to provide local protection. Antigenic stimulation with mucosa-associated microbiota, through intestinal and other mucosal surfaces, results in a rapid maturation of the SIgA system, so that adult levels of SIgA in saliva and probably other fluids are attained within several months (Cripps and Gleeson, 2005). In contrast, adult levels of plasma IgA are not reached until puberty, due to the slow maturation of the systemic IgA compartment (Heremans, 1974).

13.2. Increased Levels of IgA

13.2.1. Monoclonal Disorders of IgA Biosynthesis

13.2.1.1. IgA Multiple Myeloma

Multiple myeloma is a neoplasm of plasma cells characterized by tropism for the bone marrow and production of monoclonal Ig, which can be detected in plasma and/or urine (Mitsiades et al., 2004). The frequency of occurrence of IgA multiple myeloma reflects the relative distribution of cells producing various Ig isotypes in the bone marrow. Consequently, IgG multiple myeloma is more frequent than IgA, and IgA1 is much more frequent than IgA2 multiple myeloma (Mestecky and Russell, 1986). Levels of IgA1 or IgA2 myeloma proteins might reach several g/100mL of plasma.The availability of such large quantities of monoclonal IgA1 and IgA2 has facilitated research studies of the primary amino acid and glycan structures (see Chapter 1). In contrast to the normal bone marrow, in which IgA-producing cells are mostly J-chain-negative and secrete almost exclusively monomeric IgA (Alley et al., 1982; Kutteh et al., 1982a), almost all the neoplastic cells in IgA multiple myeloma express J-chain and and secrete predominantly polymeric IgA (Radl et al., 1974). Interestingly, the expression of J-chain by bone marrow plasma cells is also commonly seen in IgG and IgD multiple myeloma (Bast et al., 1981; Lokhorst et al., 1990; Mestecky et al., 1980, 1990a), but not in benign gammopathies, and can therefore be used as a differential marker (Bast et al., 1981). Mucosal tissues rich in IgA-producing cells are usually devoid of malignant cells in individuals with IgA multiple myeloma, and external secretions contain only trace amounts of monoclonal IgA myeloma proteins (Kubagawa et al., 1987). The fact that circulating polymeric myeloma IgA is not transported into secretions by pIgR is consistent with the physiological separation of the systemic and mucosal IgA systems in humans (see Sect. 13.1).

Spontaneous IgA multiple myeloma occurs rarely in animal species; however, it can easily be induced in mice and rats by the injection of mineral oils into the peritoneal cavity (Potter, 1972). Prior to the advent of monoclonal antibody technology, these mineral-oil-induced plasmacytomas represented a valuable source of murine monoclonal Ig for structural and functional studies.

13.2.1.2. α-Chain Disease

Immunoglibulin heavy (H)-chain diseases are relatively rare lymphoproliferative disorders of B-lymphocytes that produce incomplete H-chains devoid of L-chains; α-chain disease is the most frequent isotype of H-chain disease. In contrast to multiple myeloma, α-chain disease usually originates in the mucosal immune system rather than in the bone marrow (reviewed in Rambaud et al., 1994). Depending on the stage of the disease, the lamina propria of the small intestine and draining mesenteric lymph nodes are heavily infiltrated by IgA-producing cells of lymphoblast and plasma cell morphology. Systemic lymphoid and nonlymphoid organs (peripheral lymph nodes, spleen, stomach, colon, liver, bone marrow, and central nervous system) are involved only rarely.

Structural studies of α-chain proteins in this disease have demonstrated that they comprise the intact Fc fragment, including the N-terminal hinge region, the CH2 and CH3 domains, and the C-terminal "tail," and are predominantly of the IgA1 isotype (reviewed in Rambaud et al., 1994). Using sensitive techniques, these abnormal proteins can be detected in patient sera and intestinal secretions. In the latter fluid, α-chain proteins were found to contain J-chain and to be associated with the secretory component, indicating that they were transcytosed through intestinal epithelial cells by the polymeric Ig receptor (pIgR).

13.2.2. *Polyclonal Increases in IgA Biosynthesis*

13.2.2.1. Gastrointestinal Tract and Liver Diseases

Elevated levels of serum IgA have been reported in patients with several gastrointestinal and liver disorders, including Crohn's disease, ileojejunal bypass, chronic active hepatitis, acute hepatitis, primary biliary cirrhosis, and alcoholic liver disease (Brown, 2005; Delacroix and Vaerman, 1983; Delacroix et al., 1983; Kutteh *et al.*, 1982b; Mestecky and Russell, 1986). It appears that both increased production and decreased catabolism contribute to the increased serum IgA levels. Interestingly, immunochemical studies of serum IgA in these diseases revealed that pIgA and SIgA were significantly elevated, particularly in patients with liver cirrhosis (Delacroix and Vaerman, 1983; Delacroix et al., 1983; Kutteh et al., 1982b). With respect to IgA subclasses, the proportion of IgA1 was higher in chronic active hepatitis, whereas the proportion of IgA2 tended to be higher in patients with Crohn's disease and alcoholic liver disease (Delacroix et al., 1983). In addition to the elevated levels of serum IgA in alcoholic liver disease, there are frequent deposits of monomeric IgA1 along the margins of liver sinusoids and occasionally in the skin and kidneys (Brown, 2005).

13.2.2.2. IgA Nephropathy and Related Diseases

Immunoglobulin A nephropathy (IgAN) is an inflammatory disease—glomerulonephritis—characterized histologically by deposits of IgA and frequently by codeposition of Ig of other isotypes and components of the complement system in the glomerular mesangium. The inflammatory response leads to proliferation of mesangial cells and production of matrix proteins. IgAN is the most common form of glomerulonephritis in the world, and 20–40% of patients progress to end-stage renal failure requiring continued dialysis or kidney transplantation.

Typical cases of IgAN involve adolescent males who have recently recovered from an upper respiratory tract infection. Hallmarks of the disease include the presence of erythrocytes and protein in the urine, proliferation of mesangial cells, and glomerular deposits of IgA with complement factor 3 (C3) and often IgG and IgM. This subsection will focus on characteristic immunological features, with emphasis on the IgA system. Further epidemiological data, genetics, uniqueness of the racial distribution, clinical picture, complications, and treatment have been extensively reviewed (Donadio and Grande, 2002; Emancipator et al., 2005; Julian and Novak, 2004; Julian et al., 1999; Novak et al., 2001; Smith and Feehally, 2003).

13.2.2.2.1. IgAN: Immunological Features

Sera and urine of IgAN patients contain higher levels of total IgA than those of healthy individuals or patients with other forms of glomerulonephritis (Emancipator et al., 2005; Galla et al., 1985). In addition, most patient sera contain circulating immune complexes (CICs) and IgA rheumatoid factor (Czerkinsky et al., 1986). CICs are especially prominent in the acute phase of the disease.

Increased levels of serum IgA in IgAN are probably due to the enhanced production of IgA in the bone marrow and/or mucosal tissues as well as altered catabolism of IgA (see Sect. 13.2.2.2.5). However, elevated serum levels of IgA per se are not sufficient for formation of CICs and IgA deposition in the kidney mesangium; patients with elevated serum IgA from other causes (e.g., IgA multiple myeloma and other diseases described earlier) do not develop these pathologic sequelae. It has been speculated that the enhanced IgA production in IgAN is due to upregulation of IgA synthesis in B-cells by interactions with T-cells and their soluble products in response to mucosal infections (Emancipator et al., 2005).

13.2.2.2.2. Circulating Immune Complexes

Physicochemical and immunochemical analyses of CICs revealed that their molecular masses and corresponding sedimentation constants range from ~350 to >1000 kDa (10S to >19S). Both the size and molecular composition of CICs contribute to their pathogenic potential. CICs might contain exclusively IgA1, C3, IgG, or IgM (Czerkinsky et al., 1986; Novak et al., 2005; Tomana et al., 1997). CICs containing IgA2 have not been detected, and it is not known whether large CICs contain IgA1, IgG, and IgM in a "mixed" form.

Circulating immune complexes can be detected by several assays, including enzyme-linked immunosorbent assay (ELISA), binding to various cell types, and molecular sieve chromatography or gradient ultracentrifugation followed by the detection of Ig and C3 components by immunochemical methods. Although limited in scope, attempts to identify exogenous antigens (e.g., food antigen, viruses, and bacterial products) as components of IgA1 and C3-containing CICs or IgA1-containing mesangial deposits have yielded no convincing evidence for their presenceare (Mestecky et al., 1990b; Russell et al., 1986). Instead, there is evidence that IgAN can be considered an autoimmune disease in which IgA1 molecules with altered glycans act as antigens and bind naturally occurring antiglycan antibodies (see Sect. 13.2.2.2.3). IgA1 in CICs consists of both mIgA1 and J-chain-containing pIgA1, with a predominance of the latter form (Czerkinsky et al., 1986; Tomana et al., 1997, 1999).

13.2.2.2.3. Structural Alterations of IgA1 in IgAN

The exclusive involvement of the IgA1 subclass in IgAN suggests that it has a greater potential than IgA2 to induce an autoimmune response. In humans and hominoid primates, IgA1 and IgA2 have pronounced differences in their structure, body fluid distribution, and function (Mestecky and Russell, 1986; Mestecky et al., 2005) (see Chapter 1). Although the α1 and α2 H-chains display a very high degree of primary structure homology in their CH1, CH2, and CH3 domains, there are major differences in their hinge regions and glycosylation patterns (Mestecky et al., 2005) (Fig. 13.1).The amino

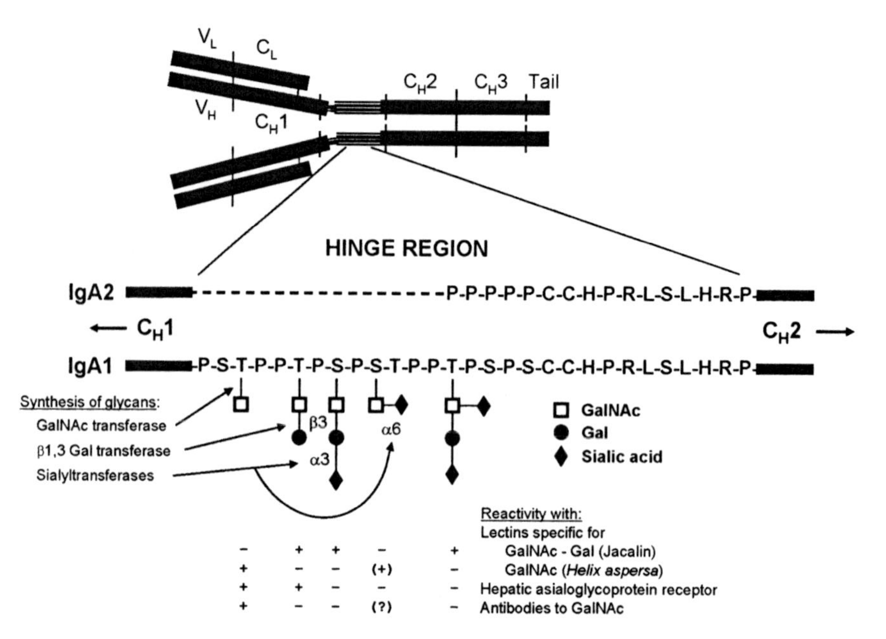

FIG. 13.1. The hinge regions of human IgA1 and IgA2 including the potential structures of O-linked glycan chains. Glycosyltransferases involved in the biosynthesis and glycosidases involved in the enzymatic removal of sugars are listed on the left and right, respectively. Reactivities of particular O-linked glycans of various structures with lectins specific for GalNAcβ1,3Gal (jacalin) and teminal GalNAc (*Helix aspersa*), the asialoglycoprotein receptor (ASGP-R) expressed on hepatocytes and specific for terminal Gal or GalNAc, and naturally occurring IgG antibodies to GalNAc are specified.

acid sequence of the IgA1 hinge region is similar to that of mucins, with repeating residues of proline (P), serine (S), and threonine (T). Furthermore, there are several (3–5) *O*-linked (to S or T) glycan side chains containing *N*-acetylgalactosamine (GalNAc), galactose (Gal), and sialic acid (SA). The hinge region of the α1-chain is also a natural substrate for IgA1 proteases produced by bacterial pathogens that cleave IgA1 into Fab and Fc fragments, with functionally important biologic consequences (Kilian and Russell, 2005) (see Chapter 1).

Comparative studies of the primary structures of IgA molecules from different vertebrate species unquestionably demonstrate that human IgA2 is phylogenetically older than IgA1 and resembles the single IgA type of many vertebrates. In contrast, the IgA1 subclass is a relatively new form that apparently emerged during primate evolution (reviewed in Mestecky et al., 2005; Peppard et al., 2005). However, the origin of the inserted gene segment encoding the hinge region of IgA1 remains obscure.

The *O*-linked glycans of the IgA1 hinge region display a considerable degree of structural heterogeneity with respect to their number, localization, and composition (Mattu et al., 1998). Alterations in the *O*-linked glycan chains have been reported in IgA1 from IgAN patients (reviewed in Emancipator et al., 2005; Julian and Novak, 2004; Smith and Feehally, 2003). Specifically, a deficiency of Gal residues has been detected by the combination of direct glycan analysis and reactivity with lectins that recognize terminal glycans and glycosidic bonds (Fig. 13.1). An absence of Gal in the glycan chain leads to the exposure of GalNAc residues linked to the polypeptide chain. Once exposed, *O*-linked GalNAc is recognized by naturally occurring antibodies of the IgG, IgA1, and perhaps IgM isotypes, resulting in the formation of CIC (Tomana et al., 1997, 1999). Indeed, all of the IgA1 in CICs of IgAN patients appears to contain Gal-deficient *O*-linked glycans.

The *O*-linked glycans are synthesized in a stepwise pathway catalyzed by a family of enzymes termed *glycosyltransferases*. In the initial step, GalNAc is linked to T and S residues by a specific *N*-acetylgalactosaminyltransferase. The linkage of Gal to GalNAc by a β1,3 glycosidic bond and the linkage of SA to Gal by α2,3 and/or to GalNAc by α2,6 bond are mediated by corresponding glycosyltransferases (Julian and Novak, 2004; Novak et al., 2001) (Fig. 13.1). At present, the biosynthetic pathways that result in the exclusion of Gal residues have not been elucidated. Proposed mechanisms include deficiency or impaired function of β1,3 Gal transferase, or premature attachment of SA to GalNAc (Julian and Novak, 2004; Novak et al., 2001). Furthermore, intracellular expression and biological activities of glycosyltransferases are regulated by several cytokines, viral infections of Ig-producing cells, and some bacterial products. It is tempting to speculate that in the initial phase of IgAN, microbial infection influences, directly or indirectly through signals delivered by soluble products of other cell types, the glycosylation process in IgA1-producing cells.

The GalNAc residue at the N-terminus of the hinge region of IgA1 is recognized by antibodies of the IgG and IgA1 isotypes. These ubiquitous antibodies are apparently induced by cross-reactive antigens of microbial origin, including Epstein–Barr virus, respiratory syncytial virus, and Gram-negative bacteria. GalNAc-specific antibodies are found in sera of healthy individuals, but are present in higher titers in IgAN patients (Tomana et al., 1999). Similar antibodies have also been detected in sera of macaques and mice (our unpublished results). *In vitro* experiments have provided evidence for the concept that CICs form when the Gal-deficient hinge region of IgA1 associates with anti-GalNAc antibodies. Reassociation of acid-dissociated CICs was competitively inhibited by GalNAc covalently linked to a carrier, asialoorosomucoid, and also by enzymatically degalactosylated IgA1 (Tomana et al., 1999). In contrast, native IgA1 with Gal-containing *O*-linked glycans did not inhibit reassociation of CICs.

13.2.2.2.4. IgA1-Containing CICs and Mesangial Deposits

Immunohistochemical detection of IgA1-containing deposits in the glomerular mesangium is the diagnostic hallmark of IgAN. However, it has not been conclusively established whether the glomerular deposits originate from CICs or are formed *in situ*. Several lines of evidence provide support for the former possibility. Gal-deficient IgA1 in CIC and mesangial deposits from the same individuals have been shown to share common idiotypic determinants, suggesting a common source (reviewed in Emancipator et al., 2005; Novak et al., 2001). The idiotypes of the IgA1 molecules were not disease-specific, indicating that the Gal-deficient *O*-glycans of the hinge region, not the variable regions, were targets of autoantibodies found in CICs and mesangial deposits. Detection of Gal-deficient IgA1 in the glomerular mesangium of IgAN patients has provided crucial evidence for the pathogenic potential of IgA1 molecules with altered *O*-glycan structure (Allen et al., 2001; Hiki et al., 2001). Further observations in other diseases has strengthened this concept. For example, the presence of Gal-deficient IgA1 in patients with a related disease, Henoch-Schöenlein purpura, was associated with the development of nephritis (Allen et al., 1998). The same researchers subsequently reported that the monoclonal IgA1 in serum from a patient with multiple myeloma and glomerulonephritis was Gal-deficient (Zickerman et al., 2000).

13.2.2.2.5. IgA Catabolism, Clearance of CICs, and Biological Consequences of Their Mesangial Deposition

Plasma IgA, which is produced in the bone marrow, spleen, lymph nodes, and, to a lesser degree, mucosal tissues, is catabolized primarily by hepatocytes (reviewed in Conley and Delacroix, 1987; Mestecky et al., 2005). IgA and other glycoproteins interact with the asialoglycoprotein receptor, which recognizes their terminal Gal and GalNAc residues (Stockert et al., 1982; Stockert, 1995). This clearance process is obviously not effective for CIC in IgAN, due to the Gal-deficiency and masking of GalNAc residues by GalNAc-specific antibodies, as described earlier. Furthermore, the impaired clearance of CICs in IgAN is probably influenced by their size, which inhibits their binding to hepatic receptors. In order to bind to hepatocytes, plasma proteins must pass from the circulation into the space of Disse, through the fenestrae of endothelial cells lining the hepatic sinusoids. However, these endothelial openings are permeable only to molecules with a molecular mass less than 1000 kDa (Socken et al., 1981). Therefore, large CICs escape hepatic clearance and deposit in extrahepatic tissues, including the kidney. It appears that the endothelial fenestrae in glomerular capillaries are significantly larger than those in the liver, thus permitting CICs to be deposited in the mesangium.

Although pathogenic IgA1 in CICs is typically derived from systemic sources such as bone marrow, reduced pIgR-mediated clearance could potentially lead to the buildup of mucosally synthesized IgA1 in the systemic circulation. In this regard, it is significant that polymorphisms in the human *PIGR* gene have been associated with increased susceptibility for IgAN (see Chapter 3).

It has been assumed that CICs deposited in the mesangium associate with a specific receptor on mesangial cells (Monteiro et al., 2002; Novak et al., 2002). This, in turn, leads to the activation and proliferation of mesangial cells, the production of cytokines, and the enhanced formation of matrix proteins (reviewed in Julian and Novak, 2004). Although several potential receptors have been reported to mediate IgA and IgA CIC binding [e.g., the asialoglycoprotein receptor, FcαRI (CD89), pIgR, and the Fcα/μR] (see Chapters 3 and 4), subsequent biochemical and molecular biological studies demonstrated that these receptors are not expressed on mesangial cells (Novak et al., 2002). Currently, a novel transferrin–IgA receptor (CD71) is considered to be the most probable candidate (Haddad et al., 2003; Moura et al., 2001).

In vitro incubation of cultured mesangial cells with size-fractionated CICs revealed that only large (>700 kDa) complexes stimulate cellular proliferation and production of soluble mediators and selected extracellular matrix proteins (Novak et al., 2005). Comparative analysis of gene expression in normal kidney tissue and renal biopsies from IgAN patients revealed specific upregulation of 13 genes, including those involved in the metabolism of extracellular matrix proteins (Waga et al., 2003). Thus, a plausible model of events that occur in IgAN begins to emerge. In genetically susceptible individuals (Julian and Novak, 2004; Novak et al., 2001), Gal-deficient IgA1 is produced as a consequence of microbial infection. Gal-deficient IgA1 with GalNAc as a terminal glycan residue is recognized by naturally occurring GalNAc-specific antibodies, resulting in the formation of large CICs. These CICs are not effectively cleared, due to the absence of Gal and masking of GalNAc normally involved in IgA catabolism, as well as their large size that prevents interactions with the hepatic asialoglycoprotein receptor. Large CICs deposited in the renal mesangium activate resident mesangial cells and induce their proliferation, and they enhance extracellular matrix protein production. It should be emphasized that this is still a working hypothesis, which, nevertheless, provides stimulus for further experimentation and potential selective intervention to prevent disease progression. Consistent with this model, a recent study of 670 Chinese IgAN patients and 494 geographically matched healthy controls has demonstrated an association between polymorphisms in the gene encoding galactosyltransferase core 1-β3-Gal-T and susceptibility to IgAN (Li et al., 2007).

13.2.2.3. Infectious Diseases

A polyclonal induction of IgA antibodies in sera and secretions accompanies a broad spectrum of systemic and mucosal diseases with inflammatory components, including infectious diseases and inflammatory conditions of the urinary, respiratory, and gastrointestinal tracts and the joints (e.g., rheumatoid arthritis), and some autoimmune diseases (reviewed in Heremans, 1974; Russell et al., 1992). The biochemical nature of antigens greatly influences the preferential IgA subclass of induced antibodies. In general, protein and glycoprotein antigens induce predominantly IgA1 antibodies, whereas polysaccharide antigens and cell wall components of bacterial origin preferentially induce IgA2 antibodies. Regardless of the route of infection or immunization, most of the antigen-induced IgA antibodies are polymeric in form, suggesting that J-chain production might be upregulated. During infections, pIgA antibodies have functional advantages over monomeric

IgA (mIgA) antibodies, including greater ability to agglutinate particulate antigens and to neutralize pathogens such as influenza virus (Renegar et al., 1998; Russell and Kilian, 2005). Furthermore, pIgA, unlike mIgA, can be transported into external secretions and exhibits greater binding to cellular IgA receptors (see Chapters 1, 3, and 4). Thus, despite their low intrinsic affinity early in the immune response, the presence of multiple antigen-binding sites endows pIgA with significantly higher avidity, resulting in enhanced elimination of pathogens and their products. Although infections or immunizations induce an increase in antigen-specific pIgA, the *total* levels of IgA in sera and secretions usually remain unaltered. However, certain infectious agents, such as the human immunodeficiency virus (HIV), stimulate a marked polyclonal increase in IgA production (Mestecky et al., 2004). Unexpectedly, and in contrast to IgG, HIV-*specific* IgA responses are either absent or low in sera and external secretions of infected individuals (see Chapter 12). A potential mechanism for HIV-stimulated polyclonal increases in IgA and IgG was suggested by the recent report that HIV-1 envelope proteins trigger polyclonal Ig class-switch recombination through interactions with BAFF and C-type lectin receptors (He et al., 2006).

13.3. Decreased Levels of IgA–IgA Deficiency

13.3.1. Incidence

Selective IgA deficiency (IgAD), defined clinically as a serum IgA level of ≤0.07 g/L, is the most common form of primary immunodeficiency in the Western world, with a prevalence of 1 in 600 individuals. It is equally common among men and women. IgAD is significantly more rare in Asian populations (Table 13.1), strongly suggesting a genetic influence on the development of the disease. Based on an extensive segregation analysis, Borecki and co-workers (1994) suggested that a single gene is responsible for the regulation of serum levels of IgA and, thus, that IgAD might be a monogenic disease.

With a few exceptions, prevalence figures are based on screening of adults; relatively few studies have evaluated the frequency of IgAD in children (Table 13.1). It is still not known whether the deficiency is manifest at birth or whether it appears later in life. However, as the IgA system matures rather late in children, a definitive diagnosis of IgAD cannot be made until the early teens. Once manifested, IgAD is usually persistent for life (Koskinen et al., 1994).

13.3.2. Clinical Manifestations of IgAD

In 1964, IgA deficiency was observed in two healthy individuals, suggesting that IgA might be dispensable in the immune system (Rockey et al., 1964). However, based on a review of 30 cases, it was later recognized that the defect frequently resulted in a propensity for infection and that IgA patients suffer from recurrent upper and, occasionally, lower respiratory tract infections

Table 13.1. Prevalence of IgA deficiency in different populations.

Population	Number	IgA cutoff	Frequency	Reference
United States	73,569	<0.10	1/650	Vyas et al. (1975)
Canada	15,500	<0.05	1/1,300	Pai et al. (1974)
England	11,000	<0.05	1/480	Hobbs (1968)
England	29,745	<0.04	1/522	Holt et al. (1977)
France	15,200	<0.05	1/2,172	Frommel et al. (1973)
France	108,000	<0.02	1/1,300	Ropars et al. (1982)
Spain[a]	1,856	<0.05	1/169	Pereira et al. (1997)
Sweden[b]	70,117	<0.05	1/626	Hammarström et al. (2006)
Finland	64,588	<0.01	1/396	Koistinen (1975)
Iceland	15,663	<0.10	1/633	Ulfarsson et al. (1982)
Hungary	2,800	<0.05	1/933	Kramer et al. (1988)
Czech Rep.	5,310	<0.05	1/408	Litzman et al. (2000)
Nigeria[a]	3,772	<0.01	1/629	Ezeoke (1988)
Brazil	11,576	<0.05	1/965	Carneiro-Sampaio et al. (1989)
China	33,171	<0.05	1/4,100	Feng et al. (1992)
China[b]	8,000	<0.05	1/2,667	Hammarström et al. (2005)
Malaysia	2,025	<0.05	0/2,025	Yadav and Lyngkaran (1979)
Japan	222,597	<0.05	1/18,550	Kanoh et al. (1986)
Japan	93,020	<0.03	1/23,255	Ozawa et al. (1986)

[a]Data on children
[b]Unpublished data.

(Ammann and Hong, 1971). The pathogens most often responsible for recurrent infections in IgAD patients are noncapsulated *Haemophilus influenzae* and *Streptococcus pneumoniae,* and exposure plays a major role in determining the frequency of infections in the patients.

The risk for gastrointestinal infections is also slightly elevated in IgAD patients, and persistence of mucosal infections is often noted. As an example, poliovirus can be detected in the stool of IgAD patients for a considerably longer period than in normal individuals after oral vaccination with live attenuated virus (Savilahti et al., 1988) and was suggested to be the cause of overt polio disease in a vaccinated IgA-deficient child (Asindi et al, 1988). However, chronic poliovirus shedding has not been observed, suggesting that IgA facilitates but is not essential for viral clearance (Halsey et al., 2004).

Selective IgAD has been associated with increased risk for several chronic immune and inflammatory diseases (reviewed in Hammarström and Smith, 2007). A few patients with IgAD who suffer from Crohn's disease have previously been described, but a detailed risk analysis has not been made due to lack of a sufficiently large patient population. Our own unpublished data suggest that IgAD is associated with a sixfold increased risk for development of Crohn's disease. A potential association between IgAD and inflammatory bowel disease highlights the importance of IgA in the maintenance of intestinal homeostasis and regulation of inflammation (see Chapter 10).

Celiac disease, a relatively common autoimmune disorder induced by the ingestion of gluten from wheat, barley, and rye, is also overrepresented among patients with IgAD. Screening of a very large number of samples in our laboratory, submitted for routine testing for antibodies against gliadin, endomysium, or transglutaminase in patients with suspected celiac disease, indicated that IgAD was four times more common among these patients than in the general population. Because demonstration of IgA antibodies against these antigens is required for diagnosis of celiac disease, surrogate markers are currently being sought in IgAD patients with suspected celiac disease. Titers of specific IgG antibodies against the antigens described earlier are often elevated in celiac disease, but the diagnostic value of these antibodies in IgAD patients is yet to be fully established.

Selective IgAD has also been suggested to be more common in patients with type 1 diabetes, systemic lupus erythematosus, and a variety of other autoimmune disorders. However, it is not clear whether IgAD per se causes a predisposition for autoimmunity; alternative explanations for a link between IgAD and autoimmunity include the increased rate of infections in IgAD patients or a shared genetic predisposition (see Sect. 13.3.6).

Theoretically, IgAD could result in a predisposition to allergic diseases, due to increased levels of circulating antigens because of increased permeability at mucosal sites. However, surprisingly few allergies have been reported in IgAD patients. An exception is the anaphylactic reaction that can occur when an IgAD patient produces IgE anti-IgA antibodies following exposure to IgA containing blood products (see Sect. 13.3.4). Clinical experience suggests that the prevalence of IgE-mediated allergic symptoms is lower in IgAD patients than in the general population, possibly reflecting the reduced rate of class switching to isotypes, including IgE, that are downstream of IgG1 (see below).

Although a high frequency of cancer, especially malignant lymphomas, has long been noted among patients with primary immunodeficiencies of the B-lymphocyte lineage, the cancer risk of IgAD patients had not been evaluated until recently. In a retrospective of 386 Danish and Swedish patients with IgAD, followed for 20 years, no increased risk for malignancy was observed (Mellemkjaer et al., 2002).

13.3.3. IgA Deficiency and Common Variable Immunodeficiency

By definition, IgAD is considered to be selective and confined to the IgA class. However, in many cases, a concurrent abnormality in the IgG subclass pattern is seen, with a lack of specific antipolysaccharide antibodies of the IgG2 subclass (Hammarström et al., 1985c) or a total lack of serum IgG2 (Oxelius et al., 1981). In other cases, deficiencies in IgG4 and IgE are also observed (Hammarström et al., 1986), reflecting a partial or complete block in class switching to Ig heavy-chain gene segments downstream of $\gamma 1$.

In approximately 1 in 50 patients, IgAD progresses over a period of 20 or so years to common variable immunodeficiency (CVID) (reviewed in Mohammadi et al., 2007). CVID is a more severe primary immunodeficiency, characterized by reduced production of both IgA and IgG and, in about half of the patients, IgM as well. The observation that IgAD can progress to CVID suggests that the two disorders might reflect facets of the same underlying defect. In support of this notion, a linkage of IgAD and CVID has been observed among family members in ~20% of Swedish multicase families, usually presenting as CVID in the parental generation and IgAD in the children (Vorechovsky et al., 1995). In 27 families, the disorders segregated as an autosomal dominant trait, affecting at least two generations, albeit with a limited penetrance. However, the fact that ~80% of cases of IgAD appear to be sporadic suggests that the familial cases are due to a genetic defect that is distinct from the underlying cause of IgAD in the majority of patients.

13.3.4. Immunological Abnormalities

The hallmark of IgAD is a low level of serum IgA with normal levels of serum IgM and IgG. However, elevated levels of IgG, mainly of the IgG1 and IgG3 subclasses, might be seen in up to a third of the patients. In the vast majority of IgAD patients, there is a concomitant lack of SIgA in external secretions; a saliva sample might thus suffice for diagnosis in children, avoiding blood sampling. Although a compensatory rise in salivary IgM has been noted in some IgAD patients, there is no definite association of elevated IgM levels with protection against infection.

The number of surface IgA-positive B-cells is usually low in patients with IgAD, and most of those IgA-positive cells have an immature phenotype with coexpression of IgD (Conley and Cooper, 1981). A large number of additional cell surface markers have been analyzed in B-cells from IgAD patients, but they currently have little or no diagnostic value.

Antibodies against IgA are a common finding in IgAD, and ~40% of patients have demonstrable titers. The etiology of this immune response is unknown, and these autoantibodies are not seen in individuals with normal levels of IgA. The antibodies are usually of the IgG class, mainly IgG1 and IgG4 (Hammarström et al., 1983), but IgM anti-IgA antibodies might also be present (Björkander et al., 1987). In exceptional cases, the antibodies have been suggested to be of the IgE class (Burks et al., 1986; Ferreira et al., 1988), which might then result in serious anaphylactic reactions upon exposure to IgA-containing blood products. Development of anti-IgA antibodies is, like many other immune responses, associated with particular major histocompatibility complex (MHC) alleles (Olerup et al., 1990; Strothman et al., 1989) and Gm allotypes (Hammarström et al., 1985a).

13.3.5. Mechanisms Involved in Development of IgAD

Manifestation of IgAD is apparent at the stem cell level; transfer of bone marrow from an IgAD donor to a normal recipient results in IgA deficiency in the recipient (Hammarström et al., 1985b), whereas transfer of bone marrow from a normal individual to an IgAD patient corrects the defect (Kurobane et al., 1991).

Several lines of evidence suggest that lack of expression of IgA is due to dysregulation of Ig class switching rather than mutations in the coding regions of the α-heavy-chain gene segments. Normal IgA expression is often seen in the children of IgAD parents (Hammarström et al., 1987), indicating that the inherited α-heavy-chain genes were not intrinsically defective. Furthermore, *in vitro* stimulation of B-cells from the vast majority of IgAD patients, using anti-CD40 antibodies or trimeric CD40L together with interleukin (IL)-10, results in normal or near-normal production of IgA (Briere et al., 1994, Friman et al., 1996). The finding of IgAD in only one of two monozygotic twins (Huntley and Stephenson, 1968, Lewkonia *et al.*, 1976, Ulfarsson *et al.*, 1982) is also consistent with a regulatory defect, which might be triggered by an environmental, possibly infectious, agent.

Selective IgAD is occasionally seen in patients treated with antirheumatic and antiepileptic drugs (reviewed in Zhou et al., 2007). IgAD is reversible after cessation of treatment in approximately half of these patients, although full recovery requires months or even years. Among a group of 350 patients with rheumatoid arthritis and other inflammatory joint diseases treated with sulfasalazine, 3% developed IgAD between 8 and 20 weeks after starting drug therapy (Farr et al., 1991), suggesting that selected individuals might be genetically predisposed to develop drug-induced IgAD.

Selective IgAD has been reported in the offspring of IgA-deficient mothers with circulating anti-IgA antibodies, suggesting that placental transmission of maternal anti-IgA antibodies could have interfered with the developing IgA system of the infants (de Laat et al., 1991; Petty et al., 1985). In support of this concept, two studies have demonstrated that anti-IgA antibodies inhibit mitogen-induced IgA production in human B-cells *in vitro* (Hammarström et al., 1983; Warrington et al., 1982).

13.3.6. Genetic Basis of IgAD

The human leukocyte antigens (HLAs), encoded by the MHC, have been implicated in the genetic susceptibility to a large number of immune-related diseases. In the late 1970s and early 1980s, associations were reported between IgAD and the MHC class I alleles HLA-A1 and HLA-B8 (Ambrus et al., 1977; Oen et al., 1982). Hammarström and Smith (1983) subsequently demonstrated an association between certain MHC class II alleles and IgAD. A later report by Wilton and co-workers (1985) suggested that genes within the MHC class III region were associated with IgAD, although the gene(s) could not be identified.

Our own work suggests that the class II region, in particular the HLA-DQ locus, is involved in mediating both susceptibility and resistance to development of IgAD (Kralovicova et al., 2003; Olerup et al., 1990, 1991). Comparison of the sequences of the polymorphic N-terminal domain of the DQ chain showed that three allelic forms associated with susceptibility to IgAD all had a neutral alanine or valine at position 57, whereas the "protective" allelic form had a negatively charged aspartic acid at this position. Recently, independent reports have suggested that two distinct loci in two different MHC haplotypes, one in the class II region and one in the class III region, confer susceptibility to development of IgAD (de la Concha et al., 2002; Gual et al., 2004;

Schroeder et al., 2004). However, the strong linkage disequilibrium within the MHC region makes it difficult to locate a specific susceptibility gene, and the question of whether IgAD is associated with genes present within the MHC class I, II, and/or III region has not yet been answered.

Although gene(s) within the MHC locus appear to be major contributor(s) to development of IgAD, additional loci have also been implicated, including chromosome 4q (Finck et al., 2006), chromosome 16q (Schaffer et al., 2006), chromosome 17q (Castigli et al., 2005), and chromosome 18 (reviewed in Hammarström and Smith, 2007). However, the pathogenetic effect of mutations in the gene-encoding transmembrane activator and calcium modulator and cyclophilin ligand interactor (TACI) on chromosome 17 has recently been challenged, and it is likely that these mutations contribute little, if anything, to IgAD in adults (Pan-Hammarström et al., 2007). The lack of IgA in mice with a targeted deletion of the gene encoding APRIL (Castigli et al., 2004), one of the ligands for TACI, has prompted a search for mutations in this gene in patients with IgAD. However, no abnormalities have been found to date.

In conclusion, induction of IgAD probably requires the presence of a predisposing gene and a triggering factor, the identities of which are still unknown. In most of the multicase families described to date, affected family members are found in more than one generation, which is compatible with an autosomal dominant inheritance with a limited penetrance. The key to understanding the true basis of IgAD is the identification of the gene(s) involved in its pathogenesis.

13.3.7. Defective Immunoglobulin Class Switching in IgAD

Germline transcripts containing the α gene are easily detected in unstimulated peripheral blood mononuclear cells from normal donors, whereas they are absent in cells from IgAD patients (Islam et al., 1994). However, they can readily be induced by the addition of transforming growth factor (TGF)-β, demonstrating that the locus is accessible for transcription as long as additional signals are supplied. Although TGF-β has been suggested as a switch factor for IgA, no difference was found in steady-state levels of TGF-β mRNA in B-cells from IgAD patients versus normal individuals. However, serum levels of the TGF-β have been reported to be markedly lower in IgAD patients as compared to controls (Muller et al., 1995), suggesting that a dysregulation in TGF-β might indeed be involved in the pathogenesis of IgAD.

In mice, IL-5 has been shown to be important for IgA production, and *in vivo* administration of IL-5 to low-responder mice results in an enhanced IgA response (Dieli et al., 1995). In humans, expression of IL-5 is regulated by genes within the MHC complex (Dieli et al., 1995), suggesting that dysregulation of IL-5 could be linked to susceptibility for IgAD. We have observed impairment of IL-5 production *in vitro* in cells from individuals carrying the extended HLA-B8, DR3 haplotype that is strongly associated with IgAD (Lio et al., 1995). However, we have failed to demonstrate any decrease in the frequency of IL-5-producing cells or in the level of IL-5 mRNA in individual cells from IgAD individuals (Smith et al., 1990). These data suggest that diminished production of IL-5 is not a part of the pathophysiological process in IgAD. However, although the timing and concentration of cytokines might play a crucial role in the switching process, there is, as yet, no strong evidence to implicate any particular cytokine or cytokine receptor in the development of IgAD.

13.3.8. Treatment of IgAD

The foundation of therapy for IgAD patients is the liberal use of antibiotics. Gamma-globulin infusions have been used for decades in patients with CVID, but they have previously been considered to be contraindicated in patients with IgAD. More recently, infection-prone patients with IgAD have been treated with gamma-globulin, with therapeutically beneficial effects (Gustafson et al., 1997). The doses employed have gradually risen, and, today, a weekly dose of 100–150 mg gamma-globulin/kg body weight by the subcutaneous route is recommended (Gardulf et al., 1991; Gardulf and Hammarström, 1996).

Oral and nasal administration of polyclonal human IgA has previously been shown to be therapeutically effective against selected infections (Hammarström et al., 1993; Lindberg et al., 1993; Tjellström et al., 1993). Although this form of therapy remains experimental, it might constitute a therapeutic option in selected cases.

Acknowledgments. Supported in part by grant AI28147 from the National Institutes of Health, and Research Project MSMT #0021620819 from the Czech Republic.

References

Allen, A. C., Bailey, E.M., Brenchley, P. E., Buck, K. S., Barratt, J., and Feehally, J. (2001). Mesangial IgA1 in IgA nephropathy exhibits aberrant *O*-glycosylation: observations in three patients. *Kidney Int.* 60:969–973.

Allen, A. C., Willis, F. R., Beattie, T. J., and Feehally, J. (1998). Abnormal IgA glycosylation in Henoch-Schönlein purpura restricted to patients with clinical nephritis. *Nephrol. Dial. Transplant.* 13:930–934.

Alley, C. D., Nash, G. S., and MacDermott, R. P. (1982). Marked *in vitro* spontaneous secretion of IgA by human rib bone marrow mononuclear cells. *J. Immunol.* 128:2604–2608.

Ambrus, M., Hernadi, E., and Bajtai, G. (1977). Prevalence of HLA-A1 and HLA-B8 antigens in selective IgA deficiency. *Clin. Immunol. Immunopathol.* 7:311–314.

Ammann, A. J., and Hong, R. (1971). Selective IgA deficiency: presentation of 30 cases and a review of the literature. *Medicine* 50:223–236.

Asindi, A. A., Bell, E. J., Browning, M. J., and Stephenson, J. B. (1988). Vaccine-induced polioencephalomyelitis in Scotland. *Scott. Med. J.* 33:306–307.

Bast, E. J., Van Camp, B., Boom, S. E., Jaspers, F. C., and Ballieux, R. E. (1981). Differentiation between benign and malignant monoclonal gammopathy by the presence of the J chain. *Clin. Exp. Immunol.* 44:375–382.

Björkander, J., Hammarström, L., Smith, C. I., Buckley, R. H., Cunningham-Rundles, C., and Hanson, L. Å. (1987). Immunoglobulin prophylaxis in patients with antibody deficiency syndromes and anti-IgA antibodies. *J. Clin. Immunol.* 7:8–15.

Borecki, I. B., McGue, M., Gerrard, J. W., Lebowitz, M. D., and Rao, D. C. (1994). Familial resemblance for immunoglobulin levels. *Hum. Genet.* 94:179–185.

Briere, F., Bridon, J. M., Chevet, D., Souillet, G., Bienvenu, F., Guret, C., Martinez-Valdez, H., and Banchereau, J. (1994). Interleukin 10 induces B lymphocytes from IgA-deficient patients to secrete IgA. *J. Clin. Invest.* 94:97–104.

Brown, W. R. (2005). Liver and biliary tract. In: Mestecky, J., Bienenstock, J., Lamm, M. E., Mayer, L., McGhee, J. R., and Strober, W. (eds.), *Mucosal Immunology*, 3rd. ed., Elsevier/Academic Press, Amsterdam, pp. 1373–1387.

Burks, A. W., Sampson, H. A., and Buckley, R. H. (1986). Anaphylactic reactions after gamma globulin administration in patients with hypogammaglobulinemia. Detection of IgE antibodies to IgA. *N. Engl. J. Med.* 314:560–564.

Carneiro-Sampaio, M. M., Carbonare, S. B., Rozentraub, R. B., de Araujo, M. N., Riberiro, M. A., and Porto, M. H. (1989). Frequency of selective IgA deficiency among Brazilian blood donors and healthy pregnant women. *Allergol. Immunopathol.* 17:213–216.

Castigli, E., Scott, S., Dedeoglu, F., Bryce, P., Jabara, H., Bhan, A. K., Mizoguchi, E., and Geha, R. S. (2004). Impaired IgA class switching in APRIL-deficient mice. *Proc. Natl. Acad. Sci. U.S.A.* 101:3903–3908.

Castigli, E., Wilson, S. A., Garibyan, L., Rachid, R., Bonilla, F., Schneider, L., and Geha, R. S. (2005). TACI is mutant in common variable immunodeficiency and IgA deficiency. *Nat. Genet.* 37:829–834.

Conley, M. E., and Cooper, M. D. (1981). Immature IgA B cells in IgA-deficient patients. *N. Engl. J. Med.* 305:495–497.

Conley, M. E., and Delacroix, D. L. (1987). Intravascular and mucosal immunoglobulin A: Two separate but related systems of immune defense? *Ann. Intern. Med.* 106:892–899.

Cripps, A. W., and Gleeson, M. (2005). Ontogeny of mucosal immunity and aging. In: Mestecky, J., Bienenstock, J., Lamm, M. E., Mayer, L., McGhee, J. R., and Strober, W. (eds.), *Mucosal Immunology*, 3rd. ed., Elsevier/Academic Press, Amsterdam, pp. 305–321.

Czerkinsky, C., Koopman, W. J., Jackson, S., Collins, J. E., Crago, S. S., Schrohenloher, R. E., Julian, B. A., Galla, J. H., and Mestecky, J. (1986). Circulating immune complexes and immunoglobulin A rheumatoid factor in patients with mesangial immunoglobulin A nephropathies. *J. Clin. Invest.* 77:1931–1938.

de la Concha, E. G., Fernandez-Arquero, M., Gual, L., Vigil, P., Martinez, A., Urcelay, E., Ferreira, A., Garcia-Rodriguez, M. C., and Fontan, G. (2002). MHC susceptibility genes to IgA deficiency are located in different regions on different HLA haplotypes. *J. Immunol.* 169:4637–4643.

de Laat, P. C., Weemaes, C. M., Bakkeren, J. A., van den Brandt, F. C., van Lith, T. G., de Graaf, R., van Munster, P. J., and Stoelinga, G. B. (1991). Familial selective IgA deficiency with circulating anti-IgA antibodies: a distinct group of patients? *Clin. Immunol. Immunopathol.* 58:92–101.

Delacroix, D. L., Elkon, K. B., Geubel, A. P., Hodgson, H. F., Dive, C., and Vaerman, J. P. (1983). Changes in size, subclass, and metabolic properties in serum immunoglobulin A in liver diseases with high serum immunoglobulin A. *J. Clin. Invest.* 71:358–367.

Delacroix, D. L., and Vaerman, J. P. (1983). Function of the human liver in IgA homeostasis in plasma. *Ann. NY Acad. Sci.* 409:383–401.

Dieli, F., Asherson, G. L., Sireci, G., Lio, D., Bonanno, C. T., and Salerno, A. (1995). IL-5 enhances *in vitro* and *in vivo* antigen-specific IgA production in MHC genetically determined low IL-5 responder mice. *Cell. Immunol.* 163:309–313.

Donadio, J. V., and Grande, J. P. (2002). IgA nephropathy. *N. Engl. J. Med.* 347:738–748.

Emancipator, S., Mestecky, J., and Lamm, M. E. (2005). IgA nephropathy and related diseases. In: Mestecky, J., Bienenstock, J., Lamm, M. E., Mayer, L., McGhee, J. R., and Strober, W. (eds.), *Mucosal Immunology*, 3rd. ed., Elsevier/Academic Press, Amsterdam, pp. 1579–1600.

Ezeoke, A. C. (1988). Selective IgA deficiency (SIgAD) in Eastern Nigeria. *Afr. J. Med. Sci.* 17:17–21.

Farr, M., Kitas, G. D., Tunn, E. J., and Bacon, P. A. (1991). Immunodeficiencies associated with sulphasalazine therapy in inflammatory arthritis. *Br. J. Rheumatol.* 30:413–417.

Feng, L. (1992). Epidemiological study of selective IgA deficiency among 6 nationalities in China. *Chin. Med. J.* 72:88–90.

Ferreira, A., Garcia Rodriguez, M. C., Lopez-Trascasa, M., Pascual Salcedo, D., and Fontan, G. (1988). Anti-IgA antibodies in selective IgA deficiency and in primary immunodeficient patients treated with γ-globulin. *Clin. Immunol. Immunopathol.* 47:199–207.

Finck, A., van der Meer, J. W. M., Schaffer, A. A., Pfannstiel, J., Fieschi, C., Plebani, A., Webster, A. D. B., Hammarström, L., and Grimbacher, B. (2006). Linkage of autosomal dominant common variable immunodeficiency to chromosome 4q. *Eur. J. Hum. Genet.* 14:867–875.

Friman, V., Hanson, L. A., Bridon, J. M., Tarkowski, A., Banchereau, J., and Briere, F. (1996). IL-10-driven immunoglobulin production by B lymphocytes from IgA-deficient individuals correlates to infection proneness. *Clin. Exp. Immunol.* 104:432–438.

Frommel, D., Moullec, J., Lambin, P., and Fine, J. M. (1973). Selective serum IgA deficiency. Frequency among 15,200 French blood donors. *Vox Sang.* 25:513–518.

Galla, J. H., Spotswood, M. F., Harrison, L. A., and Mestecky, J. (1985). Urinary IgA in IgA nephropathy and Henoch-Schoenlein purpura. *J. Clin. Immunol.* 5:298–306.

Gardulf, A., and Hammarström, L. (1996). Subcutaneous administration of immunoglobulins. What are the advantages. *Clin. Immunother.* 6:108–116.

Gardulf, A., Hammarström, L., and Smith, C. I. (1991). Home treatment of hypogammaglobulinaemia with subcutaneous gammaglobulin by rapid infusion. *Lancet* 338:162–166.

Gual, L., Martinez, A., Fernandez-Arquero, M., Garcia-Rodriguez, M. C., Ferreira, A., Fontan, G., de la Concha, E. G., and Urcelay, E. (2004). Major histocompatibility complex haplotypes in Spanish immunoglobulin A deficiency patients: A comparative fine mapping microsatellite study. *Tissue Antigens* 64:671–677.

Gustafson, R., Gardulf, A., Granert, C., Hansen, S., and Hammarström, L. (1997). Prophylactic therapy for selective IgA deficiency. *Lancet* 350:865.

Haddad, E., Moura, I. C., Arcos-Fajardo, M., Macher, M. A., Baudouin, V., Alberti, C., Loirat, C., Monteiro, R. C., and Peuchmaur, M. (2003). Enhanced expression of the CD71 mesangial IgA1 receptor in Berger disease and Henoch-Schönlein nephritis: Association between CD71 expression and IgA deposits. *J. Am. Soc. Nephrol.* 14:327–337.

Halsey, N. A., Pinto, J., Espinosa-Rosales, F., Faure-Fontenla, M. A., da Silva, E., Khan, A. J., Webster, A. D., Minor, P., Dunn, G., Asturias, E., Hussain, H., Pallansch, M. A.,

Kew, O. M., Winkelstein, J., and Sutter, R. (2004). Search for poliovirus carriers among people with primary immune deficiency diseases in the United States, Mexico, Brazil, and the United Kingdom. *Bull. World Health Org.* 82:3–8.

Hammarström, L., de Lange, G. G., and Smith, C. (1987). IgA2 allotypes determined by restriction fragment length polymorphism in IgA deficiency. Re-expression of the silent A2m(2) allotype in the children of IgA-deficient patients. *J. Immunogenet.* 14:197–201.

Hammarström, L., Grubb, R., Jakobsen, B. K., Oxelius, V., Persson, U., and Smith, C. I. E. (1986). Concomitant deficiency of IgG4 and IgE in IgA deficient donors with high titres of anti-IgA. *Monogr. Allergy* 20:234–235.

Hammarström, L., Grubb, R., and Smith, C. I. (1985a). Gm allotypes in IgA deficiency. *J. Immunogenet.* 12:125–130.

Hammarström, L., Lönnqvist, B., Ringdén, O., Smith, C.I., and Wiebe, T. (1985b). Transfer of IgA deficiency to a bone-marrow-grafted patient with aplastic anaemia. *Lancet* 1:778–781.

Hammarström, L., Persson, M. A., and Smith, C. I. (1983). Anti-IgA in selective IgA deficiency. *In vitro* effects and Ig subclass pattern of human anti-IgA. *Scand. J. Immunol.* 18:509–513.

Hammarström, L., Persson, M. A., and Smith, C. I. (1985c). Immunoglobulin subclass distribution of human anti-carbohydrate antibodies: aberrant pattern in IgA-deficient donors. *Immunology* 54:821–826.

Hammarström, L., and Smith, C. I. (1983). HLA-A, B, C and DR antigens in immunoglobulin A deficiency. *Tissue Antigens* 21:75–79.

Hammarström, L., and Smith, C. I. E. (2007). Genetic approach to common variable immunodeficiency and IgA deficiency. In: Ochs, H., Smith, C. I. E., and Puck, J. (eds.), *Primary Immunodeficiency Diseases, a Molecular and Genetic Approach.* Oxford University Press, pp. 313–325.

Hammarström, V., Smith, C. I., and Hammarström, L. (1993). Oral immunoglobulin treatment in *Campylobacter jejuni* enteritis. *Lancet* 341:1036.

He, B., Qiao, X., Klasse, P. J., Chiu, A., Chadburn, A., Knowles, D. M., Moore, J. P., and Cerutti, A. (2006). HIV-1 envelope triggers polyclonal Ig class switch recombination through a CD40-independent mechanism involving BAFF and C-type lectin receptors. *J. Immunol.* 176:3931–3941.

Heremans, J. F. (1974). Immunoglobulin A. In: Sela, M. (ed.), *The Antigens.* Academic Press, New York, pp. 365–522.

Hiki, Y., Odani, H., Takahashi, M., Yasuda, Y., Nishimoto, A., Iwase, H., Shinzato, T., Kobayashi, Y., and Maeda, K. (2001). Mass spectrometry proves under *O*-glycosylation of glomerular IgA1 in IgA nephropathy. *Kidney Int.* 59:1077–1085.

Hobbs, J. R. (1968). Immune imbalance in dysgammaglobulinaemia type IV. *Lancet* 1:110–114.

Holt, P. D., Tandy, N. P., and Anstee, D. J. (1977). Screening of blood donors for IgA deficiency: A study of the donor population of south-west England. *J. Clin. Pathol.* 30:1007–1010.

Huntley, C. C., and Stephenson, R. L. (1968). IgA deficiency: family studies. *NC Med. J.* 29:325–331.

Islam, K. B., Baskin, B., Nilsson, L., Hammarström, L., Sideras, P., and Smith, C. I. (1994). Molecular analysis of IgA deficiency. Evidence for impaired switching to IgA. *J. Immunol.*152:1442–1452.

Julian, B. A., and Novak, J. (2004). IgA nephropathy: An update. *Curr. Opin. Nephrol. Hypertens.* 13:171–179.

Julian, B. A., Tomana, M., Novak, J., and Mestecky, J. (1999). Progress in pathogenesis of IgA nephropathy. In: Grunfeld, J.-P., Bach, J. F., and Kreis, H. (eds.), *Advances in Nephrology*, Vol. 29, Mosby, St. Louis, MO, pp. 53–72.

Kanoh, T., Mizumoto, T., Yasuda, N., Koya, M., Ohno, Y., Uchino, H., Yoshimura, K., Ohkubo, Y., and Yamaguchi, H. (1986). Selective IgA deficiency in Japanese blood donors: Frequency and statistical analysis. *Vox Sang.* 50:81–86.

Kilian, M., and Russell, M. W. (2005). Microbial evasion of IgA functions. In: Mestecky, J., Bienenstock, J., Lamm, M. E., Mayer, L., McGhee, J. R., and Strober, W. (eds.), *Mucosal Immunology*, 3rd. ed., Elsevier/Academic Press, Amsterdam, pp. 291–303.

Koistinen, J. (1975). Selective IgA deficiency in blood donors. *Vox Sang.* 29:192–202.

Koskinen, S., Tolo, H., Hirvonen, M., and Koistinen, J. (1994). Long-term persistence of selective IgA deficiency in healthy adults. *J. Clin. Immunol.* 14:116–119.

Kralovicova, J., Hammarström, L., Plebani, A., Webster, A. D., and Vorechovsky, I. (2003). Fine-scale mapping at *IGAD1* and genome-wide genetic linkage analysis implicate *HLA-DQ/DR* as a major susceptibility locus in selective IgA deficiency and common variable immunodeficiency. *J. Immunol.* 170:2765–2775.

Kramer, J., Kassai, T., Medgyesi, G. A., Tauszik, T., and Fust, G. (1988). Screening of IgA deficiency by ELISA: Population frequency in Hungary. *Haematologia* 21:233–238.

Kubagawa, H., Bertoli, L. F., Barton, J. C., Koopman, W. J., Mestecky, J., and Cooper, M. D. (1987). Analysis of paraprotein transport into saliva by using anti-idiotype antibodies. *J. Immunol.* 138:435–439.

Kurobane, I., Riches, P. G., Sheldon, J., Jones, S., and Hobbs, J. R. (1991). Incidental correction of severe IgA deficiency by displacement bone marrow transplantation. *Bone Marrow Transplant.* 7:494–495.

Kutteh, W. H., Prince, S. J., and Mestecky, J. (1982a). Tissue origins of human polymeric and monomeric IgA. *J. Immunol.* 128:990–995.

Kutteh, W. H., Prince, S. J., Phillips, J. O., Spenney, J. G., and Mestecky, J. (1982b). Properties of immunoglobulin A in serum of individuals with liver diseases and in hepatic bile. *Gastroenterology* 82:184–193.

Lewkonia, R. M., Gairdner, D., and Doe, W. F. (1976). IgA deficiency in one of identical twins. *Br. Med. J.* 1:311–313.

Li, G. S., Zhang, H., Lv, J. C., Shen, Y., and Wang, H. Y. (2007). Variants of *C1GALT1* gene are associated with the genetic susceptibility to IgA nephropathy. *Kidney Int.* 71:448–453.

Lindberg, K., Samuelson, A., Rynnel-Dagöö, B., Smith, E., and Hammarström, L. (1993). Nasal administration of IgA to individuals with hypogammaglobulinemia. *Scand. J. Infect. Dis.* 25:395–397.

Lio, D., D'Anna, C., Gervasi, F., Cigna, D., Modica, M. A., Candore, G., and Caruso, C. (1995). *In vitro* impairment of interleukin-5 production in HLA-B8, DR3-positive individuals implications for immunoglobulin A synthesis dysfunction. *Hum. Immunol.* 44:170–174.

Litzman, J., Sevcikova, I., Stikarovska, D., Pikulova, Z., Pazdirkova, A., and Lokaj, J. (2000). IgA deficiency in Czech healthy individuals and selected patient groups. *Int. Arch. Allergy Immunol.* 123:177–180.

Lokhorst, H. M., Dekker, A. W., Baarlen, J. V., and Bast, E. J. E. G. (1990). J chain disease: An aggressive evolution of multiple myeloma. *Am. J. Med.* 88:417–420.

Mattu, T. S., Pleass, R. J., Willis, A. C., Kilian, M., Wormald, M. R., Lellouch, A. C., Rudd, P. M., Woof, J. M., and Dwek, R. A. (1998). The glycosylation and structure

of human serum IgA1, Fab, and Fc regions and the role of *N*-glycosylation on Fc alpha receptor interactions. *J. Biol. Chem.* 273:2260–2272.
Mellemkjaer, L., Hammarström, L., Andersen, V., Yuen, J., Heilmann, C., Barington, T., Björkander, J., and Olsen, J. H. (2002). Cancer risk among patients with IgA deficiency or common variable immunodeficiency and their relatives: A combined Danish and Swedish study. *Clin. Exp. Immunol.* 130:495–500.
Mestecky, J., Jackson, S., Moldoveanu, Z., Nesbit, L. R., Kulhavy, R., Prince, S. J., Sabbaj, S., Mulligan, M. J., and Goepfert, P. A. (2004). Paucity of antigen-specific IgA responses in sera and external secretions of HIV-type 1-infected individuals. *AIDS Res. Hum. Retroviruses* 20:972–988.
Mestecky, J., Moldoveanu, Z., Julian, B. A., and Prchal, J. T. (1990a). J chain disease: A novel form of plasma cell dyscrasia. *Am. J. Med.* 88:411–416.
Mestecky, J., Moro, I., Kerr, M. A., and Woof, J. M. (2005). Mucosal immunoglobulins. In: Mestecky, J., Bienenstock, J., Lamm, M. E., Mayer, L., McGhee, J. R., and Strober, W. (eds.), *Mucosal Immunology*, 3rd. ed., Elsevier/Academic Press, Amsterdam, pp. 153–181.
Mestecky, J., Preud'homme, J.-L., Crago, S. S., Mihaesco, E., Prchal, J. T., and Okos, A. J. (1980). Presence of J chain in human lymphoid cells. *Clin. Exp. Immunol.* 39:371–385.
Mestecky, J., and Russell, M. W. (1986). IgA Subclasses. *Monogr. Allergy* 19:277–301.
Mestecky, J., Waldo, F. B., Britt, W. J., Julian, B. A., Tomana, M., van den Wall Bake, A. W. L., Russell, M. W., Galla, J. H., Moldoveanu, Z., and Jackson, S. (1990b). Exogenous antigens deposited in the glomeruli of patients with IgA nephropathy. In: Sakai, H., Sakai, and Nomoto, Y. (eds.), *Pathogenesis of IgA Nephropathy*. Harcourt Brace Jovanovich, Tokyo, pp. 247–257.
Mitsiades, C. S., Mitsiades, N., Munshi, N. C., and Anderson, K. C. (2004). Focus on multiple myeloma. *Cancer Cell* 6:439–444.
Mohammadi, J., Hammarström, L., and Aghamohammadi, A. (2007). Induction of CVID. Case reports and a review of the literature. *Iran J. Allergy Asthma Immunol.* submitted.
Monteiro, R. C., Moura, I. C., Launay, P., Tsuge, T., Haddad, E., Benhamou, M., Cooper, M. D., and Arcos-Fajardo, M. (2002). Pathogenic significance of IgA receptor interactions in IgA nephropathy. *Trends Mol. Med.* 8:464–468.
Moura, I. C., Centelles, M. N., Arcos-Fajardo, M., Malheiros, D. M., Collawn, J. F., Cooper, M. D., and Monteiro, R. C. (2001). Identification of the transferrin receptor as a novel immunoglobulin (Ig)A1 receptor and its enhanced expression on mesangial cells in IgA nephropathy. *J. Exp. Med.* 194:417–425.
Muller, F., Aukrust, P., Nilssen, D. E., and Frøland, S. S. (1995). Reduced serum level of transforming growth factor-beta in patients with IgA deficiency. *Clin. Immunol. Immunopathol.* 76:203–208.
Novak, J., Julian, B. A., Tomana, M., and Mestecky, J. (2001). Progress in molecular and genetic studies of IgA nephropathy. *J. Clin. Immunol.* 21:310–327.
Novak, J., Tomana, M., Matousovic, K., Brown, R., Hall, S., Novak, L., Julian, B. A., Wyatt, R. J., and Mestecky, J. (2005). IgA1-containing immune complexes in IgA nephropathy differentially affect proliferation of mesangial cells. *Kidney Int.* 67:504–513.
Novak, J., Vu, H. L., Novak, L., Julian, B. A., Mestecky, J., and Tomana, M. (2002). Interactions of human mesangial cells with IgA and IgA-containing immune complexes. *Kidney Int.* 62:465–475.

Oen, K., Petty, R. E., and Schroeder, M. L. (1982). Immunoglobulin A deficiency: Genetic studies. *Tissue Antigens* 19:174–182.

Olerup, O., Smith, C. I., and Hammarström, L. (1990). Different amino acids at position 57 of the HLA-DQ beta chain associated with susceptibility and resistance to IgA deficiency. *Nature* 347:289–290.

Olerup, O., Smith, C. I., and Hammarström, L. (1991). Is selective IgA deficiency associated with central HLA genes or alleles of the DR-DQ region? *Immunol. Today* 12:134–136.

Oxelius, V. A., Laurell, A. B., Lindquist, B., Golebiowska, H., Axelsson, U., Björkander, J., and Hanson, L. Å. (1981). IgG subclasses in selective IgA deficiency: importance of IgG2-IgA deficiency. *N. Engl. J. Med.* 304:1476–1477.

Ozawa, N., Shimizu, M., Imai, M., Miyakawa, Y., and Mayumi, M. (1986). Selective absence of immunoglobulin A1 or A2 among blood donors and hospital patients. *Transfusion* 26:73–76.

Pai, M. K., Davison, M., Bedritis, I., and Zipursky, A. (1974). Selective IgA deficiency in Rh-negative women. *Vox Sang.* 27:87–91.

Pan-Hammarström, Q., Du, L., Björkander, J., Cunningham-Rundles, C., Nelson, D. L., Salzer, U., Offer, S., Behrens, T. W., Grimbacher, B., and Hammarström, L. (2007). Mutations in TNFRSF13B encoding TACI and selective IgA deficiency. *Nat. Genet.* 39:429–430.

Peppard, J. V., Kaetzel, C. S., and Russell, M. W. (2005). Phylogeny and comparative physiology of IgA. In: Mestecky, J., Bienenstock, J., Lamm, M. E., Mayer, L., McGhee, J. R., and Strober, W. (eds.), *Mucosal Immunology*, 3rd. ed., Elsevier/Academic Press, Amsterdam, pp. 195–210.

Pereira, L. F., Sapina, A. M., Arroyo, J., Vinuelas, J., Bardaji, R. M., and Prieto, L. (1997). Prevalence of selective IgA deficiency in Spain: More than we thought. *Blood* 90:893.

Petty, R. E., Sherry, D. D., and Johannson, J. (1985). Anti-IgA antibodies in pregnancy. *N. Engl. J. Med.* 313:1620–1625.

Potter, M. (1972). Immunoglobulin-producing tumors and myeloma proteins of mice. *Physiol. Rev.* 52:631–719.

Radl, J., Schuit, H. R. E., Mestecky, J., and Hijmans, W. (1974). The origin of monomeric and polymeric forms of IgA in man. *Adv. Exp. Med. Biol.* 45:57–65.

Rambaud, J.-C., Brouet, J.-C., and Seligmann, M. (1994). Alpha chain disease and related lymphoproliferative disorders. In: Ogra, P. L., Mestecky, J., Lamm, M. E., Strober, W., McGhee, J. R., and Bienenstock, J. (eds.), *Handbook of Mucosal Immunology*. Academic Press, New York, pp. 425–433.

Renegar, K. B., Jackson, G. D., and Mestecky, J. (1998). *In vitro* comparison of the biologic activities of monoclonal monomeric IgA, polymeric IgA, and secretory IgA. *J. Immunol.* 160:1219–1223.

Rockey, J. H., Hanson, L. Å., Heremans, J. F., and Kunkel, H. G. (1964). Beta-2A aglobulinemia in two healthy men. *J. Lab. Clin. Med.* 63:205–212.

Ropars, C., Muller, A., Paint, N., Beige, D., and Avenard, G. (1982). Large scale detection of IgA deficient blood donors. *J. Immunol. Methods* 54:183–189.

Russell, M. W., and Kilian, M. (2005). Biological activities of IgA. In: Mestecky, J., Bienenstock, J., Lamm, M. E., Mayer, L., McGhee, J. R., and Strober, W. (eds.), *Mucosal Immunology*, 3rd. ed., Elsevier/Academic Press, Amsterdam, pp. 267–289.

Russell, M. W., Lue, C., van den Wall Bake, A. W. L., Moldoveanu, Z., and Mestecky, J. (1992). Molecular heterogeneity of human IgA antibodies during an immune response. *Clin. Exp. Immunol.* 87:1–6.

Russell, M. W., Mestecky, J., Julian, B. A., and Galla, J. H. (1986). IgA-associated renal diseases. Antibodies to environmental antigens in sera and deposition of immunoglobulins and antigens in glomeruli. *J. Clin. Immunol.* 6:74–86.

Savilahti, E., Klemola, T., Carlsson, B., Mellander, L., Stenvik, M., and Hovi, T. (1988). Inadequacy of mucosal IgM antibodies in selective IgA deficiency: Excretion of attenuated polio viruses is prolonged. *J. Clin. Immunol.* 8:89–94.

Schaffer, A. A., Pfannstiel, J., Webster, A. D., Plebani, A., Hammarström, L., and Grimbacher, B. (2006). Analysis of families with common variable immunodeficiency (CVID) and IgA deficiency suggests linkage of CVID to chromosome 16q. *Hum. Genetics* 118:725–729.

Schroeder, H. W., Jr., Schroeder, H. W., 3rd, and Sheikh, S. M. (2004). The complex genetics of common variable immunodeficiency. *J. Investig. Med.* 52:90–103.

Smith, A. C., and Feehally, J. (2003). New insights into the pathogenesis of IgA ephropathy. Pathogenesis of IgA nephropathy. *Springer Semin. Immunopathol.* 24:477–493.

Smith, C. I., Moller, G., Severinson, E., and Hammarström, L. (1990). Frequencies of interleukin-5 mRNA-producing cells in healthy individuals and in immunoglobulin-deficient patients, measured by in situ hybridization. *Clin. Exp. Immunol.* 81:417–422.

Socken, D. J., Simms, E. S., Nagy, B. R., Fisher, M. M., and Underdown, B. J. (1981). Secretory component-dependent hepatic transport of IgA antibody-antigen complexes. *J. Immunol.* 127:316–319.

Stockert, R. J. (1995). The asialoglycoprotein receptor: Relationships between structure, function, and expression. *Physiol. Rev.* 75:591–609.

Stockert, R. J., Kressner, M. S., Collins, J. C., Sternlieb, I., and Morell, A. G. (1982). IgA interaction with the asialoglycoprotein receptor. *Proc. Natl. Acad. Sci. U.S.A.* 79:6229–6231.

Strothman, R. A., Sedestrom, L. M., Ball, M. J., and Chen, S. N. (1989). HLA association of anti-IgA antibody production. *Tissue Antigens* 34:141–144.

Tjellström, B., Stenhammar, L., Eriksson, S., and Magnusson, K. E. (1993). Oral immunoglobulin A supplement in treatment of *Clostridium difficile* enteritis. *Lancet* 341:701–702.

Tomana, M., Matousovic, K., Julian, B. A., Radl, J., Konecny, K., and Mestecky, J. (1997). Galactose-deficient IgA1 in sera of IgA nephropathy patients is present in complexes with IgG. *Kidney Int.* 52:509–516.

Tomana, M., Novak, J., Julian, B. A., Matousovic, K., Konecny, K., and Mestecky, J. (1999). Circulating immune complexes in IgA nephropathy consist of IgA1 with galactose-deficient hinge region and antiglycan antibodies. *J. Clin. Invest.* 104:73–81.

Ulfarsson, J., Gudmundsson, S., Birgisdottir, B., Kjeld, J. M., and Jensson, O. (1982). Selective serum IgA deficiency in Icelanders. Frequency, family studies and Ig levels. *Acta Med. Scand.* 211:481–487.

Vorechovsky, I., Zetterquist, H., Paganelli, R., Koskinen, S., Webster, A. D. B., Björkander, J., Smith, C. I. E., and Hammarström, L. (1995). Family and linkage study of selective IgA deficiency and common variable immunodeficiency. *Clin. Immunol. Immunopathol.* 77:214–218.

Vyas, G. N., Perkins, H. A., Yang, Y. M., and Basantani, G. K. (1975). Healthy blood donors with selective absence of immunoglobulin A: Prevention of anaphylactic transfusion reactions caused by antibodies to IgA. *J. Lab. Clin. Med.* 85:838–842.

Waga, I., Yamamoto, J., Sasai, H., Munger, W. E., Hogan, S. L., Preston, G. A., Sun, H. W., Jennette, J. C., Falk, R. J., and Alcorta, D. A. (2003). Altered mRNA expression in renal biopsy tissue from patients with IgA nephropathy. *Kidney Int.* 64:1253–1264.

Warrington, R. J., Rutherford, W. J., Sauder, P. J., and Bees, W. C. (1982). Homologous antibody to human immunoglobulin (Ig)-A suppresses *in vitro* mitogen-induced IgA synthesis. *Clin. Immunol. Immunopathol.* 23:698–704.

Wilton, A. N., Cobain, T. J., and Dawkins, R. L. (1985). Family studies of IgA deficiency. *Immunogenetics* 21:333–342.

Yadav, M., and Lyngkaran, N. (1979). Low incidence of selective IgA deficiency in normal Malaysians. *Med. J. Malaysia* 34:145–148.

Zhou, W., Pan-Hammarström, Q., Freidin, M., Aarli, J. A., Webster, A. D. B., Olerup, O., and Hammarström, L. (2007). Drug induced immunoglobulin deficiency. *Chem. Immunol.* in press.

Zickerman, A. M., Allen, A. C., Talwar, V., Olczak, S. A., Brownlee, A., Holland, M., Furness, P. N., Brunskill, N. J., and Feehally, J. (2000). IgA myeloma presenting as Henoch-Schönlein purpura with nephritis. *Am. J. Kidney Dis.* 36:E19.

14 Mucosal SIgA Enhancement: Development of Safe and Effective Mucosal Adjuvants and Mucosal Antigen Delivery Vehicles

Jun Kunisawa[1], Jerry R. McGhee[2], and Hiroshi Kiyono[1]

[1] Division of Mucosal Immunology, Department of Microbiology and Immunology, The Institute of Medical Science, The University of Tokyo and Core Research for Evolutional Science and Technology (CREST), Japan Science and Technology Corporation (JST), Tokyo, JAPAN.
[2] Departments of Oral Biology and Microbiology, The Immunobiology Vaccine Center, The University of Alabama at Birmingham, AL, USA.

14.1. Introduction

14.1.1 Unique Aspects of the Induction and Regulation of IgA Antibody Responses

The respiratory and digestive tracts represent major entry routes for pathogenic microorganisms from the lumen into an almost sterile environment of the body. Several physical and biological barriers associated with the innate immune system protect these sites from invasion and help to maintain mucosal homeostasis. The first physical defense line is a barrier structure made up of epithelial cells (ECs) joined firmly by tight junction proteins with brush-border microvilli and a dense layer of mucin (Berkes et al., 2003). Antimicrobial peptides such as defensins and type II phospholipase A2 produced by ECs and Paneth cells are additional molecules preventing the attachment and penetration of pathogenic microorganisms into mucosal tissues (Selsted and Ouellette, 2005).

In addition to these physical and biological barriers, immunological barriers, most notably secretory immunoglobulin A (SIgA) antibody (Ab), which is the predominant isotype at mucosal sites, play an important role in preventing invasion by pathogens (Kunisawa and Kiyono, 2005). A distinct feature of SIgA when compared with IgG is the ability to form polymers [polymeric IgA (pIgA)], mediated by the J-chain produced by mucosal plasma cells (Halpern and Koshland, 1970; Mestecky et al., 1971; Woof and Mestecky, 2005). The pIgA requires cooperation with mucosal ECs expressing the polymeric Ig receptor (pIgR) for transport into the lumen (Kaetzel, 2005; Kaetzel et al., 1991) (see Chapter 3). The pIgR is expressed on the basal membrane of ECs and acts as a receptor for the mucosal pIgA containing the J-chain, thereby accelerating the internalization and transport of the complex to the apical site via transcytosis. Cleavage of pIgR to the secretory component (SC) at the apical surface releases SIgA into the lumen. These SIgA Abs abolish microbial infections by inhibiting their adherence to host ECs. SIgA Abs also neutralize exotoxins by binding to their biologically active site. Thus, SIgA seems to be the sole immunological molecule exhibiting antigen specificity in the outside regions of our body. Furthermore, the SC moiety of SIgA confers novel functions that enhance mucosal immune defense (see Chapter 8).

14.1.2. Common Mucosal Immune System - Dependent IgA Induction Pathways

To induce antigen-specific IgA via mucosal sites, mucosal tissues contain a mucosal network known as the common mucosal immune system (CMIS) linking inductive and effector tissues (Fig. 14.1) (Kunisawa and Kiyono, 2005). The major inductive site for orally administered antigen is the gut-associated lymphoreticular tissues (GALT), which include the Peyer's patches (PPs), and for nasally administered antigen, it is the nasopharynx-associated lymphoreticular tissue (NALT) (Kiyono and Fukuyama, 2004; Kunisawa et al., 2005) (see Chapter 2). As an additional inductive tissue, isolated lymphoid follicles (ILFs) were identified throughout the intestine (Hamada et al., 2002). In spite of differences in their organogenesis pathway, the inductive tissues share several features for initiation of antigen-specific immune responses (Kiyono and Fukuyama, 2004; Kunisawa *et al.*, 2005). For instance, the PPs and NALT are overlaid by a follicle-associated epithelium (FAE) containing antigen-sampling M (microfold)-cells for selective antigen uptake into underlying regions containing antigen-presenting cells (APCs) such as dendritic cells (DCs) (Neutra et al., 2001). The immunological interactions among DCs, T-cells, and B-cells in these inductive tissues promotes IgA commitment of B-cells, which undergo a μ-to-α isotype class-switch recombination in the germinal centers of the inductive tissues (Brandtzaeg and Johansen, 2005; Shikina et al., 2004). In contrast to the dominant class switch to IgA in PPs, B-cell differentiation in NALT leads to the production of both IgA and IgG following sequential class switch from Cμ to Cα via Cγ (Shimoda et al., 2001). These findings may explain the equal commitment of B-cells to IgG and to IgA in NALT, but further analyses will be required to reveal the molecular mechanisms involved in the generation of mucosal B-cells that express those two different isotypes.

Following the class switch to IgA by mucosal B-cells through their interaction with T-cells and APCs, both B- and T-cells emigrate from the inductive tissue (e.g., PPs and NALT), circulate through the bloodstream, and home to distant mucosal effector compartments, especially the lamina propria (LP) regions of the gastrointestinal (GI), respiratory, and reproductive tracts (Fig. 14.1). Although immunization via one mucosal site often activates other, remote mucosal sites, immunization via certain mucosal inductive tissues can lead to the preferential induction of humoral immune responses in the same mucosal sites. The tropism of B- and T-cells is determined by the site-specific combination of adhesion molecules and chemokines. Several lines of evidence have suggested that intestinal DCs play a crucial role in determining the gut tropism of T-cells (Iwata et al., 2004Johansson-Lindbom et al., 2003; Mora et al., 2003; Stagg et al., 2002). They induce $\alpha_4\beta_7$ integrin and CCR9 on antigen-primed T-cells, which interact with the mucosal addressin cellular adhesion molecule-1 (MAdCAM-1), expressed by the endothelium in the LP, and the thymus-expressed chemokine (TECK, also known as CCL25), produced by small-intestinalECs, respectively, which determines the gut-tropism of T-cells. A similar pathway is now proposed for B-cells. On the other hand,

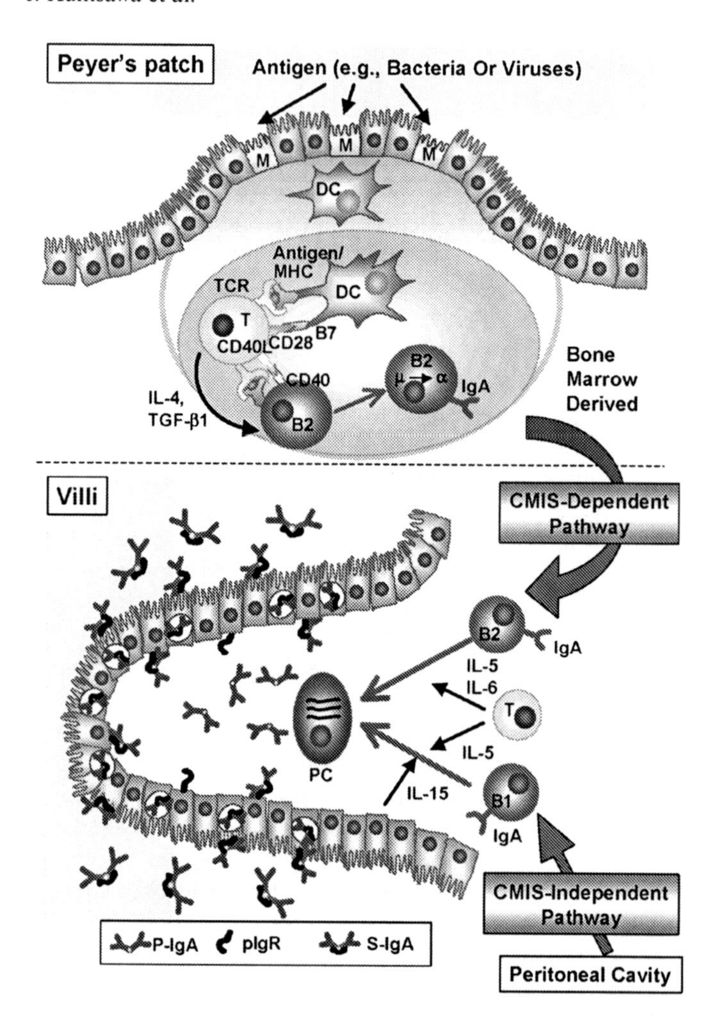

FIG. 14.1. CMIS-dependent and CMIS-independent pathways for the induction of SIgA Ab responses. In the CMIS-dependent pathway, M-cells take up the luminal antigen for transport to the DCs, which, in turn, activate naïve B-cells, also known as B2-lymphocytes, in a T-cell-dependent manner within PPs. In this step, several factors (e.g., CD40, cytokines) induce class-switch recombination from IgM to IgA. The IgA-committed B-cells exit through the lymph and home through the thoracic duct and peripheral blood to mucosal effector sites like the lamina propria of the GI tract. In these effector sites, IgA-committed B2-cells are stimulated by cytokines, including IL-5 and IL-6, resulting in their differentiation into plasma cells (PCs). The plasma cells produce IgA as a dimer joined by the J-chain. Dimeric or higher-molecular-weight forms of p-IgA bind to the pIgR on epithelial cells and are transported across the epithelium and released into the intestinal lumen as SIgA. Another lineage of B-cells, B1-cells, are mainly derived from the peritoneal cavity and act as a CMIS-independent source of intestinal SIgA.

nasal immunization induces the upregulation of $\alpha_4\beta_1$ integrin and CCR10, allowing selective trafficking of B-cells to nasal passage epithelium expressing their ligands, VCAM-1 and CCL28, respectively (Kunkel et al., 2003).

Upon arrival at effector sites, such as the LP in the aerodigestive tracts, the IgA-committed B-cells further differentiate into IgA-producing plasma cells under the influence of interleukin (IL)-5, IL-6, and IL-10 (Hiroi et al., 1999; McGhee et al., 1991; Takatsu et al., 1988). In addition to the help provided by the IgA-enhancing cytokines, IgA production into the lumen requires the expression of the J-chain and pIgR, as mentioned earlier. Along with SIgA, mucosal cytotoxic T-lymphocyte (CTL) responses are important for the clearance of enteric or respiratory viruses and intracellular parasites, such as *Listeria* (Shastri et al., 2005). In this respect, like the IgA-producing B-cells, large numbers of mucosal T-cells are continuously induced by the CMIS-dependent pathway.

14.1.3. CMIS-Independent SIgA Induction Pathways

In addition to the CMIS-dependent pathway, several CMIS-independent pathways have also been identified, especially in the GI tract. It was shown that antigen-specific IgA Ab responses can be induced even under conditions of PP deficiency and thus suggested the presence of additional antigen sampling sites for the induction of antigen-specific immune responses (Yamamoto et al., 2000; Kunisawa et al., 2002). As an alternative gatekeeper in the GI tract, we have identified M-cells on villi (we termed them villous M-cells) that are capable of taking up antigen from the intestinal lumen (Jang et al., 2004). Intestinal villous M-cells develop in various organized lymphoid tissue-defective (e.g., PP-/ILF-null) mice and are capable of taking up bacterial antigens. The discovery of villous M-cells has not only shed light on a novel gateway for antigen uptake into the GI tract but has also suggested the possibility of previously unsuspected routes of pathogen invasion.

In addition to villous M-cells, recent studies have now identified a unique DC population among intestinal ECs (Niess *et al.,* 2005; Rescigno et al., 2001). These intraepithelial DCs migrate into the epithelium via a CX3CR1-mediated pathway and express tight-junction-associated proteins (e.g., occuludin, claudin 1, and zona occuludens 1) for extending their dendrites between ECs and taking up gut luminal antigens.

Unique subsets of B-cells are also found in the intestine. Two lineages of murine B-cells, B1- and B2-cells, can be distinguished by differential expression of cell surface molecules (B220, IgM, IgD, CD5, and Mac-1), origin, growth properties, and antigen-specificity (Berland and Wortis, 2002). "Conventional" B2-cells originate from the bone marrow, recognize primarily T-dependent protein antigens, and are capable of undergoing affinity maturation and memory cell generation in mucosal inductive sites (see Chapter 2). Some of the mucosal IgA plasma cells are derived from B2-cells, which home to mucosal sites via the CMIS-mediated pathway. In contrast, B1-cells are self-renewing and recognize primarily T-independent nonprotein antigens with low affinity. Significant numbers of B1-cells are found in the pleural and peritoneal cavities, as well as mucosal effector sites like the LP of the aerodigestive tract. IL-5, a well-known IgA-enhancing cytokine, was shown to induce the differentiation of both B1- and B2-cells into IgA plasma cells (Hiroi et al., 1999). In contrast, mucosal EC-derived IL-15 promoted IgA commitment and differentiation of B1-cells but not B2-cells (Hiroi et al., 2000). Recent evidence has suggested that intestinal B1-cells migrate from noninductive sites, presumably the peritoneal cavity, to effector sites. Examination of B-cell populations in *aly*/*aly* mice, which carry a point mutation in the nuclear factor (NF)-κB-inducing kinase (NIK), demonstrated elevated B-cell levels in the peritoneal cavity but a complete absence of B-cells in the LP of the GI tract (Fagarasan et al., 2000). Those findings suggested that NIK-mediated signals are essential for the migration of B1-cells into the intestinal compartments.

Because B1-cells respond to T-cell-independent antigens, SIgA Ab production originating from B1-cells was detected in TCR β- and δ-chain-deficient mice (Macpherson et al., 2000). In another study, about 65% of fecal bacteria bound B1-derived SIgA Abs, whereas 30% of bacteria bound B2-derived SIgA, indicating that B1-cell-derived SIgA Abs recognized a large population of commensal bacteria (Bos et al., 2000). Thus, it is likely that the B1-cell-derived SIgA not only inhibits pathogenic bacterial invasion but also plays an important role in maintaining mucosal homeostasis by preventing the attachment of commensal bacteria to mucosal epithelial cells. In contrast, B2-cell-derived SIgA is a key protective Ab against pathogenic microorganisms induced in a T-cell-dependent manner.

In addition to B-cells, a unique subset of T-cells also characterizes the CMIS-independent mucosal immunity. This T-cell population is usually referred to as intraepithelial lymphocytes (IELs). The IELs occur at a frequency of 1 IEL/4–9 ECs and express either TCRαβ or TCRγδ. Thus, it has been suggested that IELs are the bridge between innate and acquired immunity. The IEL function and characteristics have been well reviewed elsewhere (Cheroutre, 2005; Guy-Grand and Vassalli, 2002).

14.2.2. Why are Mucosal Adjuvants Essential for the Induction of IgA Antibody Responses?

The fascinating characteristics of the mucosal immune system in the prevention of infections by pathogens has led to much attention for the development of mucosal (e.g., oral and nasal) vaccines. Mucosal vaccines offer numerous advantages over traditional injection-type parenteral vaccines, including needle-free, easy administration and the possibility of self-delivery. Most importantly, mucosal vaccines can induce both mucosal and systemic immune responses, whereas parenteral immunization yields only systemic immune responses. Hence, traditional parenteral immunization does not lead to the generation of mucosal immunity which would inhibit the initial attachment of pathogens to host cells in mucosal sites.

In contrast, mucosal vaccines can establish a first line of immunological defense at mucosal sites as well as provide a systemic immune surveillance to detect and destroy invading pathogens.

In general, it is often difficult to induce strong SIgA Ab responses as well as T-cell-mediated immune responses at mucosal and systemic sites by mucosal immunization with protein antigen alone. Subunit protein-based vaccines are generally safer than live attenuated or inactivated vaccine using whole microorganisms, but they are less efficacious because of the potential for degradation in the aerodigestive tracts under conditions of low pH, numerous digestive enzymes, and detergent activity by bile salts. Other problems with delivery of protein-based vaccines to mucosal inductive sites include significant dilution in the lumen and physical barriers, including mucus and epithelial tight junctions. The same issues also apply to mucosal DNA or RNA vaccines.

Another major problem in the development of mucosal vaccines is the potential for development of oral or nasal tolerance instead of SIgA Ab-mediated mucosal immunity. Both oral and nasal tolerances are naturally achieved in order to prevent or suppress the development of harmful immune responses against ingested or inhaled foreign proteins. Generally, orally or nasally administered antigen induces immunological tolerance by the induction of T-cell deletion, T-cell anergy, or regulatory T-cells, which are determined by the dose of antigen given (Dubois et al., 2005).

To overcome these obstacles in the development of mucosal vaccines, major efforts are being aimed at development of mucosal adjuvants as well as antigen delivery systems. At a minimum, these systems should protect the antigen from physical and biological elimination. In addition, a major research focus has been aimed at molecular and cellular elucidation of key immunological mechanisms for the simultaneous induction and regulation of active (e.g., SIgA) and silent (e.g., tolerance) immune responses. In this chapter, we will discuss the recent advances in the development of mucosal adjuvants and antigen delivery systems for use with prospective mucosal vaccines for successful SIgA Ab responses.

14.3. Enterotoxin-Based Mucosal Adjuvants

14.3.1. Cholera Toxin and the Escherichia coli Heat-Labile Enterotoxin Are Potent Enhancers of Mucosal SIgA Production

Perhaps the most potent mucosal adjuvants are the bacterial toxin derivatives. Among them, cholera toxin (CT) and the closely related heat-labile enterotoxin (LT) are the most effective and well-studied mucosal adjuvants, which are derived from *Vibrio cholerae* and *Escherichia coli*, respectively (Lycke, 2005; Rappuoli et al., 1999; Yamamoto et al., 2001). They are not only potent immunogens but also adjuvants that enhance both mucosal and systemic immune responses against mucosally coadministered antigens. These enterotoxins possess adjuvant activity when given by oral, nasal, or rectal routes, and the antigen should be administered by the same route as the adjuvant (Lycke and Holmgren, 1986). These findings indicate that local molecular and cellular interactions among the adjuvant, antigen, and the host mucosal immune system are all essential for maximal adjuvanticity.

Cholera toxin and LT are structurally similar (83% homology at the amino acid level) hexameric toxins composed of one A subunit (CT-A or LT-A) and a pentamer of B subunits (CT-B_5 or LT-B_5) (Spangler, 1992). The A subunit contains two distinct domains, A1 and A2, which are linked by a disulfide bond. The A1 domain possesses ADP-ribosyltransferase activity and the A2 subunit is responsible for linking the A1 with the B subunit (Spangler, 1992). The B subunit participates in the binding of CT or LT to host cells. Different binding activities between CT-B and LT-B have been reported. The CT-B binds to GM1-ganglioside, whereas the LT-B binds to GM1-ganglioside as well as asialo GM1 and GM2 (Fukuta et al., 1988; van Heyningen, 1977). They also exhibit different immunological effects. For instance, CT shows a bias for inducing IL-4-mediated Th2-type responses, whereas LT induces both Th1- and Th2-type responses associated with interferon (IFN)-γ, IL-5, IL-6, and IL-10 production (Takahashi et al., 1996; Xu-Amano et al., 1993; Yamamoto et al., 1999). In this regard, it was reported that CT inhibited IL-12 production and alternatively induced IL-10 by DCs, macrophages, and ECs, which might account for the dominant induction of Th2-type responses by CT (Braun et al., 1999; Soriani et al., 2002). Further implication of the Th1- and Th2-inducing ability of the toxins was obtained from experiments using artificially created chimeric forms of the enterotoxin adjuvants. To address which subunit plays a crucial role in determining the balance of Th1 or Th2 response, chimeras of CT-A/LT-B and LT-A/CT-B were constructed (Bowman and Clements, 2001; Boyaka et al., 2003; Kweon et al., 2002). Both chimeras induced SIgA Ab responses, and, intriguingly, CT-A/LT-B upregulated both Th1- and Th2-type responses, wherease LT-A/CT-B promoted only Th2 responses. These observations imply that the B subunit of a particular enterotoxin adjuvant is mainly responsible for their propensity to direct Th cell subset responses.

For cellular aspects of enterotoxin-induced adjuvanticity, several possible pathways operate via different subsets of immunocompetent cells. For example, both CT and LT trigger APCs such as DCs. As mentioned earlier, DCs play a pivotal role in the initiation of antigen-specific immune responses in the inductive tissues like PPs (Sato and Iwasaki, 2005). It was previously reported that CT induced migration of DCs from the subepithelial dome to T- and B-cell areas of PPs, which resulted in efficient interactions among DCs, T-cells, and B-cells (Shreedhar et al., 2003). CT additionally induced the maturation of DCs accompanied by the upregulation of major histocompatiblity complex (MHC) and costimulatory molecules (CD80 and CD86), thereby promoting potent T- and B-cell responses (Cong et al., 1997;

Yamamoto et al., 1999). CT has also been shown to promote the production of IL-1, IL-6, and IL-10 from both DCs and macrophages (Cong et al., 2001). The direct effects of CT and LT as adjuvants are not only specific for APCs, but also for the other immunocompetent cells. Binding of LT to B-cells induced upregulation of MHC class II and CD25, which was associated with the activation of extracellular signal-regulated kinase (Erk1 and Erk2) (Bone et al., 2002). In addition, CT enhances B-cell isotype switching by two different mechanisms (Lycke, 1993). In the first case, ADP-ribosyltransferase activity for cAMP induction was involved in the germline IgH-chain RNA transcripts. In the second case, the binding of CT-B to the GM1-ganglioside on B-cells promoted B-cell differentiation (Lycke, 1993). Enterotoxins also directly activate T-cells, especially $CD4^+$ T-cells (Williams et al., 1999). Thus, the adjuvant activity of CT was shown to be decreased in CD4-deficient mice (Hornquist et al., 1996).

In addition to SIgA and plasma IgG Ab responses, CTL responses are also primed by oral immunization with CT or LT (Bowen et al., 1994; Simmons et al., 1999). For the induction of CTL responses, antigen processing in the cytoplasm and endoplasmic reticulum (ER) and subsequent MHC class I-restricted antigen presentation are required (Kunisawa and Shastri, 2003; Shastri et al., 2005). In this regard, molecular studies concerning intracellular CT trafficking revealed a unique transport pathway for CT in antigen delivery (Fig. 14.2) (Lencer and Tsai, 2003). Following CT binding to the host GM1 ganglioside through CT-B, CT is internalized by both clathrin-dependent and clathrin-independent endocytosis. Interestingly, instead of transfer to the lysosome with subsequent degradation, the internalized CT is rapidly sorted into the trans-Golgi network to enter the retrograde trafficking pathway to the ER (Feng et al., 2004). Export of CT-A from the ER into the cytosol was found to be mediated by temperature-sensitive section 61p (Sec61p), a transporter on the ER membrane (Schmitz et al., 2000). Using this unique transport pathway, coadministered antigen appears to be delivered to the cytosol, subsequently leading to the MHC class I-restricted antigen presentation required for the induction of CTL responses (Fraser et al., 2003).

Intriguingly, unlike intact CT, CT-B does not activate immune responses, but instead, it enhances mucosal tolerance induction (Sun et al., 1994). The fact that CT-B does not trigger DC maturation might explain the induction of tolerance rather than adjuvant activity (Lycke, 2004). A different possibility could be the involvement of DCs in the LP because recent evidence has shown that DCs in the LP have the ability to induce tolerance by several different mechanisms, whereas DCs in PPs are capable of activating immune responses (Kelsall and Leon, 2005). Support for this pathway comes from studies demonstrating that CT-B was directly transported to the basolateral side of polarized ECs in the villi (Lencer et al., 1995). However, this issue is still controversial because some reports demonstrated that recombinant CT-B had adjuvant activity in terms of promoting DC maturation and for

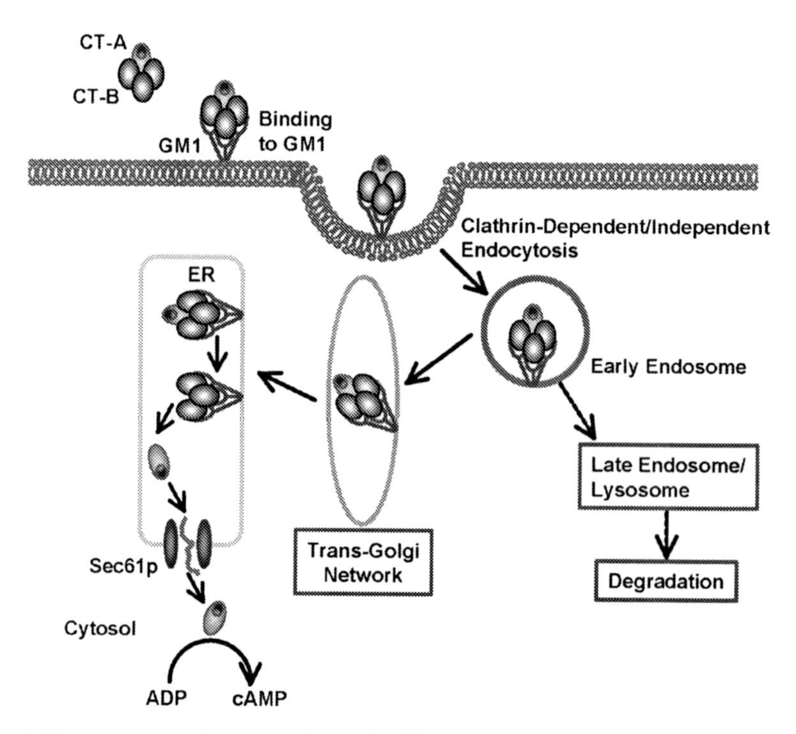

FIG. 14.2. A unique trafficking pathway for CT into the cytoplasm. The intact CT molecule binds to GM1 ganglioside via the CT-B subunit. After endocytosis of CT–GM1, the complex is transported to the endoplasmic reticulum (ER) through a retrograde pathway mediated by the trans-Golgi network. In the ER, CT-A is dissociated from CT-B, and Sec61p mediates the transport of unfolded CT-A into the cytosol, where CT-A becomes folded into the active form with ADP-ribosyltransferase activity.

enhancing mucosal SIgA and systemic IgG Ab responses (Isaka et al., 2004; Isomura et al., 2005).

14.3.2. CT- and LT-Based Mutant Nontoxic Adjuvants

The clinical use of CT and LT has been hampered by the fact that both enterotoxins induce severe diarrhea after oral administration or natural infection. Both enterotoxins also have undesirable side effects involving their entry into the central nervous system when given by the nasal route (Fujihashi et al., 2002). Thus, efforts are now focused on the development of new mucosal adjuvants that do not possess toxicity but that retain adjuvant activity. As described earlier, both CT and LT belong to the AB_5 family of toxins, comprising an A subunit with ADP-ribosylating activity and a pentamer of CT-B subunits that bind to gangliosides on the cell membrane. Thus, several groups including our own have attempted to modify CT-A in order to remove toxicity

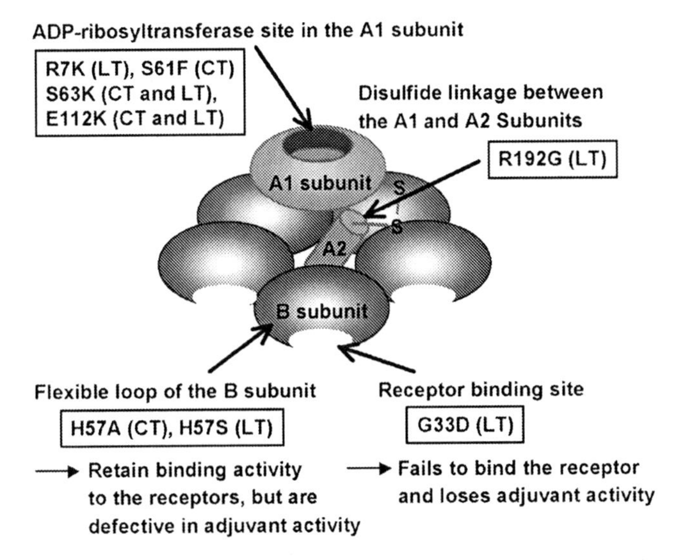

FIG. 14.3. Structure of CT and LT, showing key functional moieties and sites of mutation for generation of mutants of CT and LT as mucosal adjuvants.

(Douce et al., 1995; de Haan et al., 1996; Yamamoto et al., 1997a, 1997b) (Fig. 14.3). In our previous studies, two types of mutants were constructed from CT (Yamamoto et al., 1997a, 1997b). These mutants involved replacement of serine with phenylalanine at position 61 (S61F) or glutamic acid with lysine at position 112 (E112K). Both mutants lacked ADP-ribosyltransferase activity but retained adjuvant activity. Thus, when mutant CTs like native CT were applied nasally, high levels of SIgA Ab and Th2 cell-mediated immune responses were induced.

Subsequently, our studies demonstrated that the mutant forms of CT were effective for the induction of immune responses against tetanus toxin, *Streptococcus pneumoniae*, influenza virus, diphtheria toxin, and botulinum toxin (Kobayashi et al., 2005; Ohmura et al., 2001; Watanabe et al., 2002; Yamamoto et al., 1997a, 1998). Consistent with the numerous successes of mutant forms of CT in the murine system, we have recently reported that the mutant form of CT (E112K) was an effective adjuvant for an anti-human immunodeficiency virus (HIV) vaccine in rhesus macaques, without showing any toxicity (Yoshino et al., 2004). These results offer the possibility of the clinical application of a mutant form of CT (E112K) as a safe and effective mucosal adjuvant. We also constructed a chimeric form of mutant CT containing mutant CT-A (E112K) fused with LT-B. As mentioned earlier, the B subunit of enterotoxin adjuvants determines the direction of Th cell subset that is induced. Consistent with this, the mutant chimera, CT-A(E112K)/LT-B, induced protective immunity in both mucosal and systemic immune compartments against tetanus toxin and influenza virus (Kweon et al., 2002). Most importantly, the immune responses induced by CT-A(E112K)/LT-B were accompanied by lower IgE Ab responses when compared to mutant CT, suggesting that the chimeric form of mutant CT can induce protective mucosal and systemic immune responses without undesirable allergic responses. A second generation of double mutant CT adjuvants (dmCTs) have recently been constructed, which contain the E112K mutation in the ADP-ribosyltransferase active center as well as mutations in the C-terminal KDEL intracellular targeting motif (Hagiwara et al., 2006). These dmCTs retained strong mucosal adjuvant activity without central nervous system toxicity and may prove to be safer than other mutant forms of CT.

Mutants of LT have also been constructed in attempts to develop a safe mucosal adjuvant (de Haan et al., 1996; Douce et al., 1995; Rappuoli et al., 1995). As with mutant CTs, the main target for mutations has been the LT-A subunit, in order to disrupt ADP-ribosyltransferase activity (Douce et al., 1995). Mutant forms of LT have been demonstrated to enhance protective immunity mediated by SIgA and CTL against measles and influenza viruses, tetanus toxin, *Helicobacter pylori*, and *Streptococcus mutans*, among many others (Barchfeld et al., 1999; De Magistris et al., 1998; Marchetti et al., 1998; Nawar et al., 2007; Partidos et al., 1996). Mutant forms of LT have also shown promise in clinical trials (Peppoloni et al., 2003; Pizza et al., 2000;). Another unique approach has been to detoxify LT by modifying a protease-sensitive residue joining the A1 and A2 subunits (Dickinson and Clements, 1995). The mutant LT (R192G) was used for mucosal adjuvants against infectious diseases like rotavirus, salmonella, *Candida albicans*, and HIV (Cardenas-Freytag et al., 1999; Chong et al., 1998; Morris et al., 2000; O'Neal et al., 1998). The dissociation of the A1 from the A2 subunit is essential for the transport of the A1 subunit from the ER into the cytoplasm, where its ADP-ribosylating function is activated (Fig. 14.2) (Lencer and Tsai, 2003). Thus, the decreased toxicity of LT R192G might be due to the inhibition of LT-A transport into the cytoplasm.

In addition to the A subunit, mutations in B subunits of CT and LT have been examined (Fig. 14.3) (Aman et al., 2001; Fraser et al., 2003; Nashar *et al.*, 1996). Disruption of the binding of CT to its GM1-ganglioside receptor resulted in the disruption of adjuvant activity (Nashar et al., 1996). However, a mutant form of CT-B, in which the histidine at position 57 was replaced with alanine, retained the ability to bind GM1-ganglioside receptor but abolished adjuvant activity (Aman et al., 2001). These data indicated that an undetermined activity of CT-B, separate from its ability to bind to surface receptors, is also important in adjuvant activity. Of note, the mutation disrupted the loop structure of the CT-B subunit pentamer. It appears, therefore, that stable conformation of the toxins is important for their adjuvant activity.

14.3.3. Other Toxin-Type Adjuvants

Several other bacterial toxins have been shown to exhibit mucosal adjuvant activity. For instance, Shiga toxin 1 (STX1) and a mutant form of STX1 have recently been shown to exhibit mucosal adjuvanticity. Like CT and LT, STX1 has an AB_5-type structure. Nasal immunization with OVA and a mutant form of STX1 elicited plasma IgG and mucosal SIgA Ab responses (Ohmura-Hoshino et al., 2004). Our recent experiments revealed that mutant STX1 enhanced expression of MHC and costimulatory molecules on DCs (Ohmura et al., 2005). Further, IL-12 and TNF-α expression were enhanced, eventually leading to the effective induction of Th1 and Th2 cell responses as well as the SIgA Ab responses.

Anthrax edema toxin (EdTx), like CT and LT, is an AB_5-type toxin. The A subunit (edema factor) has adenylate cyclase activity, and the B subunits bind to anthrax toxin receptors on target cells. Nasal immunization of mice with OVA antigen plus EdTx induced OVA-specific serum IgG, salivary SIgA, and $CD4^+$ T-cells secreting Th1 and Th2 cytokines (Duverger et al., 2006). An added benefit of this vaccine was the generation of an immune response to the cell-binding B subunit of EdTx. Nontoxic derivatives of EdTx could represent an alternative to ganglioside-binding enterotoxin adjuvants and might be useful for inducing protective immunity against anthrax.

A genetically engineered pertussis toxin (PTX) was developed by removal of its ADP-ribosylating activity and was found to be an effective adjuvant for enhancing mucosal immune responses (Roberts et al., 1995). PTX recognizes N-linked glycan chains containing a branched mannose core and *N*-acetyl glucosamine, found on cell surface glycoproteins of various types of mammalian cell. Nasal immunization with tetanus toxin and PTX augmented parenteral and mucosal Ab responses (Roberts et al., 1995). It should be noted that native PTX has adjuvant activity for enhancement of plasma IgG and mucosal SIgA Ab responses, but it also stimulates production of plasma IgE Abs (Lindsay et al., 1994). However, mutant PTX did not promote IgE responses (Roberts et al., 1995), which alleviated concerns that the adjuvant would induce undesirable allergic responses.

Zonula occludens toxin (Zot) is a single polypeptide encoded by the filamentous bacteriophage infecting toxigenic strains of *V. cholerae* (Baudry et al., 1992). It is capable of disrupting epithelial tight junctions, allowing increased permeability of luminal antigens into mucosal sites via the paracellular route (Fasano et al., 1991). Thus, nasal or rectal immunization with Zot resulted in the induction of plasma IgG and mucosal SIgA Ab responses against coadministered antigens, mediated by both Th1- and Th2-type cells (Marinaro et al., 1999b, 2003).

CTA1-DD is a chimeric adjuvant composed of an enzymatically active CT-A subunit and a dimer of an Ig-binding element of *Staphylococcus aureus* protein A, which targets vaccines to B-cells (Agren et al., 1997). When CTA-DD was applied nasally, it enhanced antigen-specific immune responses in both mucosal and systemic sites without causing inflammation. A subsequent study indicated that the adjuvanticity of CTA1-DD was mediated at least in part by promoting germinal center formation (Agren et al., 2000). Surprisingly, CTA1-DD is nontoxic, although it contains the intact form of CT-A, and both ADP-ribosyltransferase activity and Ig-binding activity are required for its adjuvanticity (Agren et al., 1999). Furthermore, unlike the mutant form of CT (E112K), a similar mutation in CTA1-DD (CTA1E112K-DD) failed to retain adjuvant activity (Lycke, 2004).

It should be noted that most mutant forms of adjuvants derived from bacterial toxins retained full adjuvant activity after nasal and parenteral immunization but possessed less adjuvant activity when administered orally. The reason for different adjuvant activities after nasal versus oral delivery remains an open question and further experiments are necessary for effective oral delivery use of toxin-based mutant adjuvants.

14.4. Development of Safe and Effective Cytokine- and Chemokine-Based Mucosal Adjuvants

14.4.1. Cytokine-Based Mucosal Adjuvants

Innate-type cytokines have also been successfully used to enhance mucosal immune responses (Table 14.1). As expected from results showing that CT treatment induced IL-1 production, IL-1 itself has been shown to enhance mucosal immune responses to coadministered antigens (Staats and Ennis, 1999).

TABLE 14.1. Cytokine-, chemokine-, and TLR ligand-based mucosal adjuvants.

Adjuvant	Antibody responses	Th cell responses
Cytokines		
IFN-α/β	Plasma IgG and mucosal SIgA	Th1
IL-1	Plasma IgG and mucosal SIgA	Th2
IL-2	Plasma IgG and mucosal SIgA	Th1 and Th2
IL-12	Plasma IgG and mucosal SIgA	Th1 and Th2
IL-15	Plasma IgG and mucosal SIgA	Th1
IL-18	Plasma IgG and mucosal SIgA	Th1
Chemokines		
Lymphotactin	Plasma IgG, IgE and mucosal SIgA	Th1 and Th2
RANTES	Plasma IgG and mucosal SIgA	Th1
MIP-1α	Plasma IgG only (no mucosal SIgA)	Th1
MIP-1β	Mucosal SIgA with less plasma IgG	Th2
Defensin	Plasma IgG only (no mucosal SIgA)	Th1 and Th2
TLR ligands (receptors)		
MLA (TLR4)	Plasma IgG and mucosal SIgA	Th1 and Th2
MDP (TLR2)	Plasma IgG and mucosal SIgA	Not done
MALP-2 (TLR2/6)	Plasma IgG and mucosal SIgA	Th1 and/or Th2
CpG DNA (TLR9)	Plasma IgG and mucosal SIgA	Th1
Poly (I:C) (TLR3)	Plasma IgG and mucosal SIgA	Th1 and Th2

Abbreviations: MALP, mycoplasma-derived macrophage-activating 2kDa lipopeptide; MDP, muramyl dipeptide; MIP, macrophage inflammatory protein; MLA, monophosphoryl lipid A; RANTES, regulated on activation normal T-cell expressed and secreted; Poly (I:C), polyriboinosinic polyribocytidylic acid; TLR, Toll-like receptor.

Like CT, nasal administration of OVA or tetanus toxoid with IL-1 promoted SIgA and plasma IgG Ab responses with Th2-type helper responses.

Type I interferons (IFNs) (INF-α and IFN-β) were originally identified as antiviral cytokines. Type I IFNs are rapidly produced during bacterial or viral infections and possess multiple pathways to activate the host immune system, including enhancement of cytotoxic activity of natural killer (NK)- and T-cells, upregulation of antigen presentation, and activation of B-cells (Basler and Garcia-Sastre, 2002). Consistent with these activities, it was demonstrated that type I IFN was effective as an adjuvant to enhance systemic IgG, mucosal SIgA, and Th1-type responses (Proietti et al., 2002). Because type I IFNs have already been utilized clinically for the treatment of virus-infected hepatitis patients, they can be considered to be promising candidates for use in humans as mucosal adjuvants.

The well-characterized cytokine IL-2 is mainly produced by CD4-positive ($CD4^+$) T-cells. IL-2 is a lymphoproliferative cytokine and was reported to enhance mucosal and systemic immunity after oral administration (Wierzbicki et al., 2002). This strategy seems to be especially effective in the elderly, because their diminished immune activity is often accompanied by reduced IL-2 production by T-cells (Haynes and Eaton, 2005). In this regard, it was demonstrated that coadministration of IL-2 with papillomavirus pseudoviruses via the oral route restored mucosal and systemic immune responses in aged animals (Fayad et al., 2004). A similar approach might be applied to immunodeficient patients. Recent studies have demonstrated that nasal immunization with a plasmid encoding both IL-2 and antigen is effective for the control of HIV/SIV infection and disease progress in nonhuman primates (Bertley et al., 2004).

The cytokine IL-12 is produced by APCs (e.g., DCs) for induction of Th1-type responses. Like IL-1, IL-12 was found to promote mucosal SIgA and plasma IgG Ab responses when given by the nasal route (Boyaka et al., 1999). The difference between IL-1 and IL-12 was that nasal administration of IL-12 induced both Th1- and Th2-responses (Boyaka et al., 1999). Another approach has involved coadministration of IL-12 with CT. Nasal administration of IL-12 plus CT along with tetanus toxoid antigen preferentially induced Th1-type responses, whereas Th2-type responses were enhanced by oral delivery of CT combined with nasal delivery of IL-12 (Marinaro et al., 1999a). These findings indicate that CT and IL-12 may act through different mechanisms to enhance mucosal immunization, and the determination of Th cell subset induced depends on the route of vaccine delivery. The application of IL-12 as a nasal adjuvant has now been extended to vaccines against numerous infectious diseases, such as tetanus, influenza, the pneumococcus, and *Francisella tularensis* (Arulanandam et al., 1999, 2001; Boyaka et al., 1999; Duckett et al., 2005).

The cytokine IL-15 is produced by DCs, macrophages, and ECs. The IL-15 receptor shares β– and γ-chains with the IL-2 receptor, but it has a unique α-chain that determines its distinct activities (Kovanen and Leonard, 2004).

Both IL-15 and IL-2 activate T- and NK-cells; however, IL-15 preferentially enhances the generation of memory T-cells (Oh et al., 2003). In addition to its effects on T-cells, IL-15 also activates B-cells, particularly B1-cells. IL-15 derived from mucosal ECs was reported to enhance the proliferation and differentiation of B1-cells (but not of B2-cells) into IgA-producing plasma cells in the small intestine (Hiroi et al., 2000). Thus, treatment with IL-15 resulted in the proliferation of B1-cells, whereas anti-IL-15 mAb caused a decrease in the number of B1-cells at the effector sites of the GI and nasal tracts (Hiroi et al., 2000). In accordance with these functions of IL-15, nasal administration of a plasmid encoding IL-15 enhanced Ab and T-cell responses against coexpressed HIV or herpes simplex virus (HSV) antigen for long periods in both mucosal and systemic lymphoid compartments (Toka and Rouse, 2005; Xin et al., 1999).

The cytokine IL-18 has been shown to induce production of IL-12 by DCs and macrophages and thus promotes Th1-type responses. Nasal administration of a plasmid encoding IL-18 plus antigen enhanced SIgA responses in both rectal and vaginal mucosa and subsequently provided protective immunity against HSV-1 (Lee et al., 2003).

Although these cytokines showed adjuvant activities by themselves, studies of the adjuvant activity of cytokines in combination have revealed synergistic effects. For instance, simultaneous administration of IL-1, IL-12, and IL-18 resulted in much stronger adjuvant activity for induction of mucosal SIgA and systemic IgG Abs than that observed with each cytokine alone or with a dual combination of IL-12 plus IL-18 (Bradney et al., 2002). In contrast to the synergistic effects of IL-1, IL-12, and IL-18, co-expression of IL-15 and IL-12 did not enhance adjuvant activity (Xin et al., 1999). These findings suggest that the mechanisms by which cytokines act as immune adjuvants are complex and should be carefully optimized in the development of effective mucosal vaccines against specific infections.

14.4.2. Chemokines and Chemotactic Molecules as Mucosal Adjuvants

Chemokines interact with chemokine receptors on lymphocytes and other cells of the immune system to promote cellular migration along a concentration gradient. Several chemokines were reported to act as innate-type mucosal adjuvants (Table 14.1). For instance, lymphotactin (XCL1) is a C chemokine produced by NK-cells and $CD8^+$ T-cells, including IELs (Boismenu et al., 1996). Lymphotactin promotes chemotaxis of NK-cells and T-cells and has been shown to enhance tumor immunity (Dilloo et al., 1996). When lymphotactin was nasally delivered along with OVA antigen, marked enhancement of OVA-specific SIgA Abs in various mucosal secretions (e.g., feces, saliva, vaginal, and nasal washes) and plasma IgG Ab responses were detected and these responses were supported by both Th1 and Th2 cells (Lillard et al., 1999b).

Mucosal ECs produce RANTES, a CC chemokine, in response to bacterial and viral infections (Saito et al., 1997; Yang et al., 1997). Nasal coadministration of RANTES with OVA induced high levels of SIgA, plasma IgG. and preferential Th1-type responses (Lillard et al., 2001). *In vitro* studies further revealed that expression of CD28, CD40 ligand, and IL-12 receptor on T-cells was increased after treatment with RANTES (Lillard et al., 2001). In spite of the fact that both lymphotactin and RANTES induced mucosal SIgA and plasma IgG Abs, only lymphotactin induced IgE Ab production (Lillard et al., 1999b, 2001). This differential effect might be explained by the different cytokine profiles induced by lymphotactin and RANTES and suggests that, to avoid undesirable allergic responses mediated by IgE, RANTES is likely to be a safer adjuvant than lymphotactin.

Macrophage inflammatory protein (MIP)-1 is another CC chemokine that was analyzed for its ability to act as a mucosal adjuvant. MIP-1 contains two homologous subtypes: MIP-1α and MIP-1β. It was demonstrated that nasal administration of MIP-1α enhanced antigen-specific Ab responses to coadministered antigen in systemic but not in mucosal sites (Lillard et al., 2003). Although MIP-1β shares the same ligand (CCR5), it promoted mucosal SIgA Ab responses with less efficient induction of systemic immune responses (Lillard et al., 2003). These findings indicate that MIP-1α and MIP-1β activate distinct adjuvant pathways, although the mechanisms remain to be elucidated.

Defensins belong to a family of antimicrobial peptides produced by Paneth cells (Selsted and Ouellette, 2005). Defensins also possess chemotactic activity for T-cells and exert adjuvant activity (Lillard et al., 1999a). Coadministration of defensins with antigen promoted Th1- and Th2-mediated systemic IgG Ab responses associated with IFN-γ, IL-5, IL-6, and IL-10 production. It is interesting to note that no mucosal SIgA Ab responses were induced after nasal immunization with defensins. Thus, defensins might be unique adjuvants that enhance systemic immune responses without induction of mucosal SIgA Ab production.

14.5. TLR-Targeted Mucosal Adjuvants

Innate immunity plays a pivotal role in host defense against invading microbial pathogens at early stages of infection. A major breakthrough in our molecular understanding of innate immunity occurred after the discovery of Toll-like receptors (TLRs), which recognize pathogen-associated molecular patterns. To date, 13 mammalian TLRs have been identified (Akira and Takeda, 2004). TLR-mediated signals induce cytokine production like type I IFN, IL-1, and IL-12, as well as antimicrobial peptides like defensins, which are all known to have adjuvant activity, as discussed earlier. In addition to the important role of TLRs in innate immunity, accumulating evidence has revealed that TLRs play a pivotal role in the induction of acquired immunity. These attractive features of TLRs might now lead to their application as mucosal adjuvants for the enhancement of SIgA Ab responses (Table 14.1).

14.5.1. Surface TLR-Mediated Adjuvants

Generally, TLR families can be divided into two groups by their cell locale. TLR2 and TLR4 are expressed on the cell surface. The receptor for lipopolysaccharide (LPS) was the TLR to be discovered and is now termed TLR4 (Poltorak et al., 1998). Although LPS itself may be a potent mucosal adjuvant, the medical use of LPS is hampered by its severe toxicity. Thus, recent focus has been aimed at development of an agonist possessing useful immunomodulatory effects without undesirable toxic effects. These efforts have led to the discovery of monophosphoryl lipid A (MLA), a ligand for TLR4, which has already been shown to be a systemic adjuvant preclinically and clinically (Baldridge et al., 2004; Evans et al., 2003). Experimental studies have also demonstrated the effectiveness of MLA as a mucosal adjuvant (Baldridge et al., 2000; Doherty et al., 2002; Pinczewski et al., 2005). Nasal or oral administration of antigens like hepatitis B surface antigen, tetanus toxoid, influenza, *Mycobacterium tuberculosis*, or HIV antigens with MLA promoted SIgA Ab responses in proximal and distal mucosal sites and plasma IgG Abs in systemic sites. The primary target cells of TLR4 agonists are presumably DCs, because TLR4 expression is very low or absent on ECs (Iwasaki and Medzhitov, 2004). In support of this concept, other studies demonstrated that treatment of DCs with MLA induced IL-12 production and enhanced CD80 and CD86 (Martin et al., 2003).

TLR2 is also the target of a specific mucosal adjuvant, muramyl dipeptide (MDP), which is derived from the cell walls of mycobacteria. Prior to its identification as a TLR2 ligands, MDP had been used as an adjuvant for intravaginal and oral immunization (Thapar et al., 1990) and had been shown to stimulate PP cells for the enhancement of IgA Ab responses (Kiyono et al., 1982). Recent studies demonstrated that mycoplasma-derived macrophage-activating 2kDa lipopeptide (MALP-2) promoted Th2, plasma IgG, and mucosal SIgA responses against coadministered antigens such as β-galactosidase and HIV-1 Tat protein (Borsutzky et al., 2003; Rharbaoui et al., 2002). The adjuvant activity of MALP-2 was associated with binding to heterodimers of TLR2 and TLR6 on NALT B-cells, which induced upregulation of MHC, costimulatory molecules, and CD40 (Borsutzky et al., 2005; Rharbaoui et al., 2004).

14.5.2. Adjuvants Targeting Intracellular TLRs for use with Mucosal Vaccines

The discovery that microbial nucleic acids (e.g., DNA and RNA) have immunostimulatory activity has led to their development as mucosal adjuvants. Bacterial but not eukaryotic DNA generally contains nonmethylated "CpG motifs" and acts as a ligand for TLR9, thus enhancing innate and adaptive immunity (Hemmi et al., 2000). TLR9 is expressed in intracellular compartments of the endosome–lysosome pathway in APCs, and TLR9-mediated signals have been shown to enhance antigen presentation, expression of costimulatory molecules, including CD80, CD86, and CD40, and production cytokines such as IFN-α/β and IL-12 (Akira and Takeda, 2004). Additionally, CpG directly induces B-cells to proliferate and secrete Ig. In contrast to its intracellular expression on immune cells, it has recently been reported that TLR9 is expressed on the surface of intestinal ECs (Lee et al., 2006). Interestingly, stimulation of TLR9 from the basolateral surface of polarized ECs resulted in NF-kB activation and secretion of IL-8, whereas apical stimulation conferred intracellular tolerance to subsequent TLR challenges. These investigators suggested that the distinctive apical TLR9 signaling in intestinal ECs might contribute to the maintenance of colonic homeostasis in the presence of large numbers of commensal bacteria (see Chapter 10).

TLR9-mediated cellular activation can be achieved by the use of synthetic oligonucleotides (ODN) containing CpG motifs, such as GACGTT in mice (Krieg et al., 1995) and GTCGTT in humans (Hartmann and Krieg, 2000). In addition to the adjuvant effects on systemic immune responses, numerous studies have been focused on the use of CpG ODN as a mucosal adjuvant. Mucosal administration of antigen with CpG ODN has been shown to promote mucosal SIgA, plasma IgG, and T-cell responses, including $CD8^+$ CTLs and $CD4^+$ Th1 cells that stimulate type I IFN production by DCs. The adjuvant activity of CpG ODN has been demonstrated to enhance protective immunity against various types of mucosal infection, such as those caused by *Streptococcus pneumoniae*, HIV, HSV-2, and *Helicobacter pylori* (Chu et al., 2000; Dumais et al., 2002; Gallichan et al., 2001; Harandi and Holmgren, 2004; Horner et al., 2001; Jiang et al., 2003). Comparisons of the adjuvant effects of CpG ODN administered via different mucosal routes demonstrated that nasal and oral delivery resulted in similar enhancement of mucosal and systemic immune responses, whereas the rectal route was less effective (McCluskie and Davis, 2000). Although the clinical use of CpG ODN in mucosal vaccines has not yet been investigated, the numerous successes with CpG ODN as a mucosal adjuvant in experimental animals, as well as several studies with CpG ODN in clinical trials for systemic immunization, should accelerate the acceptance of CpG ODN as a mucosal adjuvant for human vaccines. Finally, from the view of stability, cost, and quality control, CpG ODN seems to be one of the most feasible mucosal adjuvant candidates.

TLR3 and TLR7/8, which act as receptors for double-stranded RNA (dsRNA) and single-stranded RNA, respectively, are, like TLR9, localized in intracellular compartments (Akira and Takeda, 2004). Recent studies demonstrated that nasal delivery of the synthetic dsRNA molecule polyriboinosinic polyribocytidylic acid [poly (I:C)], along with inactivated influenza virus HA, induced protective immune responses against influenza virus infection (Ichinohe et al., 2005). To date, few studies have been performed to analyze the ability of TLR3- and TLR7/8-mediated pathways when given with a mucosal vaccine. However, one can reasonably expect additional studies to examine the feasibility of TLR3- and TLR7/8-targeted mucosal adjuvants for the induction of protective immunity against various mucosal bacterial and viral infections.

14.6. Mucosal Antigen Delivery Systems

The mode of antigen delivery is another important aspect of mucosal vaccine development. Particulate antigens appear to be more effective than soluble ones, partly due to the protection of the antigen from the harsh conditions of the mucosal environment such as low pH, detergent effects of bile salts, and extensive proteolytic enzyme activity. Additionally, uptake of antigen by M-cells, a key gateway system for antigen sampling, is more effective with particulate antigens (Clark et al., 2001b) (see Chapters 2 and 9). In this section, we will outline various approaches to the development of an ideal mucosal antigen delivery system.

14.6.1. Inert Synthetic and Hybrid Delivery Systems for Mucosal Vaccines

A variety of mucosal antigen delivery systems have been developed using inert particles, including biodegradable polymer-based particles as well as lipid-based particles such as liposomes and ISCOMs (Vajdy et al., 2004). As a representative of polymer-based particles, poly-lactide coglycolide (PLG) microparticles have been extensively investigated (O'Hagan and Singh, 2003). Their biodegradability and easy regulation for controlled drug release have facilitated the application of the PLG system in humans as drug delivery vehicles (Okada and Toguchi, 1995). In the early 1990s, several groups adapted this PLG delivery system to vaccine development and demonstrated that nasal or oral immunization with PLG microparticles encapsulating protein or DNA antigens induced high levels of protective mucosal SIgA and plasma IgG Abs as well as helper T-cells and CTLs (reviewed in O'Hagan and Singh, 2003). Several approaches have been exploited to enhance antigen delivery efficacy by changing the chemical properties of microparticles. For instance, enteric coatings have been employed to protect encapsulated antigens from the acidic gastric environment and to allow the rapid release of antigen in the small intestine (Jain et al., 1996; Vogel et al., 1998). Gelatin capsules have been used for this purpose because they dissolve in the alkaline pH of the intestine but not in the acidic pH of the stomach (Moldoveanu et al., 1993). An additional example would be the chemical mucoadhesive molecules (e.g., carboxy vinyl polymer), which have been used to elongate particles containing protein antigens, thereby prolonging antigen persistence in the intestine (Kunisawa et al., 2000).

Liposomes, spherical particles with a bilayered phospholipid membrane, act as antigen delivery systems for enhancing immune responses by protecting the incorporated antigen from degradation (Somavarapu et al., 2003; Vogel et al., 1998). Liposomes can be prepared with different phospholipid compositions, thus conferring different chemical and biological properties. Because liposomes are generally unstable in acids, lipases, and bile salts in the GI tract, several efforts have been made to develop more stable forms of liposomes (reviewed in Zho and Neutra, 2002). Previous studies had demonstrated that stable liposomes could be constructed with dipalmitoyl-phosphatidylserine, dipalmitoyl-phosphatidylcholine, and cholesterol (Aramaki et al., 1993; Han et al., 1997). Antigen-specific SIgA responses were induced when mice were immunized with GM1 antigen using these more stable liposomes (Han et al., 1997). Another approach to improve the stability of liposomes in the GI tract was to create the cross-linked network among the lipid membrane by covalent bonds (Chen and Langer, 1997; Okada et al., 1995).

Additional modifications of antigen delivery systems have been attempted in order to target the antigen selectively to M-cells. Some evidence suggests that the physical properties of synthetic particles (e.g., size, hydrophobicity, and surface charge) influence the efficiency of the selective delivery of the encapsulated antigen to M-cells (Clark et al., 2001b). For example, polylactic acid (PLA) microparticles 4 μm in diameter enhanced plasma IgG Abs but not intestinal SIgA Abs, whereas 7-μm PLA microparticles enhanced SIgA Ab production. However, 26-μm microparticles were ineffective because they were too large to be taken up by M-cells in the PP (Tabata et al., 1996).

In addition to physical factors, lectins and microbial adhesins have been widely exploited to enhance access of microencapsulated antigens to M-cells (Jepson et al., 2004). *Ulex europaeus agglutinin* 1 (UEA1), a lectin specific for α-L-fucose residues, has been shown to bind selectively to the apical surface of M-cells of murine PPs and NALT (Giannasca et al., 1994; Takata et al., 2000). Incorporation of UEA1 into microparticles or liposomes resulted in selective and efficient delivery of antigens to M cells after oral administration (Chen et al., 1996; ; Clark et al., 2001a; Foster et al., 1998). The high efficacy of UEA-1-mediated antigen delivery resulted in significant enhancement of OVA-specific Ab responses when mice were immunized orally with microparticles coated with UEA-1 and OVA (Foster and Hirst, 2005). This strategy has now been applied to the induction of mucosal and systemic immune responses to HIV (Manocha et al., 2005; Wang et al., 2005).

Recent advances in biotechnology have resulted in the identification of additional candidate molecules for efficient delivery of antigen to M-cells (Higgins et al., 2004; Lambkin et al., 2003). Organic molecules and peptides that mimic the functional activity of UEA-1 were identified from mixture-based positional scanning synthetic combinatorial libraries and phage peptide libraries, respectively. Synthetic digalloyl D-lysine amide and tetragalloyl D-lLysine amide molecules were found to bind to the surface of M-cells. Coating of microparticles with these compounds resulted in their selective and efficient delivery to M-cells with high efficacy (Lambkin et al., 2003). It was subsequently demonstrated that the synthetic peptide YQCSYTMPHPPV selectively bound to the M-cell-rich subepithelial dome region of the PP and enhanced the delivery of microparticles to M-cells (Higgins et al., 2004).

Another approach has been to apply microbial adhesins for the targeted delivery of synthetic particles to M-cells. Enhanced antigen uptake was achieved by coating polystyrene nanoparticles with *Yersinia*-derived invasin, a ligand for β1 integrins on the apical side of M-cells (Clark et al., 1998). Similarly, reoviruses are known to invade through M-cells using a 45-kDa viral hemagglutinin sigma one (σ1) protein (Forrest and Dermody, 2003). Mucosal immune responses were significantly increased by mucosal immunization of antigen coupled to σ1 protein (Wang et al., 2003; Wu et al., 2001). Combining the advantages of liposomes and specific targeting molecules, we developed a "fusogenic liposome," a hybrid antigen delivery vehicle composed of a synthetic liposome and ultraviolet-inactivated Sendai virus, also known as Hemagglutinating Virus of Japan (Fig. 14.4) (Kunisawa et al., 2001a). Using the fusion activity of Sendai virus, the fusogenic liposomes effectively delivered the encapsulated antigen to NALT ECs, including M-cells, when given nasally (Kunisawa et al., 2001b). Nasal immunization with fusogenic liposomes containing OVA or HIV glycoprotein 160 (gp160) induced high levels of antigen-specific plasma IgG and mucosal SIgA in saliva, fecal extracts, nasal and vaginal washes, as well as CTL responses (Kunisawa et al., 2001b; Sakaue et al., 2003). Similar viruslike particles (e.g., influenza virus) have been developed for targeted delivery of protein antigens administered by the nasal route (Lambkin et al., 2004).

Immune stimulating complexes (ISCOMs) are another category of lipid-based vehicles that has been well studied as a mucosal delivery system. ISCOMs consist of a particle 30–40 nm in diameter composed of phospholipid, cholesterol, and Quil A or QS-21 (the purified component from Quil A). Quil A is a saponin adjuvant originally isolated from *Quillaja saponaria* (Kensil et al., 1991). The saponin intercalates with the cell membrane of APCs to form pores, thus delivering antigen into the cytoplasm where presentation by the MHC class I pathway leads to induction of CTLs (Bangham et al., 1962; Sanders et al., 2005). QS-21 has also been shown to enhance production of the Th1-type cytokines IFN-γ and IL-12 (Mikloska et al., 2000; Silla et al., 1999). ISCOMs initially were applied to parental vaccination for prevention of infectious diseases caused by influenza virus and HIV-1 (Rimmelzwaan et al., 1997; Takahashi et al., 1990). The first application of ISCOMs for mucosal vaccine delivery was a study of oral immunization with OVA antigen (Mowat et al., 1991). Oral delivery of ISCOMs containing OVA induced antigen-specific mucosal SIgA, plasma IgG, Th1/Th2, and CTL responses. Subsequent studies demonstrated the effectiveness of ISCOMs for enhancing protective immune responses against viral infections such as influenza, rotavirus, and HSV-2 (Fooks, 2000; Kazanji et al., 1994; Mohamedi et al., 2000; Simms et al., 2000; van Pinxteren et al., 1999). As a possible mechanism for the mucosal adjuvant activity of ISCOMs, one study has demonstrated that oral feeding of ISCOMs resulted in the recruitment and activation of DCs in the mesenteric lymph nodes and PPs (Furrie et al., 2002). Similarly, nasal immunization using ISCOMs was quite effective for inducing mucosal SIgA, plasma IgG, and CTL responses for protective immunity against influenza

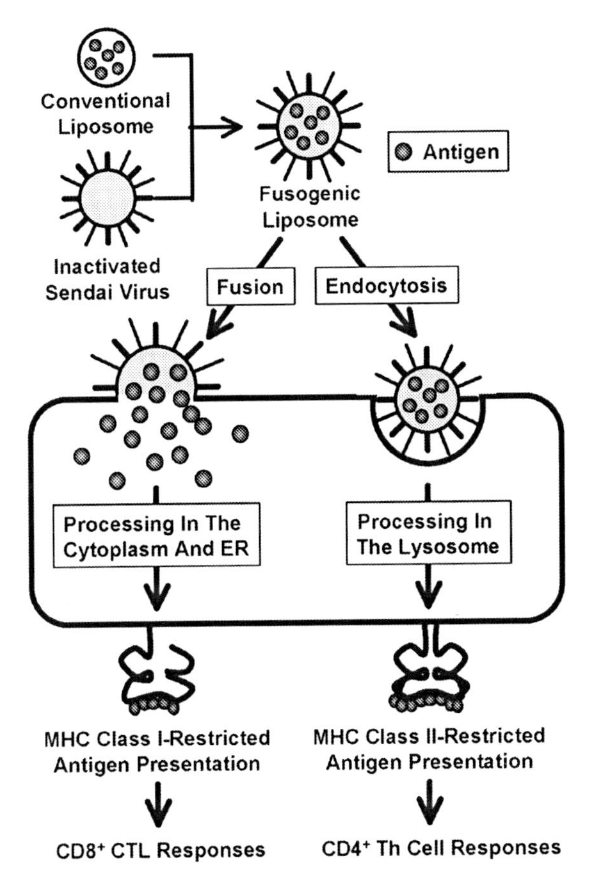

FIG. 14.4. A hybrid delivery vehicle, termed fusogenic liposome, used for the induction of CTL and SIgA Ab production via MHC class I- and II-restricted antigen presentation, respectively. Fusogenic liposomes were prepared by fusing conventional liposomes with ultraviolet-inactivated Sendai virus. Fusogenic liposomes can deliver the encapsulated antigen into the cytoplasm via its fusion with the cell membrane. The endocytosis pathway also participates in the uptake of the fusogenic liposome. The dual pathways for fusogenic liposome uptake allow antigen presentation restricted by both MHC class I and II molecules, leading to $CD8^+$ CTL and $CD4^+$ Th cell responses, respectively.

virus, HSV-2, and respiratory syncytial virus infections (Ben-Ahmeida et al., 1994; Hu et al., 1998, 2001; Ugozzoli et al., 1998).

14.6.2. Genetically Modified Live Microorganisms for Antigen Delivery

Historically, inactivated vaccines have been prepared from microorganisms inactivated by formaldehyde or β-propiolactone treatment. However, these inactivated forms of vaccine do not induce protective mucosal immune responses unless they are coadministered with mucosal adjuvants. On the other hand, live attenuated microorganisms have been shown to be highly effective as mucosal vaccines (e.g., poliovirus, *Salmonella typhi* Ty21a, and *Vibrio cholerae*) (Dietrich et al., 2003). Unlike chemical inactivation, the attenuation process may not affect the activity of naturally expressed ligands for M-cells, which promote vaccine uptake via M-cells for the effective induction of mucosal and systemic immune responses. Therefore, live attenuated microorganisms are taken up by M-cells into the PPs or NALT, which results in the efficient priming of antigen-specific immune responses.

The effective uptake of live attenuated microorganisms by M-cells has led to the development of recombinant attenuated microorganisms to deliver heterologous antigen. Recent progress in recombinant DNA technology has allowed the creation of a new generation of mucosal vaccines, in which live attenuated microorganisms are engineered to carry DNA encoding heterologous antigen (Curtiss, 2005). To ensure both safety and effectiveness, several genes determining pathogenicity are mutated or disrupted, and a gene encoding a heterologous antigen is inserted. An important consideration in the development of recombinant live attenuated vaccines for viral, parasitic, and fungal pathogens is that the vaccine antigen should be expressed in the host eukaryotic cells, as appropriate glycosylation and folding are required for immunogenicity of the expressed protein antigens (Darji et al., 1997). This can be accomplished through the use of intracellular bacteria as vaccine vectors, which invade host cells and release plasmid DNA encoding the vaccine antigen. On the other hand, for protection against bacterial infections, it is necessary that the antigens be expressed within the recombinant bacteria that are used as the vaccine vector.

Several types of recombinant attenuated bacteria have been considered as candidates for delivery vehicles of DNA encoding heterologous vaccine antigens, which can be divided into invasive and noninvasive types. The former include attenuated strains of *Salmonella typhi*, *Shigella flexneri*, and *Listeria monocytogenes,* and the latter include *Vibrio cholerae*, *Lactobacllus* spp., and *Yersinia enterocolitica* (Curtiss, 2005). Important attributes of candidate bacterial vaccine vectors include the ability to survive in the hostile environment of the GI and respiratory tracts and to bind to M-cells in order to selectively enable the effective induction of mucosal and systemic immune responses (e.g., mucosal SIgA, plasma IgG, and CTL). It is also critical that the attenuated bacterium does not revert back to a virulent form capable of triggering disease symptoms.

Among the bacteria described earlier, *Salmonella* are the best studied in regard to attenuating mutations. The licensed oral *S. typhi* vaccine strain, Ty21a, was prepared by nitrosoguanidine mutagenesis. In addition to nonspecific chemical mutagenesis, attenuation by targeted gene disruption has been investigated since the early 1970s. The *gal*E gene was the first target for site-directed mutagenesis (Germanier and Furer, 1971). This mutation decreased UDP-galactose epimerase activity but was not successful for vaccine purposes because the mutant bacteria were not immunogenic. In contrast, *Salmonella* strains with a deletion in the *aro* genes, which blocks the biosynthesis of the aromatic amino acids tyrosine, phenylalanine, and tryptophan, were both avirulent and immunogenic (Dougan *et al.*, 1987; Hoiseth and Stocker, 1981). However, administration of high doses of *S. typhimurium* strains with *aro* mutations to humans caused potentially dangerous bacteremia (Hone et al., 1992). Therefore, an additional attenuating mutation in the *htr*A gene was performed (Tacket et al., 1997b). Deletion of the *htr*A gene, which encodes a heat stress protein, results in a less virulent organism because of reduced survival and replicative ability (Johnson et al., 1991). The doubly mutated *Salmonella* strain successfully induced antigen-specific Ab production and T-cell responses in humans when given by the oral route, without causing undesirable side effects (Tacket et al., 1997b). Additional gene deletions, including those encoding adenylate cyclase, the cyclic AMP receptor protein (*cya*), and the regulatory system for phosphate sensing (*pho*), were also tested for vaccine development against *Salmonella* (Curtiss and Kelly, 1987; Hohmann et al., 1996; Tacket et al., 1997a). Similar approaches have been employed to generate mutants of other bacteria, such as *Shigella*, *L. monocytogenes,* and *Mycobacterium bovis* (also known as Bacille Calmette-Guerin, BCG) for vaccine use (Dramsi et al., 1996; Haile and Kallenius, 2005; Levine et al., 1997).

Using these mutant forms, recombinant bacteria have been engineered to express heterologous antigen by plasmid transformation. It should be noted that amplification of the transformed plasmids and expression of antigenic protein must be tightly regulated, as overexpression of antigens may cause the death of the bacteria due to a metabolic burden (Galen and Levine, 2001). One possible solution to this problem is the use of specific promoters that are active only after entry into host cells. The most extensively examined promoters include the anaerobically inducible *nir*B promoter, the temperature-inducible promoter *htr*A, and the macrophage-inducible promoter *pag*C (Chatfield et al., 1992; Everest et al., 1995; Hohmann et al., 1995). These delivery systems have proven effective for induction of antigen-specific SIgA Ab and CTL responses against various virulence determinants of pathogens like diphtheria exotoxin, HIV, HBV, and pertussis (Barry et al., 1996; Gomez-Duarte et al., 1995; Grillot-Courvalin et al., 1998; Hone et al., 2002; Karem et al., 1997; Woo et al., 2001; Wu et al., 1997).

Similarly, recombinant viruses have been developed as vehicles for mucosal vaccine delivery. Efforts have been aimed at developing mucosal vaccines that induce both mucosal SIgA Abs to prevent initial contact of pathogens with host cells in mucosal sites and antigen-specific CTLs for surveillance of virus-infected cells. Several types of virus have the advantage of their natural tropism for mucosal sites, including DNA viruses such as poxvirus, adenovirus, HSV, and adeno-associated virus, and RNA viruses such as alphaviruses (e.g., Semliki Forest virus and Sindbis virus), vesicular stomatitis virus, and poliovirus (Ertl and Xiang, 1996; Khromykh, 2000).

The poxviruses, including smallpox, vaccinia, and the avian poxvirus, comprise the best studied DNA viruses for vaccine vectors. Poxviruses have a double-stranded DNA genome ranging in size between 130 and 300 kb. The gene encoding heterologous antigen is usually inserted at nonessential sites, such as the viral thymidine kinase locus (Mackett et al., 1984). It was determined that modified or synthetic promoters were required to achieve sufficient expression of heterologous antigen, because the endogenous poxvirus promoters (P7.5 and H5) were too weak (Ourmanov et al., 2000; Wyatt et al., 1996). Modified vaccinia virus Ankara (MVA) is the best candidate in the poxvirus family for vaccines due to their relative safety. MVA is an attenuated strain generated by extensive passage in chicken embryo fibroblasts (Stickl et al., 1974). Another safe strain is NYVAC, which was generated by the deletion of 18 genes encoding virulence factors (Tartaglia et al., 1992). Both MVA and NYVAC expressing heterologous antigen elicited strong mucosal SIgA and plasma IgG Abs as well as CTL responses when they were administered via mucosal routes, leading to their testing as potential vectors for an HIV vaccine (Drexler et al., 2004; Hutchings et al., 2005; Paoletti, 1996; Sauter et al., 2005).

Adenovirus is another DNA virus that has been investigated for mucosal vaccine development and for gene therapy. Adenoviruses have several advantages over other viruses, as they can be readily obtained in high titers and stored at room temperature after freeze-drying. Most importantly, adenovirus naturally infect via mucosal routes; thus, they can deliver the antigen into mucosal sites with great efficacy. Adenoviruses are nonenveloped and contain a linear double-stranded DNA genome between 30 and 45 kb in size. Replication-defective adenovirus vectors have been constructed by insertion of genes encoding heterologous antigens into the E1 region, which is essential for viral replication (Mizuguchi et al., 2001). The E3 and E4 regions of the adenovirus genome have also been found suitable for cloning of heterologous genes (Mizuguchi et al., 2001). These recombinant adenoviruses have been shown to elicit potent Ab and cellular immune responses against antigens when they were applied by mucosal routes (e.g., oral, nasal, rectal, and vaginal), resulting in protective immunity against viral infections, including HBV, HSV, influenza virus, and HIV (Babiuk and Tikoo, 2000; Santosuosso et al., 2005). The success of experimental adenovirus-mediated mucosal vaccines against HIV has led to their use in clinical trials (Barouch and Nabel, 2005; Gomez-Roman and Robert-Guroff, 2003).

A novel adenovirus-derived delivery system has recently been described for mucosal targeting of a vaccine against botulinum neurotoxin A (BoNT/A) (Maddaloni *et al.*, 2006). A chimeric protein was constructed in which the cell-binding domain in the C terminus (Hc) of BoNT/A was fused to adenovirus 2 fiber protein (Ad2F), which binds to receptors on respiratory epithelial cells. Nasal immunization of mice with Hc-Ad2F plus CT adjuvant induced robust intestinal SIgA and plasma IgA responses, which were protective against subsequent lethal BoNT/A challenge.

Alphaviruses have received attention as a mucosal antigen delivery system due to their high efficiency of gene transduction and apparent safety (Lundstrom, 2002). In contrast to the DNA viruses like poxvirus and adenovirus, alphaviruses have an RNA genome and replicate in the host cytosol, which provides safety advantages by eliminating the possibility of integration of heterologous antigen into the host genome. In this family, Sindbis virus, Semliki Forest virus, and Venezuelan equine encephalitis virus have been well studied for expression of heterologous antigen (Karlsson and Liljestrom, 2003; Lundstrom, 2003b). These viruses contain a single-stranded positive-sense RNA genome, 11–12 kb in size, which consists of two open reading frames (ORFs). The heterologous antigen is generally inserted into the ORF encoding the viral structural protein. This cloning strategy results in the generation of mutant viruses that are incapable of producing viral progeny. Significant mucosal SIgA, plasma IgG, and CTL responses have been observed in response to mucosal administration of recombinant alphaviruses, which induced protective immunity against a broad range of viral, bacterial, and parasitic pathogens (Lundstrom, 2003a).

Poliovirus also belongs to a family of positive-stranded RNA viruses. Since attenuated poliovirus has been used clinically for many years, its safety is widely accepted. Polioviruses are resistant to the harsh conditions in the GI tract and they bind to M-cells after oral inoculation (Sicinski et al., 1990). Both replication-competent and replication-defective polioviruses have been developed as heterologous antigen delivery vehicles. Replication-competent polioviruses were engineered with a modified capsid protein expressing heterologous antigen (Andino et al., 1994). In the replication-defective poliovirus, the structural proteins (VP2 and VP3) were replaced with genes encoding heterologous antigen (Porter et al., 1993). Both systems were shown to be effective as mucosal antigen delivery systems and generated protective immunity against infectious diseases such as HIV (Crotty and Andino, 2004).

It should be noted that both bacteria-based and virus-based vaccines are not only effective for the delivery of antigens but also possess natural adjuvant activities because they express ligands for TLR family members. A potential concern, however, is that antivector immune responses may develop after repeated immunization, hampering the reuse of the recombinant vaccine for boosting of the immune response.

14.7. Concluding Remarks

Infectious diseases remain a threat to human health in both developing and advanced nations. It is crucial to develop effective mucosal vaccines against pathogens that induce both mucosal and systemic immune responses, especially SIgA and plasma IgG Abs as the first and second layers of humoral protection. In addition, the induction of mucosal CTLs is needed for protective immunity against viral pathogens. Recent advances in biomedical engineering and progress in understanding the cellular and molecular immunology of infectious diseases have allowed the development of versatile mucosal vaccine systems, based on novel mucosal adjuvants and mucosal antigen delivery systems. Creation of a new generation of mucosal vaccines requires coordination of multiple factors, including the choice of the antigen delivery system and the optimal mucosal adjuvant. In this regard, it might be necessary to compensate weak points of one with strong attributes of the other, by exploiting potential synergistic effects among different adjuvants and delivery systems. Although further investigation is clearly needed, there is no doubt that novel mucosal adjuvants and antigen delivery systems will facilitate the development of a new generation of mucosal vaccines that will contribute to human health.

References

Agren, L., Sverremark, E., Ekman, L., Schon, K., Lowenadler, B., Fernandez, C., and Lycke, N. (2000). The ADP-ribosylating CTA1-DD adjuvant enhances T cell-dependent and independent responses by direct action on B cells involving anti-apoptotic Bcl-2- and germinal center-promoting effects. *J. Immunol.* 164: 6276–6286.

Agren, L. C., Ekman, L., Lowenadler, B., and Lycke, N. Y. (1997). Genetically engineered nontoxic vaccine adjuvant that combines B cell targeting with immunomodulation by cholera toxin A1 subunit. *J. Immunol.* 158:3936–3946.

Agren, L. C., Ekman, L., Lowenadler, B., Nedrud, J. G., and Lycke, N. Y. (1999). Adjuvanticity of the cholera toxin A1-based gene fusion protein, CTA1-DD, is critically dependent on the ADP-ribosyltransferase and Ig-binding activity. *J. Immunol.* 162:2432–2440.

Akira, S., and Takeda, K. (2004). Toll-like receptor signalling. *Nat. Rev. Immunol.* 4:499–511.

Aman, A. T., Fraser, S., Merritt, E. A., Rodigherio, C., Kenny, M., Ahn, M., Hol, W. G., Williams, N. A., Lencer, W. I., and Hirst, T. R. (2001). A mutant cholera toxin B subunit that binds GM1-ganglioside but lacks immunomodulatory or toxic activity. *Proc. Natl. Acad. Sci. USA* 98:8536–8541.

Andino, R., Silvera, D., Suggett, S. D., Achacoso, P. L., Miller, C. J., Baltimore, D., and Feinberg, M. B. (1994). Engineering poliovirus as a vaccine vector for the expression of diverse antigens. *Science* 265:1448–1451.

Aramaki, Y., Tomizawa, H., Hara, T., Yachi, K., Kikuchi, H., and Tsuchiya, S. (1993). Stability of liposomes *in vitro* and their uptake by rat Peyer's patches following oral administration. *Pharm. Res.* 10:1228–1231.

Arulanandam, B. P., Lynch, J. M., Briles, D. E., Hollingshead, S., and Metzger, D. W. (2001). Intranasal vaccination with pneumococcal surface protein A and interleukin-12 augments antibody-mediated opsonization and protective immunity against *Streptococcus pneumoniae* infection. *Infect. Immun.* 69:6718–6724.

Arulanandam, B. P., O'Toole, M., and Metzger, D. W. (1999). Intranasal interleukin-12 is a powerful adjuvant for protective mucosal immunity. *J. Infect. Dis.* 180:940–949.

Babiuk, L. A., and Tikoo, S. K. (2000). Adenoviruses as vectors for delivering vaccines to mucosal surfaces. *J. Biotechnol.* 83:105–113.

Baldridge, J. R., McGowan, P., Evans, J. T., Cluff, C., Mossman, S., Johnson, D., and Persing, D. (2004). Taking a Toll on human disease: Toll-like receptor 4 agonists as vaccine adjuvants and monotherapeutic agents. *Expert. Opin. Biol. Ther.* 4:1129–1138.

Baldridge, J. R., Yorgensen, Y., Ward, J. R., and Ulrich, J. T. (2000). Monophosphoryl lipid A enhances mucosal and systemic immunity to vaccine antigens following intranasal administration. *Vaccine* 18:2416–2425.

Bangham, A. D., Horne, R. W., Glauert, A. M., Dingle, J. T., and Lucy, J. A. (1962). Action of saponin on biological cell membranes. *Nature* 196:952–955.

Barchfeld, G. L., Hessler, A. L., Chen, M., Pizza, M., Rappuoli, R., and Van Nest, G. A. (1999). The adjuvants MF59 and LT-K63 enhance the mucosal and systemic immunogenicity of subunit influenza vaccine administered intranasally in mice. *Vaccine* 17:695–704.

Barouch, D. H., and Nabel, G. J. (2005). Adenovirus vector-based vaccines for human immunodeficiency virus type 1. *Hum. Gene Ther.* 16:149–156.

Barry, E. M., Gomez-Duarte, O., Chatfield, S., Rappuoli, R., Pizza, M., Losonsky, G., Galen, J., and Levine, M. M. (1996). Expression and immunogenicity of pertussis toxin S1 subunit-tetanus toxin fragment C fusions in *Salmonella typhi* vaccine strain CVD 908. *Infect. Immun.* 64:4172–4181.

Basler, C. F., and Garcia-Sastre, A. (2002). Viruses and the type I interferon antiviral system: induction and evasion. *Int. Rev. Immunol.* 21:305–337.

Baudry, B., Fasano, A., Ketley, J., and Kaper, J. B. (1992). Cloning of a gene (zot) encoding a new toxin produced by *Vibrio cholerae. Infect. Immun.* 60:428–434.

Ben-Ahmeida, E. T., Potter, C. W., Gregoriadis, G., Adithan, C., and Jennings, R. (1994). IgG subclass response and protection against challenge following immunisation of mice with various influenza A vaccines. *J. Med. Microbiol.* 40:261–269.

Berkes, J., Viswanathan, V. K., Savkovic, S. D., and Hecht, G. (2003). Intestinal epithelial responses to enteric pathogens: Effects on the tight junction barrier, ion transport, and inflammation. *Gut* 52:439–451.

Berland, R., and Wortis, H. H. (2002). Origins and functions of B-1 cells with notes on the role of CD5. *Annu. Rev. Immunol.* 20:253–300.

Bertley, F. M., Kozlowski, P. A., Wang, S. W., Chappelle, J., Patel, J., Sonuyi, O., Mazzara, G., Montefiori, D., Carville, A., Mansfield, K. G., and Aldovini, A. (2004). Control of simian/human immunodeficiency virus viremia and disease progression after IL-2-augmented DNA-modified vaccinia virus Ankara nasal vaccination in nonhuman primates. *J. Immunol.* 172:3745–3757.

Boismenu, R., Feng, L., Xia, Y. Y., Chang, J. C., and Havran, W. L. (1996). Chemokine expression by intraepithelial gamma delta T cells. Implications for the recruitment of inflammatory cells to damaged epithelia. *J. Immunol.* 157:985–992.

Bone, H., Eckholdt, S., and Williams, N. A. (2002). Modulation of B lymphocyte signalling by the B subunit of *Escherichia coli* heat-labile enterotoxin. *Int. Immunol.* 14:647–658.

Borsutzky, S., Fiorelli, V., Ebensen, T., Tripiciano, A., Rharbaoui, F., Scoglio, A., Link, C., Nappi, F., Morr, M., Butto, S., Cafaro, A., Muhlradt, P. F., Ensoli, B., and Guzman, C. A. (2003). Efficient mucosal delivery of the HIV-1 Tat protein using the synthetic lipopeptide MALP-2 as adjuvant. *Eur. J. Immunol.* 33:1548–1556.

Borsutzky, S., Kretschmer, K., Becker, P. D., Muhlradt, P. F., Kirschning, C. J., Weiss, S., and Guzman, C. A. (2005). The mucosal adjuvant macrophage-activating lipopeptide-2 directly stimulates B lymphocytes via the TLR2 without the need of accessory cells. *J. Immunol.* 174:6308–6313.

Bos, N. A., Cebra, J. J., and Kroese, F. G. (2000). B-1 cells and the intestinal microflora. *Curr. Topics Microbiol. Immunol.* 252:211–220.

Bowen, J. C., Nair, S. K., Reddy, R., and Rouse, B. T. (1994). Cholera toxin acts as a potent adjuvant for the induction of cytotoxic T-lymphocyte responses with non-replicating antigens. *Immunology* 81:338–342.

Bowman, C. C., and Clements, J. D. (2001). Differential biological and adjuvant activities of cholera toxin and *Escherichia coli* heat-labile enterotoxin hybrids. *Infect. Immun.* 69:1528–1535.

Boyaka, P. N., Marinaro, M., Jackson, R. J., Menon, S., Kiyono, H., Jirillo, E., and McGhee, J. R. (1999). IL-12 is an effective adjuvant for induction of mucosal immunity. *J. Immunol.* 162:122–128.

Boyaka, P. N., Ohmura, M., Fujihashi, K., Koga, T., Yamamoto, M., Kweon, M. N., Takeda, Y., Jackson, R. J., Kiyono, H., Yuki, Y., and McGhee, J. R. (2003). Chimeras of labile toxin one and cholera toxin retain mucosal adjuvanticity and direct Th cell subsets via their B subunit. *J. Immunol.* 170:454–462.

Bradney, C. P., Sempowski, G. D., Liao, H. X., Haynes, B. F., and Staats, H. F. (2002). Cytokines as adjuvants for the induction of anti-human immunodeficiency virus peptide immunoglobulin G (IgG) and IgA antibodies in serum and mucosal secretions after nasal immunization. *J. Virol.* 76:517–524.

Brandtzaeg, P., and Johansen, F. E. (2005). Mucosal B cells: phenotypic characteristics, transcriptional regulation, and homing properties. *Immunol. Rev.* 206:32–63.

Braun, M. C., He, J., Wu, C. Y., and Kelsall, B. L. (1999). Cholera toxin suppresses interleukin (IL)-12 production and IL-12 receptor beta1 and beta2 chain expression. *J. Exp. Med.* 189:541–552.

Cardenas-Freytag, L., Cheng, E., Mayeux, P., Domer, J. E., and Clements, J. D. (1999). Effectiveness of a vaccine composed of heat-killed *Candida albicans* and a novel mucosal adjuvant, LT(R192G), against systemic candidiasis. *Infect. Immun.* 67:826–833.

Chatfield, S. N., Charles, I. G., Makoff, A. J., Oxer, M. D., Dougan, G., Pickard, D., Slater, D., and Fairweather, N. F. (1992). Use of the nirB promoter to direct the stable expression of heterologous antigens in Salmonella oral vaccine strains: development of a single-dose oral tetanus vaccine. *Biotechnology (N Y)* 10:888–892.

Chen, H., and Langer, R. (1997). Magnetically-responsive polymerized liposomes as potential oral delivery vehicles. *Pharm. Res.* 14:537–540.

Chen, H., Torchilin, V., and Langer, R. (1996). Lectin-bearing polymerized liposomes as potential oral vaccine carriers. *Pharm. Res.* 13:1378–1383.

Cheroutre, H. (2005). IELs: Enforcing law and order in the court of the intestinal epithelium. *Immunol. Rev.* 206:114–131.

Chong, C., Friberg, M., and Clements, J. D. (1998). LT(R192G), a non-toxic mutant of the heat-labile enterotoxin of *Escherichia coli*, elicits enhanced humoral and cellular immune responses associated with protection against lethal oral challenge with *Salmonella spp. Vaccine* 16:732–740.

Chu, R. S., McCool, T., Greenspan, N. S., Schreiber, J. R., and Harding, C. V. (2000). CpG oligodeoxynucleotides act as adjuvants for pneumococcal polysaccharide–protein conjugate vaccines and enhance antipolysaccharide immunoglobulin G2a (IgG2a) and IgG3 antibodies. *Infect. Immun.* 68:1450–1456.

Clark, M. A., Blair, H., Liang, L., Brey, R. N., Brayden, D., and Hirst, B. H. (2001a). Targeting polymerised liposome vaccine carriers to intestinal M cells. *Vaccine* 20:208–217.

Clark, M. A., Hirst, B. H., and Jepson, M. A. (1998). M-cell surface beta1 integrin expression and invasin-mediated targeting of *Yersinia pseudotuberculosis* to mouse Peyer's patch M cells. *Infect. Immun.* 66:1237–1243.
Clark, M. A., Jepson, M. A., and Hirst, B. H. (2001b). Exploiting M cells for drug and vaccine delivery. *Adv. Drug Deliv. Rev.* 50:81–106.
Cong, Y., Oliver, A. O., and Elson, C. O. (2001). Effects of cholera toxin on macrophage production of co-stimulatory cytokines. *Eur. J. Immunol.* 31:64–71.
Cong, Y., Weaver, C. T., and Elson, C. O. (1997). The mucosal adjuvanticity of cholera toxin involves enhancement of costimulatory activity by selective up-regulation of B7.2 expression. *J. Immunol.* 159:5301–5308.
Crotty, S., and Andino, R. (2004). Poliovirus vaccine strains as mucosal vaccine vectors and their potential use to develop an AIDS vaccine. *Adv. Drug Deliv. Rev.* 56:835–852.
Curtiss, R., 3rd (2005). Antigen delivery systems II: Development of live recombinant attenuated bacterial antigen and DNA vaccine delivery vector vaccines. In: Mestecky, J., Bienenstock, J., Lamm, M. E., Storober, W., and McGhee, J. R. (eds.), *Mucosal Immunology*, Academic Press, San Diego, pp. 1009–1037.
Curtiss, R., 3rd, and Kelly, S. M. (1987). *Salmonella typhimurium* deletion mutants lacking adenylate cyclase and cyclic AMP receptor protein are avirulent and immunogenic. *Infect. Immun.* 55:3035–3043.
Darji, A., Guzman, C. A., Gerstel, B., Wachholz, P., Timmis, K. N., Wehland, J., Chakraborty, T., and Weiss, S. (1997). Oral somatic transgene vaccination using attenuated *S. typhimurium*. *Cell* 91:765–775.
de Haan, L., Verweij, W. R., Feil, I. K., Lijnema, T. H., Hol, W. G., Agsteribbe, E., and Wilschut, J. (1996). Mutants of the *Escherichia coli* heat-labile enterotoxin with reduced ADP-ribosylation activity or no activity retain the immunogenic properties of the native holotoxin. *Infect. Immun.* 64:5413–5416.
De Magistris, M. T., Pizza, M., Douce, G., Ghiara, P., Dougan, G., and Rappuoli, R. (1998). Adjuvant effect of non-toxic mutants of *E. coli* heat-labile enterotoxin following intranasal, oral and intravaginal immunization. *Dev. Biol. Stand.* 92:123–126.
Dickinson, B. L., and Clements, J. D. (1995). Dissociation of *Escherichia coli* heat-labile enterotoxin adjuvanticity from ADP-ribosyltransferase activity. *Infect. Immun.* 63:1617–1623.
Dietrich, G., Spreng, S., Favre, D., Viret, J. F., and Guzman, C. A. (2003). Live attenuated bacteria as vectors to deliver plasmid DNA vaccines. *Curr. Opin. Mol. Ther.* 5:10–19.
Dilloo, D., Bacon, K., Holden, W., Zhong, W., Burdach, S., Zlotnik, A., and Brenner, M. (1996). Combined chemokine and cytokine gene transfer enhances antitumor immunity. *Nat. Med.* 2:1090–1095.
Doherty, T. M., Olsen, A. W., van Pinxteren, L., and Andersen, P. (2002). Oral vaccination with subunit vaccines protects animals against aerosol infection with *Mycobacterium tuberculosis*. *Infect. Immun.* 70:3111–3121.
Douce, G., Turcotte, C., Cropley, I., Roberts, M., Pizza, M., Domenghini, M., Rappuoli, R., and Dougan, G. (1995). Mutants of *Escherichia coli* heat-labile toxin lacking ADP-ribosyltransferase activity act as nontoxic, mucosal adjuvants. *Proc. Natl. Acad. Sci. USA* 92:1644–1648.
Dougan, G., Maskell, D., Pickard, D., and Hormaeche, C. (1987). Isolation of stable aroA mutants of *Salmonella typhi* Ty2: Properties and preliminary characterisation in mice. *Mol. Gen. Genet.* 207:402–405.

Dramsi, S., Lebrun, M., and Cossart, P. (1996). Molecular and genetic determinants involved in invasion of mammalian cells by *Listeria monocytogenes. Curr. Topics Microbiol. Immunol.* 209:61–77.

Drexler, I., Staib, C., and Sutter, G. (2004). Modified vaccinia virus Ankara as antigen delivery system: how can we best use its potential? *Curr. Opin. Biotechnol.* 15:506–512.

Dubois, B., Goubier, A., Joubert, G., and Kaiserlian, D. (2005). Oral tolerance and regulation of mucosal immunity. *Cell. Mol. Life Sci.* 62:1322–1332.

Duckett, N. S., Olmos, S., Durrant, D. M., and Metzger, D. W. (2005). Intranasal interleukin-12 treatment for protection against respiratory infection with the *Francisella tularensis* live vaccine strain. *Infect. Immun.* 73:2306–2311.

Dumais, N., Patrick, A., Moss, R. B., Davis, H. L., and Rosenthal, K. L. (2002). Mucosal immunization with inactivated human immunodeficiency virus plus CpG oligodeoxynucleotides induces genital immune responses and protection against intravaginal challenge. *J. Infect. Dis.* 186:1098–1105.

Duverger, A., Jackson, R. J., van Ginkel, F. W., Fischer, R., Tafaro, A., Leppla, S. H., Fujihashi, K., Kiyono, H., McGhee, J. R., and Boyaka, P. N. (2006). Bacillus anthracis edema toxin acts as an adjuvant for mucosal immune responses to nasally administered vaccine antigens. *J. Immunol.* 176:1776–1783.

Ertl, H. C., and Xiang, Z. (1996). Novel vaccine approaches. *J. Immunol.* 156:3579–3582.

Evans, J. T., Cluff, C. W., Johnson, D. A., Lacy, M. J., Persing, D. H., and Baldridge, J. R. (2003). Enhancement of antigen-specific immunity via the TLR4 ligands MPL adjuvant and Ribi.529. *Expert. Rev. Vaccines* 2:219–229.

Everest, P., Frankel, G., Li, J., Lund, P., Chatfield, S., and Dougan, G. (1995). Expression of LacZ from the htrA, nirB and groE promoters in a *Salmonella* vaccine strain: influence of growth in mammalian cells. *FEMS Microbiol. Lett.* 126:97–101.

Fagarasan, S., Shinkura, R., Kamata, T., Nogaki, F., Ikuta, K., Tashiro, K., and Honjo, T. (2000). Alymphoplasia (aly)-type nuclear factor kappaB-inducing kinase (NIK) causes defects in secondary lymphoid tissue chemokine receptor signaling and homing of peritoneal cells to the gut-associated lymphatic tissue system. *J. Exp. Med.* 191:1477–1486.

Fasano, A., Baudry, B., Pumplin, D. W., Wasserman, S. S., Tall, B. D., Ketley, J. M., and Kaper, J. B. (1991). *Vibrio cholerae* produces a second enterotoxin, which affects intestinal tight junctions. *Proc. Natl. Acad. Sci. USA* 88:5242–5246.

Fayad, R., Zhang, H., Quinn, D., Huang, Y., and Qiao, L. (2004). Oral administration with papillomavirus pseudovirus encoding IL-2 fully restores mucosal and systemic immune responses to vaccinations in aged mice. *J. Immunol.* 173:2692–2698.

Feng, Y., Jadhav, A. P., Rodighiero, C., Fujinaga, Y., Kirchhausen, T., and Lencer, W. I. (2004). Retrograde transport of cholera toxin from the plasma membrane to the endoplasmic reticulum requires the trans-Golgi network but not the Golgi apparatus in Exo2-treated cells. *EMBO Rep.* 5:596–601.

Fooks, A. R. (2000). Development of oral vaccines for human use. *Curr. Opin. Mol. Ther.* 2:80–86.

Forrest, J. C., and Dermody, T. S. (2003). Reovirus receptors and pathogenesis. *J. Virol.* 77:9109–9115.

Foster, N., Clark, M. A., Jepson, M. A., and Hirst, B. H. (1998). *Ulex europaeus* 1 lectin targets microspheres to mouse Peyer's patch M-cells *in vivo. Vaccine* 16:536–541.

Foster, N., and Hirst, B. H. (2005). Exploiting receptor biology for oral vaccination with biodegradable particulates. *Adv. Drug Deliv. Rev.* 57:431–450.

Fraser, S. A., de Haan, L., Hearn, A. R., Bone, H. K., Salmond, R. J., Rivett, A. J., Williams, N. A., and Hirst, T. R. (2003). Mutant *Escherichia coli* heat-labile toxin B subunit that separates toxoid-mediated signaling and immunomodulatory action from trafficking and delivery functions. *Infect. Immun.* 71:1527–1537.

Fujihashi, K., Koga, T., van Ginkel, F. W., Hagiwara, Y., and McGhee, J. R. (2002). A dilemma for mucosal vaccination: Efficacy versus toxicity using enterotoxin-based adjuvants. *Vaccine* 20:2431–2438.

Fukuta, S., Magnani, J. L., Twiddy, E. M., Holmes, R. K., and Ginsburg, V. (1988). Comparison of the carbohydrate-binding specificities of cholera toxin and *Escherichia coli* heat-labile enterotoxins LTh-I, LT-IIa, and LT-IIb. *Infect. Immun.* 56:1748–1753.

Furrie, E., Smith, R. E., Turner, M. W., Strobel, S., and Mowat, A. M. (2002). Induction of local innate immune responses and modulation of antigen uptake as mechanisms underlying the mucosal adjuvant properties of immune stimulating complexes (ISCOMS). *Vaccine* 20:2254–2262.

Galen, J. E., and Levine, M. M. (2001). Can a 'flawless' live vector vaccine strain be engineered? *Trends Microbiol.* 9:372–376.

Gallichan, W. S., Woolstencroft, R. N., Guarasci, T., McCluskie, M. J., Davis, H. L., and Rosenthal, K. L. (2001). Intranasal immunization with CpG oligodeoxynucleotides as an adjuvant dramatically increases IgA and protection against herpes simplex virus-2 in the genital tract. *J. Immunol.* 166:3451–3457.

Germanier, R., and Furer, E. (1971). Immunity in experimental salmonellosis. II. Basis for the avirulence and protective capacity of gal E mutants of *Salmonella typhimurium*. *Infect. Immun.* 4:663–673.

Giannasca, P. J., Giannasca, K. T., Falk, P., Gordon, J. I., and Neutra, M. R. (1994). Regional differences in glycoconjugates of intestinal M cells in mice: Potential targets for mucosal vaccines. *Am. J. Physiol.* 267:G1108–G1121.

Gomez-Duarte, O. G., Galen, J., Chatfield, S. N., Rappuoli, R., Eidels, L., and Levine, M. M. (1995). Expression of fragment C of tetanus toxin fused to a carboxyl-terminal fragment of diphtheria toxin in *Salmonella typhi* CVD 908 vaccine strain. *Vaccine* 13:1596–1602.

Gomez-Roman, V. R., and Robert-Guroff, M. (2003). Adenoviruses as vectors for HIV vaccines. *AIDS Rev.* 5:178–185.

Grillot-Courvalin, C., Goussard, S., Huetz, F., Ojcius, D. M., and Courvalin, P. (1998). Functional gene transfer from intracellular bacteria to mammalian cells. *Nat. Biotechnol.* 16:862–866.

Guy-Grand, D., and Vassalli, P. (2002). Gut intraepithelial lymphocyte development. *Curr. Opin. Immunol.* 14:255–259.

Hagiwara, Y., Kawamura, Y. I., Kataoka, K., Rahima, B., Jackson, R. J., Komase, K., Dohi, T., Boyaka, P. N., Takeda, Y., Kiyono, H., McGhee, J. R., and Fujihashi, K. (2006). A second generation of double mutant cholera toxin adjuvants: Enhanced immunity without intracellular trafficking. *J. Immunol.* 177:3045–3054.

Haile, M., and Kallenius, G. (2005). Recent developments in tuberculosis vaccines. *Curr. Opin. Infect. Dis.* 18:211–215.

Halpern, M. S., and Koshland, M. E. (1970). Noval subunit in secretory IgA. *Nature* 228:1276–1278.

Hamada, H., Hiroi, T., Nishiyama, Y., Takahashi, H., Masunaga, Y., Hachimura, S., Kaminogawa, S., Takahashi-Iwanaga, H., Iwanaga, T., Kiyono, H., Yamamoto, H., and Ishikawa, H. (2002). Identification of multiple isolated lymphoid follicles on the antimesenteric wall of the mouse small intestine. *J. Immunol.* 168:57–64.

Han, M., Watarai, S., Kobayashi, K., and Yasuda, T. (1997). Application of liposomes for development of oral vaccines: study of *in vitro* stability of liposomes and antibody response to antigen associated with liposomes after oral immunization. *J. Vet. Med. Sci.* 59:1109–1114.

Harandi, A. M., and Holmgren, J. (2004). CpG DNA as a potent inducer of mucosal immunity: Implications for immunoprophylaxis and immunotherapy of mucosal infections. *Curr. Opin. Invest. Drugs* 5:141–145.

Hartmann, G., and Krieg, A. M. (2000). Mechanism and function of a newly identified CpG DNA motif in human primary B cells. *J. Immunol.* 164:944–953.

Haynes, L., and Eaton, S. M. (2005). The effect of age on the cognate function of $CD4^+$ T cells. *Immunol. Rev.* 205:220–228.

Hemmi, H., Takeuchi, O., Kawai, T., Kaisho, T., Sato, S., Sanjo, H., Matsumoto, M., Hoshino, K., Wagner, H., Takeda, K., and Akira, S. (2000). A Toll-like receptor recognizes bacterial DNA. *Nature* 408:740–745.

Higgins, L. M., Lambkin, I., Donnelly, G., Byrne, D., Wilson, C., Dee, J., Smith, M., and O'Mahony, D. J. (2004). *In vivo* phage display to identify M cell-targeting ligands. *Pharm. Res.* 21:695–705.

Hiroi, T., Yanagita, M., Iijima, H., Iwatani, K., Yoshida, T., Takatsu, K., and Kiyono, H. (1999). Deficiency of IL-5 receptor alpha-chain selectively influences the development of the common mucosal immune system independent IgA-producing B-1 cell in mucosa-associated tissues. *J. Immunol.* 162:821–828.

Hiroi, T., Yanagita, M., Ohta, N., Sakaue, G., and Kiyono, H. (2000). IL-15 and IL-15 receptor selectively regulate differentiation of common mucosal immune system-independent B-1 cells for IgA responses. *J. Immunol.* 165:4329–4337.

Hohmann, E. L., Oletta, C. A., Killeen, K. P., and Miller, S. I. (1996). phoP/phoQ-deleted *Salmonella typhi* (Ty800) is a safe and immunogenic single-dose typhoid fever vaccine in volunteers. *J. Infect. Dis.* 173:1408–1414.

Hohmann, E. L., Oletta, C. A., Loomis, W. P., and Miller, S. I. (1995). Macrophage-inducible expression of a model antigen in *Salmonella typhimurium* enhances immunogenicity. *Proc. Natl. Acad. Sci. USA* 92:2904–2908.

Hoiseth, S. K., and Stocker, B. A. (1981). Aromatic-dependent *Salmonella typhimurium* are non-virulent and effective as live vaccines. *Nature* 291:238–239.

Hone, D. M., DeVico, A. L., Fouts, T. R., Onyabe, D. Y., Agwale, S. M., Wambebe, C. O., Blattner, W. A., Gallo, R. C., and Lewis, G. K. (2002). Development of vaccination strategies that elicit broadly neutralizing antibodies against human immunodeficiency virus type 1 in both the mucosal and systemic immune compartments. *J. Hum. Virol.* 5:17–23.

Hone, D. M., Tacket, C. O., Harris, A. M., Kay, B., Losonsky, G., and Levine, M. M. (1992). Evaluation in volunteers of a candidate live oral attenuated *Salmonella typhi* vector vaccine. *J. Clin. Invest.* 90:412–420.

Horner, A. A., Datta, S. K., Takabayashi, K., Belyakov, I. M., Hayashi, T., Cinman, N., Nguyen, M. D., Van Uden, J. H., Berzofsky, J. A., Richman, D. D., and Raz, E. (2001). Immunostimulatory DNA-based vaccines elicit multifaceted immune responses against HIV at systemic and mucosal sites. *J. Immunol.* 167:1584–1591.

Hornquist, E., Grdic, D., Mak, T., and Lycke, N. (1996). CD8-deficient mice exhibit augmented mucosal immune responses and intact adjuvant effects to cholera toxin. *Immunology* 87:220–229.
Hu, K. F., Elvander, M., Merza, M., Akerblom, L., Brandenburg, A., and Morein, B. (1998). The immunostimulating complex (ISCOM) is an efficient mucosal delivery system for respiratory syncytial virus (RSV) envelope antigens inducing high local and systemic antibody responses. *Clin. Exp. Immunol.* 113:235–243.
Hu, K. F., Lovgren-Bengtsson, K., and Morein, B. (2001). Immunostimulating complexes (ISCOMs) for nasal vaccination. *Adv. Drug Deliv. Rev.* 51:149–159.
Hutchings, C. L., Gilbert, S. C., Hill, A. V., and Moore, A. C. (2005). Novel protein and poxvirus-based vaccine combinations for simultaneous induction of humoral and cell-mediated immunity. *J. Immunol.* 175:599–606.
Ichinohe, T., Watanabe, I., Ito, S., Fujii, H., Moriyama, M., Tamura, S., Takahashi, H., Sawa, H., Chiba, J., Kurata, T., Sata, T., and Hasegawa, H. (2005). Synthetic double-stranded RNA poly(I:C) combined with mucosal vaccine protects against influenza virus infection. *J. Virol.* 79:2910–2919.
Isaka, M., Komiya, T., Takahashi, M., Yasuda, Y., Taniguchi, T., Zhao, Y., Matano, K., Matsui, H., Maeyama, J., Morokuma, K., Ohkuma, K., Goto, N., and Tochikubo, K. (2004). Recombinant cholera toxin B subunit (rCTB) as a mucosal adjuvant enhances induction of diphtheria and tetanus antitoxin antibodies in mice by intranasal administration with diphtheria-pertussis-tetanus (DPT) combination vaccine. *Vaccine* 22:3061–3068.
Isomura, I., Yasuda, Y., Tsujimura, K., Takahashi, T., Tochikubo, K., and Morita, A. (2005). Recombinant cholera toxin B subunit activates dendritic cells and enhances antitumor immunity. *Microbiol. Immunol.* 49:79–87.
Iwasaki, A., and Medzhitov, R. (2004). Toll-like receptor control of the adaptive immune responses. *Nat. Immunol.* 5:987–995.
Iwata, M., Hirakiyama, A., Eshima, Y., Kagechika, H., Kato, C., and Song, S. Y. (2004). Retinoic acid imprints gut-homing specificity on T cells. *Immunity* 21:527–538.
Jain, S. L., Barone, K. S., and Michael, J. G. (1996). Activation patterns of murine T cells after oral administration of an enterocoated soluble antigen. *Cell. Immunol.* 167:170–175.
Jang, M. H., Kweon, M. N., Iwatani, K., Yamamoto, M., Terahara, K., Sasakawa, C., Suzuki, T., Nochi, T., Yokota, Y., Rennert, P. D., Hiroi, T., Tamagawa, H., Iijima, H., Kunisawa, J., Yuki, Y., and Kiyono, H. (2004). Intestinal villous M cells: An antigen entry site in the mucosal epithelium. *Proc. Natl. Acad. Sci. USA* 101:6110–6115.
Jepson, M. A., Clark, M. A., and Hirst, B. H. (2004). M cell targeting by lectins: A strategy for mucosal vaccination and drug delivery. *Adv. Drug Deliv. Rev.* 56:511–525.
Jiang, W., Baker, H. J., and Smith, B. F. (2003). Mucosal immunization with helicobacter, CpG DNA, and cholera toxin is protective. *Infect. Immun.* 71:40–46.
Johansson-Lindbom, B., Svensson, M., Wurbel, M. A., Malissen, B., Marquez, G., and Agace, W. (2003). Selective generation of gut tropic T cells in gut-associated lymphoid tissue (GALT): Requirement for GALT dendritic cells and adjuvant. *J. Exp. Med.* 198:963–969.
Johnson, K., Charles, I., Dougan, G., Pickard, D., O'Gaora, P., Costa, G., Ali, T., Miller, I., and Hormaeche, C. (1991). The role of a stress-response protein in *Salmonella typhimurium* virulence. *Mol. Microbiol.* 5:401–407.

Kaetzel, C. S. (2005). The polymeric immunoglobulin receptor: Bridging innate and adaptive immune responses at mucosal surfaces. *Immunol. Rev.* 206:83–99.

Kaetzel, C. S., Robinson, J. K., Chintalacharuvu, K. R., Vaerman, J. P., and Lamm, M. E. (1991). The polymeric immunoglobulin receptor (secretory component) mediates transport of immune complexes across epithelial cells: A local defense function for IgA. *Proc. Natl. Acad. Sci. USA* 88:8796–8800.

Karem, K. L., Bowen, J., Kuklin, N., and Rouse, B. T. (1997). Protective immunity against herpes simplex virus (HSV) type 1 following oral administration of recombinant *Salmonella typhimurium* vaccine strains expressing HSV antigens. *J. Gen. Virol.* 78(Pt. 2):427–434.

Karlsson, G. B., and Liljestrom, P. (2003). Live viral vectors: Semliki Forest virus. *Methods Mol. Med.* 87:69–82.

Kazanji, M., Laurent, F., and Pery, P. (1994). Immune responses and protective effect in mice vaccinated orally with surface sporozoite protein of *Eimeria falciformis* in ISCOMs. *Vaccine* 12:798–804.

Kelsall, B. L., and Leon, F. (2005). Involvement of intestinal dendritic cells in oral tolerance, immunity to pathogens, and inflammatory bowel disease. *Immunol. Rev.* 206:132–148.

Kensil, C. R., Patel, U., Lennick, M., and Marciani, D. (1991). Separation and characterization of saponins with adjuvant activity from *Quillaja saponaria* Molina cortex. *J. Immunol.* 146:431–437.

Khromykh, A. A. (2000). Replicon-based vectors of positive strand RNA viruses. *Curr. Opin. Mol. Ther.* 2:555–569.

Kiyono, H., and Fukuyama, S. (2004). NALT- versus Peyer's-patch-mediated mucosal immunity. *Nat. Rev. Immunol.* 4:699–710.

Kiyono, H., Michalek, S. M., Mosteller, L. M., Torii, M., Hamada, S., and McGhee, J. R. (1982). Enhancement of murine immune responses to orally administered haptenated *Streptococcus mutans*. *Scand. J. Immunol.* 16:455–463.

Kobayashi, R., Kohda, T., Kataoka, K., Ihara, H., Kozaki, S., Pascual, D. W., Staats, H. F., Kiyono, H., McGhee, J. R., and Fujihashi, K. (2005). A novel neurotoxoid vaccine prevents mucosal botulism. *J. Immunol.* 174:2190–2195.

Kovanen, P. E., and Leonard, W. J. (2004). Cytokines and immunodeficiency diseases: critical roles of the gamma(c)-dependent cytokines interleukins 2, 4, 7, 9, 15, and 21, and their signaling pathways. *Immunol. Rev.* 202:67–83.

Krieg, A. M., Yi, A. K., Matson, S., Waldschmidt, T. J., Bishop, G. A., Teasdale, R., Koretzky, G. A., and Klinman, D. M. (1995). CpG motifs in bacterial DNA trigger direct B-cell activation. *Nature* 374:546–549.

Kunisawa, J., Fukuyama, S., and Kiyono, H. (2005). Mucosa-associated lymphoid tissues in aerodigestive tract: Their shared and divergent traits and their importance to the orchestration of the mucosal immune system. *Curr. Mol. Med.* 5:557–572.

Kunisawa, J., and Kiyono, H. (2005). A marvel of mucosal T cells and secretory antibodies for the creation of first lines of defense. *Cell. Mol. Life Sci.* 62:1308–1321.

Kunisawa, J., Nakagawa, S., and Mayumi, T. (2001a). Pharmacotherapy by intracellular delivery of drugs using fusogenic liposomes: Application to vaccine development. *Adv. Drug Deliv. Rev.* 52:177–186.

Kunisawa, J., Nakanishi, T., Takahashi, I., Okudaira, A., Tsutsumi, Y., Katayama, K., Nakagawa, S., Kiyono, H., and Mayumi, T. (2001b). Sendai virus fusion protein mediates simultaneous induction of MHC class I/II-dependent mucosal and

systemic immune responses via the nasopharyngeal-associated lymphoreticular tissue immune system. *J. Immunol.* 167:1406–1412.
Kunisawa, J., Okudaira, A., Tsutsumi, Y., Takahashi, I., Nakanishi, T., Kiyono, H., and Mayumi, T. (2000). Characterization of mucoadhesive microspheres for the induction of mucosal and systemic immune responses. *Vaccine* 19:589–594.
Kunisawa, J., and Shastri, N. (2003). The group II chaperonin TRiC protects proteolytic intermediates from degradation in the MHC class I antigen processing pathway. *Mol. Cell* 12:565–576.
Kunisawa, J., Takahashi, I., Okudaira, A., Hiroi, T., Katayama, K., Ariyama, T., Tsutsumi, Y., Nakagawa, S., Kiyono, H., and Mayumi, T. (2002). Lack of antigen-specific immune responses in anti-IL-7 receptor alpha chain antibody-treated Peyer's patch-null mice following intestinal immunization with microencapsulated antigen. *Eur. J. Immunol.* 32:2347–2355.
Kunkel, E. J., Kim, C. H., Lazarus, N. H., Vierra, M. A., Soler, D., Bowman, E. P., and Butcher, E. C. (2003). CCR10 expression is a common feature of circulating and mucosal epithelial tissue IgA Ab-secreting cells. *J. Clin. Invest.* 111:1001–1010.
Kweon, M. N., Yamamoto, M., Watanabe, F., Tamura, S., Van Ginkel, F. W., Miyauchi, A., Takagi, H., Takeda, Y., Hamabata, T., Fujihashi, K., McGhee, J. R., and Kiyono, H. (2002). A nontoxic chimeric enterotoxin adjuvant induces protective immunity in both mucosal and systemic compartments with reduced IgE antibodies. *J. Infect. Dis.* 186:1261–1269.
Lambkin, I., Pinilla, C., Hamashin, C., Spindler, L., Russell, S., Schink, A., Moya-Castro, R., Allicotti, G., Higgins, L., Smith, M., Dee, J., Wilson, C., Houghten, R., and O'Mahony, D. (2003). Toward targeted oral vaccine delivery systems: Selection of lectin mimetics from combinatorial libraries. *Pharm. Res.* 20:1258–1266.
Lambkin, R., Oxford, J. S., Bossuyt, S., Mann, A., Metcalfe, I. C., Herzog, C., Viret, J. F., and Gluck, R. (2004). Strong local and systemic protective immunity induced in the ferret model by an intranasal virosome-formulated influenza subunit vaccine. *Vaccine* 22:4390–4396.
Lee, J., Mo, J. H., Katakura, K., Alkalay, I., Rucker, A. N., Liu, Y. T., Lee, H. K., Shen, C., Cojocaru, G., Shenouda, S., Kagnoff, M., Eckmann, L., Ben-Neriah, Y., and Raz, E. (2006). Maintenance of colonic homeostasis by distinctive apical TLR9 signalling in intestinal epithelial cells. *Nature Cell Biol.* 8:1327–1336.
Lee, S., Gierynska, M., Eo, S. K., Kuklin, N., and Rouse, B. T. (2003). Influence of DNA encoding cytokines on systemic and mucosal immunity following genetic vaccination against herpes simplex virus. *Microbes Infect.* 5:571–578.
Lencer, W. I., Moe, S., Rufo, P. A., and Madara, J. L. (1995). Transcytosis of cholera toxin subunits across model human intestinal epithelia. *Proc. Natl. Acad. Sci. USA* 92:10,094–10,098.
Lencer, W. I., and Tsai, B. (2003). The intracellular voyage of cholera toxin: Going retro. *Trends Biochem. Sci.* 28:639–645.
Levine, M. M., Galen, J., Barry, E., Noriega, F., Tacket, C., Sztein, M., Chatfield, S., Dougan, G., Losonsky, G., and Kotloff, K. (1997). Attenuated *Salmonella typhi* and *Shigella* as live oral vaccines and as live vectors. *Behring. Inst. Mitt.* 120–123.
Lillard, J. W., Jr., Boyaka, P. N., Chertov, O., Oppenheim, J. J., and McGhee, J. R. (1999a). Mechanisms for induction of acquired host immunity by neutrophil peptide defensins. *Proc. Natl. Acad. Sci. USA* 96:651–656.

Lillard, J. W., Jr., Boyaka, P. N., Hedrick, J. A., Zlotnik, A., and McGhee, J. R. (1999b). Lymphotactin acts as an innate mucosal adjuvant. *J. Immunol.* 162:1959–1965.

Lillard, J. W., Jr., Boyaka, P. N., Taub, D. D., and McGhee, J. R. (2001). RANTES potentiates antigen-specific mucosal immune responses. *J. Immunol.* 166:162–169.

Lillard, J. W., Jr., Singh, U. P., Boyaka, P. N., Singh, S., Taub, D. D., and McGhee, J. R. (2003). MIP-1alpha and MIP-1beta differentially mediate mucosal and systemic adaptive immunity. *Blood* 101:807–814.

Lindsay, D. S., Parton, R., and Wardlaw, A. C. (1994). Adjuvant effect of pertussis toxin on the production of anti-ovalbumin IgE in mice and lack of direct correlation between PCA and ELISA. *Int. Arch. Allergy Immunol.* 105:281–288.

Lundstrom, K. (2002). Alphavirus-based vaccines. *Curr. Opin. Mol. Ther.* 4:28–34.

Lundstrom, K. (2003a). Alphavirus vectors for vaccine production and gene therapy. *Expert. Rev. Vaccines* 2:447–459.

Lundstrom, K. (2003b). Semliki Forest virus vectors for gene therapy. *Expert Opin. Biol. Ther.* 3:771–777.

Lycke, N. (2004). ADP-ribosylating bacterial enzymes for the targeted control of mucosal tolerance and immunity. *Ann. NY Acad. Sci.* 1029:193–208.

Lycke, N. (2005). Targeted vaccine adjuvants based on modified cholera toxin. *Curr. Mol. Med.* 5:591–597.

Lycke, N., and Holmgren, J. (1986). Strong adjuvant properties of cholera toxin on gut mucosal immune responses to orally presented antigens. *Immunology* 59:301–308.

Lycke, N. Y. (1993). Cholera toxin promotes B cell isotype switching by two different mechanisms. cAMP induction augments germ-line Ig H-chain RNA transcripts whereas membrane ganglioside GM1-receptor binding enhances later events in differentiation. *J. Immunol.* 150:4810–4821.

Mackett, M., Smith, G. L., and Moss, B. (1984). General method for production and selection of infectious vaccinia virus recombinants expressing foreign genes. *J. Virol.* 49:857–864.

Macpherson, A. J., Gatto, D., Sainsbury, E., Harriman, G. R., Hengartner, H., and Zinkernagel, R. M. (2000). A primitive T cell-independent mechanism of intestinal mucosal IgA responses to commensal bacteria. *Science* 288:2222–2226.

Maddaloni, M., Staats, H. F., Mierzejewska, D., Hoyt, T., Robinson, A., Callis, G., Kozaki, S., Kiyono, H., McGhee, J. R., Fujihashi, K., and Pascual, D. W. (2006). Mucosal vaccine targeting improves onset of mucosal and systemic immunity to botulinum neurotoxin A. *J. Immunol.* 177:5524–5532.

Manocha, M., Pal, P. C., Chitralekha, K. T., Thomas, B. E., Tripathi, V., Gupta, S. D., Paranjape, R., Kulkarni, S., and Rao, D. N. (2005). Enhanced mucosal and systemic immune response with intranasal immunization of mice with HIV peptides entrapped in PLG microparticles in combination with *Ulex europaeus*-I lectin as M cell target. *Vaccine* 23:5599–5617.

Marchetti, M., Rossi, M., Giannelli, V., Giuliani, M. M., Pizza, M., Censini, S., Covacci, A., Massari, P., Pagliaccia, C., Manetti, R., Telford, J. L., Douce, G., Dougan, G., Rappuoli, R., and Ghiara, P. (1998). Protection against *Helicobacter pylori* infection in mice by intragastric vaccination with *H. pylori* antigens is achieved using a non-toxic mutant of *E. coli* heat-labile enterotoxin (LT) as adjuvant. *Vaccine* 16:33–37.

Marinaro, M., Boyaka, P. N., Jackson, R. J., Finkelman, F. D., Kiyono, H., Jirillo, E., and McGhee, J. R. (1999a). Use of intranasal IL-12 to target predominantly Th1 responses to nasal and Th2 responses to oral vaccines given with cholera toxin. *J. Immunol.* 162:114–121.

Marinaro, M., Di Tommaso, A., Uzzau, S., Fasano, A., and De Magistris, M. T. (1999b). *Zonula occludens* toxin is a powerful mucosal adjuvant for intranasally delivered antigens. *Infect. Immun.* 67:1287–1291.

Marinaro, M., Fasano, A., and De Magistris, M. T. (2003). *Zonula occludens* toxin acts as an adjuvant through different mucosal routes and induces protective immune responses. *Infect. Immun.* 71:1897–1902.

Martin, M., Michalek, S. M., and Katz, J. (2003). Role of innate immune factors in the adjuvant activity of monophosphoryl lipid A. *Infect. Immun.* 71:2498–2507.

McCluskie, M. J., and Davis, H. L. (2000). Oral, intrarectal and intranasal immunizations using CpG and non-CpG oligodeoxynucleotides as adjuvants. *Vaccine* 19:413–422.

McGhee, J. R., Fujihashi, K., Beagley, K. W., and Kiyono, H. (1991). Role of interleukin-6 in human and mouse mucosal IgA plasma cell responses. *Immunol. Res.* 10:418–422.

Mestecky, J., Zikan, J., and Butler, W. T. (1971). Immunoglobulin M and secretory immunoglobulin A: Presence of a common polypeptide chain different from light chains. *Science* 171:1163–1165.

Mikloska, Z., Ruckholdt, M., Ghadiminejad, I., Dunckley, H., Denis, M., and Cunningham, A. L. (2000). Monophosphoryl lipid A and QS21 increase CD8 T lymphocyte cytotoxicity to herpes simplex virus-2 infected cell proteins 4 and 27 through IFN-gamma and IL-12 production. *J. Immunol.* 164:5167–5176.

Mizuguchi, H., Kay, M. A., and Hayakawa, T. (2001). Approaches for generating recombinant adenovirus vectors. *Adv. Drug Deliv. Rev.* 52:165–176.

Mohamedi, S. A., Brewer, J. M., Alexander, J., Heath, A. W., and Jennings, R. (2000). Antibody responses, cytokine levels and protection of mice immunised with HSV-2 antigens formulated into NISV or ISCOM delivery systems. *Vaccine* 18:2083–2094.

Moldoveanu, Z., Novak, M., Huang, W. Q., Gilley, R. M., Staas, J. K., Schafer, D., Compans, R. W., and Mestecky, J. (1993). Oral immunization with influenza virus in biodegradable microspheres. *J. Infect. Dis.* 167:84–90.

Mora, J. R., Bono, M. R., Manjunath, N., Weninger, W., Cavanagh, L. L., Rosemblatt, M., and Von Andrian, U. H. (2003). Selective imprinting of gut-homing T cells by Peyer's patch dendritic cells. *Nature* 424:88–93.

Morris, C. B., Cheng, E., Thanawastien, A., Cardenas-Freytag, L., and Clements, J. D. (2000). Effectiveness of intranasal immunization with HIV-gp160 and an HIV-1 env CTL epitope peptide (E7) in combination with the mucosal adjuvant LT(R192G). *Vaccine* 18:1944–1951.

Mowat, A. M., Donachie, A. M., Reid, G., and Jarrett, O. (1991). Immune-stimulating complexes containing Quil A and protein antigen prime class I MHC-restricted T lymphocytes *in vivo* and are immunogenic by the oral route. *Immunology* 72:317–322.

Nashar, T. O., Webb, H. M., Eaglestone, S., Williams, N. A., and Hirst, T. R. (1996). Potent immunogenicity of the B subunits of *Escherichia coli* heat-labile enterotoxin: Receptor binding is essential and induces differential modulation of lymphocyte subsets. *Proc. Natl. Acad. Sci. USA* 93:226–230.

Nawar, H. F., Arce, S., Russell, M. W., and Connell, T. D. (2007). Mutants of type II heat-labile enterotoxin LT-IIa with altered ganglioside-binding activities and diminished toxicity are potent mucosal adjuvants. *Infect. Immun.* 75:621–633.

Neutra, M. R., Mantis, N. J., and Kraehenbuhl, J. P. (2001). Collaboration of epithelial cells with organized mucosal lymphoid tissues. *Nat. Immunol.* 2:1004–1009.

Niess, J. H., Brand, S., Gu, X., Landsman, L., Jung, S., McCormick, B. A., Vyas, J. M., Boes, M., Ploegh, H. L., Fox, J. G., Littman, D. R., and Reinecker, H. C. (2005). CX3CR1-mediated dendritic cell access to the intestinal lumen and bacterial clearance. *Science* 307:254–258.

Oh, S., Berzofsky, J. A., Burke, D. S., Waldmann, T. A., and Perera, L. P. (2003). Coadministration of HIV vaccine vectors with vaccinia viruses expressing IL-15 but not IL-2 induces long-lasting cellular immunity. *Proc. Natl. Acad. Sci. USA* 100:3392–3397.

O'Hagan, D. T., and Singh, M. (2003). Microparticles as vaccine adjuvants and delivery systems. *Expert. Rev. Vaccines* 2:269–283.

Ohmura, M., Yamamoto, M., Kiyono, H., Fujihashi, K., Takeda, Y., and McGhee, J. R. (2001). Highly purified mutant E112K of cholera toxin elicits protective lung mucosal immunity to diphtheria toxin. *Vaccine* 20:756–762.

Ohmura, M., Yamamoto, M., Tomiyama-Miyaji, C., Yuki, Y., Takeda, Y., and Kiyono, H. (2005). Nontoxic Shiga toxin derivatives from *Escherichia coli* possess adjuvant activity for the augmentation of antigen-specific immune responses via dendritic cell activation. *Infect. Immun.* 73:4088–4097.

Ohmura-Hoshino, M., Yamamoto, M., Yuki, Y., Takeda, Y., and Kiyono, H. (2004). Non-toxic Stx derivatives from *Escherichia coli* possess adjuvant activity for mucosal immunity. *Vaccine* 22:3751–3761.

Okada, H., and Toguchi, H. (1995). Biodegradable microspheres in drug delivery. *Crit. Rev. Ther. Drug Carrier Syst.* 12:1–99.

Okada, J., Cohen, S., and Langer, R. (1995). *In vitro* evaluation of polymerized liposomes as an oral drug delivery system. *Pharm. Res.* 12:576–582.

O'Neal, C. M., Clements, J. D., Estes, M. K., and Conner, M. E. (1998). Rotavirus 2/6 viruslike particles administered intranasally with cholera toxin, *Escherichia coli* heat-labile toxin (LT), and LT-R192G induce protection from rotavirus challenge. *J. Virol.* 72:3390–3393.

Ourmanov, I., Brown, C. R., Moss, B., Carroll, M., Wyatt, L., Pletneva, L., Goldstein, S., Venzon, D., and Hirsch, V. M. (2000). Comparative efficacy of recombinant modified vaccinia virus Ankara expressing simian immunodeficiency virus (SIV) Gag-Pol and/or Env in macaques challenged with pathogenic SIV. *J. Virol.* 74:2740–2751.

Paoletti, E. (1996). Applications of pox virus vectors to vaccination: an update. *Proc. Natl. Acad. Sci. USA* 93:11349–11353.

Partidos, C. D., Pizza, M., Rappuoli, R., and Steward, M. W. (1996). The adjuvant effect of a non-toxic mutant of heat-labile enterotoxin of *Escherichia coli* for the induction of measles virus-specific CTL responses after intranasal co-immunization with a synthetic peptide. *Immunology* 89:483–487.

Peppoloni, S., Ruggiero, P., Contorni, M., Morandi, M., Pizza, M., Rappuoli, R., Podda, A., and Del Giudice, G. (2003). Mutants of the *Escherichia coli* heat-labile enterotoxin as safe and strong adjuvants for intranasal delivery of vaccines. *Expert. Rev. Vaccines* 2:285–293.

Pinczewski, J., Zhao, J., Malkevitch, N., Patterson, L. J., Aldrich, K., Alvord, W. G., and Robert-Guroff, M. (2005). Enhanced immunity and protective efficacy against SIVmac251 intrarectal challenge following ad-SIV priming by multiple mucosal routes and gp120 boosting in MPL-SE. *Viral Immunol.* 18:236–243.

Pizza, M., Giuliani, M. M., Fontana, M. R., Douce, G., Dougan, G., and Rappuoli, R. (2000). LTK63 and LTR72, two mucosal adjuvants ready for clinical trials. *Int. J. Med. Microbiol.* 290:455–461.

Poltorak, A., He, X., Smirnova, I., Liu, M. Y., Van Huffel, C., Du, X., Birdwell, D., Alejos, E., Silva, M., Galanos, C., Freudenberg, M., Ricciardi-Castagnoli, P., Layton, B., and Beutler, B. (1998). Defective LPS signaling in C3H/HeJ and C57BL/10ScCr mice: mutations in Tlr4 gene. *Science* 282:2085–2088.

Porter, D. C., Ansardi, D. C., Choi, W. S., and Morrow, C. D. (1993). Encapsidation of genetically engineered poliovirus minireplicons which express human immunodeficiency virus type 1 Gag and Pol proteins upon infection. *J. Virol.* 67:3712–3719.

Proietti, E., Bracci, L., Puzelli, S., Di Pucchio, T., Sestili, P., De Vincenzi, E., Venditti, M., Capone, I., Seif, I., De Maeyer, E., Tough, D., Donatelli, I., and Belardelli, F. (2002). Type I IFN as a natural adjuvant for a protective immune response: lessons from the influenza vaccine model. *J. Immunol.* 169:375–383.

Rappuoli, R., Douce, G., Dougan, G., and Pizza, M. (1995). Genetic detoxification of bacterial toxins: A new approach to vaccine development. *Int. Arch. Allergy Immunol.* 108:327–333.

Rappuoli, R., Pizza, M., Douce, G., and Dougan, G. (1999). Structure and mucosal adjuvanticity of cholera and *Escherichia coli* heat-labile enterotoxins. *Immunol. Today* 20:493–500.

Rescigno, M., Urbano, M., Valzasina, B., Francolini, M., Rotta, G., Bonasio, R., Granucci, F., Kraehenbuhl, J. P., and Ricciardi-Castagnoli, P. (2001). Dendritic cells express tight junction proteins and penetrate gut epithelial monolayers to sample bacteria. *Nat. Immunol.* 2:361–367.

Rharbaoui, F., Drabner, B., Borsutzky, S., Winckler, U., Morr, M., Ensoli, B., Muhlradt, P. F., and Guzman, C. A. (2002). The Mycoplasma-derived lipopeptide MALP-2 is a potent mucosal adjuvant. *Eur. J. Immunol.* 32:2857–2865.

Rharbaoui, F., Westendorf, A., Link, C., Felk, S., Buer, J., Gunzer, M., and Guzman, C. A. (2004). The mycoplasma-derived macrophage-activating 2-kilodalton lipopeptide triggers global immune activation on nasal mucosa-associated lymphoid tissues. *Infect. Immun.* 72:6978–6986.

Rimmelzwaan, G. F., Baars, M., van Beek, R., van Amerongen, G., Lovgren-Bengtsson, K., Claas, E. C., and Osterhaus, A. D. (1997). Induction of protective immunity against influenza virus in a macaque model: Comparison of conventional and iscom vaccines. *J. Gen. Virol.* 78(Pt. 4):757–765.

Roberts, M., Bacon, A., Rappuoli, R., Pizza, M., Cropley, I., Douce, G., Dougan, G., Marinaro, M., McGhee, J., and Chatfield, S. (1995). A mutant pertussis toxin molecule that lacks ADP-ribosyltransferase activity, PT-9K/129G, is an effective mucosal adjuvant for intranasally delivered proteins. *Infect. Immun.* 63:2100–2108.

Saito, T., Deskin, R. W., Casola, A., Haeberle, H., Olszewska, B., Ernst, P. B., Alam, R., Ogra, P. L., and Garofalo, R. (1997). Respiratory syncytial virus induces selective production of the chemokine RANTES by upper airway epithelial cells. *J. Infect. Dis.* 175:497–504.

Sakaue, G., Hiroi, T., Nakagawa, Y., Someya, K., Iwatani, K., Sawa, Y., Takahashi, H., Honda, M., Kunisawa, J., and Kiyono, H. (2003). HIV mucosal vaccine: nasal immunization with gp160-encapsulated hemagglutinating virus of Japan-liposome induces antigen-specific CTLs and neutralizing antibody responses. *J. Immunol.* 170:495–502.

Sanders, M. T., Brown, L. E., Deliyannis, G., and Pearse, M. J. (2005). ISCOM-based vaccines: The second decade. *Immunol. Cell. Biol.* 83:119–128.

Santosuosso, M., McCormick, S., and Xing, Z. (2005). Adenoviral vectors for mucosal vaccination against infectious diseases. *Viral Immunol.* 18:283–291.

Sato, A., and Iwasaki, A. (2005). Peyer's patch dendritic cells as regulators of mucosal adaptive immunity. *Cell. Mol. Life Sci.* 62:1333–1338.

Sauter, S. L., Rahman, A., and Muralidhar, G. (2005). Non-replicating viral vector-based AIDS vaccines: Interplay between viral vectors and the immune system. *Curr. HIV Res.* 3:157–181.

Schmitz, A., Herrgen, H., Winkeler, A., and Herzog, V. (2000). Cholera toxin is exported from microsomes by the Sec61p complex. *J. Cell. Biol.* 148:1203–1212.

Selsted, M. E., and Ouellette, A. J. (2005). Mammalian defensins in the antimicrobial immune response. *Nat. Immunol.* 6:551–557.

Shastri, N., Cardinaud, S., Schwab, S. R., Serwold, T., and Kunisawa, J. (2005). All the peptides that fit: The beginning, the middle, and the end of the MHC class I antigen processing pathway. *Immunol. Rev.* 207:31–41.

Shikina, T., Hiroi, T., Iwatani, K., Jang, M. H., Fukuyama, S., Tamura, M., Kubo, T., Ishikawa, H., and Kiyono, H. (2004). IgA class switch occurs in the organized nasopharynx- and gut-associated lymphoid tissue, but not in the diffuse lamina propria of airways and gut. *J. Immunol.* 172:6259–6264.

Shimoda, M., Nakamura, T., Takahashi, Y., Asanuma, H., Tamura, S., Kurata, T., Mizuochi, T., Azuma, N., Kanno, C., and Takemori, T. (2001). Isotype-specific selection of high affinity memory B cells in nasal-associated lymphoid tissue. *J. Exp. Med.* 194:1597–1607.

Shreedhar, V. K., Kelsall, B. L., and Neutra, M. R. (2003). Cholera toxin induces migration of dendritic cells from the subepithelial dome region to T- and B-cell areas of Peyer's patches. *Infect. Immun.* 71:504–509.

Sicinski, P., Rowinski, J., Warchol, J. B., Jarzabek, Z., Gut, W., Szczygiel, B., Bielecki, K., and Koch, G. (1990). Poliovirus type 1 enters the human host through intestinal M cells. *Gastroenterology* 98:56–58.

Silla, S., Fallarino, F., Boon, T., and Uyttenhove, C. (1999). Enhancement by IL-12 of the cytolytic T lymphocyte (CTL) response of mice immunized with tumor-specific peptides in an adjuvant containing QS21 and MPL. *Eur. Cytokine Netw.* 10:181–190.

Simmons, C. P., Mastroeni, P., Fowler, R., Ghaem-maghami, M., Lycke, N., Pizza, M., Rappuoli, R., and Dougan, G. (1999). MHC class I-restricted cytotoxic lymphocyte responses induced by enterotoxin-based mucosal adjuvants. *J. Immunol.* 163:6502–6510.

Simms, J. R., Heath, A. W., and Jennings, R. (2000). Use of herpes simplex virus (HSV) type 1 ISCOMS 703 vaccine for prophylactic and therapeutic treatment of primary and recurrent HSV-2 infection in guinea pigs. *J. Infect. Dis.* 181:1240–1248.

Somavarapu, S., Bramwell, V. W., and Alpar, H. O. (2003). Oral plasmid DNA delivery systems for genetic immunisation. *J. Drug Target* 11:547–553.

Soriani, M., Bailey, L., and Hirst, T. R. (2002). Contribution of the ADP-ribosylating and receptor-binding properties of cholera-like enterotoxins in modulating cytokine secretion by human intestinal epithelial cells. *Microbiology* 148:667–676.

Spangler, B. D. (1992). Structure and function of cholera toxin and the related *Escherichia coli* heat-labile enterotoxin. *Microbiol. Rev.* 56:622–647.

Staats, H. F., and Ennis, F. A., Jr. (1999). IL-1 is an effective adjuvant for mucosal and systemic immune responses when coadministered with protein immunogens. *J. Immunol.* 162:6141–6147.

Stagg, A. J., Kamm, M. A., and Knight, S. C. (2002). Intestinal dendritic cells increase T cell expression of alpha4beta7 integrin. *Eur. J. Immunol.* 32:1445–1454.

Stickl, H., Hochstein-Mintzel, V., Mayr, A., Huber, H. C., Schafer, H., and Holzner, A. (1974). [MVA vaccination against smallpox: Clinical tests with an attenuated live vaccinia virus strain (MVA) (author's transl)]. *Dtsch. Med. Wochenschr.* 99:2386–2392.

Sun, J. B., Holmgren, J., and Czerkinsky, C. (1994). Cholera toxin B subunit: An efficient transmucosal carrier-delivery system for induction of peripheral immunological tolerance. *Proc. Natl. Acad. Sci. USA* 91:10,795–10,799.

Tabata, Y., Inoue, Y., and Ikada, Y. (1996). Size effect on systemic and mucosal immune responses induced by oral administration of biodegradable microspheres. *Vaccine* 14:1677–1685.

Tacket, C. O., Kelly, S. M., Schodel, F., Losonsky, G., Nataro, J. P., Edelman, R., Levine, M. M., and Curtiss, R., 3rd (1997a). Safety and immunogenicity in humans of an attenuated *Salmonella typhi* vaccine vector strain expressing plasmid-encoded hepatitis B antigens stabilized by the Asd-balanced lethal vector system. *Infect. Immun.* 65:3381–3385.

Tacket, C. O., Sztein, M. B., Losonsky, G. A., Wasserman, S. S., Nataro, J. P., Edelman, R., Pickard, D., Dougan, G., Chatfield, S. N., and Levine, M. M. (1997b). Safety of live oral *Salmonella typhi* vaccine strains with deletions in htrA and aroC aroD and immune response in humans. *Infect. Immun.* 65:452–456.

Takahashi, H., Takeshita, T., Morein, B., Putney, S., Germain, R. N., and Berzofsky, J. A. (1990). Induction of $CD8^{+}$ cytotoxic T cells by immunization with purified HIV-1 envelope protein in ISCOMs. *Nature* 344:873–875.

Takahashi, I., Marinaro, M., Kiyono, H., Jackson, R. J., Nakagawa, I., Fujihashi, K., Hamada, S., Clements, J. D., Bost, K. L., and McGhee, J. R. (1996). Mechanisms for mucosal immunogenicity and adjuvancy of *Escherichia coli* labile enterotoxin. *J. Infect. Dis.* 173:627–635.

Takata, S., Ohtani, O., and Watanabe, Y. (2000). Lectin binding patterns in rat nasal-associated lymphoid tissue (NALT) and the influence of various types of lectin on particle uptake in NALT. *Arch. Histol. Cytol.* 63:305–312.

Takatsu, K., Tominaga, A., Harada, N., Mita, S., Matsumoto, M., Takahashi, T., Kikuchi, Y., and Yamaguchi, N. (1988). T cell-replacing factor (TRF)/interleukin 5 (IL-5): molecular and functional properties. *Immunol. Rev.* 102:107–135.

Tartaglia, J., Perkus, M. E., Taylor, J., Norton, E. K., Audonnet, J. C., Cox, W. I., Davis, S. W., van der Hoeven, J., Meignier, B., Riviere, M., et al. (1992). NYVAC: A highly attenuated strain of vaccinia virus. *Virology* 188:217–232.

Thapar, M. A., Parr, E. L., and Parr, M. B. (1990). Secretory immune responses in mouse vaginal fluid after pelvic, parenteral or vaginal immunization. *Immunology* 70:121–125.

Toka, F. N., and Rouse, B. T. (2005). Mucosal application of plasmid-encoded IL-15 sustains a highly protective anti-Herpes simplex virus immunity. *J. Leuk. Biol.* 78:178–186.

Ugozzoli, M., O'Hagan, D. T., and Ott, G. S. (1998). Intranasal immunization of mice with herpes simplex virus type 2 recombinant gD2: The effect of adjuvants on mucosal and serum antibody responses. *Immunology* 93:563–571.

Vajdy, M., Srivastava, I., Polo, J., Donnelly, J., O'Hagan, D., and Singh, M. (2004). Mucosal adjuvants and delivery systems for protein-, DNA- and RNA-based vaccines. *Immunol. Cell. Biol.* 82:617–627.

van Heyningen, S. (1977). Cholera toxin. *Biol. Rev. Camb. Phil. Soc.* 52:509.

van Pinxteren, L. A., Bruce, M. G., Campbell, I., Wood, A., Clarke, C. J., Bellman, A., Morein, B., and Snodgrass, D. R. (1999). Effect of oral rotavirus/iscom vaccines on immune responses in gnotobiotic lambs. *Vet. Immunol. Immunopathol.* 71:53–67.

Vogel, K., Kantor, J., Wood, L., Rivera, R., and Schlom, J. (1998). Oral immunization with enterocoated microbeads induces antigen-specific cytolytic T-cell responses. *Cell. Immunol.* 190:61–67.

Wang, X., Hone, D. M., Haddad, A., Shata, M. T., and Pascual, D. W. (2003). M cell DNA vaccination for CTL immunity to HIV. *J. Immunol.* 171:4717–4725.

Wang, X., Kochetkova, I., Haddad, A., Hoyt, T., Hone, D. M., and Pascual, D. W. (2005). Transgene vaccination using *Ulex europaeus* agglutinin I (UEA-1) for targeted mucosal immunization against HIV-1 envelope. *Vaccine* 23:3836–3842.

Watanabe, I., Hagiwara, Y., Kadowaki, S. E., Yoshikawa, T., Komase, K., Aizawa, C., Kiyono, H., Takeda, Y., McGhee, J. R., Chiba, J., Sata, T., Kurata, T., and Tamura, S. (2002). Characterization of protective immune responses induced by nasal influenza vaccine containing mutant cholera toxin as a safe adjuvant (CT112K). *Vaccine* 20:3443–3455.

Wierzbicki, A., Kiszka, I., Kaneko, H., Kmieciak, D., Wasik, T. J., Gzyl, J., Kaneko, Y., and Kozbor, D. (2002). Immunization strategies to augment oral vaccination with DNA and viral vectors expressing HIV envelope glycoprotein. *Vaccine* 20:1295–1307.

Williams, N. A., Hirst, T. R., and Nashar, T. O. (1999). Immune modulation by the cholera-like enterotoxins: From adjuvant to therapeutic. *Immunol. Today* 20:95–101.

Woo, P. C., Wong, L. P., Zheng, B. J., and Yuen, K. Y. (2001). Unique immunogenicity of hepatitis B virus DNA vaccine presented by live-attenuated *Salmonella typhimurium*. *Vaccine* 19:2945–2954.

Woof, J. M., and Mestecky, J. (2005). Mucosal immunoglobulins. *Immunol. Rev.* 206:64–82.

Wu, S., Pascual, D. W., Lewis, G. K., and Hone, D. M. (1997). Induction of mucosal and systemic responses against human immunodeficiency virus type 1 glycoprotein 120 in mice after oral immunization with a single dose of a *Salmonella*-HIV vector. *AIDS Res. Hum. Retroviruses* 13:1187–1194.

Wu, Y., Wang, X., Csencsits, K. L., Haddad, A., Walters, N., and Pascual, D. W. (2001). M cell-targeted DNA vaccination. *Proc. Natl. Acad. Sci. USA* 98:9318–9323.

Wyatt, L. S., Shors, S. T., Murphy, B. R., and Moss, B. (1996). Development of a replication-deficient recombinant vaccinia virus vaccine effective against parainfluenza virus 3 infection in an animal model. *Vaccine* 14:1451–1458.

Xin, K. Q., Hamajima, K., Sasaki, S., Tsuji, T., Watabe, S., Okada, E., and Okuda, K. (1999). IL-15 expression plasmid enhances cell-mediated immunity induced by an HIV-1 DNA vaccine. *Vaccine* 17:858–866.

Xu-Amano, J., Kiyono, H., Jackson, R. J., Staats, H. F., Fujihashi, K., Burrows, P. D., Elson, C. O., Pillai, S., and McGhee, J. R. (1993). Helper T cell subsets for immunoglobulin A responses: Oral immunization with tetanus toxoid and cholera toxin as adjuvant selectively induces Th2 cells in mucosa-associated tissues. *J. Exp. Med.* 178:1309–1320.

Yamamoto, M., Briles, D. E., Yamamoto, S., Ohmura, M., Kiyono, H., and McGhee, J. R. (1998). A nontoxic adjuvant for mucosal immunity to pneumococcal surface protein A. *J. Immunol.* 161:4115–4121.

Yamamoto, M., Kiyono, H., Yamamoto, S., Batanero, E., Kweon, M. N., Otake, S., Azuma, M., Takeda, Y., and McGhee, J. R. (1999). Direct effects on antigen-presenting cells and T lymphocytes explain the adjuvanticity of a nontoxic cholera toxin mutant. *J. Immunol.* 162:7015–7021.

Yamamoto, M., McGhee, J. R., Hagiwara, Y., Otake, S., and Kiyono, H. (2001). Genetically manipulated bacterial toxin as a new generation mucosal adjuvant. *Scand. J. Immunol.* 53:211–217.

Yamamoto, M., Rennert, P., McGhee, J. R., Kweon, M. N., Yamamoto, S., Dohi, T., Otake, S., Bluethmann, H., Fujihashi, K., and Kiyono, H. (2000). Alternate mucosal immune system: Organized Peyer's patches are not required for IgA responses in the gastrointestinal tract. *J. Immunol.* 164:5184–5191.

Yamamoto, S., Kiyono, H., Yamamoto, M., Imaoka, K., Fujihashi, K., Van Ginkel, F. W., Noda, M., Takeda, Y., and McGhee, J. R. (1997a). A nontoxic mutant of cholera toxin elicits Th2-type responses for enhanced mucosal immunity. *Proc. Natl. Acad. Sci. USA* 94:5267–5272.

Yamamoto, S., Takeda, Y., Yamamoto, M., Kurazono, H., Imaoka, K., Yamamoto, M., Fujihashi, K., Noda, M., Kiyono, H., and McGhee, J. R. (1997b). Mutants in the ADP-ribosyltransferase cleft of cholera toxin lack diarrheagenicity but retain adjuvanticity. *J. Exp. Med.* 185:1203–1210.

Yang, S. K., Eckmann, L., Panja, A., and Kagnoff, M. F. (1997). Differential and regulated expression of C-X-C, C-C, and C-chemokines by human colon epithelial cells. *Gastroenterology* 113:1214–1223.

Yoshino, N., Lu, F. X., Fujihashi, K., Hagiwara, Y., Kataoka, K., Lu, D., Hirst, L., Honda, M., van Ginkel, F. W., Takeda, Y., Miller, C. J., Kiyono, H., and McGhee, J. R. (2004). A novel adjuvant for mucosal immunity to HIV-1 gp120 in nonhuman primates. *J. Immunol.* 173:6850–6857.

Zho, F., and Neutra, M. R. (2002). Antigen delivery to mucosa-associated lymphoid tissues using liposomes as a carrier. *Biosci. Rep.* 22:355–369.

15
Recombinant IgA Antibodies

Esther M. Yoo[1], Koteswara R. Chintalacharuvu[1], and Sherie L. Morrison[1]

15.1. Introduction

The production of monoclonal antibodies and the development of recombinant antibody technology have made antibodies one of the largest classes of drugs in development for prophylactic, therapeutic and diagnostic

[1] Department of Microbiology, Immunology and Molecular Genetics and the Molecular Biology Institute, University of California Los Angeles, Los Angeles, CA 90095, USA.

purposes. Currently, all of the Food and Drug Administration (FDA)-approved antibodies are immunoglobulin Gs (IgGs). However, more than 95% of the infections are initiated at the mucosal surfaces, where IgA is the primary immune effector antibody.

Immunoglobulin A has many properties that make it an attractive therapeutic agent. At mucosal surfaces, IgA is polymeric, making it more efficient at agglutinating antigens than IgG, thereby facilitating immune exclusion. IgA is anti-inflammatory due to both its inability to activate complement and its ability to inhibit complement activation by IgG and IgM (Griffiss and Goroff, 1983; Russell et al., 1998). When associated with the secretory component (SC), IgA is more resistant to proteolysis than IgG (Chintalacharuvu and Morrison, 1997; Lindh, 1975).

Studies in animals and humans suggest that passive administration of IgA could provide protection against a wide range of pathogens, including bacteria and their toxins and viruses such as HIV and respiratory syncytial virus (RSV) (Gupta et al., 2005; Hemming et al., 1987; Kozlowski and Neutra, 2003). A recombinant secretory IgA (SIgA) against *Streptococcus mutans* has been approved in Europe for topical administration to prevent dental caries (Ma et al., 1998). Antitransferrin IgA has been shown to be effective against cancers when administered intravenously (Brooks et al., 1995). However, topical administration of an anti-RSV IgA did not show significant efficacy as a prophylactic agent in clinical trials, although an IgG anti-RSV has been shown to be effective when administered intramuscularly and is approved by the FDA for use in humans (Subramanian et al., 1998). Thus, although there is much potential for IgA-based therapeutic and prophylactic agents, it is essential to use recombinant antibody technology to define the correlations between IgA structure and function, to study the biology of IgA and determine the mechanisms by which IgA and SIgA provide protection. This knowledge will facilitate the design and production of effective IgA-based immunotherapeutics for a wide range of applications.

Production of recombinant IgA poses many challenges. IgA exists as two different isotypes, with one isotype, IgA2, having three different allotypic forms. A decision must be made as to which IgA form is most appropriate for the intended application. Like IgG, production of IgA requires the expression of heavy and light chains. However, to make polymeric IgA (pIgA) a third chain, the J-chain, must also be expressed. If SIgA is needed, a fourth polypeptide, SC, must also be expressed. These chains must be appropriately assembled and secreted. IgA is a glycoprotein with two to five N-linked glycans present on the heavy chain; O-Linked glycans are present in the hinge region of IgA1. The N-linked glycans present on the α-heavy-chain are exposed on the surface of the molecule (Mattu et al., 1998; Merry et al., 1992). The J-chain and SC are also glycoproteins. Because glycans on IgA play an important role in its function, particular attention must be paid to the structure of the glycans attached by different expression systems. The cost of the product is also an important consideration. Thus, the identification of the appropriate expression system is a significant challenge when producing recombinant IgA.

15.2. Expression Systems for the Production of Recombinant IgA

Recombinant IgA has been produced in several different expression systems, including insect, plant, and mammalian cells and transgenic animals. Each system has its advantages, limitations, and potential applications. Large quantities of proteins can be produced in insect cells and in agricultural amounts in plants, making these expression systems an economical alternative to producing therapeutic and diagnostic reagents. Normal posttranslational modifications such as signal peptide cleavage, intrachain and interchain disulfide bond formation, and the addition of O-linked and N-linked carbohydrates are observed in plant and insect cells. However, the *N*-glycans attached by insect and plant cells differ from those normally found on human IgA (Butters et al., 1981; Jarvis and Finn, 1995, 1996 Martin et al., 1988; Ogonah et al., 1995). The *N*-glycans present on IgA produced in transgenic animals will depend on the species used. These differences in glycosylation might influence *in vivo* biologic properties such as biodistribution, half-life, and effector functions. Intact and fully assembled IgA has been produced in mammalian cells. However, in this case, cost is an important consideration. Regardless of which expression system is used, careful characterization of the *in vivo* biologic properties of the resulting IgA will be essential.

15.2.1. Baculovirus

Antibodies have been produced in insect cells. However, insect cells do not appear to contain all of the enzymes required to produce sialylated complex carbohydrates (Jarvis and Finn, 1995) although expression of glycosyltransferases can lead to proteins with more mammalianlike glycans (Jarvis and Finn, 1996). Insect cells also produce proteins with glycans such as core α1→6 fucosylated oligomannose (Man3GlcNAc2) that are not found on mammalian cells (Ogonah et al., 1995).

Recombinant IgA1 has been produced in *Spodoptera frugiperada* (Sf9) insect cells by double infection with baculovirus containing chimeric mouse–human H-chain and κ L-chain. Yields of $0.75\,\mu g/10^6$ cells per 72h were obtained. The secreted antibody was a H_2L_2 heterodimer containing O-linked Galβ(1–3) GalNAc disaccharides, and N-linked high mannose glycans. It bound antigen (*p*-azophenylarsonate), FcαR on HL-60 promyelocytic leukemia cells, and complement component C3. IgA dimers were assembled inefficiently when the human J-chain was also expressed, with the majority of extracellular IgA being monomers associated with the J-chain (Carayannopoulos et al., 1994). It is not clear if the instability of the covalent bond between IgA and the J-chain is due to differences in carbohydrate structure or the production level of the J-chain in the baculovirus system. In contrast to what was found for IgA produced in Chinese hamster ovary (CHO) cells (Mattu et al., 1998), insect-cell-produced IgA1 lacking the N-linked carbohydrate in C_H2 did not bind to cells expressing FcαR. Human SC has also been produced in Sf9 cells with yields of 50 mg/L. Although the recombinant SC was not fully glycosylated, it was shown to associate with mouse dimeric IgA (Rindisbacher et al., 1995).

15.2.2. Plants

Transgenic tobacco, maize, soybean, and alfalfa are expected to yield over 10 kg of therapeutic antibody per acre, with costs of about one-tenth of what would be required for the production of antibodies using mammalian cells (Larrick et al., 2001). Although plants produce proteins with both high mannose and complex N-linked glycans, they attach β(1→2)-linked xylose and α(1→3)-linked core fucose instead of the α(1→6)-linked core fucose found in mammals. In addition, plants do not normally add Gal to their complex N-glycans because they lack β(1→4) galactosyltransferase and the addition of sialic acid is also absent. Although Gal was added to the complex carbohydrate of a mouse IgG1 when human β(1→4) galactosyltransferase was expressed in tobacco plants, xylosylation and fucosylation were not affected (Bakker et al., 2001). The age of the leaf from which the antibodies are purified also affects the structure of N-glycans (Elbers et al., 2001). Even when a KDEL endoplasmic reticulum anchorage domain was added to a mouse IgG, *N*-glycans containing β(1→2) xylose and α(1→3) fucose were present (Ramirez et al., 2003).

A significant concern is whether the glycans attached by plants will be immunogenic. Although IgE specific for plant glycoallergens β(1→2) xylose and α(1→3) fucose has been detected in serum from allergic patients (van Ree et al., 2000), only two of six mice immunized subcutaneously with a mouse IgG1 produced in tobacco plants had a detectable immune response. However, it should be noted that the assay used detected only IgG2a, IgG2b, and IgG3, not other isotypes including IgE (Chargelegue et al., 2000).

Immunoglobulin A specific for the *Streptococcus mutans* and *S. sobrinus* antigen I/II cell surface adhesion molecule has been produced in tobacco plants (Ma et al., 1995). Hybrid IgA-producing plants were generated by crossing plants expressing a murine–human α–γ H-chain and murine κ L-chain. Dimeric IgA was produced by subsequent crossing with plants expressing the murine J-chain. Finally, crossing with rabbit SC-producing plants resulted in the expression of SIgA in a single plant cell. However, only about 50% of IgA was covalently associated with SC. In addition, the SC bound to IgA was 50 kDa, which is smaller than the expected size of 66.5 kDa. Antibody yields of 10–80 mg of antibody/kg of fresh plant material were obtained (Ma et al., 1998).

15.2.3. Transgenic Animals

Transgenic animals are attractive expression systems for the large-scale production of antibodies. However, mammals differ in their sialic acid structure, ability to add bisecting GlcNAc and Gal, and core fucosylation, making the choice of species an important consideration (Raju et al., 2000). Transgenic mice secreting a chimeric murine–porcine IgA against transmissible gastroenteritis coronavirus (TGEV) have been engineered (Sola et al., 1998). β-Lactoglobin regulatory sequences were used to target IgA synthesis to the mammary glands. The concentration of porcine IgA in the milk was 0.01–6 mg/mL, with no detectable levels in the serum of the transgenic animals. Transgenic mice overexpressing murine pIgR in mammary gland epithelial cells were shown to have increased levels of IgA in milk as compared with control mice (De Groot et al., 2000). Mouse IgA against phosphorylcholine has been expressed in transgenic mice, pigs, and sheep (Lo et al., 1991).

15.2.4. Mammalian Cells

Mammalian expression systems have been used extensively for the production of IgA and SIgA. To date, nonimmunoglobulin producing murine myeloma and CHO cell lines have been the expression systems of choice. However, the glycans attached by these expression systems are not identical to those present on normal human IgA. In addition, we have observed incomplete assembly of IgA2 produced in CHO cells (Chintalacharuvu et al., unpublished data).

Murine carbohydrates differ from those present in the human. Proteins produced in murine myelomas lack the bisecting GlcNAc linked to the trimannosyl core of N-linked glycans (Fukuta et al., 2000), which has been implicated as being important for antibody function (Davies et al., 2001; Lifely et al., 1995; Umana *et al.*, 1999). In addition, sialic acid structure varies among mammalian species, with human cells attaching *N*-acetylneuraminic acid (NANA), whereas mouse cells attach *N*-glycolylneuraminic acid (NGNA) (Raju et al., 2000). Mouse cells also synthesize Galα1→3Galβ1→4GlcNAc. A significant amount of circulating IgG in humans is against Galα1→3Galβ1→4GlcNAc, an epitope abundant on glycoproteins from nonprimate mammals, prosimians, and New World monkey, but absent from Old World monkeys, apes, and man (Galili, 1989). The presence of this carbohydrate structure on therapeutic proteins would have significant consequences. The presence of this epitope on recombinant antibodies is quite variable, with a humanized IgG1 produced by GS-NS0 cells containing no Galα1→3Galβ1 residues (Hills et al., 2001) and H-chain glycans produced in J558 possessing the epitope (Lund et al., 1990, 1993). However, much of the N-linked glycans on human IgA1 and IgA2 produced in murine myeloma cells appear to contain Galα1→3Galβ1 (Morrison et al., unpublished observation). The exposed nature of the glycans on IgA might facilitate processing to this glycan form.

Chinese hamster ovary cells appear to lack the glycosyltransferases necessary for generating both the bisecting GlcNAc and α(2→6) linked sialic acid on N-linked glycans (Routier et al., 1997). Transfection of exogenous glycosyltransferases into CHO cells has made it possible to express proteins that contain more humanized glycans with less glycan heterogeneity (Lee et al., 1989; Weikert *et al.*, 1999). *In vitro* galactosylation and sialylation has also been used for glyco-engineering of antibody molecules (Raju *et al.*, 2001). In addition, CHO cells have been engineered to overexpress protein disulfide isomerase, facilitating the assembly of recombinant proteins (Davis et al., 2000).

15.3. Vectors for Expression of Recombinant IgA

The general strategy has been to design immunoglobulin expression vectors as cassettes to facilitate manipulation of the antibody genes. The vectors for expression of H- and L-chains usually contain different selectable markers, although bicistronic expression vectors containing human κ L-and J-chain genes and human IgA1 and murine dihydrofolate reductase (DHFR) genes have also been described (Wolbank et al., 2003). Because the amount of L-chain produced by a transfectant appears to limit the amount of antibody produced, an effective method for generating transfectants is to first isolate good L-chain-producing transfectants. A second round of transfections with the H-chain and selection using a different marker yields transfectants producing large amounts of antibody (Chintalacharuvu et al., 1994 Chintalacharuvu and Morrison, 1996; Morton et al., 1993). Alternatively, cotransfection of L- and H-chains can be performed, but only one marker is usually used for selection (Berdoz *et al.*, 1999). Variable and constant region genes were initially obtained by genomic cloning but now are more routinely amplified by the polymerase chain reaction (PCR) from cDNA made from antibody-producing cells (Berdoz et al., 1995; Campbell et al., 1992; Coloma et al., 1992; Gavilondo-Cowley *et al.*, 1990; Gillies et al., 1989; Larrick et al., 1989; Orlandi et al., 1989). PCR has been used successfully for rapid cloning and modification of V regions from antibodies of many different specificities. If the sequence of the variable region to be cloned is known, specific PCR primers can be produced. If the sequence is unknown, degenerate primer sets that anneal to the relatively conserved leader or first framework sequences at the 5′ end are used. A limited number of primers will amplify all of the heavy-chain isotypes; there are only two isotypes of the light chain. By including appropriate restriction sites in the primers, it is straightforward to clone the PCR product into an expression vector. In addition, fully human IgAs have also been produced by cloning V regions from single-chain Fv fragments selected by phage display (Berdoz and Corthesy, 2004; Boel et al., 2000; Huls et al., 1999b).

The SC is produced by proteolytic cleavage of pIgR during epithelial transcytosis (see Chapter 4). Initially, SIgA was produced by coculturing

IgA-producing hybridomas with Madin-Darby canine kidney (MDCK) cells expressing pIgR (Hirt et al., 1993). In another approach, polymeric murine IgA produced in mice ascites was injected into Lewis rats, where it associated with the rat pIgR, and SIgA with rat SC was collected from bile (Renegar et al., 1998). As an alternative approach, SC was produced by introducing a stop codon into the position corresponding to the natural cleavage site in the pIgR (Rindisbacher et al., 1995). Purified pIgA from hybridoma cell lines can then combine with the purified recombinant SC in the test tube (Crottet et al., 1999; Lullau et al., 1996; Rindisbacher et al., 1995). These SIgA complexes had the same affinity for antigen as the native molecule, and formed the correct covalent bonds between the two molecules (Lullau et al., 1996). Although SIgA is normally a product of two different cell types, recombinant SIgA has been produced in a single cell. When the gene for human SC was transfected into a myeloma cell producing chimeric mouse–human pIgA containing the endogenous murine J-chain, SIgA was produced. Virtually all of the SC was covalently bound to dimeric IgA, and SIgA assembly appeared to take place in the Golgi apparatus (Chintalacharuvu and Morrison, 1997). Subsequently, CHO (Berdoz et al., 1999; Johansen et al., 1999) and Baby Hamster Kidney (BHK) (Vidarsson et al., 2001) cells have been used to produce SIgA by transfecting in genes for the human J-chain and SC into cells already expressing chimeric IgA.

The controlling elements used to direct IgA expression depend on the cell type to be transfected. In B-lineage cell lines, the expression of the recombinant H- and L-chain genes is usually driven by the immunoglobulin promoter and enhancer elements (Bruggemann et al., 1987; Chintalacharuvu et al., 1994; Chintalacharuvu and Morrison, 1996; Knight et al., 1988). Expression of IgA in B-lineage cell lines is advantageous in that the endogenous J-chain is incorporated into the recombinant IgA. Strong heterologous promoters such as the human cytomegalovirus (CMV) promoter can also be used to direct IgA expression. These viral controlling elements have led to high antibody yields and have the advantage that they are versatile and can function in a variety of cell types such as CHO, BHK, and COS cells as well as lymphoid cells. Chimeric murine–human IgA1 and IgA2 have been produced in these cell lines (Berdoz et al., 1999; Berdoz and Corthesy, 2004; Chintalacharuvu et al., 1994; Chintalacharuvu and Morrison, 1996; Morton et al., 1993; Trill et al., 1995; Vidarsson et al., 2001). Chimeric murine–porcine IgA (Sola et al., 1998) and 12 isotypes of chimeric murine–rabbit IgA (Schneiderman et al., 1989) have been produced in Sp2/0 cells; IgA2m(1) has also been produced in plasmacytoma J558L cells (Braathen et al., 2002; Bruggemann et al., 1987) and chimeric murine–bovine IgA has been produced in the murine hybridoma 27/44 (Knight et al., 1988).

Vaccinia virus expression systems are advantageous in that the extended host range of the virus makes it possible to express the heterologous gene in almost any mammalian cell type. Viral promotors such as the vaccinia late promoter P11K flanked by the vaccinia thymidine kinase sequence, which directs recombination into the viral genome, and viral or mammalian transcriptional and translational elements have been used to regulate expression. In the case of human SC, equivalent amounts of protein (5–10 mg/L) were produced when human SC or vaccinia 11K regulatory sequences were used in HeLa and HeLaS3 (human epithelial), CV-1 (monkey fibroblast) and TK^- 143B (human fibroblast) cell lines (Rindisbacher et al., 1995). The recombinant SC produced was capable of associating with dimeric mouse IgA and contained complex *N*-glycans with cell-line-specific processing. Rabbit SC (Corthesy et al., 1996) and murine SC have also been produced using a vaccinia virus expression system in a non-Ig-producing hybridoma cell line (Crottet *et al.*, 1999). The glycosylation profile for the recombinant murine SC was the same as that found on SC from murine milk or bile.

15.4. Purification and Characterization of IgA and SIgA Proteins

A variety of methods have been used to purify recombinant IgA and SIgA. Purification protocols based on selection by antigen are widely used. In addition, a synthetic peptide of 50 residues derived from a streptococcal M protein has been shown to bind to human IgA. The peptide, designated Sap for streptococcal IgA-binding peptide, was shown to deplete >99% of IgA from human serum but only 45% of SIgA from saliva. Sap appears to bind both IgA1 and IgA2 with similar affinities (Sandin et al., 2002), making it an attractive reagent for isolation and purification of recombinant human IgA. Lectins such as concavalin A have been used to purify SC (Rindisbacher et al., 1995) and the lectin Jacalin binds the O-linked glycans in the hinge of IgA1. Tags can also be added to aid the purification of the recombinant proteins; if separated from the protein by a protease cleavage site, these tags can then be removed following protein purification. A 6xHis tag at the C-terminus of human SC did not alter its secretion by either the baculovirus or vaccinia virus expression systems (Rindisbacher et al., 1995) and a FLAG tag has been used to purify murine SC (Crottet et al., 1999).

15.4.1. Assembly of Human IgA and SIgA

Recombinant technology has made it possible to shed light on the requirements necessary for IgA assembly and secretion, polymer formation, and association with other proteins such as the J-chain and pIgR/SC. Studies using site-directed mutagenesis and domain-exchanged proteins have identified some of the structural elements important for assembly of IgA and SIgA.

The α H-chains and L-chains, which are synthesized on polysomes, are translocated into the lumen of the endoplasmic reticulum, where they are assembled into monomers and higher polymers. The H- and L-chains first associate noncovalently and then covalently through disulfide bonds. The predominant covalent assembly pathway for human IgA1, IgA2m(2), and IgA2(n) molecules is through an HL intermediate in which the H- and L-chains are first disulfide bonded before forming the covalently associated H_2L_2 monomer. In contrast, the major intermediate in the assembly of IgA2m(1) is H_2, and the majority of the IgA2m(1) molecules lack disulfide bonds between the H- and L-chains (Chintalacharuvu and Morrison, 1996). Cys133 in C_H1, which is absent in IgA2, participates in disulfide bond formation with the L-chain in IgA1 (Chintalacharuvu and Morrison, 1996). In IgA2, the L-chain is covalently associated to the hinge proximal Cys241 or Cys242 in C_H2, with the structure of the C_H1 influencing the efficiency of disulfide bond formation (Chintalacharuvu et al., 2002).

Immunoglobulin A possesses an 18-amino-acid extension in the C-terminus called the tailpiece, which participates in polymerization; IgM, which also forms polymers, also possesses a tailpiece. The tailpiece contains an N-linked carbohydrate addition site, which is required for IgA dimer formation (Atkin et al., 1996), and a conserved penultimate Cys residue, which is involved in polymerization. Studies using domain-exchanged proteins have revealed that in addition to the tailpiece, structural motifs in C_H are critical for polymer assembly and J-chain incorporation. Assembly of dimers containing the J-chain requires $C_\alpha3$ and the α tailpiece, with more efficient dimer assembly in the presence of $C_\alpha2$ (Braathen et al., 2002; Chintalacharuvu et al., 2001; Yoo et al., 1999).

Only polymers containing the J-chain can bind to the pIgR (Brandtzaeg and Prydz, 1984) (see Chapter 4). Both covalent and noncovalent interactions are found between pIgA and pIgR. Human pIgR forms disulfide bonds with human, rat, and rabbit dimeric IgA (Tamer et al., 1995); 11 rabbit IgA-f isotypes were shown to bind to rabbit pIgR covalently, whereas 1 rabbit IgA-g isotype was noncovalently associated (Schneiderman et al., 1989). Disulfide bond formation does not appear to facilitate transport but, rather, might be important in stabilizing pIgA–SC complexes. Although the J-chain promotes IgA polymer assembly, human IgA1 and IgA2m(1) produced in CHO cells were able to form dimers in its absence (Morton et al., 1993). The J-chain contains eight Cys residues, two of which (Cys15 and 69) form disulfide bonds with the α-chain and six of which (Cys13, 72, 92, 101, 109, and 134) form intrachain disulfide bridges. The C-terminus of the J-chain as well as two of the three intrachain disulfide bridges were found to be dispensable for formation of IgA polymers but were required for binding to SC (Johansen et al., 2001). Either Cys15 or Cys69 of the J-chain was required for IgA polymerization, but these polymers were stabilized by noncovalent interactions. Although the noncovalently associated polymers bound free SC, transcytosis by pIgR-expressing MDCK cells was greatly diminished (Johansen et al., 2001).

Because the J-chain is required for pIgR binding and both $C_\alpha3$ and the tailpieces are required for J-chain incorporation, it is not surprising that studies using domain-exchanged proteins showed that $C_\alpha3$ is required for binding and for transport by pIgR-expressed on MDCK cells (Braathen et al., 2002;

Hexham et al., 1999). An exposed loop of the $C_\alpha3$ domain containing amino acids 402–410 (QEPSQGTTT) was predicted to be the binding site for pIgR (Hexham et al., 1999), with amino acids 430–443 also participating in pIgR binding (White and Capra, 2002). However, a naturally occurring mutant that lacks 36 amino acids in $C_\alpha3$, including amino acids 402–410, was found to associate with rat SC and be transported from blood into bile in a manner indistinguishable from pIgA (Switzer et al., 1992). Nevertheless, green fluorescent protein (GFP) fusion proteins containing monomeric and dimeric forms of amino acids 402–410 and a peptide from a phage display library resembling amino acids 402–410 were transported by the pIgR (Hexham et al., 1999), suggesting that such peptides might be used to target and deliver therapeutics to mucosal sites. Although the addition of $C_\alpha3$ from IgA2m(2) to IgG1 resulted in proteins that incorporated the J-chain and bound to pIgR, the hybrid molecules were transported by pIgR-expressing MDCK cells only with low efficiency (Chintalacharuvu et al., 2001).

15.4.2. Stability of IgA and SIgA

The environment at mucosal surfaces is a hostile one, with the pH in the gastrointestinal tract ranging from 2.0 in the stomach to 8.0 in the intestinal lumen. IgA has evolved to function in this hostile environment, and the addition of $C_\alpha3$ of IgA2m(2) to IgG1 resulted in proteins that were more stable than wild-type IgG1 at most pHs. Replacement of $C_\gamma1$ with $C_\alpha1$ further increased the stability of the hybrid antibodies, especially at low pH (Chintalacharuvu *et al.,* 2001). SC serves to increase the stability of IgA (see Chapter 8). In mice, SIgA1 was cleared less readily than dimeric IgA (Chintalacharuvu and Morrison, 1997), and SIgA was more resistant to degradation than dimeric IgA when exposed to intestinal washes (Berdoz et al., 1999; Chintalacharuvu and Morrison, 1997; Crottet and Corthesy, 1998). In addition, recombinant SIgA produced in plants was found to be more stable in the human oral cavity than an IgG containing identical V regions (Ma et al., 1998).

Many bacteria that cause disease at mucosal surfaces secrete proteases that cleave IgA1 but not IgA2. Type 1 proteases cleave between Pro-Ser residues, whereas type 2 cleave between Pro-Thr residues. Streptococcal species produce type 2 metalloproteases, whereas *Neisseria meningitides, N. gonorrhoeae*, and *Haemophilus influenzae* produce both type 1 and type 2 proteases. The O-linked glycans in the hinge appear to be important in sensitivity to streptococcal proteases (Batten et al., 2003). The length of the hinge has been shown to be important for recognition and cleavage by proteases (Senior and Woof, 2005b) (see Chapter 1). Studies using human IgA1 hinge mutants revealed that the bacterial proteases require specific amino acid sequences for cleavage (Batten *et al.,* 2003), with many but not all proteases able to cleave a hybrid IgA2/IgA1 with a shortened hinge (Senior et al., 2000). The proteases are able to cleave alternative peptide bonds if the wild-type sites are not present (Senior and Woof, 2005a). In addition to sequence requirements in the hinge, both C_H2 and C_H3 of IgA1 are required for the cleavage by proteases from *H. influenzae* and *N. gonorrhoeae* (Chintalacharuvu et al., 2003).

15.4.3. Binding to FcαRI (CD89)

Immunoglobulin A-mediated immune effector mechanisms such as phagocytosis, antibody-dependent cell-mediated cytotoxicity (ADCC), respiratory burst, and cytokine release are mediated through FcαRI, an IgA-specific receptor present on human macrophages, monocytes, eosinophils, and neutrophils (Monteiro et al., 1990) (see Chapter 4). No comparable receptor is expressed in the mouse. Both monomeric and dimeric IgA can bind the FcαR and activate the immune effector cells (Vidarsson et al., 2001). IgA forms a 2:1 complex with the receptor by binding to each C_H2/C_H3 interface (Herr et al., 2003). Initial studies using recombinant human IgA1/IgG1 domain-exchanged proteins revealed that $C_\alpha2$ and $C_\alpha3$ are necessary and sufficient for binding to FcαRI (Carayannopoulos et al., 1996). Specifically, residues Leu257 and Leu258 in the C_H2 loop and Pro440, Leu441, Ala442, and Phe443 in the C_H3 loop are important for binding and signaling through the receptor (Carayannopoulos et al., 1996; Pleass et al., 1999). Electrostatic interactions of acidic residues Glu254, Asp255, and Glu437 on IgA1 do not seem to play a major role in binding to FcαRI because mutation of these residues had little affect on affinity for the receptor (Pleass et al., 2003a). The crystal structure of FcαRI in complex with the Fc of IgA has confirmed and extended these conclusions (Herr et al., 2003). The FcαRI:Fcα interface is composed of a central hydrophobic core composed of Leu257, Leu258, Met433, Leu441, Ala442, Phe443, and the aliphatic portion of the Arg382 side chain flanked by charged residues (Arg382, Glu389, and Glu437). The *N*-glycans approach within 8 Å of FcαRI but do not directly contact the receptor. Fcα is bound to FcαRI in an "upright" orientation such that its C-terminus would be oriented near the cell membrane when bound to cell-associated FcαRI. Bound SC would appear to occlude the binding site, and SIgA does not enhance or stimulate phagocytosis (Vidarsson et al., 2001). Indeed, activation of neutrophil effector functions can be prevented by SIgA (Motegi and Kita, 1998).

15.4.4. Complement Fixation

Whereas IgA does not fix complement via the classical pathway, it is debatable whether it can activate the alternative pathway and might depend on the species producing the IgA (see Chapter 6). Chimeric murine–human IgA1 against *Pseudomonas aeruginosa* lipopolysaccharide (LPS) did not mediate deposition of C3 onto the surface of purified LPS or whole bacteria (Preston et al., 1998); although chimeric mouse–human IgA1 specific for the hapten dansyl bound C3, it did not appear to activate the alternative pathway because subsequent steps in the complement cascade (factor B cleavage and terminal complex binding) were not observed (Chuang and Morrison, 1997). However, human serum IgA has been shown to activate the alternative pathway to enhance the killing of bacteria under some circumstances (Janoff et al., 1999), and 12 recombinant rabbit IgA isotypes activated the alternative pathway (Schneiderman et al., 1990). The absence of glycosylation on IgA resulted in a reduction in C3 binding (Zhang and Lachmann, 1994).

15.4.5. Role of Carbohydrates in IgA and SIgA

Immunoglobulin A and SIgA are heavily glycosylated proteins with oligosaccharide side chains accounting for >10% of the molecular mass of the IgA H chain (Tomana et al., 1976), 8% of the J-chain (Baenziger, 1979; Royle *et al.*, 2003), and 15–20% of SC (Hughes et al., 1999; Phalipon et al., 2002). The presence or absence of glycans as well as their structure can affect IgA assembly, secretion, and function. Deletion of N-linked glycans in C_H2 and the tailpiece did not affect assembly and secretion of human IgA1 (Chuang and Morrison, 1997) but did inhibit the secretion of murine IgA (Taylor and Wall, 1988). The absence of N-linked carbohydrate in human IgA1 did not affect its binding to pIgR (Chuang and Morrison, 1997). When IgA was produced in insect cells, glycosylation in C_H2 was found to be critical for interaction with FcαRI (Carayannopoulos et al., 1996); however, this was not the case when the IgA was produced in CHO cells (Mattu et al., 1998), possibly reflecting structural differences that occur because of the altered structure of either the N-linked carbohydrate at Asn459 or the O-linked glycans in the hinge.

The N- and/or O-linked oligosaccharides of IgA and SC have been proposed to function as ligands for bacterial adhesions and appear to be important in providing protection by serving as receptor analogues in innate immunity (Dallas and Rolfe, 1998; Mantis et al., 2004; Royle et al., 2003) (see Chapter 8). In addition, the N-linked glycans on recombinant human SC were shown to be critical for directing the appropriate localization of IgA and conferring protection against *Shigella flexneri* infection in mice (Phalipon et al., 2002). Secretion of recombinant human SC requires that it be glycosylated because cells grown in the presence of tunicamycin, which inhibits N-linked glycosylation, were unable to secrete recombinant human SC (Cottet and Corthesy, 1997).

Carbohydrate structure also influences the pharmacokinetic properties of glycoproteins. The asialoglycoprotein-receptor (ASGPR) mediates the rapid clearance of glycoproteins bearing terminal β-linked *N*-acetylgalactosamine (GalNAc) or Gal. In mice, all three allotypes of IgA2 were rapidly cleared from the serum by ASGPR expressed in the liver, whereas much less IgA1 was cleared from circulation by this route (Rifai et al., 2000). IgA1 has been found to be highly sialylated (Mattu et al., 1998) and increasing the amount of sialic acid content on the carbohydrate was found to decrease the interaction of IgA1 and IgA2 with the ASGPR (Basset et al., 1999). Greater serum levels of IgA1 than IgA2 might be explained in part by this rapid clearance of IgA2 by ASGPR.

15.5. Applications of Recombinant IgA

A growing public health concern is the increased presence of antibiotic-resistant disease-causing microbes. The increased number of these bacteria has led to renewed interest in the use of passive antibodies for the treatment of infectious disease. Passive antibodies were widely used in the preantibiotic era but were abandoned when the widespread use of antibiotics became common (reviewed in Casadevall et al., 2004). Currently, most of the antibodies in use and in clinical trails are for systemic administration. However, possibly the greatest use for antibodies might lie in their administrations for both the prevention and treatment of infections at the mucosal surfaces, the major portal of entry for infectious agents. From a public health perspective, prevention is very important and direct application of antibody to the site of infection can block pathogen entry. Local administration of antibodies provides effective concentrations of drug immediately, and topical application is an easy and attractive mode of administration. Therapeutics based on IgA would appear to hold great promise for this route of treatment. SIgA can confer protection when administered passively, as in breast milk. In addition, recombinant IgAs have been used in a variety of disease models.

15.5.1. Bacterial Diseases

Recombinant IgA specific for a wide range of bacterial products has been produced. Fully human dimeric and pIgA and SIgA produced in CHO cells bind *Helicobacter pylori* urease (Berdoz and Corthesy, 2004) and a chimeric murine–human IgA1 recognizes *Pseudomonas aeruginosa* serogroup O6 lipopolysaccharide (Preston et al., 1998). Recombinant IgG (1–4), IgA1, and IgA2 against porin A from *Neisseria meningitidis* have been produced in BHK cells (Vidarsson et al., 2001). A mixture of monomeric and dimeric IgA2 containing the J-chain was shown to trigger a respiratory burst in polymorphonuclear leukocytes (PMNs) and monomeric IgA but not SIgA was shown to activate phagocytosis of heat-killed *Neisseria*. Polymeric chimeric IgA2m(1) specific for the enterotoxin A of *Clostridium difficile* showed increased avidity, which enhanced both the efficacy and duration of protection of human colonic epithelial T84 cell monolayers against the destructive effects of toxin A in comparison to IgG and monomeric IgA containing the same V regions (Stubbe et al., 2000). The protection in *Shigella flexneri* infection conferred by IgA specific for serotype 5a LPS was shown to be enhanced when pIgA is associated with SC. SC effectively localized IgA to the epithelial surface of the nasal cavity and the bronchial mucus, whereas IgA by itself was distributed diffusely in tissue. (Phalipon et al., 2002).

The first human trial of monoclonal secretory antibody has been performed using tobacco plant-derived SIgA against *Streptococcus mutans*

(Ma et al., 1998). The SIgA/IgG hybrid antibody was found to be more stable than murine IgG in the human oral cavity and SIgA was shown to prevent oral colonization by *S. mutans* for at least 4 months.

15.5.2. Viral Diseases

About 75% of human immunodeficiency virus (HIV)-resistant women expressed HIV-1-specific IgA in their genital tract as opposed to 26% of infected women (Kaul et al., 1999) (see Chapter 12). IgA appears to play an important role in preventing sexual transmission of HIV-1 because the IgA from these persistently seronegative individuals not only neutralizes HIV-1 but also prevents its transcytosis across tight epithelial cell layers (Devito et al., 2000a, 2000b). Human IgA1 against a nonimmunodominant epitope on the HIV-1 gp41 produced in CHO cells was shown to neutralize transepithelial HIV-1 entry *in vitro* using Me180 or HT-29 polarized epithelial monolayers, although it appears to be less effective than IgG and IgM at virus neutralization through complement and protection of human peripheral blood mononuclear cells (PBMCs) with HIV primary isolates (Wolbank et al., 2003).

Transmissible gastroenteritis coronavirus (TGEV) infects both enteric and respiratory tissues and causes close to 100% mortality in newborn pigs. Recombinant chimeric murine–porcine IgA produced in Sp2/0 cells neutralized TGEV 50-fold more efficiently than IgG1 and presumably due to the increased avidity of dimeric IgA. Transgenic mice expressing this same porcine IgA in the mammary gland were shown to secrete virus neutralizing antibodies into milk, suggesting that transgenic technology can be used to prevent neonatal infection and disease in livestock (Sola et al., 1998).

15.5.3. Parasitic Diseases

Recombinant IgA also shows promise in the treatment of parasitic diseases such as malaria. Therapeutics based on IgA mighty be advantageous over IgG because parasites initiate hypergammaglobulinemia as a smokescreen to evade the immune system (Miller et al., 1994); IgG also binds to the inhibitory receptor FcγRIIb, thereby interfering with the immune response to the parasite (Clynes et al., 2000). Pleass et al (2003b) produced a recombinant murine–human chimeric IgA1 and IgG1 against $MSP1_{19}$, merozoite surface protein 1, an antigen on *Plasmodium yoelii*. IgA1 was more effective than IgG1 in stimulating an oxidative burst in human neutrophils *in vitro*. However, in a murine model of malaria, IgA1 did not provide protection against lethal parasite infection, as did mouse IgG2b containing the identical V region, presumably because human IgA1 fails to interact with mouse Fc receptor(s). In humans, FcαRI is expressed on monocytes and neutrophils, potent effector cells in merozoite killing, as well as on Kupffer and dendritic cells, which are involved in $MSP1_{19}$ antigen presentation. Stimulation of FcαRI on human monocytes inhibits tumor necrosis factor (TNF)-α and interleukin (IL)-6 synthesis, which when elevated has shown to correlate with poor prognosis in malaria. In addition, high titers of *Plasmodium*-specific IgA in serum (Biswas et al., 1995) and breast milk (Leke et al., 1992) have been found in humans living in endemic areas, suggesting that IgA plays a crucial role in immunity against malaria and that recombinant IgA might be an effective therapeutic agent against malaria.

A chimeric murine–human IgA1 against a major ragweed pollen antigen (Sun et al., 1995) was shown to prevent increased airway hyperresponsiveness and lung eosinophilia in sensitized mice (Schwarze et al., 1998).

15.5.4. SIgA as a Mucosal Vaccine Delivery System

Studies suggest that SIgA can also serve as a mucosal vaccine delivery system. An eight-amino-acid sequence in the loop connecting the E and F β-strands of domain I of rabbit pIgR/SC (residues 79–86) was replaced with a nine-amino-acid linear epitope from *S. flexneri* invasin B to engineer an "antigenized" SC. Although this mutation resulted in a significantly lower rate of secretion and, when expressed on MDCK cells, failed to transport dimeric IgA, the mutated SC was able to associate with murine dimeric IgA. When the antigenized SIgA together with cholera toxin as adjuvant was orally administered to mice, it was able to elicit production of invasin B- and rabbit SC-specific antibodies (Corthesy et al., 1996). In addition, when the *H. pylori* GroES chaperonin was fused at the C-terminus of human SC to serve as a mucosal vaccine carrier, the antigenized SC was shown to associate with polymeric and dimeric IgA isolated from murine hybridoma HNK20 (Favre et al., 2003).

15.5.5. Cancer Therapy

Recombinant IgAs have shown promise as anticancer therapeutics. Recombinant human IgA1 produced in BHK cells specific for Ep-CAM, a tumor-associated antigen, was shown to mediate the killing of tumor cells by unstimulated PMNs. It also mediated phagocytosis of tumor cells by macrophages, although not as efficiently as human IgG1 containing the same V regions (Huls et al., 1999a). In addition, chimeric murine–human IgA1 and IgA2 against the HLA class II molecule expressed on B-cell lymphomas were effective in directing the killing of a variety of target cell lines by PMNs. Target cell lysis was inhibited by anti-FcαRI, indicating that FcαRI is the receptor responsible for antitumor activity (Dechant et al., 2002). Neutrophils can exert a potent cytolytic capacity against a variety of tumor cells in the presence of antitumor antibody. Immature neutrophils mobilized from the bone marrow by granulocyte colony-stimulating factor treatment have been shown to efficiently trigger tumor cell lysis via FcαRI, but not through the FcγR (Otten et al., 2005). Carcinoembryonic antigen (CEA) is expressed on the apical side of carcinoma cells (Buchegger et al., 1989; Burtin et *al.*, 1973), and it is thought that much of the radiolabeled anti-CEA IgG antibody used for detection (Baum et al., 1994) or therapy of colon carcinomas (Breitz et al., 1992) is unable to reach it due to its particular histological localization. Because chimeric IgA2m(2) against CEA can be transcytosed by the pIgR, it might be a more effective agent (Terskikh et al., 1994). A novel mouse–human chimeric dimeric IgA specific for the human transferrin receptor (hTfR) has been shown to inhibit proliferation of and induce apoptosis in myeloma cell lines ARH-77 and IM-9 (Chintalacharuvu et al., unpublished results; Prost et al., 1998; Shinohara et al., 2000).

15.6. Concluding Remarks

Immunoglobulin A mediates immune protection at mucosal surfaces, the major portal of entry for many pathogens, and therefore holds great potential as a therapeutic and prophylatic agent. One of the challenges in producing therapeutic IgAs is to ensure that this complex recombinant protein has the desired functional properties. Advances in molecular biology techniques have allowed for rapid cloning, expression, and characterization of antibodies, resulting in a greater understanding of protein assembly and secretion pathways, ligand–receptor interactions, the role of carbohydrates in glycoprotein function, and antibody effector functions. The challenge is to now use the resulting information to produce effective IgA-based therapeutic molecules.

Posttranslational modifications are important for correct antibody function. An important posttranslational modification is glycosylation, and consistent and correct glycosylation is especially important for producing functional IgA and SIgA in which the H-chain, J-chain, and SC all have significant amounts of surface-exposed carbohydrates. Murine myelomas and CHO cells have been the expression system of choice for the production of antibodies. However, these expression systems do not add carbohydrates identical to those found on human IgA. The use of human cell lines for expression might help to resolve this problem. Transgenic animals are an attractive expression system, especially for the production of SIgA in milk (Houdebine, 2000). However, the choice of species becomes important because there are species-specific differences in glycan addition. However, these differences might be less of an issue when the recombinant protein will be used for topical administration.

The recombinant antibody industry is in its infancy and shows great promise for rapid and extensive growth. It is expected that IgA will take its place alongside IgG as an important therapeutic molecule. However, for this to take place, certain issues and problems must be addressed. These include the issue of the cost of these complex biologics that require the use of expensive media along with the high costs associated with the purification process. The availability and cost of bovine serum in many growth media are issues that can be addressed by the use of serum-free growth media. Batch-to-batch variation and heterogeneity in the final product are also important considerations (Stoll et al., 1996). The design of novel bioreactor configurations for large-scale culture in suspension (McKinney et al., 1995; Sauer et al., 2000), careful monitoring and control of culture conditions (Monica et al., 1993; Nyberg et al., 1999; Schneider et al., 1996), and engineering of the production cell lines to improve glycosylation and assembly might solve some of these problems (Davies et al., 2001; Weikert et al., 1999).

Recombinant IgAs have been shown to be effective against pathogens such as bacteria and viruses. In addition, IgA might also be protective against cancer cells. Although mice do not express an FcαRI homologue, the use of human FcαRI transgenic mice (van Egmond et al., 1999) might facilitate the study of the role of IgA in immune protection in various disease models. IgA is important in mucosal secretions and SC has an important role in anchoring SIgA to the mucus; however, most clinical trials have been done using oral or nasal administration of pIgA lacking SC (Zeitlin et al., 1999). With all things considered, we can expect recombinant IgA and SIgA to assume an ever increasing role in the clinic.

References

Atkin, J. D., Pleass, R. J., Owens, R. J., and Woof, J. M. (1996). Mutagenesis of the human IgA1 heavy chain tailpiece that prevents dimer assembly. *J. Immunol.* 157:156–159.

Baenziger, J. U. (1979). Structure of the oligosaccharide of human J chain. *J. Biol. Chem.* 254:4063–4071.

Bakker, H., Bardor, M., Molthoff, J. W., Gomord, V., Elbers, I., Stevens, L. H., Jordi, W., Lommen, A., Faye, L., Lerouge, P., and Bosch, D. (2001). Galactose-extended glycans of antibodies produced by transgenic plants. *Proc. Natl. Acad. Sci. USA* 98:2899–2904.

Basset, C., Devauchelle, V., Durand, V., Jamin, C., Pennec, Y. L., Youinou, P., and Dueymes, M. (1999). Glycosylation of immunoglobulin A influences its receptor binding. *Scand. J. Immunol.* 50:572–579.

Batten, M. R., Senior, B. W., Kilian, M., and Woof, J. M. (2003). Amino acid sequence requirements in the hinge of human immunoglobulin A1 (IgA1) for cleavage by streptococcal IgA1 proteases. *Infect. Immun.* 71:1462–1469.

Baum, R. P., Niesen, A., Hertel, A., Nancy, A., Hess, H., Donnerstag, B., Sykes, T. R., Sykes, C. J., Suresh, M. R., Noujaim, A. A., et al. (1994). Activating anti-idiotypic human anti-mouse antibodies for immunotherapy of ovarian carcinoma. *Cancer* 73:1121–1125.

Berdoz, J., Blanc, C. T., Reinhardt, M., Kraehenbuhl, J. P., and Corthesy, B. (1999). In vitro comparison of the antigen-binding and stability properties of the various molecular forms of IgA antibodies assembled and produced in CHO cells. *Proc. Natl. Acad. Sci. USA* 96:3029–3034.

Berdoz, J., and Corthesy, B. (2004). Human polymeric IgA is superior to IgG and single-chain Fv of the same monoclonal specificity to inhibit urease activity associated with *Helicobacter pylori*. *Mol. Immunol.* 41:1013–1022.

Berdoz, J., Monath, T. P., and Kraehenbuhl, J. P. (1995). Specific amplification by PCR of rearranged genomic variable regions of immunoglobulin genes from mouse hybridoma cells. *PCR Methods Appl.* 4:256–264.

Biswas, S., Saxena, Q. B., Roy, A., and Kubilan, L. (1995). Naturally occurring *Plasmodium* specific IgA antibodies in humans from a malaria endemic area. *J. Biosci.* 20:453–460.

Boel, E., Verlaan, S., Poppelier, M. J., Westerdaal, N. A., Van Strijp, J. A., and Logtenberg, T. (2000). Functional human monoclonal antibodies of all isotypes constructed from phage display library-derived single-chain Fv antibody fragments. *J. Immunol. Methods* 239:153–166.

Braathen, R., Sorensen, V., Brandtzaeg, P., Sandlie, I., and Johansen, F. E. (2002). The carboxyl-terminal domains of IgA and IgM direct isotype-specific polymerization and interaction with the polymeric immunoglobulin receptor. *J. Biol. Chem.* 277:42,755–42,762.

Brandtzaeg, P., and Prydz, H. (1984). Direct evidence for an integrated function of J chain and secretory component in epithelial transport of immunoglobulins. *Nature* 311:71–73.

Breitz, H. B., Weiden, P. L., Vanderheyden, J. L., Appelbaum, J. W., Bjorn, M. J., Fer, M. F., Wolf, S. B., Ratliff, B. A., Seiler, C. A., Foisie, D. C., et al. (1992). Clinical experience with rhenium-186-labeled monoclonal antibodies for radioimmunotherapy: results of phase I trials. *J. Nucl. Med.* 33:1099–1109.

Brooks, D., Taylor, C., Dos Santos, B., Linden, H., Houghton, A., Hecht, T. T., Kornfeld, S., and Taetle, R. (1995). Phase Ia trial of murine immunoglobulin A antitransferrin receptor antibody 42/6. *Clin. Cancer Res.* 1:1259–1265.

Bruggemann, M., Williams, G. T., Bindon, C. I., Clark, M. R., Walker, M. R., Jefferis, R., Waldmann, H., and Neuberger, M. S. (1987). Comparison of the effector functions of human immunoglobulins using a matched set of chimeric antibodies. *J. Exp. Med.* 166:1351–1361.

Buchegger, F., Pfister, C., Fournier, K., Prevel, F., Schreyer, M., Carrel, S., and Mach, J. P. (1989). Ablation of human colon carcinoma in nude mice by 131I-labeled monoclonal anti-carcinoembryonic antigen antibody F(ab')2 fragments. *J. Clin. Invest.* 83:1449–1456.

Burtin, P., von Kleist, S., Sabine, M. C., and King, M. (1973). Immunohistological localization of carcinoembryonic antigen and nonspecific cross-reacting antigen in gastrointestinal normal and tumoral tissues. *Cancer Res.* 33:3299–3305.

Butters, T. D., Hughes, R. C., and Vischer, P. (1981). Steps in the biosynthesis of mosquito cell membrane glycoproteins and the effects of tunicamycin. *Biochim. Biophys. Acta* 640:672–686.

Campbell, M. J., Zelenetz, A. D., Levy, S., and Levy, R. (1992). Use of family specific leader region primers for PCR amplification of the human heavy chain variable region gene repertoire. *Mol. Immunol.* 29:193–203.

Carayannopoulos, L., Hexham, J. M., and Capra, J. D. (1996). Localization of the binding site for the monocyte immunoglobulin (Ig) A-Fc receptor (CD89) to the domain boundary between Calpha2 and Calpha3 in human IgA1. *J. Exp. Med.* 183:1579–1586.

Carayannopoulos, L., Max, E. E., and Capra, J. D. (1994). Recombinant human IgA expressed in insect cells. *Proc. Natl. Acad. Sci. USA* 91:8348–8352.

Casadevall, A., Dadachova, E., and Pirofski, L. A. (2004). Passive antibody therapy for infectious diseases. *Nat. Rev. Microbiol.* 2:695–703.

Chargelegue, D., Vine, N. D., van Dolleweerd, C. J., Drake, P. M., and Ma, J. K. (2000). A murine monoclonal antibody produced in transgenic plants with plant-specific glycans is not immunogenic in mice. *Transgenic Res.* 9:187–194.

Chintalacharuvu, K. R., Chuang, P. D., Dragoman, A., Fernandez, C. Z., Qiu, J., Plaut, A. G., Trinh, K. R., Gala, F. A., and Morrison, S. L. (2003). Cleavage of the human immunoglobulin A1 (IgA1) hinge region by IgA1 proteases requires structures in the Fc region of IgA. *Infect. Immun.* 71:2563–2570.

Chintalacharuvu, K. R., and Morrison, S. L. (1996). Residues critical for H-L disulfide bond formation in human IgA1 and IgA2. *J. Immunol.* 157:3443–3449.

Chintalacharuvu, K. R., and Morrison, S. L. (1997). Production of secretory immunoglobulin A by a single mammalian cell. *Proc. Natl. Acad. Sci. USA* 94:6364–6368.

Chintalacharuvu, K. R., Raines, M., and Morrison, S. L. (1994). Divergence of human alpha-chain constant region gene sequences. A novel recombinant alpha 2 gene. *J. Immunol.* 152:5299–5304.

Chintalacharuvu, K. R., Vuong, L. U., Loi, L. A., Larrick, J. W., and Morrison, S. L. (2001). Hybrid IgA2/IgG1 antibodies with tailor-made effector functions. *Clin. Immunol.* 101:21–31.

Chintalacharuvu, K. R., Yu, L. J., Bhola, N., Kobayashi, K., Fernandez, C. Z., and Morrison, S. L. (2002). Cysteine residues required for the attachment of the light chain in human IgA2. *J. Immunol.* 169:5072–5077.

Chuang, P. D., and Morrison, S. L. (1997). Elimination of N-linked glycosylation sites from the human IgA1 constant region: effects on structure and function. *J. Immunol.* 158:724–732.

Clynes, R. A., Towers, T. L., Presta, L. G., and Ravetch, J. V. (2000). Inhibitory Fc receptors modulate in vivo cytoxicity against tumor targets. *Nat. Med.* 6:443–446.

Coloma, M. J., Hastings, A., Wims, L. A., and Morrison, S. L. (1992). Novel vectors for the expression of antibody molecules using variable regions generated by polymerase chain reaction. *J. Immunol. Methods* 152:89–104.

Corthesy, B., Kaufmann, M., Phalipon, A., Peitsch, M., Neutra, M. R., and Kraehenbuhl, J. P. (1996). A pathogen-specific epitope inserted into recombinant secretory immunoglobulin A is immunogenic by the oral route. *J. Biol. Chem.* 271:33,670–677.

Cottet, S., and Corthesy, B. (1997). Cellular processing limits the heterologous expression of secretory component in mammalian cells. *Eur. J. Biochem.* 246:23–31.

Crottet, P., and Corthesy, B. (1998). Secretory component delays the conversion of secretory IgA into antigen-binding competent F(ab')2: A possible implication for mucosal defense. *J. Immunol.* 161:5445–5453.

Crottet, P., Cottet, S., and Corthesy, B. (1999). Expression, purification and biochemical characterization of recombinant murine secretory component: a novel tool in mucosal immunology. *Biochem. J.* 341(Pt. 2):299–306.

Dallas, S. D., and Rolfe, R. D. (1998). Binding of *Clostridium difficile* toxin A to human milk secretory component. *J. Med. Microbiol.* 47:879–888.

Davies, J., Jiang, L., Pan, L. Z., LaBarre, M. J., Anderson, D., and Reff, M. (2001). Expression of GnTIII in a recombinant anti-CD20 CHO production cell line: Expression of antibodies with altered glycoforms leads to an increase in ADCC through higher affinity for Fc gamma RIII. *Biotechnol. Bioeng.* 74:288–294.

Davis, R., Schooley, K., Rasmussen, B., Thomas, J., and Reddy, P. (2000). Effect of PDI overexpression on recombinant protein secretion in CHO cells. *Biotechnol. Prog.* 16:736–743.

Dechant, M., Vidarsson, G., Stockmeyer, B., Repp, R., Glennie, M. J., Gramatzki, M., van De Winkel, J. G., and Valerius, T. (2002). Chimeric IgA antibodies against HLA class II effectively trigger lymphoma cell killing. *Blood* 100:4574–4580.

De Groot, N., Van Kuik-Romeijn, P., Lee, S. H., and De Boer, H. A. (2000). Increased immunoglobulin A levels in milk by over-expressing the murine polymeric immunoglobulin receptor gene in the mammary gland epithelial cells of transgenic mice. *Immunology* 101:218–224.

Devito, C., Broliden, K., Kaul, R., Svensson, L., Johansen, K., Kiama, P., Kimani, J., Lopalco, L., Piconi, S., Bwayo, J. J., et al. (2000a). Mucosal and plasma IgA from HIV-1-exposed uninfected individuals inhibit HIV-1 transcytosis across human epithelial cells. *J. Immunol.* 165:5170–5176.

Devito, C., Hinkula, J., Kaul, R., Lopalco, L., Bwayo, J. J., Plummer, F., Clerici, M., and Broliden, K. (2000b). Mucosal and plasma IgA from HIV-exposed seronegative individuals neutralize a primary HIV-1 isolate. *AIDS* 14:1917–1920.

Elbers, I. J., Stoopen, G. M., Bakker, H., Stevens, L. H., Bardor, M., Molthoff, J. W., Jordi, W. J., Bosch, D., and Lommen, A. (2001). Influence of growth conditions and developmental stage on N-glycan heterogeneity of transgenic immunoglobulin G and endogenous proteins in tobacco leaves. *Plant Physiol.* 126:1314–1322.

Favre, L. I., Spertini, F., and Corthesy, B. (2003). Simplified procedure to recover recombinant antigenized secretory IgA to be used as a vaccine vector. *J. Chromatogr. B: Anal. Technol. Biomed. Life Sci.* 786:143–151.

Fukuta, K., Abe, R., Yokomatsu, T., Omae, F., Asanagi, M., and Makino, T. (2000). Control of bisecting GlcNAc addition to N-linked sugar chains. *J. Biol. Chem.* 275:23,456–461.

Galili, U. (1989). Abnormal expression of alpha-galactosyl epitopes in man. A trigger for autoimmune processes? *Lancet* 2:358–361.

Gavilondo-Cowley, J. V., Coloma, M. J., Vazquez, J., Ayala, M., Macias, A., Fry, K. E., and Larrick, J. W. (1990). Specific amplification of rearranged immunoglobulin variable region genes from mouse hybridoma cells. *Hybridoma* 9:407–417.

Gillies, S. D., Lo, K. M., and Wesolowski, J. (1989). High-level expression of chimeric antibodies using adapted cDNA variable region cassettes. *J. Immunol. Methods* 125:191–202.

Griffiss, G. M., and Goroff, D. K. (1983). IgA blocks IgM and IgG-initiated immune lysis by separate molecular mechanisms. *J. Immunol.* 130:2882–2885.

Gupta, N., Arthos, J., Khazanie, P., Steenbeke, T. D., Censoplano, N. M., Chung, E. A., Cruz, C. C., Chaikin, M. A., Daucher, M., Kottilil, S., Mavilio, D., Schuck, P., Sun, P. D., Rabin, R. L., Radaev, S., Van Ryk, D., Cicala, C., and Fauci, A. S. (2005). Targeted lysis of HIV-infected cells by natural killler cells armed and triggered by a recombinant immunoglobulin fusion protein: Implication for immunotherapy. *Vaccine* 332:491–497.

Hemming, V. G., Rodriguez, W., Kim, H. M., Brandt, C. D., Parrott, R. H., Burch, B., Prince, G. A., Baron, P. A., Fink, R. J., and Reaman, G. (1987). Intravenous immunoglobulin treatment of respiratory syncytial virus infection in infants and young children. *Antimicrob. Agents Chemother.* 31:1882–1886.

Herr, A. B., Ballister, E. R., and Bjorkman, P. J. (2003). Insights into IgA-mediated immune response from the cystal structures of human Fc alpha RI and its complex with IgA1-Fc. *Nature* 423:614–620.

Hexham, J. M., White, K. D., Carayannopoulos, L. N., Mandecki, W., Brisette, R., Yang, Y. S., and Capra, J. D. (1999). A human immunoglobulin (Ig)A C alpha 3 domain motif directs polymeric Ig receptor-mediated secretion. *J. Exp. Med.* 189:747–752.

Hills, A. E., Patel, A., Boyd, P., and James, D. C. (2001). Metabolic control of recombinant monoclonal antibody N-glycosylation in GS-NS0 cells. *Biotechnol. Bioeng.* 75:239–251.

Hirt, R. P., Hughes, G. J., Frutiger, S., Michetti, P., Perregaux, C., Poulain-Godefroy, O., Jeanguenat, N., Neutra, M. R., and Kraehenbuhl, J. P. (1993). Transcytosis of the polymeric Ig receptor requires phosphorylation of serine 664 in the absence but not the presence of dimeric IgA. *Cell* 74:245–255.

Houdebine, L. M. (2000). Transgenic animal bioreactors. *Transgenic Res.* 9:305–320.

Hughes, G. J., Reason, A. J., Savoy, L., Jaton, J., and Frutiger-Hughes, S. (1999). Carbohydrate moieties in human secretory component. *Biochim. Biophys. Acta* 1434:86–93.

Huls, G., Heijnen, I. A., Cuomo, E., van der Linden, J., Boel, E., van de Winkel, J. G., and Logtenberg, T. (1999a). Antitumor immune effector mechanisms recruited by phage display-derived fully human IgG1 and IgA1 monoclonal antibodies. *Cancer Res.* 59:5778–5784.

Huls, G. A., Heijnen, I. A., Cuomo, M. E., Koningsberger, J. C., Wiegman, L., Boel, E., van der Vuurst de Vries, A. R., Loyson, S. A., Helfrich, W., van Berge Henegouwen, G. P., van Meijer, M., de Kruif, J., and Logtenberg, T. (1999b). A recombinant, fully human monoclonal antibody with antitumor activity constructed from phage-displayed antibody fragments. *Nat. Biotechnol.* 17:276–281.

Janoff, E. N., Fasching, C., Orenstein, J. M., Rubins, J. B., Opstad, N. L., and Dalmasso, A. P. (1999). Killing of *Streptococcus pneumoniae* by capsular polysaccharide-specific polymeric IgA, complement, and phagocytes. *J. Clin. Invest.* 104:1139–1147.

Jarvis, D. L., and Finn, E. E. (1995). Biochemical analysis of the N-glycosylation pathway in baculovirus-infected lepidopteran insect cells. *Virology* 212:500–511.

Jarvis, D. L., and Finn, E. E. (1996). Modifying the insect cell N-glycosylation pathway with immediate early baculovirus expression vectors. *Nat. Biotechnol.* 14:1288–1292.

Johansen, F. E., Braathen, R., and Brandtzaeg, P. (2001). The J chain is essential for polymeric Ig receptor-mediated epithelial transport of IgA. *J. Immunol.* 167:5185–5192.

Johansen, F. E., Natvig Norderhaug, I., Roe, M., Sandlie, I., and Brandtzaeg, P. (1999). Recombinant expression of polymeric IgA: Incorporation of J chain and secretory component of human origin. *Eur. J. Immunol.* 29:1701–1708.

Kaul, R., Trabattoni, D., Bwayo, J. J., Arienti, D., Zagliani, A., Mwangi, F. M., Kariuki, C., Ngugi, E. N., MacDonald, K. S., Ball, T. B., Clerici, M., and Plummer, F. A. (1999). HIV-1-specific mucosal IgA in a cohort of HIV-1-resistant Kenyan sex workers. *AIDS* 13:23–29.

Knight, K. L., Suter, M., and Becker, R. S. (1988). Genetic engineering of bovine Ig. Construction and characterization of hapten-binding bovine/murine chimeric IgE, IgA, IgG1, IgG2, and IgG3 molecules. *J. Immunol.* 140:3654–3659.

Kozlowski, P. A., and Neutra, M. R. (2003). The role of mucosal immunity in preventing of HIV transmission. *Curr. Mol. Med.* 3:217–228.

Larrick, J. W., Danielsson, L., Brenner, C. A., Abrahamson, M., Fry, K. E., and Borrebaeck, C. A. (1989). Rapid cloning of rearranged immunoglobulin genes from human hybridoma cells using mixed primers and the polymerase chain reaction. *Biochem. Biophys. Res. Commun.* 160:1250–1256.

Larrick, J. W., Yu, L., Naftzger, C., Jaiswal, S., and Wycoff, K. (2001). Production of secretory IgA antibodies in plants. *Biomol. Eng.* 18:87–94.

Lee, E. U., Roth, J., and Paulson, J. C. (1989). Alteration of terminal glycosylation sequences on N-linked oligosaccharides of Chinese hamster ovary cells by expression of beta-galactoside alpha 2,6-sialyltransferase. *J. Biol. Chem.* 264:13,848–855.

Leke, R. G., Ndansi, R., Southerland, N. J., Quakyi, I. A., and Taylor, D. W. (1992). Identification of anti-*Plasmodium falciparum* antibodies in human breast milk. *Scand. J. Immunol.* 11(Suppl.):17–22.

Lifely, M. R., Hale, C., Boyce, S., Keen, M. J., and Phillips, J. (1995). Glycosylation and biological activity of CAMPATH-1H expressed in different cell lines and grown under different culture conditions. *Glycobiology* 5:813–822.

Lindh, E. (1975). Increased resistance of immunoglobulin A dimers to proteolytic degradation after binding of secretory component. *J. Immunol.* 114:284–286.

Lo, D., Pursel, V., Linton, P. J., Sandgren, E., Behringer, R., Rexroad, C., Palmiter, R. D., and Brinster, R. L. (1991). Expression of mouse IgA by transgenic mice, pigs and sheep. *Eur. J. Immunol.* 21:1001–1006.

Lullau, E., Heyse, S., Vogel, H., Marison, I., von Stockar, U., Kraehenbuhl, J. P., and Corthesy, B. (1996). Antigen binding properties of purified immunoglobulin A and reconstituted secretory immunoglobulin A antibodies. *J. Biol. Chem.* 271:16,300–16,309.

Lund, J., Takahashi, N., Hindley, S., Tyler, R., Goodall, M., and Jefferis, R. (1993). Glycosylation of human IgG subclass and mouse IgG2b heavy chains secreted by mouse J558L transfectoma cell lines as chimeric antibodies. *Hum. Antibodies Hybridomas* 4:20–25.

Lund, J., Tanaka, T., Takahashi, N., Sarmay, G., Arata, Y., and Jefferis, R. (1990). A protein structural change in aglycosylated IgG3 correlates with loss of huFc gamma R1 and huFc gamma R111 binding and/or activation. *Mol. Immunol.* 27:1145–1153.

Ma, J. K., Hiatt, A., Hein, M., Vine, N. D., Wang, F., Stabila, P., van Dolleweerd, C., Mostov, K., and Lehner, T. (1995). Generation and assembly of secretory antibodies in plants. *Science* 268:716–719.

Ma, J. K., Hikmat, B. Y., Wycoff, K., Vine, N. D., Chargelegue, D., Yu, L., Hein, M. B., and Lehner, T. (1998). Characterization of a recombinant plant monoclonal secretory antibody and preventive immunotherapy in humans. *Nat. Med.* 4:601–606.

Mantis, N. J., Farrant, S. A., and Mehta, S. (2004). Oligosaccharide side chains on human secretory IgA serve as receptors for ricin. *J. Immunol.* 172:6838–6845.

Martin, B. M., Tsuji, S., LaMarca, M. E., Maysak, K., Eliason, W., and Ginns, E. I. (1988). Glycosylation and processing of high levels of active human glucocerebrosidase in invertebrate cells using a baculovirus expression vector. *DNA* 7:99–106.

Mattu, T. S., Pleass, R. J., Willis, A. C., Kilian, M., Wormald, M. R., Lellouch, A. C., Rudd, P. M., Woof, J. M., and Dwek, R. A. (1998). The glycosylation and structure of human serum IgA1, Fab, and Fc regions and the role of N-glycosylation on Fc alpha receptor interactions. *J. Biol. Chem.* 273:2260–2272.

McKinney, K. L., Dilwith, R., and Belfort, G. (1995). Optimizing antibody production in batch hybridoma cell culture. *J. Biotechnol.* 40:31–48.

Merry, A. H., Morton, C., Bruce, J., Kerr, M., and Woof, J. M. (1992). Glycosylation of recombinant chimeric and human serum IgA1. *Biochem. Soc. Trans.* 20:92S.

Miller, L. H., Good, M. F., and Milon, G. (1994). Malaria pathogenesis. *Science* 264:1878–1883.

Monica, T. J., Goochee, C. F., and Maiorella, B. L. (1993). Comparative biochemical characterization of a human IgM produced in both ascites and in vitro cell culture. *Biotechnology (NY)* 11:512–515.

Monteiro, R. C., Kubagawa, H., and Cooper, C. (1990). Cellular distribution, regulation, and biochemical nature of an Fc alpha receptor in human. *J. Exp. Med.* 171:597–613.

Morton, H. C., Atkin, J. D., Owens, R. J., and Woof, J. M. (1993). Purification and characterization of chimeric human IgA1 and IgA2 expressed in COS and Chinese hamster ovary cells. *J. Immunol.* 151:4743–4752.

Motegi, Y., and Kita, H. (1998). Interaction with secretory component stimulates effector functions of human eosinophils but not of neutrophils. *J. Immunol.* 161:4340–4346.

Nyberg, G. B., Balcarcel, R. R., Follstad, B. D., Stephanopoulos, G., and Wang, D. I. (1999). Metabolic effects on recombinant interferon-gamma glycosylation in continuous culture of Chinese hamster ovary cells. *Biotechnol. Bioeng.* 62:336–347.

Ogonah, O. W., Freedman, R. B., Jenkins, N., and Rooney, B. C. (1995). Analysis of human interferon-gamma glycoforms produced in baculovirus infected insect cells by matrix assisted laser desorption spectrometry. *Biochem. Soc. Trans.* 23:100S.

Orlandi, R., Gussow, D. H., Jones, P. T., and Winter, G. (1989). Cloning immunoglobulin variable domains for expression by the polymerase chain reaction. *Proc. Natl. Acad. Sci. USA* 86:3833–3837.

Otten, M. A., Rudolph, E., Dechant, M., Tuk, C. W., Reijmers, R. M., Beelen, R. H., van de Winkel, J. G., and van Egmond, M. (2005). Immature neutrophils mediate tumor cell killing via IgA but not IgG Fc receptors. *J. Immunol.* 174:5472–5480.

Phalipon, A., Cardona, A., Kraehenbuhl, J. P., Edelman, L., Sansonetti, P. J., and Corthesy, B. (2002). Secretory component: A new role in secretory IgA-mediated immune exclusion in vivo. *Immunity* 17:107–115.

Pleass, R. J., Dehal, P. K., Lewis, M. J., and Woof, J. M. (2003a). Limited role of charge matching in the interaction of human immunoglobulin A with the immunoglobulin A Fc receptor (Fc alpha RI) CD89. *Immunology* 109:331–335.

Pleass, R. J., Dunlop, J. I., Anderson, C. M., and Woof, J. M. (1999). Identification of residues in the CH2/CH3 domain interface of IgA essential for interaction with the human fcalpha receptor (FcalphaR) CD89. *J. Biol. Chem.* 274:23,508–514.

Pleass, R. J., Ogun, S. A., McGuinness, D. H., van de Winkel, J. G., Holder, A. A., and Woof, J. M. (2003b). Novel antimalarial antibodies highlight the importance of the antibody Fc region in mediating protection. *Blood* 102:4424–4430.

Preston, M. J., Gerceker, A. A., Reff, M. E., and Pier, G. B. (1998). Production and characterization of a set of mouse-human chimeric immunoglobulin G (IgG) subclass and IgA monoclonal antibodies with identical variable regions specific for *Pseudomonas aeruginosa* serogroup O6 lipopolysaccharide. *Infect. Immun.* 66:4137–4142.

Prost, A. C., Menegaux, F., Langlois, P., Vidal, J. M., Koulibaly, M., Jost, J. L., Duron, J. J., Chigot, J. P., Vayre, P., Aurengo, A., Legrand, J. C., Rosselin, G., and Gespach, C. (1998). Differential transferrin receptor density in human colorectal cancer: A potential probe for diagnosis and therapy. *Int. J. Oncol.* 13:871–875.

Raju, T. S., Briggs, J. B., Borge, S. M., and Jones, A. J. (2000). Species-specific variation in glycosylation of IgG: evidence for the species-specific sialylation and branch-specific galactosylation and importance for engineering recombinant glycoprotein therapeutics. *Glycobiology* 10:477–486.

Raju, T. S., Briggs, J. B., Chamow, S. M., Winkler, M. E., and Jones, A. J. (2001). Glycoengineering of therapeutic glycoproteins: in vitro galactosylation and sialylation of glycoproteins with terminal N-acetylglucosamine and galactose residues. *Biochemistry* 40:8868–8876.

Ramirez, N., Rodriguez, M., Ayala, M., Cremata, J., Perez, M., Martinez, A., Linares, M., Hevia, Y., Paez, R., Valdes, R., et al. (2003). Expression and characterization of an anti-(hepatitis B surface antigen) glycosylated mouse antibody in transgenic tobacco (*Nicotiana tabacum*) plants and its use in the immunopurification of its target antigen. *Biotechnol. Appl. Biochem.* 38:223–230.

Renegar, K. B., Jackson, G. D., and Mestecky, J. (1998a). In vitro comparision of the biologic activities of monoclonal monomeric IgA, polymeric IgA, and secretory IgA. *J. Immunol.* 160:1219–1223.

Rifai, A., Fadden, K., Morrison, S. L., and Chintalacharuvu, K. R. (2000). The N-glycans determine the differential blood clearance and hepatic uptake of human immunoglobulin (Ig)A1 and IgA2 isotypes. *J. Exp. Med.* 191:2171–2182.

Rindisbacher, L., Cottet, S., Wittek, R., Kraehenbuhl, J. P., and Corthesy, B. (1995). Production of human secretory component with dimeric IgA binding capacity using viral expression systems. *J. Biol. Chem.* 270:14,220–14,228.

Routier, F. H., Davies, M. J., Bergemann, K., and Hounsell, E. F. (1997). The glycosylation pattern of humanized IgGI antibody (D1.3) expressed in CHO cells. *Glycoconj. J.* 14:201–207.

Royle, L., Roos, A., Harvey, D. J., Wormald, M. R., van Gijlswijk-Janssen, D., Redwan el, R. M., Wilson, I. A., Daha, M. R., Dwek, R. A., and Rudd, P. M. (2003). Secretory IgA N- and O-glycans provide a link between the innate and adaptive immune systems. *J. Biol. Chem.* 278:20,140–153.

Russell, M. W., Reinholdt, J., and Kilian, M. (1998). Anti-inflammatory activity of human IgA antibodies and their Fab alpha fragments: Inhibition of IgG-mediated complement activation. *Eur. J. Immunol.* 24:1211–1217.

Sandin, C., Linse, S., Areschoug, T., Woof, J. M., Reinholdt, J., and Lindahl, G. (2002). Isolation and detection of human IgA using a streptococcal IgA-binding peptide. *J. Immunol.* 169:1357–1364.

Sauer, P. W., Burky, J. E., Wesson, M. C., Sternard, H. D., and Qu, L. (2000). A high-yielding, generic fed-batch cell culture process for production of recombinant antibodies. *Biotechnol. Bioeng.* 67:585–597.

Schneider, M., Marison, I. W., and von Stockar, U. (1996). The importance of ammonia in mammalian cell culture. *J. Biotechnol.* 46:161–185.

Schneiderman, R. D., Hanly, W. C., and Knight, K. L. (1989). Expression of 12 rabbit IgA C alpha genes as chimeric rabbit-mouse IgA antibodies. *Proc. Natl. Acad. Sci. USA* 86:7561–7565.

Schneiderman, R. D., Lint, T. F., and Knight, K. L. (1990). Activation of the alternative pathway of complement by twelve different rabbit-mouse chimeric transfectoma IgA isotypes. *J. Immunol.* 145:233–237.

Schwarze, J., Cieslewicz, G., Joetham, A., Sun, L. K., Sun, W. N., Chang, T. W., Hamelmann, E., and Gelfand, E. W. (1998). Antigen-specific immunoglobulin-A prevents increased airway responsiveness and lung eosinophilia after airway challenge in sensitized mice. *Am. J. Respir. Crit. Care Med.* 158:519–525.

Senior, B. W., Dunlop, J. I., Batten, M. R., Kilian, M., and Woof, J. M. (2000). Cleavage of a recombinant human immunoglobulin A2 (IgA2) –IgA1 hybrid antibody by certain bacterial IgA1 proteases. *Infect. Immun.* 68:463–469.

Senior, B. W., and Woof, J. M. (2005a). Effect of mutations in the human immunoglobulin A1 (IgA1) hinge on its susceptibility to cleavage by diverse bacterial IgA1 proteases. *Infect. Immun.* 73:1515–1522.

Senior, B. W., and Woof, J. M. (2005b). The influences of hinge length and composition on the susceptibility of human IgA to cleavage by diverse bacterial IgA1 proteases. *J. Immunol.* 174:7792–7799.

Shinohara, H., Fan, D., Ozawa, S., Yano, S., Van Arsdell, M., Viner, J. L., Beers, R., Pastan, I., and Fidler, I. J. (2000). Site-specific expression of transferrin receptor by human colon cancer cells directly correlates with eradication by antitransferrin recombinant immunotoxin. *Int. J. Oncol.* 17:643–651.

Sola, I., Castilla, J., Pintado, B., Sanchez-Morgado, J. M., Whitelaw, C. B., Clark, A. J., and Enjuanes, L. (1998). Transgenic mice secreting coronavirus neutralizing antibodies into the milk. *J. Virol.* 72:3762–3772.

Stoll, T. S., Muhlethaler, K., von Stockar, U., and Marison, I. W. (1996). Systematic improvement of a chemically-defined protein-free medium for hybridoma growth and monoclonal antibody production. *J. Biotechnol.* 45:111–123.

Stubbe, H., Berdoz, J., Kraehenbuhl, J. P., and Corthesy, B. (2000). Polymeric IgA is superior to monomeric IgA and IgG carrying the same variable domain in preventing *Clostridium difficile* toxin A damaging of T84 monolayers. *J. Immunol.* 164:1952–1960.

Subramanian, K. N., Weismann, L. E., Rhodes, T., Ariagno, R., Sanchez, P. J., Steichen, J., Givner, L. B., Jennings, T. L., Top, F. H. J., Carlin, D., and Connor, E. (1998). Safety, tolerance and pharmacokinetics of a humanized monoclonal antibody to respiratory syncytial virus in premature infants and infants with bronchopulmonary dysplasia. *MEDI-493 Study Group* 17:110–115.

Sun, L. K., Fung, M. S., Sun, W. N., Sun, C. R., Chang, W. I., and Chang, T. W. (1995). Human IgA monoclonal antibodies specific for a major ragweed pollen antigen. *Biotechnology (NY)* 13:779–786.

Switzer, I. C., Loney, G. M., Yang, D. S., and Underdown, B. J. (1992). Binding of secretory component to protein 511, a pIgA mouse protein lacking 36 amino acid residues of the C alpha 3 domain. *Mol. Immunol.* 29:31–35.

Tamer, C. M., Lamm, M. E., Robinson, J. K., Piskurich, J. F., and Kaetzel, C. S. (1995). Comparative studies of transcytosis and assembly of secretory IgA in Madin-Darby canine kidney cells expressing human polymeric Ig receptor. *J. Immunol.* 155:707–714.

Taylor, A. K., and Wall, R. (1988). Selective removal of alpha heavy-chain glycosylation sites causes immunoglobulin A degradation and reduced secretion. *Mol. Cell. Biol.* 8:4197–4203.

Terskikh, A., Couty, S., Pelegrin, A., Hardman, N., Hunziker, W., and Mach, J. P. (1994). Dimeric recombinant IgA directed against carcino-embryonic antigen, a novel tool for carcinoma localization. *Mol. Immunol.* 31:1313–1319.

Tomana, M., Niedermeier, W., Mestecky, J., and Skvaril, F. (1976). The differences in carbohydrate composition between the subclasses of IgA immunoglobulins. *Immunochemistry* 13:325–328.

Trill, J. J., Shatzman, A. R., and Ganguly, S. (1995). Production of monoclonal antibodies in COS and CHO cells. *Curr. Opin. Biotechnol.* 6:553–560.

Umana, P., Jean-Mairet, J., Moudry, R., Amstutz, H., and Bailey, J. E. (1999). Engineered glycoforms of an antineuroblastoma IgG1 with optimized antibody-dependent cellular cytotoxic activity. *Nat. Biotechnol.* 17:176–180.

van Egmond, M., van Vuuren, A. J., Morton, H. C., van Spriel, A. B., Shen, L., Hofhuis, F. M., Saito, T., Mayadas, T. N., Verbeek, J. S., and van de Winkel, J. G. (1999). Human immunoglobulin A receptor (FcαRI, CD89) function in transgenic mice requires both FcR γ chain and CR3 (CD11b/CD18). *Blood* 93:4387–4394.

van Ree, R., Cabanes-Macheteau, M., Akkerdaas, J., Milazzo, J. P., Loutelier-Bourhis, C., Rayon, C., Villalba, M., Koppelman, S., Aalberse, R., Rodriguez, R., Faye, L., and Lerouge, P. (2000). Beta(1,2)-xylose and alpha(1,3)-fucose residues have a strong contribution in IgE binding to plant glycoallergens. *J. Biol. Chem.* 275:11,451–11,458.

Vidarsson, G., van Der Pol, W. L., van Den Elsen, J. M., Vile, H., Jansen, M., Duijs, J., Morton, H. C., Boel, E., Daha, M. R., Corthesy, B., and van De Winkel, J. G. (2001). Activity of human IgG and IgA subclasses in immune defense against *Neisseria meningitidis* serogroup B. *J. Immunol.* 166:6250–6256.

Weikert, S., Papac, D., Briggs, J., Cowfer, D., Tom, S., Gawlitzek, M., Lofgren, J., Mehta, S., Chisholm, V., Modi, N., et al. (1999). Engineering Chinese hamster ovary cells to maximize sialic acid content of recombinant glycoproteins. *Nat. Biotechnol.* 17:1116–1121.

White, K. D., and Capra, J. D. (2002). Targeting mucosal sites by polymeric immunoglobulin receptor-directed peptides. *J. Exp. Med.* 196:551–555.

Wolbank, S., Kunert, R., Stiegler, G., and Katinger, H. (2003). Characterization of human class-switched polymeric (immunoglobulin M [IgM] and IgA) anti-human immunodeficiency virus type 1 antibodies 2F5 and 2G12. *J. Virol.* 77:4095–4103.

Yoo, E. M., Coloma, M. J., Trinh, K. R., Nguyen, T. Q., Vuong, L. U., Morrison, S. L., and Chintalacharuvu, K. R. (1999). Structural requirements for polymeric immunoglobulin assembly and association with J chain. *J. Biol. Chem.* 274:33,771–777.

Zeitlin, L., Cone, R. A., and Whaley, K. J. (1999). Using monoclonal antibodies to prevent mucosal transmission of epidemic infectious diseases. *Emerg. Infect. Dis.* 5:54–64.

Zhang, W., and Lachmann, P. J. (1994). Glycosylation of IgA is required for optimal activation of the alternative complement pathway by immune complexes. *Immunology* 81:137–141.

Index

Page numbers followed by f, indicate figures; t, tables.

N

Q

Printed in the United States
97374LV00002B/29/A

9 780387 722313